VECTOR CALCULUS

Peter Baxandall

Hans Liebeck

DOVER PUBLICATIONS, INC.
Mineola, New York

Bibliographical Note

This Dover edition, first published in 2008, is an unabridged republication of the
work originally published in 1986 by Oxford University Press, Oxford, England.

Library of Congress Cataloging-in-Publication Data

Baxandall, P. R.
    Vector calculus / Peter Baxandall, Hans Liebeck. — Dover ed.
      p. cm.
    Originally published: Oxford [Oxfordshire] : Clarendon Press ; New York :
Oxford University Press, 1986.
    Includes bibliographical references and index.
    ISBN-13: 978-0-486-46620-0
    ISBN-10: 0-486-46620-5
    1. Calculus. 2. Vector analysis. I. Liebeck, H. (Hans) II. Title.

QA303.2.B35 2008
515'.63—dc22

               2007048507

Manufactured in the United States of America
Dover Publications, Inc., 31 East 2nd Street, Mineola, N.Y. 11501

About the Authors

Hans Liebeck (my father) was born in Berlin in 1927. When he was nine years old his family emigrated to South Africa where he later obtained a BSc at the University of Cape Town in Mathematics and Physics and an MSc in Physics. He was awarded a Rhodes Scholarship to study Mathematics at Oxford University, moved on to take a PhD at Cambridge University, and then was appointed to his first lecturing post back at Cape Town University. In the early 1960's he returned to the UK with his young family to take up a lectureship at the University of Keele where he remained for the rest of his career, retiring as Reader in Pure Mathematics. Hans was a well respected research algebraist publishing around 20 articles in research journals, also writing 'Algebra for Scientists and Engineers' Wiley (1969) and (with Peter Baxandall) 'Differential Vector Calculus' Longmans (1981). An outstanding teacher he also contributed many popular articles to periodicals including The Mathematical Gazette and American Mathematical Monthly.

Hans died in March 2007. He was survived by his wife the mathematics education specialist Pamela Liebeck and their three children, to each of whom they kindly passed on their mathematical genes.

Martin Liebeck, Imperial College, London.

Peter Baxandall was born and grew up in the north of England and studied mathematics at the universities of Manchester and Cambridge specializing in geometric topology and the work of his supervisor E.C. (later Sir Christopher) Zeeman. He was Lecturer in Mathematics at the universities of Cape Town and Keele for some 20 years and then moved on to teach mathematics at Bryanston School in Dorset, England for 16 years. He has been an enthusiastic tutor of the remarkable students of the UK's Open University since it began in 1971 and it was through his work there that his interest in linear mathematics and the history of mathematics developed. He has been concerned throughout his teaching career both at Keele and at Bryanston with the presentation of the subject to non-specialist audiences being much influenced by Dick Tahta and other members of the UK's ground breaking and innovative Leapfrog Group. He also wrote 'Differential Vector Calculus with Hans Liebeck (Longmans Mathematical Texts, 1981) and contributed chapters to F. R. Watson's 'Proof in Mathematics' (Keele Mathematics Education Publications, 1978) as well as a number of articles and reviews for several journals. Peter Baxandall married the potter and sculptor Jane Surgey in 1963. They have two sons, both of whom seem grateful not to have inherited any mathematical genes at all.

PRB Blandford Forum, Dorset.

Preface

This book provides an introduction to the calculus of functions of several variables set against a background of linear algebra. It is designed as a first or second year course for undergraduates who have some knowledge of linear algebra and real analysis.

Traditionally linear algebra, vector analysis and calculus of functions of several variables are taught as separate subjects. They are, however, closely related. The underlying links are established by defining the differentiability of a vector function at a point in its domain in terms of the existence of an approximating linear transformation (differential). Many of the important classical results of vector calculus are essentially concerned with the geometrical properties of this differential and of its image and graph.

In the study of vector functions from $\mathbb{R}^m$ to $\mathbb{R}^n$ the cases $m = 1$ and $n = 1$ warrant separate discussion. After a short introductory chapter we proceed in Chapter 2 to consider functions from $\mathbb{R}$ to $\mathbb{R}^n$ (including the study of curves, differential geometry and dynamics). Chapter 3 deals with functions from $\mathbb{R}^m$ to $\mathbb{R}$ (real-valued functions of many variables, Taylor's Theorem and applications). The general theory of functions from $\mathbb{R}^m$ to $\mathbb{R}^n$ is covered in Chapter 4. In particular we prove the important Inverse Function and Implicit Function theorems. Our approach has the advantage of introducing the concept of differentiability of vector functions in easy stages. Certain theorems (the Chain Rule in particular) appear at a number of points in progressively more general form.

The remainder of the book is concerned with the classical integral theorems of vector calculus, including the Fundamental Theorem of Calculus and the theorems of Green, Stokes, and Gauss. The treatment is rather more rigorous than is usually given in a first course, but it avoids the difficulties of the more sophisticated approach adopted in advanced books. Special attention is paid to the relationship between different parametrizations of curves and surfaces and their application in line integrals and surface integrals. Full use is made of the relationship between differentiability and linearity, thereby explaining the presence of the Jacobian in integration techniques and theorems.

We have tried to give readable yet rigorous proofs (often omitted from introductory texts) of some of the important classical theorems. The theory is amply illustrated with figures and worked examples, and numerous exercises are provided, ranging from routine calculations to problems which suggest further development. Many of the exercises explain by way of counter-example the significance of the hypotheses of the theorems. Applications to physics are given both in the text and in the exercises.

It is our hope that when the student has gained familiarity with the material of this book he will approach with confidence advanced texts on calculus on manifolds and differential geometry.

We wish to express our thanks to Professor Alan Jeffrey for encouraging us to write this book. We are particularly grateful to Professor John Pym of the University of Sheffield and to the son of one of the authors, Dr Martin Liebeck of Imperial College, London, both of whom carefully read the typescript and suggested many improvements. We also acknowledge the contribution of many colleagues and friends at universities both in the United Kingdom and the USA.

Finally, very special thanks are due to Miss Sue Sutton for cheerfully typing the manuscript.

Keele   H. L.
Bryanston   P. R. B.
October 1985

Contents

1

Basic linear algebra and analysis

1.1 Introduction

Basic analysis and linear algebra play a fundamental role in the generalization of elementary calculus to the theory of vector-valued functions. We shall assume familiarity with the analysis and linear algebra usually found in a first course. Most of this background material is summarized in this chapter.

A first course in calculus deals with real-valued functions of one real variable. Such a function $f : S \subseteq \mathbb{R} \to \mathbb{R}$ is defined on a domain S which is a subset of the real line $\mathbb{R}$. The value that f takes at $x \in S$ is a real number denoted by $f(x) \in \mathbb{R}$. For example, the rule

$$f(x) = \sqrt{(1 - x^2)}$$

can be taken to define a real-valued function $f : S \subseteq \mathbb{R} \to \mathbb{R}$, where S is the closed interval $[-1, 1] = \{x \in \mathbb{R} | -1 \leqslant x \leqslant 1\}$.

Let $\mathbb{R}^m$ denote the set of all m-tuples of real numbers $(x_1, x_2, \ldots, x_m)$, $x_i \in \mathbb{R}$, $i = 1, \ldots, m$. (In particular, $\mathbb{R}^1 = \mathbb{R}$.) We shall be concerned with the study of functions $f : S \subseteq \mathbb{R}^m \to \mathbb{R}^n$ defined on a subset S of $\mathbb{R}^m$ and taking values in $\mathbb{R}^n$. For example, the rule

1.1.1 $\qquad f(x_1, x_2, x_3) = (\sqrt{(1 - x_3^2)}, x_1 x_2 x_3)$

can be taken to define a function $f : S \subseteq \mathbb{R}^3 \to \mathbb{R}^2$, where S is the subset of $\mathbb{R}^3$ given by

1.1.2 $\qquad S = \{(x_1, x_2, x_3) \in \mathbb{R}^3 | -1 \leqslant x_3 \leqslant 1\}.$

The set $\mathbb{R}^m$ defined above is given the structure of a real vector space by defining an addition of m-tuples

1.1.3 $\qquad (x_1, \ldots, x_m) + (y_1, \ldots, y_m) = (x_1 + y_1, \ldots, x_m + y_m)$

and a multiplication by scalars in $\mathbb{R}$,

1.1.4 $\qquad k(x_1, \ldots, x_m) = (kx_1, \ldots, kx_m), \qquad k \in \mathbb{R}.$

Viewed in this way, the rule 1.1.1 defines a function f on a subset S

of the vector space $\mathbb{R}^3$ such that the values that f takes lie in the vector space $\mathbb{R}^2$.

With $\mathbb{R}^m$ interpreted as a vector space, the study of functions $f : S \subseteq \mathbb{R}^m \to \mathbb{R}^n$ is appropriately called vector calculus. (The alternative title 'calculus of functions of several variables' is sometimes preferred when $m > 1$.) We observe that vector calculus includes the case $m = n = 1$ and so is a generalization of elementary calculus.

Exercise 1.1

1. Suggest a possible subset $S \subseteq \mathbb{R}^3$ as the domain of a function $f : S \subseteq \mathbb{R}^3 \to \mathbb{R}^2$ which is given by the rule
   (a) $f(x_1, x_2, x_3) = (x_1/x_2, \sqrt{(1 - x_3^2)})$;
   (b) $f(x, y, z) = (e^{1/x}\tan(xyz), x^2 + y^2 + z^2)$;
   (c) $f(x, y, z) = (1/(x^2 - z^2), \ln(xyz))$.

   Answers:
   (a) $\{(x_1, x_2, x_3) \in \mathbb{R}^3 | x_2 \neq 0, x_3 \in [-1, 1]\}$;
   (b) $\{(x, y, z) \in \mathbb{R}^3 | x \neq 0, xyz \neq (k + \frac{1}{2})\pi, \text{ for integers } k\}$;
   (c) $\{(x, y, z) \in \mathbb{R}^3 | x \neq \pm z, xyz > 0\}$.

1.2   The vector space $\mathbb{R}^m$

In section 1.1 we pointed out that the set $\mathbb{R}^m$ of all m-tuples $(x_1, \ldots, x_m)$, $x_i \in \mathbb{R}$, $i = 1, \ldots, m$, can be regarded as a vector space over $\mathbb{R}$ if we impose the rules 1.1.3, 1.1.4 of addition and scalar multiplication. We shall often denote a vector of $\mathbb{R}^m$ by a single letter in bold-face, thus: $\mathbf{x} = (x_1, \ldots, x_m)$. In $\mathbb{R}^2$ and $\mathbb{R}^3$ the familiar notation $\mathbf{r} = (x, y)$ and $\mathbf{r} = (x, y, z)$ is useful.

Consider the vectors in $\mathbb{R}^m$

$$\mathbf{e}_1 = (1, 0, \ldots, 0), \mathbf{e}_2 = (0, 1, \ldots, 0), \ldots, \mathbf{e}_m = (0, 0, \ldots, 1),$$

where $\mathbf{e}_i$ has 1 in the ith place and 0 elsewhere. The set $\{\mathbf{e}_1, \ldots, \mathbf{e}_m\}$ is clearly a basis of the vector space $\mathbb{R}^m$, since any $\mathbf{x} = (x_1, \ldots, x_m) \in \mathbb{R}^m$ has the unique expression $\mathbf{x} = x_1\mathbf{e}_1 + \cdots + x_m\mathbf{e}_m$ as a linear combination of the $\mathbf{e}_i$'s. In view of its simple form we call the set $\{\mathbf{e}_1, \ldots, \mathbf{e}_m\}$ the *standard* (or *natural*) *basis* of $\mathbb{R}^m$. No other bases of $\mathbb{R}^m$ are used in this book.

The vector spaces $\mathbb{R}^1 = \mathbb{R}$, $\mathbb{R}^2$ and $\mathbb{R}^3$ are conveniently pictured as a number line, as a plane and as three-dimensional space. For example, we picture $\mathbb{R}^2$ by choosing an ordered pair of perpendicular axes and a unit of length, and associating the vector $\mathbf{x} = (x_1, x_2) \in \mathbb{R}^2$

with the point in the plane whose coordinates relative to the axes are (x_1, x_2) in the usual way.

An important alternative way of picturing the vector $\mathbf{x} = (x_1, x_2)$ in the plane is by an arrow joining the origin $(0, 0)$ to the point labelled (x_1, x_2). Both ways of picturing vectors will frequently be used – sometimes in the same diagram. The arrow representation is particularly important in physical applications, for example when we wish to picture velocities, accelerations or forces.

In the arrow representation, the rule 1.1.3 of vector addition is the well known parallelogram law of vector addition. See Fig. 1.1(i). The arrows joining O to P and Q to R are identical in all respects except for their position in the plane. We therefore agree to picture $\mathbf{x} = (x_1, x_2)$ not only by the arrow $\overrightarrow{OP}$ but also by an arrow joining (y_1, y_2) to $(x_1 + y_1, x_2 + y_2)$, where y_1, y_2 are arbitrarily chosen real numbers.

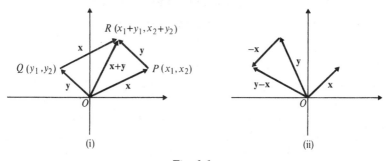

Fig. 1.1

Vector subtraction is illustrated in Fig. 1.1(ii). We have, by definition, $\mathbf{y} - \mathbf{x} = \mathbf{y} + (-\mathbf{x})$, where $-\mathbf{x} = (-x_1, -x_2)$.

Considerations similar to the above apply to a pictorial representation of $\mathbb{R}^3$ relative to three mutually perpendicular axes.

1.2.1 Definition.   *Given vectors* $\mathbf{x} = (x_1, \ldots, x_m)$ *and* $\mathbf{y} = (y_1, \ldots, y_m)$ *in* $\mathbb{R}^m$, *the* dot product *(or* scalar product*) of* $\mathbf{x}$ *and* $\mathbf{y}$ *is defined to be the real number*

$$\mathbf{x} \cdot \mathbf{y} = x_1 y_1 + \cdots + x_m y_m.$$

The dot product has the following important properties

1.2.2   *Symmetry*: $\mathbf{x}\cdot\mathbf{y} = \mathbf{y}\cdot\mathbf{x}$, $\mathbf{x}, \mathbf{y} \in \mathbb{R}^m$

1.2.3   *Linearity*: $(k\mathbf{x} + l\mathbf{y})\cdot\mathbf{z} = k(\mathbf{x}\cdot\mathbf{z}) + l(\mathbf{y}\cdot\mathbf{z})$,
          $\mathbf{x}, \mathbf{y}, \mathbf{z} \in \mathbb{R}^m$, $k, l \in \mathbb{R}$

1.2.4   *Positivity*: $\mathbf{x}\cdot\mathbf{x} > 0$, for all $\mathbf{x} \neq \mathbf{0}$, $\mathbf{x} \in \mathbb{R}^m$.

1.2.5 Definition. *The* norm *or* length *of* $\mathbf{x} \in \mathbb{R}^m$ *is the non-negative real number* $\|\mathbf{x}\| \geq 0$ *such that*

$$\|\mathbf{x}\|^2 = \mathbf{x}\cdot\mathbf{x} = x_1^2 + \ldots + x_m^2.$$

1.2.6 Example. If $\mathbf{a} = (2, 1)$, then $\|\mathbf{a}\|^2 = 2^2 + 1^2 = 5$ and $\|\mathbf{a}\| = \sqrt{5}$. Note that $\|\mathbf{x}\| = 0$ if and only if $\mathbf{x} = \mathbf{0}$.

1.2.7 Example. It follows from Definition 1.2.5 that for $\mathbf{x} \in \mathbb{R}^m$, $k \in \mathbb{R}$, $\|k\mathbf{x}\| = |k|\,\|\mathbf{x}\|$. For example, $\|-2\mathbf{x}\| = 2\|\mathbf{x}\|$.

The following theorem will be used extensively in this book.

1.2.8 Theorem. The Cauchy–Schwarz inequality. *For any vectors* $\mathbf{x}, \mathbf{y} \in \mathbb{R}^m$

1.2.9 $|\mathbf{x}\cdot\mathbf{y}| \leq \|\mathbf{x}\|\,\|\mathbf{y}\|$,

with equality if and only if the vectors $\mathbf{x}, \mathbf{y}$ *are linearly dependent.*

Proof. [*i*] If $\mathbf{y} = \mathbf{0}$, then $\mathbf{x}$ and $\mathbf{y}$ are linearly dependent, and 1.2.9 is an equality, both sides being zero.

[*ii*] If $\mathbf{y} \neq \mathbf{0}$ and $\mathbf{x}$ and $\mathbf{y}$ are linearly dependent, then there exists $k \in \mathbb{R}$ such that $\mathbf{x} = k\mathbf{y}$. In this case 1.2.9 is an equality, both sides being equal to $|k|\,\|\mathbf{y}\|^2$.

[*iii*] The remaining case is that $\mathbf{x}$ and $\mathbf{y}$ are linearly independent. Then, for any $k \in \mathbb{R}$, $\mathbf{x} + k\mathbf{y} \neq \mathbf{0}$ and, by the properties 1.2.2–1.2.4 of the dot product,

$$0 < \|\mathbf{x} - k\mathbf{y}\|^2 = (\mathbf{x} + k\mathbf{y})\cdot(\mathbf{x} + k\mathbf{y})$$
$$= \|\mathbf{x}\|^2 + 2k(\mathbf{x}\cdot\mathbf{y}) + k^2\|\mathbf{y}\|^2.$$

With $k = -(\mathbf{x}\cdot\mathbf{y})/\|\mathbf{y}\|^2$, a simple calculation results in 1.2.9 as a strict inequality.

The Cauchy–Schwarz inequality is often stated in the form

1.2.10 $$\left(\sum_{i=1}^{m} x_i y_i\right)^2 \leq \sum_{i=1}^{m} x_i^2 \sum_{i=1}^{m} y_i^2.$$

This follows by squaring 1.2.9 and applying the definitions of dot product and norm.

1.2.11 *Theorem.* The Triangle Inequality. *For any vectors* **x**, **y** *in* $\mathbb{R}^m$,

$$\|\mathbf{x} + \mathbf{y}\| \leqslant \|\mathbf{x}\| + \|\mathbf{y}\|.$$

Proof

1.2.12
$$
\begin{aligned}
\|\mathbf{x} + \mathbf{y}\|^2 &= (\mathbf{x} + \mathbf{y})\cdot(\mathbf{x} + \mathbf{y}) = \|\mathbf{x}\|^2 + 2\mathbf{x}\cdot\mathbf{y} + \|\mathbf{y}\|^2 \\
&\leqslant \|\mathbf{x}\|^2 + 2|\mathbf{x}\cdot\mathbf{y}| + \|\mathbf{y}\|^2 \\
&\leqslant \|\mathbf{x}\|^2 + 2\|\mathbf{x}\|\,\|\mathbf{y}\| + \|\mathbf{y}\|^2 \quad \text{by (1.2.9)} \\
&= (\|\mathbf{x}\| + \|\mathbf{y}\|)^2.
\end{aligned}
$$

Taking square roots on both sides, we obtain the Triangle Inequality.

A picture for the cases $\mathbb{R}^2$ and $\mathbb{R}^3$ shows why Theorem 1.2.11 is called the Triangle Inequality, for it is related to the property of triangles that the sum of the lengths $\|\mathbf{x}\|$ and $\|\mathbf{y}\|$ of two sides of a triangle is not smaller than the length $\|\mathbf{x} + \mathbf{y}\|$ of the third side. (See Fig. 1.1.)

1.2.13 *Corollary.* [i] *For any* **x** *and* **y** *in* $\mathbb{R}^m$

1.2.14
$$\big|\,\|\mathbf{x}\| - \|\mathbf{y}\|\,\big| \leqslant \|\mathbf{x} - \mathbf{y}\| \leqslant \|\mathbf{x}\| + \|\mathbf{y}\|.$$

[ii] *For any n vectors* $\mathbf{a}_1, \ldots, \mathbf{a}_n \in \mathbb{R}^m$ *and scalars* $x_1, \ldots, x_n \in \mathbb{R}$,

1.2.15
$$\|x_1\mathbf{a}_1 + \cdots + x_n\mathbf{a}_n\| \leqslant |x_1|\,\|\mathbf{a}_1\| + \cdots + |x_n|\,\|\mathbf{a}_n\|.$$

[iii] *For any* $\mathbf{x} = (x_1, \ldots, x_m) \in \mathbb{R}^m$,

1.2.16
$$\|(x_1, \ldots, x_m)\| \leqslant |x_1| + \cdots + |x_m|.$$

Proof. Exercise.

The well-known cosine rule applied to the triangle OPR of Fig. 1.1 gives

$$\|\mathbf{x} + \mathbf{y}\|^2 = \|\mathbf{x}\|^2 + \|\mathbf{y}\|^2 + 2\|\mathbf{x}\|\,\|\mathbf{y}\|\cos\theta,$$

where θ is the angle between the non-zero vectors **x** and **y**. Comparing this expression with 1.2.12 we obtain

1.2.17
$$\cos\theta = \frac{\mathbf{x}\cdot\mathbf{y}}{\|\mathbf{x}\|\,\|\mathbf{y}\|}.$$

In particular, the vector **x** is perpendicular (or orthogonal) to **y** if and only if $\mathbf{x} \cdot \mathbf{y} = 0$. The following generalization applies to $\mathbb{R}^m$.

1.2.18 Definition.   *The vector* $\mathbf{x} \in \mathbb{R}^m$ *is* orthogonal *to the vector* $\mathbf{y} \in \mathbb{R}^m$ *if and only if* $\mathbf{x} \cdot \mathbf{y} = 0$.

Note that by the symmetry of the dot product, **x** is orthogonal to **y** if and only if **y** is orthogonal to **x**.

1.2.19 Example.   *Equation of a plane in* $\mathbb{R}^3$. A plane in $\mathbb{R}^3$ is specified by a point $\mathbf{q} \in \mathbb{R}^3$ in the plane and a non-zero vector $\mathbf{n} \in \mathbb{R}^3$ perpendicular (or *normal*) to the plane.

The point $\mathbf{r} = (x, y, z) \in \mathbb{R}^3$ lies in the plane if and only if $\mathbf{r} - \mathbf{q}$ is orthogonal to **n**, that is, if and only if

$$(\mathbf{r} - \mathbf{q}) \cdot \mathbf{n} = 0.$$

This equation is called the equation of the plane containing the point **q** and having normal **n**.

For example, the equation of the plane containing $\mathbf{q} = (1, 1, 1)$ with normal $\mathbf{n} = (2, 4, 6)$ is $(x, y, z) \cdot (2, 4, 6) = (1, 1, 1) \cdot (2, 4, 6)$, that is,

$$2x + 4y + 6z = 12.$$

The dot product is defined on vectors in $\mathbb{R}^m$, where m is arbitrary. The following vector product is defined in $\mathbb{R}^3$ only.

1.2.20 Definition.   *The* cross product (*or* vector product) *of* $\mathbf{b} = (b_1, b_2, b_3)$ *and* $\mathbf{c} = (c_1, c_2, c_3)$ *in* $\mathbb{R}^3$ *is the vector*

$$\mathbf{b} \times \mathbf{c} = (b_2 c_3 - b_3 c_2, b_3 c_1 - b_1 c_3, b_1 c_2 - b_2 c_1).$$

The formula for $\mathbf{b} \times \mathbf{c}$ is conveniently obtained by expanding the formal determinant

1.2.21
$$\mathbf{b} \times \mathbf{c} = \det \begin{bmatrix} \mathbf{e}_1 & \mathbf{e}_2 & \mathbf{e}_3 \\ b_1 & b_2 & b_3 \\ c_1 & c_2 & c_3 \end{bmatrix},$$

where $\mathbf{e}_1, \mathbf{e}_2, \mathbf{e}_3$ is the standard basis of $\mathbb{R}^3$. An alternative common notation for the standard basis of $\mathbb{R}^3$ is $\mathbf{i} = (1, 0, 0), \mathbf{j} = (0, 1, 0),$ $\mathbf{k} = (0, 0, 1)$ and we shall use it occasionally.

We state the following standard results about the vector product without proof.

1.2.22 Theorem.   *Let* **b** *and* **c** *be vectors in* $\mathbb{R}^3$. *Then*
   [**i**] *the vector* $\mathbf{b} \times \mathbf{c}$ *is orthogonal to* **b** *and to* **c**;

[ii] **b** *and* **c** *are linearly dependent if and only if* $\mathbf{b} \times \mathbf{c} = \mathbf{0}$;

[iii] *if* **b** *and* **c** *are linearly independent then relative to a right-handed coordinate system of* $\mathbb{R}^3$ *the vectors* **b**,**c**, $\mathbf{b} \times \mathbf{c}$ *from a right-handed triple of vectors*;

[iv] $\|\mathbf{b} \times \mathbf{c}\| = \|\mathbf{b}\| \, \|\mathbf{c}\| \sin \phi$,

where ϕ *is the angle between* **b** *and* **c**. *Thus* $\|\mathbf{b} \times \mathbf{c}\|$ *measures the area of the parallelogram with* **b** *and* **c** *as adjacent sides.*

[v] *the volume of the parallelopiped with adjacent sides* $\mathbf{a} = (a_1, a_2, a_3)$, $\mathbf{b} = (b_1, b_2, b_3)$, $\mathbf{c} = (c_1, c_2, c_3)$ *is equal to the absolute value of the triplet product*

$$\mathbf{a} \cdot (\mathbf{b} \times \mathbf{c}) = \det \begin{bmatrix} a_1 & a_2 & a_3 \\ b_1 & b_2 & b_3 \\ c_1 & c_2 & c_3 \end{bmatrix}$$

As a corollary to Theorem 1.2.22 *(iv)*, we have the following result.

1.2.23   Theorem.   *In* $\mathbb{R}^2$ *the area of the parallelogram with adjacent sides* (b_1, b_2) *and* (c_1, c_2) *is the absolute value of the determinant*

$$\det \begin{bmatrix} b_1 & b_2 \\ c_1 & c_2 \end{bmatrix}.$$

Proof.   In $\mathbb{R}^3$ put $\mathbf{b} = (b_1, b_2, 0)$, $\mathbf{c} = (c_1, c_2, 0)$. The required area is equal to the norm of $\mathbf{b} \times \mathbf{c}$, and this is equal to $|b_1 c_2 - b_2 c_1|$.

The following example illustrates a physical application of the vector product.

1.2.24 *Example.*   Consider a rigid body rotating about a fixed axis A with angular speed ω radians/sec. See Fig. 1.2. Choose a right-handed coordinate system for $\mathbb{R}^3$ with origin O on the axis of rotation A. Let $\boldsymbol{\omega}$ be the angular velocity vector, of magnitude ω and pointing in the direction that a right-handed screw would advance under the given rotation. Let $\mathbf{r} = \overrightarrow{OP}$ be the position vector of the point P in the body. Then the velocity **v** of P is orthogonal both to $\boldsymbol{\omega}$ and to **r**, and so is in the direction of $\boldsymbol{\omega} \times \mathbf{r}$. The speed of P is

$$\|\mathbf{v}\| = (\|\mathbf{r}\| \sin \phi)\omega = \|\boldsymbol{\omega}\| \, \|\mathbf{r}\| \sin \phi = \|\boldsymbol{\omega} \times r\|.$$

Hence the velocity of P is given by

$$\mathbf{v} = \boldsymbol{\omega} \times \mathbf{r}.$$

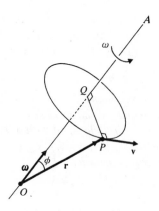

Fig. 1.2

Now let $\mathbf{r'} = \overrightarrow{OP}$ be the position vector of any point P' in the body. The velocity of P' is given by

$$\mathbf{v'} = \boldsymbol{\omega} \times \mathbf{r'} = \boldsymbol{\omega} \times (\mathbf{r'} - \mathbf{r}) + \boldsymbol{\omega} \times \mathbf{r}$$

$$= \boldsymbol{\omega} \times (\mathbf{r'} - \mathbf{r}) + \mathbf{v}.$$

So the velocity of P' is the vector sum of a rotation $\boldsymbol{\omega}$ relative to an axis through P and the velocity of P.

Exercises 1.2

1. Prove Corollary 1.2.13. (Hint: to prove the right-hand inequality 1.2.14 apply Theorem 1.2.11 to $\mathbf{x} + (-\mathbf{y})$; for the left-hand inequality put $\mathbf{x} - \mathbf{y} = \mathbf{z}$.)

2. (a) Prove that the zero vector in $\mathbb{R}^m$ is orthogonal to every vector $\mathbf{x} \in \mathbb{R}^m$.
   (b) Prove that if $\mathbf{x} \in \mathbb{R}^m$ is orthogonal to the vectors $\mathbf{y}_1, \ldots, \mathbf{y}_r$ in $\mathbb{R}^m$, then $\mathbf{x}$ is orthogonal to every vector in the subspace of $\mathbb{R}^m$ spanned by $\mathbf{y}_1, \ldots, \mathbf{y}_r$.

3. Find the angle between (a) the vectors $(1, -1, 0)$ and $(-4, 1, 1)$; (b) the vectors $(1, -1, 0)$ and $(4, -1, -1)$. Compare.
   Answer: (a) $\cos \theta = -\frac{5}{6}$;   (b) $\cos \phi = \frac{5}{6}$. $\theta + \phi = \pi$.

4. Prove that vectors $\mathbf{x} - \mathbf{y}$ and $\mathbf{x} + \mathbf{y}$ in $\mathbb{R}^m$ are orthogonal if and only if $\|\mathbf{x}\| = \|\mathbf{y}\|$. Illustrate this result in $\mathbb{R}^2$, and deduce that the diagonals of a parallelogram intersect at right angles if and only if the parallelogram is a rhombus.

5. Find the equation of the plane in $\mathbb{R}^3$
   (a) containing the point $(1, -1, 1)$ and with normal $\mathbf{n} = (0, 1, 1)$;
   (b) containing the points $(1, 0, 1)$, $(0, 1, 1)$ and $(1, 1, 0)$.

Answers: (a) $y + z = -2$;   (b) $x + y + z = 2$.

6. Prove from the definition of the cross product that
   (a) $\mathbf{b} \times \mathbf{c} = -\mathbf{c} \times \mathbf{b}$;   (b) $\mathbf{a} \times (\mathbf{b} \times \mathbf{c}) = (\mathbf{a} \cdot \mathbf{c})\mathbf{b} - (\mathbf{a} \cdot \mathbf{b})\mathbf{c}$.

7. Find a vector orthogonal to both $\mathbf{a} = (1, -1, 2)$ and $\mathbf{b} = (2, 0, 1)$.

Answer: any scalar multiple of the vector product $\mathbf{a} \times \mathbf{b} = (-1, 3, 2)$.

1.3   Linear functions

1.3.1 *Definition.* *A* linear *function* $L : \mathbb{R}^m \to \mathbb{R}^n$ *with domain* $\mathbb{R}^m$ *and codomain* $\mathbb{R}^n$ *is a rule that assigns to each* $\mathbf{x} \in \mathbb{R}^m$ *a unique vector* $L(\mathbf{x}) \in \mathbb{R}^n$ *such that for any* $\mathbf{x}, \mathbf{y} \in \mathbb{R}^m$ *and* $k, l \in \mathbb{R}$,

1.3.2 $L(k\mathbf{x} + l\mathbf{y}) = kL(\mathbf{x}) + lL(\mathbf{y}).$

It follows from 1.3.2 (by induction on r) that for any $\mathbf{a}_1, \ldots, \mathbf{a}_r \in \mathbb{R}^m$ and $k_1, \ldots, k_r \in \mathbb{R}$,

1.3.3 $L(k_1\mathbf{a}_1 + \cdots + k_r\mathbf{a}_r) = k_1 L(\mathbf{a}_1) + \cdots + k_r L(\mathbf{a}_r).$

1.3.4 *Theorem.* *A linear function* $L : \mathbb{R}^m \to \mathbb{R}^n$ *is completely determined by its effect on the standard basis* $\mathbf{e}_1, \ldots, \mathbf{e}_m$ *of* $\mathbb{R}^m$. *Moreover, an arbitrary choice of vectors* $L(\mathbf{e}_1), \ldots, L(\mathbf{e}_m)$ *in* $\mathbb{R}^n$ *determines a linear function* $L : \mathbb{R}^m \to \mathbb{R}^n$.

Proof.   Firstly, if $L : \mathbb{R}^m \to \mathbb{R}^n$ is linear then for any $\mathbf{x} = (x_1, \ldots, x_m) \in \mathbb{R}^m$,

$$L(\mathbf{x}) = L(x_1\mathbf{e}_1 + \cdots + x_m\mathbf{e}_m)$$
$$= x_1 L(\mathbf{e}_1) + \cdots + x_m L(\mathbf{e}_m), \qquad \text{by 1.3.3.}$$

Therefore $L(\mathbf{x})$ is determined by $L(\mathbf{e}_1), \ldots, L(\mathbf{e}_m)$.

Secondly, for any choice of $\mathbf{b}_1, \ldots, \mathbf{b}_m$ in $\mathbb{R}^n$ the expression

$$L(\mathbf{x}) = x_1\mathbf{b}_1 + \cdots + x_m\mathbf{b}_m, \qquad \mathbf{x} \in \mathbb{R}^m$$

determines a linear function $L : \mathbb{R}^m \to \mathbb{R}^n$ such that $L(\mathbf{e}_i) = \mathbf{b}_i$ for all $i = 1, \ldots, m$.

Note that a linear function has a vector space $\mathbb{R}^m$ as its domain, whereas a non-linear function may be defined on a subset of $\mathbb{R}^m$. See, for example, 1.1.1.

The *image* of a function $L : \mathbb{R}^m \to \mathbb{R}^n$ is defined as the set of all image vectors:

1.3.5 $\operatorname{im} L = \{L(\mathbf{x}) \in \mathbb{R}^n \mid \mathbf{x} \in \mathbb{R}^m\}.$

We say that L maps $\mathbb{R}^m$ *onto* $\mathbb{R}^n$ if $\operatorname{im} L = \mathbb{R}^n$.

The *kernel* of $L : \mathbb{R}^m \to \mathbb{R}^n$ is defined as the set of all $\mathbf{x} \in \mathbb{R}^m$ that are mapped by L to zero:

1.3.6 $\ker L = \{\mathbf{x} \in \mathbb{R}^m \mid L(\mathbf{x}) = \mathbf{0} \in \mathbb{R}^n\}.$

When $L : \mathbb{R}^m \to \mathbb{R}^n$ is linear, then $\operatorname{im} L$ is a subspace of $\mathbb{R}^n$, $\ker L$ is a subspace of $\mathbb{R}^m$, and the dimensions of $\operatorname{im} L$ and $\ker L$ are related by the celebrated formula (which we leave unproved)

1.3.7 $\dim \operatorname{im} L + \dim \ker L = \dim \mathbb{R}^m = m.$

1.3.8 Definition.   *A function f defined on a domain D is said to be 1–1 (one-to-one) on D if distinct elements of D have distinct images under f; that is if, for any $\mathbf{x} \in D$, $\mathbf{y} \in D$, $\mathbf{x} \neq \mathbf{y}$ implies that $f(\mathbf{x}) \neq f(\mathbf{y})$.*

1.3.9 Theorem.   *A linear function $L : \mathbb{R}^m \to \mathbb{R}^n$ is 1–1 if and only if $\ker L = \{\mathbf{0}\}$.*

Proof.   Exercise.

We shall require (in Section 4.6) the following definition and result concerning a linear function $L : \mathbb{R}^n \to \mathbb{R}^n$ whose domain and co-domain are the same space $\mathbb{R}^n$.

1.3.10 Definition.   *An* isomorphism *on $\mathbb{R}^n$ is a linear function $L : \mathbb{R}^n \to \mathbb{R}^n$ mapping $\mathbb{R}^n$ onto itself.*

1.3.11 Theorem.   *A linear function $L : \mathbb{R}^n \to \mathbb{R}^n$ is an isomorphism on $\mathbb{R}^n$ if and only if L is 1–1.*

Proof.   By Definition 1.3.10, L is an isomorphism on $\mathbb{R}^n$ if and only if $\operatorname{im} L = \mathbb{R}^n$. By 1.3.7 (applied to $\mathbb{R}^n$ as domain) this is so if and only if $\ker L = \{\mathbf{0}\}$. The theorem follows from Theorem 1.3.9.

We now outline the procedure for representing a linear function $L : \mathbb{R}^m \to \mathbb{R}^n$ by a matrix. By Theorem 1.3.4, the function L is determined by the images under L of the standard basis $\mathbf{e}_1, \ldots, \mathbf{e}_m$ of its domain $\mathbb{R}^m$. We must avoid confusing the standard bases of $\mathbb{R}^m$ and of $\mathbb{R}^n$, so let us denote the standard basis of $\mathbb{R}^n$ by $\mathbf{e}_1^*, \ldots, \mathbf{e}_n^*$. Suppose that for each $j = 1, \ldots, m$,

$$L(\mathbf{e}_j) = a_{1j}\mathbf{e}_1^* + \cdots + a_{nj}\mathbf{e}_n^*, \qquad a_{ij} \in \mathbb{R}, \quad i = 1, \ldots, n.$$

Then for any $\mathbf{x} = (x_1, \ldots, x_m) \in \mathbb{R}^m$, if $L(\mathbf{x}) = \mathbf{y} = (y_1, \ldots, y_n) \in \mathbb{R}^n$,

1.3.12
$$y_i = \sum_{j=1}^{m} a_{ij}x_j, \qquad i = 1, \ldots, n.$$

Formula 1.3.12 is conveniently written in matrix form

1.3.13
$$\begin{bmatrix} y_1 \\ \vdots \\ y_n \end{bmatrix} = \begin{bmatrix} a_{11} & \cdots & a_{1m} \\ \vdots & & \vdots \\ a_{n1} & \cdots & a_{nm} \end{bmatrix} \begin{bmatrix} x_1 \\ \vdots \\ x_m \end{bmatrix},$$

the evaluation being performed by the usual matrix multiplication.

We say that a vector $\mathbf{z} = (z_1, \ldots, z_q) \in \mathbb{R}^q$ is represented relative to the standard basis of $\mathbb{R}^q$ by the column ($q \times 1$ matrix)

$$[\mathbf{z}] = \begin{bmatrix} z_1 \\ \vdots \\ z_q \end{bmatrix}$$

and that the linear function $L : \mathbb{R}^m \to \mathbb{R}^n$ discussed above is represented relative to the standard bases of $\mathbb{R}^m$ and $\mathbb{R}^n$ by the $n \times m$ matrix

$$[L] = A = \begin{bmatrix} a_{11} & \cdots & a_{1m} \\ \vdots & & \\ a_{n1} & \cdots & a_{nm} \end{bmatrix}.$$

Then the matrix representation 1.3.13 of the relation $\mathbf{y} = L(\mathbf{x})$ can be expressed in the form

1.3.14
$$[\mathbf{y}] = A[\mathbf{x}],$$

the evaluation being performed by the usual matrix multiplication. The matrix A is sometimes denoted by its general term: $A = [a_{ij}]$.

Note that the columns of the matrix $[L]$ are simply the representatives of $L(\mathbf{e}_1), \ldots, L(\mathbf{e}_m)$ relative to the standard basis of $\mathbb{R}^n$.

1.3.15 *Example.* The linear function $L : \mathbb{R}^3 \to \mathbb{R}^2$ specified by

$$L(1, 0, 0) = (2, 0), \, L(0, 1, 0) = (-1, 3), \, L(0, 0, 1) = (0, 2)$$

is represented relative to the standard bases of $\mathbb{R}^3$ and $\mathbb{R}^2$ by the 2×3 matrix

$$A = \begin{bmatrix} 2 & -1 & 0 \\ 0 & 3 & 2 \end{bmatrix}.$$

By the linearity of L we have

1.3.16 $(y_1, y_2) = L(x_1, x_2, x_3) = (2x_1 - x_2, 3x_2 + 2x_3)$.

The relation 1.3.16 is conveniently remembered in the form 1.3.14.

1.3.17 *Example*.   The identity function $I : \mathbb{R}^n \to \mathbb{R}^n$ defined by

$$I(\mathbf{x}) = \mathbf{x}, \qquad \mathbf{x} \in \mathbb{R}^n$$

is a linear function. We denote the matrix representing it by I.

Given linear functions $L : \mathbb{R}^m \to \mathbb{R}^n$ and $M : \mathbb{R}^n \to \mathbb{R}^q$, the composite function $(M \circ L) : \mathbb{R}^m \to \mathbb{R}^q$ is defined by the rule

1.3.18 $(M \circ L)(\mathbf{x}) = M(L(\mathbf{x})), \qquad \mathbf{x} \in \mathbb{R}^m.$

The $q \times m$ matrix representing $M \circ L$ relative to the standard bases of $\mathbb{R}^m$ and $\mathbb{R}^q$ is given by the matrix product

1.3.19 $[M \circ L] = [M][L],$

which is evaluated in the usual way.

Exercises 1.3

1. (a)   Which of the following functions $f : \mathbb{R}^3 \to \mathbb{R}^2$ are linear? If non-linear, demonstrate a breakdown of the conditions 1.3.2 or 1.3.3.
    (i) $f(x_1, x_2, x_3) = (3x_1 - x_3, x_1 + x_3)$;
    (ii) $f(x_1, x_2, x_3) = (x_1^2 - x_2, x_2 + x_3)$;
    (iii) $f(x_1, x_2, x_3) = (\sin(x_1 x_2 x_3), 0)$.

Answers: Only (i) is linear. (ii) $f(\mathbf{e}_1 + \mathbf{e}_1) \neq f(\mathbf{e}_1) + f(\mathbf{e}_1)$.
    (iii) $f(\mathbf{e}_1 + \mathbf{e}_2 + \mathbf{e}_3) \neq f(\mathbf{e}_1) + f(\mathbf{e}_2) + f(\mathbf{e}_3)$.

    (b)   Prove that a transformation $f : \mathbb{R}^3 \to \mathbb{R}^2$ is linear if and only if there exist $a_{ij} \in \mathbb{R}$, $i = 1, 2$, and $j = 1, 2, 3$, such that, for all $(x_1, x_2, x_3) \in \mathbb{R}^3$

$$f(x_1, x_2, x_3) = (a_{11}x_1 + a_{12}x_2 + a_{13}x_3, a_{21}x_1 + a_{22}x_2 + a_{23}x_3).$$

2. Let $L : \mathbb{R}^m \to \mathbb{R}^n$ be a linear function. Prove that $\ker L$ is a subspace of $\mathbb{R}^m$ and that $\operatorname{im} L$ is a subspace of $\mathbb{R}^n$.

3. A linear function $L : \mathbb{R}^4 \to \mathbb{R}^3$ is defined by the rule

$$L(x_1, x_2, x_3, x_4)$$
$$= (x_1 + 2x_2 - 2x_4, x_3 + x_4, x_1 + 2x_2 + x_3 - x_4).$$

(a) Write down the 3×4 matrix A representing L relative to the standard bases of $\mathbb{R}^4$ and $\mathbb{R}^3$.

(b) Calculate im L. (*Hint*: im L is spanned by $L(\mathbf{e}_j)$, $j = 1, 2, 3, 4$.)

(c) Calculate ker L.

(d) Verify that dim im $L +$ dim ker $L = 4$.

Answers: (a) $A = \begin{bmatrix} 1 & 2 & 0 & -2 \\ 0 & 0 & 1 & 1 \\ 1 & 2 & 1 & -1 \end{bmatrix}$,

(b) im $L =$ span $\{(1, 0, 1), (0, 1, 1)\} \subseteq \mathbb{R}^3$.

This can be read off from the first and third columns of A, which represent $L(\mathbf{e}_1)$ and $L(\mathbf{e}_3)$. It will be seen that $L(\mathbf{e}_2) = 2L(\mathbf{e}_1)$ and $L(\mathbf{e}_4) = -2L(\mathbf{e}_1) + L(\mathbf{e}_3)$; (c) ker $L =$ span $\{(2, 0, -1, 1), (-2, 1, 0, 0)\} \subseteq \mathbb{R}^4$.

4. Prove that if $L : \mathbb{R} \to \mathbb{R}^n$ is a linear function, then there exists a vector $\mathbf{c} \in \mathbb{R}^n$ such that $L(t) = t\mathbf{c}$ for all $t \in \mathbb{R}$.
    Hint: consider $L(1)$.

5. Prove that a linear function $L : \mathbb{R}^m \to \mathbb{R}^n$ is 1–1 if and only if ker $L = \{\mathbf{0}\}$.
    Hint: If L is linear, then $L(\mathbf{x}) = L(\mathbf{y})$ if and only if $L(\mathbf{x} - \mathbf{y}) = \mathbf{0}$.

1.4 Quadratic forms

The material in this section is used in Section 3.12. We do little more than state the required results. Familiarity with diagonalizing a symmetric matrix is assumed.

1.4.1 *Definition*. *A function* $q : \mathbb{R}^m \to \mathbb{R}$ *is called a* quadratic form *on* $\mathbb{R}^m$ *if there exists an* $m \times m$ *real symmetric matrix* $A = [a_{ij}]$ *(so* $a_{ij} = a_{ji}$, *for all* $i, j = 1, \ldots, m$*) such that*

1.4.2
$$q(\mathbf{h}) = [\mathbf{h}]^t A [\mathbf{h}] \qquad \mathbf{h} \in \mathbb{R}^m$$

(evaluated using matrix multiplication), where $[\mathbf{h}]^t$ *denotes the transpose of* $[\mathbf{h}]$. *We call* q *the* quadratic form *corresponding to the symmetric matrix* A.

Multiplying out 1.4.2, we obtain the *quadratic polynomial*

$$q(\mathbf{h}) = \sum_{i=1}^{m} \sum_{j=1}^{m} a_{ij} h_i h_j = a_{11} h_1^2 + 2a_{12} h_1 h_2 + \cdots.$$

1.4.3 *Example*. The quadratic form $q : \mathbb{R}^3 \to \mathbb{R}$ given by

$$q(\mathbf{h}) = -h_1^2 + 2h_2^2 - h_3^2 + 4h_1 h_2 + 8h_1 h_3 - 4h_2 h_3$$

corresponds to the symmetric matrix

$$A = \begin{bmatrix} -1 & 2 & 4 \\ 2 & 2 & -2 \\ 4 & -2 & -1 \end{bmatrix}.$$

1.4.4 Definition.   *A quadratic form q on $\mathbb{R}^m$ is* (i) positive definite,
(ii) negative definite, (iii) indefinite *if*
  [i] $q(\mathbf{h}) > 0$ *for all* $\mathbf{h} \in \mathbb{R}^m$, $\mathbf{h} \neq \mathbf{0}$,
  [ii] $q(\mathbf{h}) < 0$ *for all* $\mathbf{h} \in \mathbb{R}^m$, $\mathbf{h} \neq \mathbf{0}$,
  [iii] $q(\mathbf{h})$ *takes both positive and negative values.*

  *We also say that a real symmetric matrix A is positive definite,
negative definite or indefinite if the quadratic form corresponding to A
has this property.*

1.4.5 Example.   The quadratic form of Example 1.4.3 is indefinite, since
$q(\mathbf{h}) = 27$ when $\mathbf{h} = (1, -2, 2)$ and $q(\mathbf{h}) = -54$ when $\mathbf{h} = (-2, 1, 2)$.

  The following theorem provides the machinery for determining the
nature of quadratic forms.

1.4.6 Theorem.   *Let A be an $m \times m$ real symmetric matrix. Then*
  [i]   *the eigenvalues $\lambda_1, \ldots, \lambda_m$ of A are real;*
  [ii] *there exist pairwise orthogonal unit vectors $\mathbf{c}_1, \ldots, \mathbf{c}_m$ in $\mathbb{R}^m$
such that $A[\mathbf{c}_i] = \lambda_i[\mathbf{c}_i]$, $i = 1, \ldots, m$;*
  [iii] *if C is the $m \times m$ matrix whose columns are $[\mathbf{c}_1], \ldots, [\mathbf{c}_m]$,
then C is orthogonal (that is, $C^{\mathrm{t}} = C^{-1}$) and*

1.4.7 $$C^{\mathrm{t}}AC = \begin{bmatrix} \lambda_1 & & 0 \\ & \vdots & \\ 0 & & \lambda_m \end{bmatrix}.$$

  [iv] *If $\lambda_{\min}$ and $\lambda_{\max}$ are the minimum and the maximum eigenvalues
of A, then*

1.4.8 $\|\mathbf{h}\|^2 \lambda_{\min} \leqslant [\mathbf{h}]^{\mathrm{t}} A [\mathbf{h}] \leqslant \|\mathbf{h}\|^2 \lambda_{\max}$, $\mathbf{h} \in \mathbb{R}^m$.

  A proof of parts (i)–(iii) will be found in most textbooks on linear
algebra.

Proof of part (**iv**).   Put $[\mathbf{h}] = C[\mathbf{k}]$. Then

1.4.9
$$\begin{cases} [\mathbf{h}]^t A[\mathbf{h}] = [\mathbf{k}]^t C^t A C[\mathbf{k}] = \lambda_1 k_1^2 + \cdots + \lambda_m k_m^2, \\[2mm] \|\mathbf{k}\|^2 = [\mathbf{h}]^t C^t C[\mathbf{h}] = [\mathbf{h}]^t[\mathbf{h}] = \|\mathbf{h}\|^2. \end{cases}$$

Now,

$$\|\mathbf{k}\|^2 \lambda_{\min} = (k_1^2 + \cdots + k_m^2)\lambda_{\min} \leqslant \lambda_1 k_1^2 + \cdots + \lambda_m k_m^2 \leqslant \|\mathbf{k}\|^2 \lambda_{\max}.$$

The inequality 1.4.8 follows from 1.4.9.

1.4.10 *Example*.   Consider the matrix A of Example 1.4.3. It will be found that the eigenvalues of A are 3, 3, -6 (solve the determinantal equation $\det(A - tI) = 0$) and that the orthogonal matrix

$$C = \frac{1}{3}\begin{bmatrix} 1 & 2 & -2 \\ -2 & 2 & 1 \\ 2 & 1 & 2 \end{bmatrix} \quad \text{is such that} \quad C^t A C = \begin{bmatrix} 3 & 0 & 0 \\ 0 & 3 & 0 \\ 0 & 0 & -6 \end{bmatrix}.$$

It follows that

1.4.11 $[\mathbf{h}]^t A[\mathbf{h}] = 3k_1^2 + 3k_2^2 - 6k_3^2,$ where $[\mathbf{h}] = C[\mathbf{k}]$.

The inequality 1.4.8 becomes

$$-6\|\mathbf{h}\|^2 \leqslant [\mathbf{h}]^t A[\mathbf{h}] \leqslant 3\|\mathbf{h}\|^2.$$

The formula 1.4.11 is useful in obtaining values of $\mathbf{h}$ for which $[\mathbf{h}]^t A[\mathbf{h}]$ takes a prescribed value. For example it is clear from 1.4.11 that $[\mathbf{h}]^t A[\mathbf{h}]$ takes the value -54 when $k_1 = k_2 = 0$, $k_3 = 3$. The corresponding value of $\mathbf{h}$ is given by

$$\begin{bmatrix} h_1 \\ h_2 \\ h_3 \end{bmatrix} = C\begin{bmatrix} 0 \\ 0 \\ 3 \end{bmatrix} = \begin{bmatrix} -2 \\ 1 \\ 2 \end{bmatrix}.$$

This calculation was used in Example 1.4.5.

1.4.12 *Theorem*.   *The real symmetric matrix A is* [**i**] *positive definite*, [**ii**] *negative definite*, [**iii**] *indefinite if and only if the eigenvalues of A are* [**i**] *all positive*, [**ii**] *all negative*, [**iii**] *some positive and some negative*.

Proof.   Apply 1.4.9.

Exercises 1.4

1.   Decide whether the following matrices are positive definite, negative

definite or indefinite.

(a) $\begin{bmatrix} 2 & 1 & 0 \\ 1 & 2 & 0 \\ 0 & 0 & 2 \end{bmatrix}$,   (b) $\begin{bmatrix} 0 & 1 & 2 \\ 1 & 0 & 2 \\ 2 & 2 & 0 \end{bmatrix}$,   (c) $\begin{bmatrix} 0 & -1 & -1 \\ -1 & 0 & 0 \\ -1 & 0 & 0 \end{bmatrix}$,

(d) $\begin{bmatrix} -5 & 2 \\ 2 & -2 \end{bmatrix}$.

Answers: (a) positive-definite;   (b) indefinite;   (c) indefinite;
(d) negative-definite.

2. Let A be an $m \times m$ real symmetric matrix with eigenvalues $\lambda_1, \ldots, \lambda_m$. Prove that the determinant of A is equal to the product of its eigenvalues: $\det A = \lambda_1 \cdots \lambda_m$.

Hint: apply Theorem 1.4.6(iii).

3. A quadratic form $q : \mathbb{R}^m \to \mathbb{R}$ and its associated symmetric matrix A are said to be *positive semi-definite* (*negative semi-definite*) if $q(\mathbf{h}) \geqslant 0$ $(q(\mathbf{h}) \leqslant 0)$ for all $\mathbf{h} \in \mathbb{R}^m$, but there exists $\mathbf{h} \neq \mathbf{0}$ such that $q(\mathbf{h}) = 0$.

   Prove that q is positive semi-definite (negative semi-definite) if and only if the eigenvalues of the associated matrix A are all non-negative (non-positive) and at least one is zero.

   Show that the symmetric matrix

$$A = \begin{bmatrix} 3 & -2 & 2 \\ -2 & 4 & 0 \\ 2 & 0 & 2 \end{bmatrix}$$

   is positive semi-definite. Find $\mathbf{h} \neq \mathbf{0}$ such that $[\mathbf{h}]^t A [\mathbf{h}] = 0$.

Answer: $\mathbf{h} = (2, 1, -2)$, for example.

1.5   Functions from $\mathbb{R}^m$ to $\mathbb{R}^n$

A *vector-valued function* $f : S \subseteq \mathbb{R}^m \to \mathbb{R}^n$ with domain $S \subseteq \mathbb{R}^m$ and codomain $\mathbb{R}^n$ is a rule that assigns to each $\mathbf{p} \in S$ a unique $f(\mathbf{p}) \in \mathbb{R}^n$. Putting

$$f(\mathbf{p}) = (f_1(\mathbf{p}), \ldots, f_n(\mathbf{p})), \qquad \mathbf{p} \in S$$

the real-valued functions $f_i : S \subseteq \mathbb{R}^m \to \mathbb{R}$ so defined are called the *coordinate functions* of f. The function f is then conveniently expressed as $f = (f_1, \ldots, f_n)$. The *image* of f is the subset $f(S)$ of $\mathbb{R}^n$

consisting of the vectors $f(\mathbf{p})$ for all $\mathbf{p} \in S$. The expressions

$$x_1 = f_1(\mathbf{p}), \ldots, x_n = f_n(\mathbf{p}), \qquad \mathbf{p} \in S$$

are said to give a *parametrization* of the image of f.

1.5.1 *Example.* The function $f : \mathbb{R}^2 \to \mathbb{R}^3$ defined for $c > 0$ by

1.5.2 $f(\phi, \theta) = (c \sin \phi \cos \theta, c \sin \phi \sin \theta, c \cos \phi),$ $\qquad (\phi, \theta) \in \mathbb{R}^2$

determines a parametrization of the sphere $x^2 + y^2 + z^2 = c^2$, centre the origin O and radius c. The parameters ϕ and θ are related to the geometry of the sphere as follows. Consider a point P on the sphere and let OQ denote the perpendicular projection of OP on to the x, y plane (see Fig. 1.3). If the angle between OP and the z-axis is ϕ and the angle between OQ and the x-axis is θ then the coordinates of P are given by $f(\phi, \theta)$.

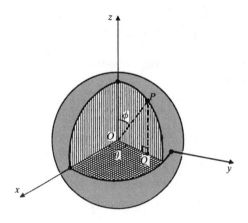

Fig. 1.3

The parametrization 1.5.2 is not 1–1. For example, $f(\phi, \theta + 2\pi) = f(\phi, \theta)$. By restricting the domain of f by the conditions

$$0 < \phi < \pi, \qquad 0 \leqslant \theta < 2\pi$$

we obtain a 1–1 parametrization of the sphere excluding the poles $(0, 0, \pm c)$.

In Chapter 2, where we consider functions $f : D \subseteq \mathbb{R} \to \mathbb{R}^n$, the domain D of f is usually taken to be an interval in $\mathbb{R}$. An *open interval* in $\mathbb{R}$ is of one of the following forms:

$$] - \infty, a[, \qquad]a, b[, \qquad]b, \infty[\qquad \text{and} \qquad \mathbb{R},$$

where, for example, $]a, b[= \{t \in \mathbb{R} | a < t < b\}$. A *closed interval* has the form

$$] - \infty, a], \qquad [a, b], \qquad [b, \infty[\qquad \text{and} \qquad \mathbb{R},$$

where, for example, $[a, b] = \{t \in \mathbb{R} | a \leqslant t \leqslant b\}$. The half-open intervals $]a, b]$ and $[a, b[$ are defined similarly. In later chapters we study functions whose domains are subsets of $\mathbb{R}^m$, $m \geqslant 2$. The most usual domains then are *open sets*. These are defined in Section 3.2. The domain of a function is often conveniently expressed as a Cartesian product. Given $A \subseteq \mathbb{R}^p$, $B \subseteq \mathbb{R}^q$, the *Cartesian product* of A and B is the set

$$A \times B = \{(\mathbf{a}, \mathbf{b}) \in \mathbb{R}^{p+q} | \mathbf{a} \in A, \mathbf{b} \in B\}.$$

1.5.3 Example. [i] $\mathbb{R} \times \mathbb{R} = \mathbb{R}^2$, and more generally $\mathbb{R}^p \times \mathbb{R}^q = \mathbb{R}^{p+q}$
  [ii] The set D given by 1.1.2 is equal to $\mathbb{R}^2 \times [-1, 1]$.

The following definition gives a number of important ways of combining functions.

1.5.4 Definition. *Given* $f : S \subseteq \mathbb{R}^m \to \mathbb{R}^n$, $g : S \subseteq \mathbb{R}^m \to \mathbb{R}^n$ *and* $\phi : S \subseteq \mathbb{R}^m \to \mathbb{R}$, *we define*
  [i] *the sum and difference functions* $f \pm g : S \subseteq \mathbb{R}^m \to \mathbb{R}^n$ *by*

$$(f \pm g)(\mathbf{x}) = f(\mathbf{x}) \pm g(\mathbf{x}), \qquad \mathbf{x} \in S;$$

  [ii] *the product* $\phi f : S \subseteq \mathbb{R}^m \to \mathbb{R}^n$ *by*

$$(\phi f)(\mathbf{x}) = \phi(\mathbf{x}) f(\mathbf{x}) = (\phi(\mathbf{x}) f_1(\mathbf{x}), \dots, \phi(\mathbf{x}) f_n(\mathbf{x})), \qquad \mathbf{x} \in S;$$

  [iii] *the dot product* $f \cdot g : S \subseteq \mathbb{R}^m \to \mathbb{R}$ *by*

$$(f \cdot g)(\mathbf{x}) = f_1(\mathbf{x}) g_1(\mathbf{x}) + \cdots + f_n(\mathbf{x}) g_n(\mathbf{x}), \qquad \mathbf{x} \in S,$$

where $f_i : S \subseteq \mathbb{R}^m \to \mathbb{R}$ *and* $g_i : S \subseteq \mathbb{R}^m \to \mathbb{R}$, $i = 1, \dots, n$ *are the coordinate functions of* f *and* g;
  [iv] *for the case* $f : S \subseteq \mathbb{R}^m \to \mathbb{R}^3$, $g : S \subseteq \mathbb{R}^m \to \mathbb{R}^3$, *the cross product function* $h = f \times g : S \subseteq \mathbb{R}^m \to \mathbb{R}^3$ *by its coordinate functions*

$$h_1 = f_2 g_3 - f_3 g_2, \qquad h_2 = f_3 g_1 - f_1 g_3, \qquad h_3 = f_1 g_2 - f_2 g_1.$$

1.5.5 Example. Given $f : \mathbb{R}^2 \to \mathbb{R}^3$ and $g : \mathbb{R}^2 \to \mathbb{R}^3$ such that

$$f(x, y) = (x, y, xy) \quad \text{and} \quad g(x, y) = (-y, x, 0), \qquad (x, y) \in \mathbb{R}^2$$

then

$$(f \cdot g)(x, y) = -xy + yx + 0 = 0$$

and

$$(f \times g)(x, y) = (0 - x^2 y, -xy^2 - 0, x^2 + y^2)$$
$$= (-x^2 y, -xy^2, x^2 + y^2).$$

1.5.6 Definition. *Let* $g : E \subseteq \mathbb{R}^l \to \mathbb{R}^m$ *and* $f : D \subseteq \mathbb{R}^m \to \mathbb{R}^n$ *be such that the domain* D *of* f *contains the image* $g(E)$ *of* g. *The* composite *function* $f \circ g : E \subseteq \mathbb{R}^l \to \mathbb{R}^n$ *is defined by*

$$(f \circ g)(\mathbf{x}) = f(g(\mathbf{x})), \qquad \mathbf{x} \in E.$$

1.5.7 Example. Define functions $g : \mathbb{R}^3 \to \mathbb{R}^2$ and $f : \mathbb{R}^2 \to \mathbb{R}^2$ by

$$g(t_1, t_2, t_3) = (t_1 t_2, t_2 t_3) \qquad (t_1, t_2, t_3) \in \mathbb{R}^3,$$
$$f(x_1, x_2) = (\sin x_1, x_1 x_2) \qquad (x_1, x_2) \in \mathbb{R}^2.$$

Then

$$(f \circ g)(t_1, t_2, t_3) = (\sin(t_1 t_2), t_1 t_2^2 t_3).$$

We conclude with a short list describing further notation used in the book.

$\mathbb{N}, \mathbb{Z}$	the natural numbers, integers
$A \backslash B$	$\{x \in A \mid x \notin B\}$
$f : U \to V$	function on domain U, codomain V, with $f(U) \subseteq V$.

Exercises 1.5

1. Let $f_i : S \subseteq \mathbb{R}^m \to \mathbb{R}$, $i = 1, \ldots, n$ be the coordinate functions of the function $f : S \subseteq \mathbb{R}^m \to \mathbb{R}^n$. Prove that

$$f_i(\mathbf{x}) = f(\mathbf{x}) \cdot \mathbf{e}_i, \qquad \mathbf{x} \in S^m, \quad i = 1, \ldots, n,$$

where $\mathbf{e}_1, \ldots, \mathbf{e}_n$ is the standard basis of $\mathbb{R}^n$.

2. Let $f : \mathbb{R}^3 \to \mathbb{R}^3$ and $g : \mathbb{R}^3 \to \mathbb{R}^3$ be defined by

$$f(x_1, x_2, x_3) = (x_2, -x_3, x_1),$$
$$g(x_1, x_2, x_3) = (x_1 x_3, x_1 x_2, x_2 x_3).$$

Determine the functions $f \cdot g, f \times g, g \times f, f \circ g$ and $g \circ f$. (Note that $f \circ g$ and $g \circ f$ are both defined. In Example 1.5.7 this is not the case.)

Answers: $(f \cdot g)(\mathbf{x}) = x_1 x_2 x_3$;
   $(f \times g)(\mathbf{x}) = (-x_2(x_1^2 + x_3^2), x_3(x_1^2 - x_2^2), x_1(x_2^2 + x_3^2))$;
   $g \times f = -f \times g$; $(f \circ g)(t_1, t_2, t_3) = (t_1 t_2, -t_2 t_3, t_1 t_3)$;
   $(g \circ f)(t_1, t_2, t_3) = (t_1 t_2, -t_2 t_3, -t_1 t_3)$.

3.   Identify and sketch the images of the following functions f from $\mathbb{R}^2$ to
     $\mathbb{R}^3$. Which of the functions are 1–1?
     (a) $f(u, v) = (2u, 4v, u + v - 2)$, $(u, v) \in \mathbb{R}^2$;
     (b) $f(\phi, \theta) = (\sin \phi \cos \theta, 2 \sin \phi \sin \theta, 3 \cos \phi)$, $(\phi, \theta) \in \mathbb{R}^2$;
     (c) $f(u, v) = (2 \cos u, 2 \sin u, v)$, $(u, v) \in \mathbb{R}^2$;
     (d) $f(u, v) = (v \cos u, v \sin u, v)$, $(u, v) \in \mathbb{R}^2$;
     (e) $f(x, y) = (x, y, \sqrt{(x^2 + y^2)})$, $(x, y) \in \mathbb{R}^2$;
     (f) $f(x, y) = (x, y, x^2 + y^2)$, $(x, y) \in \mathbb{R}^2$;
     (g) $f(x, y) = (x, y, \sqrt{(9 - x^2 - y^2)})$, $x^2 + y^2 \leqslant 9$.

Answers: (a) plane $2x + y - 4z = 8$; (b) ellipsoid $x^2 + \frac{1}{4}y^2 + z^2 = 1$; (c)
   circular cylinder; (d) right circular cone; (e) half-cone; (f) parabolic
   bowl; (g) hemispherical bowl.

1.6   Elementary real analysis and calculus

In this section we summarize the material from elementary analysis
on which we base our generalizations.

1.6.1 Definition.   *A set E of real numbers is* bounded above (below)
*if there exists $b \in \mathbb{R}$ such that $b \geqslant x$ ($b \leqslant x$) for all $x \in E$. The smallest
(largest)number b with this property is called the* least upper bound
or supremum *of E (greatest lower bound or infimum of E).*

1.6.2 Example.   The open interval $D =]-\infty, 5[$ is bounded above but not
below. Its least upper bound is 5. Note that $5 \notin D$. In general, an interval is
open if and only if it contains neither its greatest lower bound nor its least
upper bound (if these exist).

   We next consider sequences $a_1, a_2, a_3, \ldots$ of real numbers, which
we denote in short by (a_k).

1.6.3 Definition.   [i] *The sequence (a_k) of real numbers* converges *to
$a \in \mathbb{R}$ if to each $\varepsilon > 0$ there corresponds $K \in \mathbb{N}$ such that*

$$|a_k - a| < \varepsilon \quad \text{whenever} \quad k > K.$$

In this case we write $a_k \to a$ and also $\lim_{k \to \infty} a_k = a$.

[ii] *The sequence (a_k) of real numbers is a* Cauchy *sequence if to each $\varepsilon > 0$ there corresponds $K \in \mathbb{N}$ such that*

$$|a_k - a_l| < \varepsilon \quad whenever \quad k, l > K.$$

1.6.4 Theorem.   *A Cauchy sequence is convergent.*

An interval in $\mathbb{R}$ is said to be *compact* if it is closed and bounded. Thus the compact intervals are of the form $[a, b]$.

1.6.5 Theorem.   (*Bolzano–Weierstrass*) *Every sequence of real numbers in a compact interval $[a, b]$ has a subsequence that converges to a limit in $[a, b]$.*

We shall find that functions which are 1–1 on their domains play an important role in vector calculus. Suppose that $f : D \subseteq \mathbb{R}^m \to \mathbb{R}^n$ is 1–1 on D. Then we can define a (unique) function $g : f(D) \to D$ which reverses the effect of f, by the rule

$$g(f(\mathbf{x})) = \mathbf{x}, \qquad f(\mathbf{x}) \in f(D).$$

Thus if f maps $\mathbf{x}$ to $\mathbf{y}$ then g maps $\mathbf{y}$ back to $\mathbf{x}$. The function g is called the *inverse* of f and the function f is said to be *invertible*. For example, the function $f : [-1, 0] \subseteq \mathbb{R} \to \mathbb{R}$ given by the rule

$$f(x) = x^2, \qquad x \in [-1, 0]$$

is invertible, and the inverse function is $g : [0, 1] \to [-1, 0]$, where

$$g(y) = -\sqrt{y}, \qquad y \in [0, 1].$$

In contrast, the rule $f(x) = x^2, x \in [-1, 1]$ does not define an invertible function, since f is not 1–1 on the extended domain. Note that if f is invertible (on its domain) then the inverse function g is also invertible and the inverse of g is f.

We shall study inverse functions in some detail in Sections 4.5 and 4.6. The following results are needed for applications in Chapter 2.

1.6.6 Definition.   *Let D be an interval in $\mathbb{R}$. A function $f : D \subseteq \mathbb{R} \to \mathbb{R}$ is strictly increasing (strictly decreasing) if $f(b) > f(a)$ $(f(b) < f(a))$ whenever $b > a$, where $a, b \in D$. A function is strictly monotonic if it is either strictly increasing or strictly decreasing.*

1.6.7 Theorem.   *Let $f : D \subseteq \mathbb{R} \to \mathbb{R}$ be a differentiable function on an interval D such that f has positive (negative) derivative throughout D.*

Then f is 1–1 and strictly increasing (strictly decreasing). Its inverse function $f^{-1} : f(D) \subseteq \mathbb{R} \to \mathbb{R}$ is also differentiable and has positive (negative) derivative throughout $f(D)$.

Our next results concern continuity and differentiability. These concepts are defined for functions $f : D \subseteq \mathbb{R} \to \mathbb{R}^n$ in Chapter 2. We will require the following classical theorems about real-valued functions.

1.6.8 *The Intermediate-Value Theorem.* *Let $f : [a, b] \subseteq \mathbb{R} \to \mathbb{R}$ be a continuous function. If t is any point between the points $f(a)$ and $f(b)$ then there exists a point $x \in [a, b]$ such that $f(x) = t$.*

1.6.9 *The Mean-Value Theorem.* *Let p and $h > 0$ be real numbers, and suppose that $f : [p, p + h] \to \mathbb{R}$ is a real-valued function such that (a) f is continuous on $[p, p + h]$, and (b) f is differentiable at every point of the open interval $]p, p + h[$. Then there exists a point c in the open interval $]p, p + h[$ such that*

$$f(p + h) = f(p) + hf'(c).$$

1.6.10 *The Generalized Mean-Value Theorem* (Taylor's Theorem). *Let $f : D \subseteq \mathbb{R} \to \mathbb{R}$ be a real-valued function on an open interval D in $\mathbb{R}$, and suppose that f is n times differentiable at every point of D. If p and $p + h$ are points of D, then there exists in D a point $c = p + \theta h, 0 < \theta < 1$, such that*

$$f(p + h) = f(p) + hf'(p) + \frac{h^2}{2!}f''(p) + \cdots$$

$$+ \frac{h^{n-1}}{(n-1)!}f^{(n-1)}(p) + \frac{h^n}{n!}f^{(n)}(c).$$

1.6.11 *Theorem.* The Chain Rule. *Let $g : E \subseteq \mathbb{R} \to \mathbb{R}$ and $f : D \subseteq \mathbb{R} \to \mathbb{R}$ be defined on open intervals E and D such that $g(E) \subseteq D$. Suppose that g is differentiable at $a \in E$ and that f is differentiable at $g(a) \in D$. Then the composite function $F = f \circ g : E \subseteq \mathbb{R} \to \mathbb{R}$ is differentiable at $a \in E$ and*

$$F'(a) = (f \circ g)'(a) = f'(g(a))g'(a).$$

1.6.12 *Remark.* The Chain Rule is sometimes presented in the following form 1.6.13, which is easily remembered but lacks the precision of the above statement of the rule. In the expression

$$F(t) = (f \circ g)(t) = f(g(t)), \qquad t \in E,$$

put $u = g(t)$. Then $F(t) = f(u)$ and

1.6.13
$$\frac{\mathrm{d}F}{\mathrm{d}t} = \frac{\mathrm{d}f}{\mathrm{d}u}\frac{\mathrm{d}u}{\mathrm{d}t}.$$

1.6.14 *Example*.   Consider the function $F : \mathbb{R} \to \mathbb{R}$ defined by $F(t) = \sin(t^2)$, $t \in \mathbb{R}$. Then $F = f \circ g$, where $f(u) = \sin u$ and $g(t) = t^2$. By the Chain Rule

$$F'(a) = \cos(g(a))2a = 2a\cos(a^2).$$

Alternatively, Formula 1.6.13 can be applied by putting $u = t^2$ and $f(u) = \sin u$.

We shall need the following two classical theorems of elementary calculus.

1.6.15 *The Fundamental Theorem of Calculus*.   Let $f : [p, q] \subseteq \mathbb{R} \to \mathbb{R}$ *be such that the derivative f' is continuous. Then*

$$\int_p^q f'(x)\mathrm{d}x = f(q) - f(p).$$

1.6.16 *The Integral Mean-Value Theorem*.   *If* $f : [a, b] \subseteq \mathbb{R} \to \mathbb{R}$ *is continuous then there exists $\xi \in {]}a, b{[}$ such that*

$$\int_a^b f(x)\mathrm{d}x = (b - a)f(\xi).$$

The concept of uniform continuity is also important. Specifically we shall require the following result and its generalization.

1.6.17 *Definition*.   *A function $f : D \subset \mathbb{R} \to \mathbb{R}$ is* uniformly continuous *on D if to each $\varepsilon > 0$ there corresponds $\delta > 0$ (depending only on ε) such that*

$$|f(t) - f(s)| < \varepsilon \quad whenever \quad |t - s| < \delta, \quad t, s \in D.$$

1.6.18 *Theorem*.   *A function $f : [a, b] \subseteq \mathbb{R} \to \mathbb{R}$ which is continuous on the closed bounded interval $[a, b]$ is uniformly continuous on $[a, b]$.*

We shall use the following well-known comparison theorem.

1.6.19 Theorem. *Let $g : [c, d] \subseteq \mathbb{R} \to \mathbb{R}$ be a bounded function and suppose that $m \leqslant g(y) \leqslant M$ for all $y \in [c, d]$. If g is integrable over $[c, d]$, then*

$$m(d - c) \leqslant \int_c^d g(y)\mathrm{d}y \leqslant M(d - c).$$

Exercises 1.6

1. Prove that (a) $\lim_{k \to \infty} (1/k)\sin k = 0$;   (b) $\lim_{k \to \infty} k \sin (1/k) = 1$.

2. Prove that if (a_k) and (b_k) are convergent sequences of real numbers such that $a_k \to a$ and $b_k \to b$, then the sequences $(a_k + b_k)$ and $(a_k b_k)$ are convergent and $a_k + b_k \to a + b$, $a_k b_k \to ab$.

    Suppose $b_k \neq 0$ for all $k \in \mathbb{N}$. What can you say about the sequence (a_k/b_k)?

Hint: Consider for example the sequences (b_k) where
    (a) $b_k = (1 + k)/k$ and (b) $b_k = 1/k$.

3. Suppose that $f :]a, b[\subseteq \mathbb{R} \to \mathbb{R}$ is differentiable, and $f'(x) = 0$ for all $x \in]a, b[$. Prove that f is constant on $]a, b[$.

Hint: take $p, q \in]a, b[$ and use the Mean-Value Theorem to prove that $f(p) = f(q)$.

4. Prove that when $a > 0$ and $|h| < a, \sqrt{(a^2 + h)}$ differs from $a + \frac{1}{2}(h/a)$ by an amount less than $h^2/8a^3$. Show in particular that

$$10.392 < \sqrt{108} < 10.400.$$

Hint: Apply Taylor's Theorem.

5. Apply the Chain Rule to find $F'(\pi)$, given that

$$F(t) = (\sin t)^2 + e^{\sin t}, \qquad t \in \mathbb{R}.$$

Answer: -1.

6. Let $f :]0, 1[\subseteq \mathbb{R} \to \mathbb{R}$ be defined by $f(t) = 1/t, t \in]0, 1[$.
    (a) Prove that f is continuous at every point $p \in]0, 1[$. (Show that given $\varepsilon > 0$, however small, there exists $\delta > 0$ such that $|f(t) - f(p)| < \varepsilon$ whenever $t \in]0, 1[$ and $|t - p| < \delta$.)
    (b) Prove that f is not uniformly continuous on $]0, 1[$.

Answer: for a given $\varepsilon > 0$, the choice of δ in part (a) depends on the point p. A possible choice is, for example, $\delta = \min \{\frac{1}{2}p, \frac{1}{2}\varepsilon p^2\}$, for then if $|t - p| < \delta$ it follows that $t > \frac{1}{2}p$ and

$$|f(t) - f(p)| = |t - p|/|tp| < \frac{1}{2}\varepsilon p^2/\frac{1}{2}p^2 = \varepsilon.$$

(b) Take $\varepsilon = \frac{1}{2}$, for example. Then it is not possible to choose a $\delta > 0$ to satisfy the continuity condition for all $p \in]0, 1[$. However δ is chosen, there exists an integer $n > 1$ such that $\delta > 1/n$. Consider $p = 1/n$, $t = 1/(n + 1)$. Then $|t - p| < \delta$, but $|f(t) - f(p)| = 1 > \varepsilon$.)

7. Prove the following generalization of the Fundamental Theorem of Calculus 1.6.15. Let $f : [a, b] \subseteq \mathbb{R} \to \mathbb{R}$ be continuous and suppose that there exists a subdivision $a = p_0 < p_1 < \cdots < p_r = b$ of $[a, b]$ such that for each $k = 1, \ldots, r$ the function f restricted to the subinterval $[p_{k-1}, p_k]$ has continuous derivative. Prove that

$$\int_a^b f'(x)\mathrm{d}x = f(b) - f(a),$$

where the integral is given the natural interpretation

$$\int_a^b f'(x)\mathrm{d}x = \sum_{k=1}^{r} \int_{p_{k-1}}^{p_k} f'(x)\mathrm{d}x.$$

(Note that $f'(x)$ is not necessarily defined at $x = p_k$, but the left-hand and right-hand derivatives of f exist at such points. The function f is said to be *piecewise continuously differentiable*.)

8. (a) Illustrate Exercise 7 with the continuous function $f : [-1, 2] \subseteq \mathbb{R} \to \mathbb{R}$ defined by $f(x) = |x|$, $x \in [-1, 2]$.

Answer: $\displaystyle\int_{-1}^{2} f'(x)\mathrm{d}x = f(2) - f(-1) = 1.$

(b) Comment on the condition that f be continuous in Exercise 7. (Consider, for example, the discontinuous function $f : [-1, 2] \subseteq \mathbb{R} \to \mathbb{R}$ defined by

$$f(x) = \begin{cases} -x + 1, & \text{when } -1 \leqslant x \leqslant 0 \\ x, & \text{when } 0 < x \leqslant 2. \end{cases})$$

Vector-valued functions of $\mathbb{R}$

2.1 Introduction

Linear algebra is the study of linear functions from one vector space into another. In this book we shall be concerned with functions from $\mathbb{R}^m$ into $\mathbb{R}^n$ that can be approximated (in a sense to be made precise) by linear functions. This is the main theme of differential vector calculus.

In the present chapter we confine ourselves to the study of vector-valued functions $f: D \subseteq \mathbb{R} \to \mathbb{R}^n$ *where D is an interval in $\mathbb{R}$* and the dimension n of the codomain is open to choice. We shall see that such functions are significant in both geometry and dynamics. Geometrically, if f is 'continuous' on an interval D, then the image of f is a curve and the 'derivatives' of f describe the way the curve twists and turns in $\mathbb{R}^n$. In dynamics, on the other hand, the position of a particle moving in space is a function of time, so that with respect to suitable frames of reference we have a corresponding function $f: D \subseteq \mathbb{R} \to \mathbb{R}^3$ where at time t the position vector of the particle is $f(t)$. The 'derivatives' of f relate to the velocity and acceleration of the particle. Before having a closer look at geometry and dynamics, however, we must explain precisely what is meant by the continuity and differentiability of vector-valued functions of $\mathbb{R}$.

In the following definitions we introduce three fundamental ideas and then explore their relationship by considering some simple examples.

2.1.1 Definition. *Corresponding to any function $f: D \subseteq \mathbb{R} \to \mathbb{R}^n$ we define* coordinate functions $f_i: D \subseteq \mathbb{R} \to \mathbb{R}$, $i = 1, \ldots, n$, *by means of the expression*

$$f(t) = (f_1(t), \ldots, f_n(t)), \qquad t \in D.$$

Our knowledge of calculus applied to the real-valued functions f_i will lead to the calculus of the vector-valued function f.

2.1.2 Definition. *The* image *of a function $f: D \subseteq \mathbb{R} \to \mathbb{R}^n$ is the* *subset of* $\mathbb{R}^n$ *given by* $\{f(t)\in\mathbb{R}^n|t\in D\}$. *The expressions*

$$x_1 = f_1(t), \ldots, x_n = f_n(t), \qquad t\in D$$

are said to give a parametrization *of the image of f with parameter t.*

2.1.3 Definition. *The* graph *of a function* $f : D \subseteq \mathbb{R} \to \mathbb{R}^n$ *is the subset of* $\mathbb{R}^{n+1}$ *given by* $\{(t, f_1(t), \ldots, f_n(t))|t\in D\}$. *We could also express the graph as the set* $\{(t, f(t))|t\in D\} \subseteq \mathbb{R}^{n+1}$.

For the case $n = 1$, Definition 2.1.3 reduces to the usual high-school concept of the graph of $f : \mathbb{R} \to \mathbb{R}$ as a subset of $\mathbb{R}^2$. For example, the graph of the sine function consists of all points $(t, \sin t)$ in $\mathbb{R}^2$, that is, all points (t, y) such that $y = \sin t$.

2.1.4 Example. The rule

$$f(t) = (1, 2, -3) + t(2, 0, 1), \qquad t\in\mathbb{R}$$

defines a function $f : \mathbb{R} \to \mathbb{R}^3$ whose image is the straight line in $\mathbb{R}^3$ through $(1, 2, -3)$ in direction $(2, 0, 1)$. The coordinate functions of f are given by

$$f_1(t) = 1 + 2t, \qquad f_2(t) = 2, \qquad f_3(t) = t - 3, \qquad t\in\mathbb{R},$$

and the corresponding parametrization of the straight line is $x = 1 + 2t$, $y = 2, z = t - 3$.

2.1.5 Example. The function $g : \mathbb{R} \to \mathbb{R}^2$ defined by

$$g(t) = (\cos t, \sin t), \qquad t\in\mathbb{R}$$

has image the unit circle $x^2 + y^2 = 1$ in $\mathbb{R}^2$ (see Fig. 2.1).

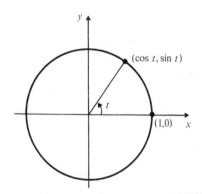

Fig. 2.1 Image of $g(t) = (\cos t, \sin t)$

The coordinate functions of g are given by $g_1(t) = \cos t$ and $g_2(t) = \sin t$, and the corresponding parametrization of the unit circle is $x = \cos t$, $y = \sin t$. The function g is periodic for, since $g(t + 2\pi) = g(t)$ for all t in $\mathbb{R}$, the values of $g(t)$ repeat regularly. The *period* of g is 2π, since this is the smallest positive number p such that $g(t + p) = g(t)$ for all t.

2.1.6 Example. Compare the function g of Example 2.1.5 with the function $h : \mathbb{R} \to \mathbb{R}^2$ defined by

$$h(t) = (\cos 2t, \sin 2t), \qquad t \in \mathbb{R}.$$

The image of h is the unit circle of Fig. 2.1 and yet g and h are different functions (for example $g(\pi) \neq h(\pi)$). Like g, the function h is periodic but unlike g its period is π. As t passes from 0 to 2π, $g(t)$ performs one revolution of the circle whereas $h(t)$ performs two revolutions. Although the functions g and h have the same image, they can be distinguished pictorially by sketching their respective graphs.

2.1.7 Example. The graphs of the functions g and h of Example 2.1.5 and 2.1.6 are respectively the set of points $(t, \cos t, \sin t)$ and $(t, \cos 2t, \sin 2t)$, for all t in $\mathbb{R}$. Just as the graph of a real-valued function (such as the sine function) can be sketched in the plane, so we can also sketch the graph of our function $g : \mathbb{R} \to \mathbb{R}^2$ (see Fig. 2.2). The graph of g is called a *circular helix*. The graph of h is also a circular helix, the windings being around the same circular cylinder but with half the pitch.

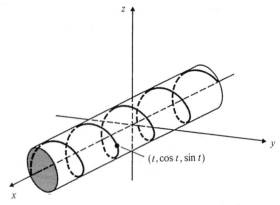

Fig. 2.2 Graph of $g(t) = (\cos t, \sin t)$

We have seen that two different functions can have the same image. It is not difficult to see that there is an infinity of different parametrizations of any non-empty subset of $\mathbb{R}^n$. The unit circle in

$\mathbb{R}^2$, for example, has parametrization

$$x = \cos(kt + c), \qquad y = \sin(kt + c), \qquad t \in \mathbb{R},$$

for any c and $k \neq 0$ in $\mathbb{R}$.

In contrast, a function is uniquely determined by its graph. That is, two different functions have different graphs (Exercise 2.1.8). The following theorem provides a useful way of thinking about the graph of a function.

2.1.8 Theorem. *The graph of a function* $f : D \subseteq \mathbb{R} \to \mathbb{R}^n$ *is the image of the function* $f^* : D \subseteq \mathbb{R} \to \mathbb{R}^{n+1}$ *given by*

$$f^*(t) = (t, f_1(t), \ldots, f_n(t)) = (t, f(t)) \in \mathbb{R}^{n+1}, \qquad t \in D.$$

Proof. The theorem follows immediately from Definitions 2.1.2 and 2.1.3.

Exercises 2.1

1. Find a vectorial equation of the straight line in $\mathbb{R}^3$ through $(1, 0, -4)$ in the direction $(-1, 2, 3)$. Prove that the point (x, y, z) lies on the line if and only if
$$\frac{x-1}{-1} = \frac{y}{2} = \frac{z+4}{3}.$$

2. The function $g : \mathbb{R} \to \mathbb{R}^2$ defined by
$$g(t) = (2\cos t, \sin t)$$
has as its image the ellipse $\frac{1}{4}x^2 + y^2 = 1$. Sketch the ellipse. Find a second parametrization of the ellipse that corresponds to a function of period 4π.

   Does there exist a non-periodic function $h : \mathbb{R} \to \mathbb{R}^2$ whose image is the ellipse?

Hint: consider, for example, $h(t) = (2\cos(t^2), \sin(t^2))$.

3. Prove that there is an infinity of different parametrizations of any non-empty subset of $\mathbb{R}^n$.

4. Describe the largest possible domain of the function f, where $f(t) = (\sqrt{(4 - t^2)}, t)$. Sketch its image. Compare with the image of g, where $g(t) = (-\sqrt{(4 - t^2)}, t)$.

5. Sketch (a) the image and (b) the graph of $f : \mathbb{R} \to \mathbb{R}^2$ given by
$$f(t) = (t, t^2).$$
Note: a rough sketch only is possible for the graph of f, since it is a subset of $\mathbb{R}^3$.

6. Find the period of the function $g : \mathbb{R} \to \mathbb{R}^2$ defined by
    $g(t) = (\cos 2t, \cos 3t)$.

7. Is it true that if the coordinate functions of $f : D \subseteq \mathbb{R} \to \mathbb{R}^n$ are periodic,
    then f is periodic?

8. Prove that a function $f : D \subseteq \mathbb{R} \to \mathbb{R}^n$ is uniquely determined by its
    graph, that is that two different functions on D have different graphs.

Hint: If f and g are different functions on D then there exists $t \in D$ such that
$f(t) \neq g(t)$.

2.2   Sequences and limits

The reader will find in Chapter 1 a definition (1.6.3) of convergence
for sequences of real numbers. The following theorem is an immediate consequence of that definition. Recall that we denote the
sequence $a_1, a_2, a_3, \ldots$, by (a_k).

2.2.1 Theorem.   *The sequence (a_k) of real numbers converges to $a \in \mathbb{R}$
if and only if* $\lim_{k \to \infty} |a_k - a| = 0$.

This theorem suggests that we can define the convergence of a
sequence of vectors in $\mathbb{R}^n$ by generalizing the idea of absolute value
to vectors in $\mathbb{R}^n$. The required generalization is the norm or length
$\|x\|$ of a vector $\mathbf{x} \in \mathbb{R}^n$ which we defined (Definition 1.2.5) by

$$\|\mathbf{x}\| = \|(x_1, \ldots, x_n)\| = (x_1^2 + \cdots + x_n^2)^{1/2}.$$

We note that in $\mathbb{R}^n$

2.2.2 $$\|\mathbf{x} - \mathbf{y}\|^2 = \sum_{r=1}^{n} (x_r - y_r)^2, \qquad \mathbf{x}, \mathbf{y} \in \mathbb{R}^n$$

and that accordingly in $\mathbb{R}$

2.2.3 $$\|x - y\| = |x - y|, \qquad x, y \in \mathbb{R}.$$

We refer to the real number $\|\mathbf{x} - \mathbf{y}\|$ as the *distance* between the
vectors $\mathbf{x}$ and $\mathbf{y}$ in $\mathbb{R}^n$. This generalizes the familiar concept of
distance in physical space. If $\mathbf{x}$ and $\mathbf{y}$ are the position vectors of
physical points P and Q relative to a rectangular coordinate system,
then $\mathbf{x} - \mathbf{y}$ is the vector $\overrightarrow{QP}$ (see Fig. 2.3(i)). The distance between P
and Q is $\|\mathbf{x} - \mathbf{y}\|$ units.

Intuitively a sequence of vectors $(\mathbf{a}_k)$ in $\mathbb{R}^n$ converges to $\mathbf{a} \in \mathbb{R}^n$ if the

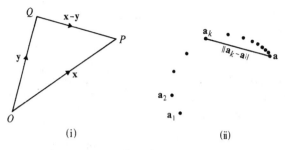

Fig. 2.3

distance between $\mathbf{a}_k$ and $\mathbf{a}$ tends to 0 as k tends to infinity (Fig. 2.3(ii)). The following definition captures this idea and is a natural extension of Theorem 2.2.1.

2.2.4 Definition. *The sequence $(\mathbf{a}_k)$ in $\mathbb{R}^n$ is said to* converge *to $\mathbf{a} \in \mathbb{R}^n$ if $\lim_{k \to \infty} \|\mathbf{a}_k - \mathbf{a}\| = 0$ in $\mathbb{R}$. The convergence of $(\mathbf{a}_k)$ to $\mathbf{a}$ is denoted by $\lim_{k \to \infty} \mathbf{a}_k = \mathbf{a}$ or by $\mathbf{a}_k \to \mathbf{a}$. If the sequence $(\mathbf{a}_k)$ does not converge to $\mathbf{a}$ we write $\mathbf{a}_k \not\to \mathbf{a}$.*

2.2.5 Example. The sequence defined by $\mathbf{a}_k = (1, k/(k+1), 1/k)$, $k \in \mathbb{N}$ converges to $(1, 1, 0)$ in $\mathbb{R}^3$ since

$$\|\mathbf{a}_k - (1, 1, 0)\| = \left\| \left(0, \frac{1}{k+1}, \frac{1}{k}\right) \right\| = \left(\left(\frac{1}{k+1}\right)^2 + \left(\frac{1}{k}\right)^2 \right)^{1/2} \to 0.$$

The following theorem is an immediate consequence of the definition of convergence in $\mathbb{R}$.

2.2.6 Theorem. *The sequence $(\mathbf{a}_k)$ in $\mathbb{R}^n$ converges to $\mathbf{a} \in \mathbb{R}^n$ if and only if to each $\varepsilon > 0$ there corresponds a natural number K such that*

$$\|\mathbf{a}_k - \mathbf{a}\| < \varepsilon \quad \text{whenever} \quad k > K.$$

The techniques we have for studying convergence in $\mathbb{R}$ can be used to study the more general situation by appealing to the following fundamental theorem.

2.2.7 Theorem. *Let $(\mathbf{a}_k)$ be a sequence in $\mathbb{R}^n$ with $\mathbf{a}_k = (a_{k1}, \ldots, a_{kn})$ for all $k \in \mathbb{N}$ and let $\mathbf{a} = (a_1, \ldots, a_n) \in \mathbb{R}^n$. Then $\mathbf{a}_k \to \mathbf{a}$ if and only if $a_{ki} \to a_i$ for each $i = 1, \ldots, n$. In other words, the convergence of $(\mathbf{a}_k)$ is equivalent to the convergence of each of its coordinate sequences.*

Proof.   It follows from 2.2.2 that for each $k \in \mathbb{N}$

2.2.8
$$\|\mathbf{a}_k - \mathbf{a}\|^2 = \sum_{i=1}^{n} (a_{ki} - a_i)^2.$$

Suppose that $a_{ki} \to a_i$ for each $i = 1, \ldots, n$. Choose any $\varepsilon > 0$.

To each $i = 1, \ldots, n$ there corresponds a number $K_i \in \mathbb{N}$ which depends on ε, such that

$$|a_{ki} - a_i| < \left(\frac{\varepsilon}{n}\right)^{1/2} \qquad \text{whenever} \quad k > K_i.$$

Let $K = \max \{K_1, \ldots, K_n\}$. By 2.2.8,

$$\|\mathbf{a}_k - \mathbf{a}\|^2 = \sum_{i=1}^{n} (a_{ki} - a_i)^2 < n\left(\frac{\varepsilon}{n}\right) = \varepsilon \qquad \text{whenever} \quad k > K.$$

This establishes that $\|\mathbf{a}_k - \mathbf{a}\|^2 \to 0$ and hence that $\|\mathbf{a}_k - \mathbf{a}\| \to 0$. Therefore $\mathbf{a}_k \to \mathbf{a}$.

The proof of the converse is left as an exercise.

2.2.9 *Example*.   [i] $\lim_{k \to \infty} (k/1 + k), k/(1 + k^2)) = (1, 0)$.

[ii] The sequence $(\mathbf{a}_k)$ where $\mathbf{a}_k = ((-1)^k, 1/k)$ does not converge in $\mathbb{R}^2$.

Exercises 2.2

1.   Show that $\lim_{k \to \infty} (k \sin (1/k), (1/k) \sin k) = (1, 0)$.

2.   Prove that a convergent sequence $(\mathbf{a}_k)$ in $\mathbb{R}^n$ has a unique limit in $\mathbb{R}^n$.

3.   Let (a_k) be the sequence of real numbers defined by $a_k = 10^{-k}[10^k \sqrt{2}]$, where $[r]$ denotes the integer part of the real number r. Verify that the first four terms of the sequence are $a_1 = 1, a_2 = 1.4, a_3 = 1.41, a_4 = 1.414$.

   Prove that (a_k) is a convergent sequence of rational numbers with limit $\sqrt{2}$.

Hint: given $\varepsilon > 0$, prove that $|a_k - \sqrt{2}| < \varepsilon$ for all k such that $1/10^k < \varepsilon$.

4.   Let $(\mathbf{a}_k)$ and $(\mathbf{b}_k)$ be convergent sequences in $\mathbb{R}^n$. Prove that if $\mathbf{a}_k \to \mathbf{a}$ and $\mathbf{b}_k \to \mathbf{b}$ then $\mathbf{a}_k + \mathbf{b}_k \to \mathbf{a} + \mathbf{b}$. Show also that for any constant c, the sequence $(c\mathbf{a}_k)$ is convergent, and $c\mathbf{a}_k \to c\mathbf{a}$.

5.   Prove that the definition of distance between $\mathbf{x}$ and $\mathbf{y}$ is symmetric in $\mathbf{x}$ and $\mathbf{y}$. (Show that $\|\mathbf{x} - \mathbf{y}\| = \|\mathbf{y} - \mathbf{x}\|$.)

6.   A sequence $(\mathbf{a}_k)$ in $\mathbb{R}^n$ is a Cauchy sequence if to each $\varepsilon > 0$ there corresponds $K \in \mathbb{N}$ such that

$$\|\mathbf{a}_k - \mathbf{a}_l\| < \varepsilon \qquad \text{whenever } k, l > K.$$

Prove that a Cauchy sequence in $\mathbb{R}^n$ is convergent.

Hint: Apply the ideas in the proof of Theorem 2.2.7 and Theorem 1.6.4.

2.3   Continuity

A function $f : D \subseteq \mathbb{R} \to \mathbb{R}$ is *discontinuous* at $p \in D$ if we can find a sequence (a_k) in D such that $a_k \to p$, but $f(a_k) \not\to f(p)$. If no such sequence can be found we say that f is continuous at p. Thus we have the following definition of continuity, generalized to functions with codomain $\mathbb{R}^n$ for arbitrary dimension n.

2.3.1 Definition. *The function $f : D \subseteq \mathbb{R} \to \mathbb{R}^n$ is said to be* continuous *at $p \in D$ if $f(a_k) \to f(p)$ whenever $a_k \to p$, where $a_k \in D$ for all $k \in \mathbb{N}$.*

The following theorem about continuity corresponds to Theorem 2.2.7 about limits.

2.3.2 Theorem. *The function $f : D \subseteq \mathbb{R} \to \mathbb{R}^n$ is continuous at $p \in D$ if and only if for each $i = 1, \ldots, n$, the coordinate function $f_i : D \subseteq \mathbb{R} \to \mathbb{R}$ is continuous at p.*

Proof. Let (a_k) be a sequence in D with $a_k \to p$. By definition of coordinate functions $f(a_k) = (f_1(a_k), \ldots, f_n(a_k))$ for each $k \in \mathbb{N}$, and $f(p) = (f_1(p), \ldots, f_n(p))$. By Theorem 2.2.7, $f(a_k) \to f(p)$ if and only if $f_i(a_k) \to f_i(p)$ for each $i = 1, \ldots, n$. The result follows.

2.3.3 Example. The function $f : \mathbb{R} \to \mathbb{R}^2$ given by $f(t) = (t, t^2)$ is continuous at all points in $\mathbb{R}$.

2.3.4 Example. The function $f : \mathbb{R} \to \mathbb{R}^2$ given by

$$f(t) = \begin{cases} (2t - 1, 2t - 1) & \text{when } t \leq 2, \\ (3 - t, 5 - t) & \text{when } t > 2, \end{cases}$$

is not continuous at 2 since the coordinate function f_1 is not continuous there. On the other hand the coordinate function f_2 is continuous everywhere.

2.3.5 Corollary. *Consider functions $f : D \subseteq \mathbb{R} \to \mathbb{R}^n$, $g : D \subseteq \mathbb{R} \to \mathbb{R}^n$ and $\phi : D \subseteq \mathbb{R} \to \mathbb{R}$. If f, g and ϕ are continuous at $p \in D$, then so are the functions ϕf, $f + g$, $f \cdot g$ and, provided $n = 3$, so is $f \times g$.*

Proof.   Exercise.

2.3.6 Definition.   *A function* $f : D \subseteq \mathbb{R} \to \mathbb{R}^n$ *is said to be* continuous *if it is continuous at p for every* $p \in D$.

Theorem 2.3.2 establishes that f is continuous if and only if all its coordinate functions are continuous.

Exercises 2.3

1. (a) Let $f : \mathbb{R} \to \mathbb{R}$ be the linear function defined by $f(t) = ct$, where c is a constant. Prove from first principles that f is continuous.
    (b) Prove that a linear function $f : \mathbb{R} \to \mathbb{R}^n$ is continuous.

Hint: consider the coordinate functions of f.

2. Prove that the function $f : \mathbb{R} \to \mathbb{R}^2$ defined by

$$f(t) = \begin{cases} (t, 3t - 2) & \text{when } t \leqslant 1 \\ (2t - 1, t + a) & \text{when } t > 1 \end{cases}$$

is continuous on $\mathbb{R}$ if and only if $a = 0$.

3. Prove Corollary 2.3.5.
Hint: consider the coordinate functions, and apply Exercise 1.6.2.

2.4   Limits and continuity

We have seen that the continuity of the function $f : D \subseteq \mathbb{R} \to \mathbb{R}^n$ at $p \in D$ depends upon how f behaves near to p as well as upon the value that f takes at p. In this section we study how f behaves near p by considering what f does to sequences (a_k) in D which converge to p but are such that $a_k \neq p$ for all $k \in \mathbb{N}$. *Remember that we are restricting ourselves to functions whose domain is an interval of some sort.*

2.4.1 Definition.   *The* cluster points *of an interval D in* $\mathbb{R}$ *are the points of D together with its end points.*

2.4.2. Example
  (i) The cluster points of $[-2, 1[$ form $[-2, 1]$.
  (ii) The cluster points of $]-\infty, 3[$ form $]-\infty, 3]$.

2.4.3 Theorem.   *A point* $p \in \mathbb{R}$ *is a cluster point of an interval D if* *and only if there is a sequence (a_k) in D such that $a_k \to p$ but $a_k \neq p$ for all $k \in \mathbb{N}$.*

Proof.   Exercise.

2.4.4 Example.   Let $D = [-2, 1[$. The sequences $(-2 + k^{-1})$, (k^{-1}) and $(1 - k^{-1})$ in D converge to cluster points -2, 0, and 1 respectively.

2.4.5 Definition.   *Consider a function $f : D \subseteq \mathbb{R} \to \mathbb{R}^n$, a cluster point p of an interval D and a point $\mathbf{q}$ in $\mathbb{R}^n$. We write $\lim_{t \to p} f(t) = \mathbf{q}$ if $f(a_k) \to \mathbf{q}$ whenever (a_k) is a sequence in D such that $a_k \to p$ and $a_k \neq p$ for all $k \in \mathbb{N}$.*

For this type of limit we have the coordinate result that we have come to expect.

2.4.6 Theorem.   *Given $f : D \subseteq \mathbb{R} \to \mathbb{R}^n$ and p a cluster point of D, then $\lim_{t \to p} f(t) = \mathbf{q}$ if and only if $\lim_{t \to p} f_i(t) = q_i$ for each $i = 1, \ldots, n$.*

Proof.   From Theorem 2.2.7, for any sequence (a_k) in D, $f(a_k) \to \mathbf{q}$ if and only if $f_i(a_k) \to q_i$ for all $i = 1, \ldots, n$. This equivalence still holds if we confine ourselves to sequences such that $a_k \to p$ and $a_k \neq p$ for all $k \in \mathbb{N}$.

2.4.7 Example.   Define $f : \mathbb{R} \to \mathbb{R}^2$ by

$$f(t) = \begin{cases} (1 + t, 2 + t) & \text{when } t \geq 0, \\ (1 + t, 1 - t) & \text{when } t < 0. \end{cases}$$

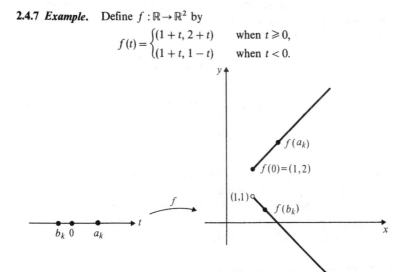

Fig. 2.4 Image of function f (Example 2.4.7)

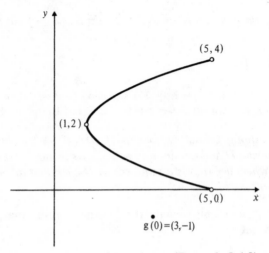

Fig. 2.5 Image of function g (Example 2.4.8)

Let $a_k = k^{-1}$ and $b_k = -k^{-1}$. Then $a_k \to 0$ and $b_k \to 0$ but $\lim_{k \to \infty} f(a_k) = (1, 2)$ and $\lim_{k \to \infty} f(b_k) = (1, 1)$. Hence $\lim_{t \to 0} f(t)$ does not exist. The situation is sketched in Fig. 2.4. In particular $\lim_{t \to 0} f_1(t) = 1$ but $\lim_{t \to 0} f_2(t)$ does not exist.

2.4.8 *Example.* Figure 2.5 is a sketch of the image of the function $g:] -2, 2[\to \mathbb{R}^2$ defined by $g(0) = (3, -1)$ and $g(t) = (t^2 + 1, t + 2)$ when $t \neq 0$. The image of 0 is separated from the rest of the image, but $\lim_{t \to 0} g(t)$ exists and, by Theorem 2.4.6,

$$\lim_{t \to 0} g(t) = \left(\lim_{t \to 0} t^2 + 1, \lim_{t \to 0} t + 2 \right) = (1, 2).$$

Notice also that, even though 2 is not in the domain of g, $\lim_{t \to 2} g(t)$ exists and

$$\lim_{t \to 2} g(t) = \left(\lim_{t \to 2} t^2 + 1, \lim_{t \to 2} t + 2 \right) = (5, 4).$$

The fact that the domains of the functions considered in Examples 2.4.7 and 2.4.8 are intervals but that their images are not in one piece indicates that the functions are not continuous. We can explore this property further by using the following theorem relating the limit concept and continuity.

2.4.9 *Theorem.* *The function* $f : D \subseteq \mathbb{R} \to \mathbb{R}^n$ *is continuous at* $p \in D$ *if and only if* $\lim_{t \to p} f(t) = f(p)$.

Proof. It is clear that if $f(a_k) \to f(p)$ whenever $a_k \to p$, with $a_k \neq p$ for all $k \in \mathbb{N}$, then $f(a_k) \to f(p)$ whenever $a_k \to p$ without that restriction.

2.4.10 Example. [i] The function g of Example 2.4.8 is not continuous at 0 since $\lim_{t \to 0} g(t) = (1, 2) \neq (3, -1) = g(0)$. If, however, we redefine the value of g at 0 to be $(1, 2)$, then the new function is continuous at 0.

[ii] The function f of Example 2.4.7 is not continuous at 0 since $\lim_{t \to 0} f(t)$ does not exist. Furthermore, f cannot be made continuous by changing its value at 0.

The following result provides an important tool for manipulating with limits.

2.4.11 Theorem. *Consider a function $f : D \subseteq \mathbb{R} \to \mathbb{R}^n$, a cluster point p of D and a point $\mathbf{q} \in \mathbb{R}^n$. Then $\lim_{t \to p} f(t) = \mathbf{q}$ if and only if to each $\varepsilon > 0$ there corresponds $\delta > 0$ such that*

2.4.12 $\qquad \|f(t) - \mathbf{q}\| < \varepsilon$ *whenever* $t \in D$ *and* $0 < |t - p| < \delta$.

Proof. Suppose firstly that the ε, δ condition is satisfied. Let (a_k) be any sequence in D such that $a_k \to p$ and $a_k \neq p$ for all $k \in \mathbb{N}$. Given any $\varepsilon > 0$ there corresponds, by 2.4.12, a $\delta > 0$ such that

2.4.13 $\qquad \|f(a_k) - \mathbf{q}\| < \varepsilon$ whenever $0 < |a_k - p| < \delta$.

But $a_k \to p$ and $a_k \neq p$ for all $k \in \mathbb{N}$, and so there exists $k \in \mathbb{N}$ such that

2.4.14 $\qquad 0 < |a_k - p| < \delta$ whenever $k > K$.

Expressions 2.4.13 and 2.4.14 together establish that $\lim_{k \to \infty} f(a_k) = \mathbf{q}$. Since this is true for all suitable sequences (a_k) it follows from Definition 2.4.5 that $\lim_{t \to p} f(t) = \mathbf{q}$.

To prove the converse we will show that the failure of the ε, δ condition 2.4.12 implies that $\lim_{t \to \infty} f(t) \neq \mathbf{q}$. Failure of the condition implies that there exists an $\varepsilon > 0$ such that no corresponding δ can be found. In particular $\delta = 1/k$ will not do for any $k \in \mathbb{N}$. For any $k \in \mathbb{N}$, therefore, there exists $a_k \in D$ such that

2.4.15 $\qquad 0 < |a_k - p| < 1/k$ but $\|f(a_k) - \mathbf{q}\| \geqslant \varepsilon$.

The first part of 2.4.15 tells us that the sequence (a_k) converges to p and that $a_k \neq p$ for all $k \in \mathbb{N}$. The second part of 2.4.15 tells us that $\lim_{k \to \infty} f(a_k) \neq \mathbf{q}$. This is the required contradiction.

The ε, δ condition of Theorem 2.4.11 means that given any required level of approximation, the value of $f(t)$ is approximately equal to $\mathbf{q}$ for all $t \neq p$ sufficiently close to p.

2.4.16 Corollary.   *Given* $f : D \subseteq \mathbb{R} \to \mathbb{R}^n$ *and* $p \in D$, *then* f *is continuous at* p *if and only if to each* $\varepsilon > 0$ *there corresponds* $\delta > 0$ *such that* $\| f(t) - f(p) \| < \varepsilon$ *whenever* $t \in D$ *and* $|t - p| < \delta$.

Proof.   Immediate from Theorem 2.4.9 and Theorem 2.4.11.

2.4.17 Remark.   The limit $\lim_{t \to p} f(t)$ defined in Definition 2.4.5 is related to a function f whose domain is an interval D in $\mathbb{R}$. The definition still makes sense if D is *any* non-empty subset of $\mathbb{R}$ and $p \in \mathbb{R}$ is any point such that there is a sequence (a_k) in D for which $a_k \to p$ and $a_k \neq p$ for all $k \in \mathbb{N}$.

2.4.18 Example.   Define $g : \mathbb{R} \backslash \{0\} \to \mathbb{R}$ by $g(t) = (\sin t)/t$. Then $\lim_{t \to 0} g(t)$ is defined and in fact $\lim_{t \to 0} g(t) = 1$ since, for all small t,

$$\left| t - \frac{t^3}{6} \right| \leqslant |\sin t| \leqslant |t|.$$

Exercises 2.4

1. Apply the test for continuity of Corollary 2.4.16 to prove that
    (a) the function $f : \mathbb{R} \to \mathbb{R}^n$ given by $f(t) = (t, t^2)$ $t \in \mathbb{R}$, is continuous at 0;
    (b) a linear function $f : \mathbb{R} \to \mathbb{R}^2$ is continuous at $p \in \mathbb{R}$.

2. Prove that the real-valued function $f : \mathbb{R} \to \mathbb{R}$ defined by $f(t) = \sin(1/t)$, $t \neq 0$, $f(0) = 0$, is discontinuous at 0 by showing that the test for continuity of Corollary 2.4.16 breaks down.

    (*Hint*: choose $\varepsilon = \frac{1}{2}$ and show that for any $\delta > 0$ there exists $0 < x < \delta$ such that $|f(x) - f(0)| > \varepsilon$. In fact, it is possible to choose $0 < x < \delta$ such that $f(x) = 1$.)

3. A function $f : D \subseteq \mathbb{R} \to \mathbb{R}^n$ is said to have a *removable discontinuity* at $p \in D$ if f is not continuous at p but there exists $\mathbf{q} \in \mathbb{R}^n$ such that $\lim_{t \to p} f(t) = \mathbf{q}$. This means that f can be made continuous at p by changing its value there. Show that 0 is a removable discontinuity of the function g of Example 2.4.8 but 0 is not a removable discontinuity of the function f of Example 2.4.7.

4. Given a function $f : D \subseteq \mathbb{R} \to \mathbb{R}^n$, consider the function $\| f \| : D \subseteq \mathbb{R} \to \mathbb{R}$ where $\| f \|(t) = \| f(t) \|$, $t \in D$. Prove that if f is continuous then $\| f \|$ is continuous. Is the converse true?

2.5   Differentiability and tangent lines

The derivative of a function $\phi : \mathbb{R} \to \mathbb{R}$ at $p \in \mathbb{R}$ is defined to be

2.5.1
$$\phi'(p) = \lim_{h \to 0} \frac{\phi(p + h) - \phi(p)}{h},$$

if this limit exists.

In generalizing this definition to a function $f : D \subseteq \mathbb{R} \to \mathbb{R}^n$ at a point $p \in \mathbb{R}$ we consider vectors of the form

2.5.2
$$\frac{1}{h}(f(p + h) - f(p)) \in \mathbb{R}^n,$$

where $h \in \mathbb{R}$ is non-zero and is such that $p + h \in D$. Since the vector given in 2.5.2 is a multiple of $f(p + h) - f(p)$ it is parallel to the line joining $f(p)$ and $f(p + h)$. It is pictured (for the case $h > 0$) in Fig. 2.4(i) as lying along this line and based at $f(p)$.

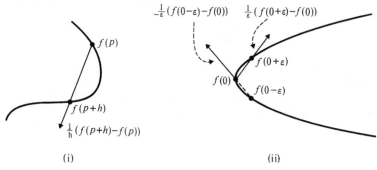

Fig. 2.6

2.5.3 *Example.* Consider the function $f : \mathbb{R} \to \mathbb{R}^2$ defined by $f(t) = (t^2 + 1, t + 2)$. Taking $p = 0$, for any $\varepsilon > 0$ the two vectors in $\mathbb{R}^2$ obtained from 2.5.2 by taking $h = \varepsilon$ and $h = -\varepsilon$ are respectively

$$\frac{1}{\varepsilon}(f(0 + \varepsilon) - f(0)) = (\varepsilon, 1) \quad \text{and} \quad \frac{1}{-\varepsilon}(f(0 - \varepsilon) - f(0)) = (-\varepsilon, 1).$$

These vectors are drawn, together with the image of f, in Fig. 2.6(ii). The image of f is the parabola $(y - 2)^2 = (x - 1)$ in $\mathbb{R}^2$.

2.5.4 *Definition.* *Let p be a point in an interval D. Then the interval D_p is defined by*

$$D_p = \{h \in \mathbb{R} \mid p + h \in D\}.$$

The interval D_p is merely the translation of D through $(-p)$. Since p lies in D, 0 lies in D_p. In view of Remark 2.4.17 the following definition makes sense.

2.5.5 Definition.   *The function* $f : D \subseteq \mathbb{R} \to \mathbb{R}^n$ *is said to be* differentiable *at* $p \in D$ *if the limit*

$$\lim_{h \to 0} \frac{1}{h}(f(p+h) - f(p)), \qquad h \in D_p \backslash \{0\}$$

exists in $\mathbb{R}^n$. *If the limit does exist then it is called the* derivative *of* f *at* p *and is denoted by* $f'(p)$.

We can say, as in the elementary case, that $f'(p)$ measures the rate of change of f at p. Notice, however, that in this case $f'(p)$ is a vector in $\mathbb{R}^n$.

2.5.6 *Example.*   For the function f of Example 2.5.3 with $p = 0$ and $h \neq 0$,

$$\frac{1}{h}(f(p+h) - f(p)) = (h, 1)$$

and Definition 2.5.5 gives $f'(0) = \lim_{h \to 0}(h, 1) = (0, 1) \in \mathbb{R}^2$. This vector is drawn in Fig. 2.7(i) based at $f(0) = (1, 2)$. Compare this figure with Fig. 2.6(ii). In particular, note that as ε tends to zero, the two vectors drawn in Fig. 2.6(ii) tend to the vector $f'(0)$ drawn in Fig. 2.7(i).

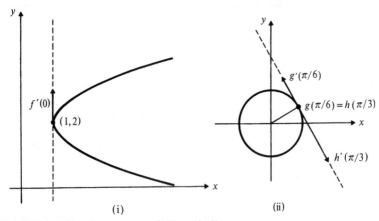

Fig. 2.7 (i) Tangent vector $f'(0) = (0, 1)$;
(ii) Point $g(\pi/6) = h(\pi/3) = (\frac{1}{2}\sqrt{3}, \frac{1}{2})$
Tangent vectors $g'(\pi/6) = \frac{1}{2}(-1, \sqrt{3})$ and $h'(\pi/3) = (1, -\sqrt{3})$

2.5.7 Theorem.   *The function $f : D \subseteq \mathbb{R} \to \mathbb{R}^n$ is differentiable at $p \in D$ if and only if for each $i = 1, \ldots, n$ the coordinate function $f_i : D \subseteq \mathbb{R} \to \mathbb{R}$ is differentiable at p. Furthermore, if f is differentiable at p, then*

$$f'(p) = (f_1{}'(p), \ldots, f_n{}'(p)).$$

Proof.   Apply Theorem 2.4.6.

2.5.8 Example.   (i) The function $g : \mathbb{R} \to \mathbb{R}^2$ defined by $g(t) = (\cos t, \sin t)$ has as its image the unit circle centre $(0, 0)$ in $\mathbb{R}^2$. The function is differentiable at $\pi/6$ with derivative

$$g'\left(\frac{\pi}{6}\right) = \left(-\sin\frac{\pi}{6}, \cos\frac{\pi}{6}\right) = \left(-\frac{1}{2}, \frac{\sqrt{3}}{2}\right).$$

This vector is drawn in Fig. 2.7(ii) based at $g(\pi/6) = (\sqrt{3}/2, 1/2)$.

   (ii) The function $h : \mathbb{R} \to \mathbb{R}^2$ defined by $h(u) = (\sin 2u, \cos 2u)$ has image the same unit circle in $\mathbb{R}^2$. The function h is differentiable at $\pi/3$ with derivative

$$h'\left(\frac{\pi}{3}\right) = \left(2\cos\frac{2\pi}{3}, -2\sin\frac{2\pi}{3}\right) = (1, -\sqrt{3}).$$

This derivative is also drawn in Fig. 2.7(ii) based, as was $g'(\pi/6)$, at $h(\pi/3) = (\sqrt{3}/2, 1/2)$.

   In Example 2.5.8 the vectors $g'(\pi/6)$ and $h'(\pi/3)$ are linearly dependent and are orthogonal to the radius vector $g(\pi/6) = h(\pi/3) = (\sqrt{3}/2, 1/2)$. Both derivatives are therefore in the direction of the tangent to the unit circle at $(\sqrt{3}/2, 1/2)$ (see Fig. 2.7(ii)). Similarly, in Example 2.5.3, the vector $f'(0)$ is in the direction of the tangent to the parabola $(y - 2)^2 = (x - 1)$ at $f(0) = (1, 2)$ (see Fig. 2.7(i)). Such geometrical considerations suggest the following definitions.

2.5.9 Definition.   *Let $f : D \subseteq \mathbb{R} \to \mathbb{R}^n$ be differentiable at $p \in D$.*

   [i]   *The derivative $f'(p)$ will also be called the* tangent vector *to f at p.*

   [ii]   *If $f'(p) \neq \mathbf{0}$, then the straight line in $\mathbb{R}^n$ through $f(p)$ in direction $f'(p)$ will be called the* tangent line *to f at p. It is the set* $\{f(p) + sf'(p) \mid s \in \mathbb{R}\} \subseteq \mathbb{R}^n$.

   By excluding the case $f'(p) = 0$ in Definition 2.5.9(ii) we avoid the tangent line degenerating to a single point. Notice that the tangent vector and tangent line are defined in terms of a function and not in terms of its image.

2.5.10 *Example.*   The tangent vector to the function f of Example 2.5.6 at
0 is $(0, 1) \in \mathbb{R}^2$. The tangent line at 0 is the line $\{(1, 2) + s(0, 1) | s \in \mathbb{R}\} \subseteq \mathbb{R}^2$. It is
the line $x = 1$ sketched as a broken line in Fig. 2.7(i).

2.5.11 *Example.*   The functions g and h of Example 2.5.8 both have image
the unit circle centre $\mathbf{0}$ in $\mathbb{R}^2$. Although $g(\pi/6) = h(\pi/3)$ the two tangent
vectors $g'(\pi/6)$ and $h'(\pi/3)$ are different. The tangent line to g at $\pi/6$ is the
set

$$\{(\sqrt{3}/2, 1/2) + s(-1/2, \sqrt{3}/2) | s \in \mathbb{R}\} \subseteq \mathbb{R}^2.$$

It is the straight line $y + \sqrt{3}x = 2$ sketched as a broken line in Fig. 2.7(ii).
Since $h(\pi/3) = g(\pi/6)$ and since $h'(\pi/3)$ is a non-zero multiple of $g'(\pi/6)$, the
tangent line to h at $\pi/3$ is the same straight line in $\mathbb{R}^2$. Our earlier work on
the geometry of Fig. 2.7(ii) shows that this tangent line is the tangent to the
circle at $(\sqrt{3}/2, 1/2)$.

We shall have a second look at the relationship between tangent
lines to a function and the image of the function in the next section.

2.5.12 *Definition.*   *The function* $f : D \subseteq \mathbb{R} \to \mathbb{R}^n$ *is said to be* differentiable *if it is differentiable at each point* $p \in D$. *The function*
$f' : D \subseteq \mathbb{R} \to \mathbb{R}^n$ *whose image at* $p \in D$ *is the vector* $f'(p) \in \mathbb{R}^n$ *is called the*
derivative *of* f. *If* f' *is itself differentiable, then its derivative is*
denoted by $f'' : D \subseteq \mathbb{R} \to \mathbb{R}^n$.

With the help of Theorem 2.5.7 many theorems of elementary
calculus are readily generalized.

2.5.13 *Theorem.*   *Consider functions* $f : D \subseteq \mathbb{R} \to \mathbb{R}^n$, $g : D \subseteq \mathbb{R} \to \mathbb{R}^n$
and $\phi : D \subseteq \mathbb{R} \to \mathbb{R}$. *If* f, g *and* ϕ *are differentiable, then so are the*
following functions with the stated derivatives.
   [i]   *The sum function:* $(f + g)' = f' + g'$.
   [ii]   *The dot product* $(f \cdot g)' = (f' \cdot g) + (f \cdot g')$
   [iii]   *(For the case* $n = 3$) *the cross product*:
         $(f \times g)' = (f' \times g) + (f \times g')$
   [iv]   *The product function:* $(\phi f)' = \phi' f + \phi f'$.

Proof.   For each $t \in D$ express $(f + g)(t)$, $(f \cdot g)(t)$, $(f \times g)(t)$ and
$(\phi f)(t)$ in terms of coordinate functions of f and of g. Now apply
Theorem 2.5.7 and use results from elementary calculus to obtain the
required results.

The following theorem is an important application of
Theorem 2.5.13(ii).

2.5.14 *Theorem.* *Let* $h: D \subseteq \mathbb{R} \to \mathbb{R}^n$ *be a differentiable function such that* $\|h(t)\| = 1$ *for all* $t \in D$. *Then* $h(t)$ *and* $h'(t)$ *are orthogonal for all* $t \in D$.

Proof. The image of h lies on the unit sphere centre $\mathbf{0}$ in $\mathbb{R}^n$. For all $t \in D$

2.5.15 $$h(t) \cdot h(t) = \|h(t)\|^2 = 1.$$

Using Theorem 2.5.13(ii) we find, on differentiating 2.5.15, that

$$h(t) \cdot h'(t) + h'(t) \cdot h(t) = 2h(t) \cdot h'(t) = 0, \qquad \text{for all } t \in D.$$

Therefore the tangent vector $h'(t)$ is orthogonal to the radius vector $h(t)$.

In particular, if $n = 2$ and if $h'(p) \neq \mathbf{0}$ then the tangent line to h at p is just the tangent to the unit circle centre $\mathbf{0} \in \mathbb{R}^2$ at $h(p)$. We have already seen this to be true in the particular cases considered in Example 2.5.11.

Despite their geometrical significance, differentiability, tangent vectors and tangent lines are defined in terms of functions rather than in terms of their images. The following two examples should serve to warn the reader against jumping to conclusions about a function merely on the basis of looking at its image.

2.5.16 *Example.* Consider the continuous functions f, g and h from $\mathbb{R}$ into $\mathbb{R}^2$ defined for each $t \in \mathbb{R}$ by

$$g(t) = (t, t^2), \quad f(t) = (t^3, t^6), \quad h(t) = \begin{cases} (t, t^2) & \text{for } t \geq 0, \\ (t^3, t^6) & \text{for } t < 0. \end{cases}$$

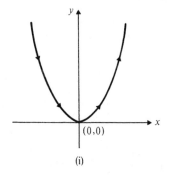

(i)

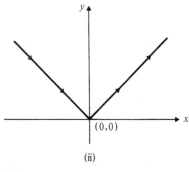

(ii)

Fig. 2.8

All three functions have the same image (the parabola $y = x^2$ in $\mathbb{R}^2$ — see Fig. 2.8(i)) and all take the value $(0, 0)$ at 0. Furthermore they cover the points of this curve in the same order as t increases (indicated by the arrows in Fig. 2.8(i)). The function g is differentiable at 0 with tangent vector $g'(0) = (1, 0)$ and tangent line $y = 0$ there. The function f is also differentiable but f has no tangent line at 0 since $f'(0) = (0, 0)$. Even more striking is the fact that, since the coordinate function h_1 is not differentiable at 0, the function h is not differentiable at 0. This is so despite the smoothness of the image of h.

2.5.17 Example.   The image of a continuous function $g : \mathbb{R} \to \mathbb{R}^2$ where $g(t) = (t|t|, t^2), t \in \mathbb{R}$, is sketched in Fig. 2.8(ii). It is the set $y = |x|$ in $\mathbb{R}^2$. The coordinate functions of g are given by

$$g_1(t) = \begin{cases} t^2 & \text{when } t \geq 0 \\ -t^2 & \text{when } t < 0 \end{cases} \quad \text{and} \quad g_2(t) = t^2, \qquad t \in \mathbb{R}.$$

Both g_1 and g_2 are differentiable at 0. Therefore g is differentiable at 0 and has a tangent vector there despite the fact that the image has a 'corner' at $g(0) = (0, 0)$. Since the tangent vector at 0 is $g'(0) = (0, 0)$, there is no associated tangent line.

In elementary calculus the differentiability of a function $\phi : \mathbb{R} \to \mathbb{R}$ is usually thought of in terms of the smoothness of the graph of ϕ. It is very important in deducing properties of $f : D \subseteq \mathbb{R} \to \mathbb{R}^n$ from a sketch of its image to remember that it is just the image and not the graph of f. This point is amply supported in Examples 2.5.16 and 2.5.17.

Exercises 2.5

1.   Find the tangent vector (a) at 0 (b) at $\frac{1}{2}\pi$ to the function $g : \mathbb{R} \to \mathbb{R}^2$ defined by $g(t) = (2 \cos t, \sin t)$. Determine the corresponding tangent lines. Sketch these and the image of g (the ellipse of Exercise 2.1.2).

2.   Sketch the images of the following differentiable functions $f : \mathbb{R} \to \mathbb{R}^2$ and indicate the tangent vectors and tangent lines (if they exist) to f
(i) at $t = 0$; (ii) at $t = 1$.
(a) $f(t) = (t + t^2, t - t^2)$
(b) $f(t) = (t^2, t^3)$

3.   Illustrate Theorem 2.5.14 and its proof with the function $h : \mathbb{R} \to \mathbb{R}^2$ defined by $h(t) = (\cos(t^2), \sin(t^2))$. Calculate $h'(t)$ and sketch the corresponding tangent lines for various values of t.

4.   Sketch the image of the differentiable function $g : \mathbb{R} \to \mathbb{R}^2$ given by $g(t) = (t, t^3), t \in \mathbb{R}$. Sketch the tangent vector and tangent line at $t = 0$ and at $t = 1$.

Using the idea of Example 2.5.16, construct a function h with the same image as g and such that $g(0) = h(0)$, but h is not differentiable at 0.

Answer: for example $h(t) = (t^{1/3}, t)$, $t \in \mathbb{R}$.

5. Prove that the function $f : \mathbb{R} \to \mathbb{R}$ defined by $f(t) = |t|$ is continuous but not differentiable at 0.

6. Verify Theorem 2.5.13 for the functions $f : \mathbb{R} \to \mathbb{R}^3$, $g : \mathbb{R} \to \mathbb{R}^3$ and $\phi : \mathbb{R} \to \mathbb{R}$, where

$$f(t) = (e^t, t, 1), \quad g(t) = (0, -t, t^2 + 1), \quad \phi(t) = t^2.$$

7. Sketch the image of the differentiable function $f : \mathbb{R} \to \mathbb{R}^2$ defined by

$$f(t) = (e^{kt} \cos t, e^{kt} \sin t), \qquad t \in \mathbb{R},$$

where k is a constant. Prove that

$$\frac{f'(t) \cdot f(t)}{\|f'(t)\| \, \|f(t)\|} = \frac{k}{\sqrt{(1 + k^2)}} \qquad t \in \mathbb{R}.$$

Deduce that the angle between the tangent vector $f'(t)$ and the line joining the origin and the point $f(t)$ is the same for all t. The image of f is called an *equiangular spiral*.

2.6 Curves and simple arcs. Orientation

Cartesian geometry and calculus developed during the first half of the seventeenth century through the study of some special curves and in particular through attempts to solve problems concerning their tangents, their length and the areas associated with them.

Intuitively one thinks of a curve as a wavy copy of an interval, with possible self-intersections and corners. It seems reasonable to define a curve in $\mathbb{R}^n$ as the image of an interval D under a continuous function $f : D \subseteq \mathbb{R} \to \mathbb{R}^n$. The point $f(t) \in \mathbb{R}^n$ would then trace the curve as the parameter t moves along the interval. This definition, however, is too general to be useful. There is, for example, such a continuous function defined on a compact interval $[a, b]$ whose image fills out a square in $\mathbb{R}^2$. For an illustration, see Liu Wen, *American Math. Monthly* **90** (1983) p. 283. We therefore impose extra conditions on f, although many of our results are also true if they are weakened.

2.6.1 Definition. *A function $f : D \subseteq \mathbb{R} \to \mathbb{R}^n$ is said to be* continuously differentiable, *or a C^1 function, if f is differentiable and if*

f' is continuous. In general a C^k function is one whose kth derivative exists and is continuous.

2.6.2 Definition.   A subset C of $\mathbb{R}^n$ is a curve if there is a C^1 function $f : D \subseteq \mathbb{R} \to \mathbb{R}^n$, where D is an interval, such that $f(D) = C$. The function f is called a C^1 parametrization of the curve.

2.6.3 Example.   The function $f(t) = (t^2 + 1, t + 2)$ of Example 2.5.3 defines a C^1 parametrization of the curve (parabola) $(y - 2)^2 = (x - 1)$ in $\mathbb{R}^2$. The functions $g(t) = (\cos t, \sin t)$ and $h(t) = (\sin 2t, \cos 2t)$ of Example 2.5.8 define two different C^1 parametrizations of the curve (circle) $x^2 + y^2 = 1$ in $\mathbb{R}^2$. Notice that g and h trace the circle in opposite directions.

2.6.4 Example.   The function $f : \mathbb{R} \to \mathbb{R}^3$ defined by $f(t) = (t, \cos t, \sin t)$ is a C^1 parametrization of the circular helix curve sketched in Fig. 2.2.

2.6.5 Example.   The function $g : \mathbb{R} \to \mathbb{R}^2$ defined by $g(t) = (\sin t, \sin 2t)$ is a C^1 parametrization of the 'figure 8' curve sketched in Fig. 2.9. The arrows in Fig. 2.9 indicate the direction in which $g(t)$ traces the curve as t increases. Clearly g has period 2π and indeed the curve is traced just once in each interval of length 2π.

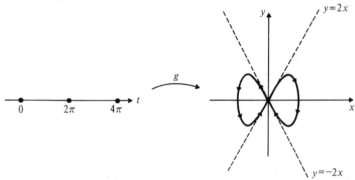

Fig. 2.9 $g(t) = (\sin t, \sin 2t)$, $t \in \mathbb{R}$

In this section we shall consider the possibility of using a parametrization of a curve to define the tangent lines to it as we did in Examples 2.5.10 and 2.5.11. Example 2.5.16 indicates that care is needed, as do the following examples.

2.6.6 Example.   The function $g : \mathbb{R} \to \mathbb{R}^2$ defined by $g(t) = (t|t|, t^2)$ as in Example 2.5.17 is a C^1 parametrization of the curve $y = |x|$ in $\mathbb{R}^2$ sketched in Fig. 2.8(ii). There is no tangent line at 0.

2.6.7 *Example.* Consider the C^1 parametrization $g(t) = (\sin t, \sin 2t)$ of the 'figure 8' curve defined in Example 2.6.5. In considering the properties of this curve at the cross-over point $(0, 0)$ we observe that $g(0) = g(\pi) = (0, 0)$. The derivative of g is given by

$$g'(t) = (\cos t, 2\cos 2t), \qquad t \in \mathbb{R}.$$

Hence $g'(0) = (1, 2)$ and the tangent line to g at 0 is $y = 2x$. On the other hand $g'(\pi) = (-1, 2)$ and the tangent line to g at π is $y = -2x$. These two lines are sketched in Fig. 2.9 as dotted lines through $g(0) = g(\pi) = (0, 0)$.

In general, $y = 2x$ is the tangent line to g at points of the form $2k\pi \in \mathbb{R}$, $k \in \mathbb{Z}$, and $y = -2x$ is the tangent line to g at points of the form $(2k + 1)\pi$, $k \in \mathbb{Z}$.

2.6.8 *Example.* (i) The function g of Example 2.6.7 restricted to the open interval $]-\pi, \pi[$ leads to a 1–1 C^1 function $f :]-\pi, \pi[\subseteq \mathbb{R} \to \mathbb{R}^2$ defined by $f(t) = (\sin t, \sin 2t)$. The image of f is the same 'figure 8' curve sketched in Fig. 2.9. However, the only value of t for which $f(t) = (0, 0)$ is $t = 0$. The tangent line to f at 0 is $y = 2x$. The other significant line through $(0, 0)$, $y = -2x$, is not revealed by this parametrization of the curve. See Fig. 2.10 (i).

(ii) By restricting g to the open interval $]0, 2\pi[$, we obtain a 1–1 C^1 function $h :]0, 2\pi[\subseteq \mathbb{R} \to \mathbb{R}^2$ defined by $h(u) = (\sin u, \sin 2u)$. Again the image of h is the 'figure 8' curve. The only value of u for which $h(u) = (0, 0)$ is $u = \pi$. The tangent line to h at π is $y = -2x$. This time the line $y = 2x$ is hidden by the parametrization. See Fig. 2.10(ii).

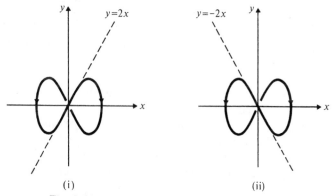

Fig. 2.10 (i) $f(t) = (\sin t, \sin 2t)$, $t \in]-\pi, \pi[$;
(ii) $h(u) = (\sin u, \sin 2u)$, $u \in]0, 2\pi[$

Nevertheless, we would expect that if two parametrizations of a curve are related in some way then they will associate the same tangent lines with points of the curve. Accordingly we define the following equivalence relation between parametrizations.

2.6.9. Definition.   Let $f : D \subseteq \mathbb{R} \to \mathbb{R}^n$ and $h : E \subseteq \mathbb{R} \to \mathbb{R}^n$ be C^1
parametrizations of a curve C. Then h is equivalent to f if there is a
differentiable function ϕ from E onto D such that
   [i] $h = f \circ \phi$ *and*
   [ii] *either* $\phi'(u) > 0$ *for all* $u \in E$ *or* $\phi'(u) < 0$ *for all* $u \in E$.

Condition [ii] of Definition 2.6.9 ensures that ϕ is strictly
monotonic and hence 1–1 from E onto D. By Theorem 1.6.7 the
inverse function ϕ^{-1} from D onto E is dfferentiable. It follows that
if h is equivalent to f via the differentiable function ϕ then f is
equivalent to h via the differentiable function ϕ^{-1}. We leave it to
the reader to complete the argument that Definition 2.6.9
establishes an equivalence relation on the class of all C^1
parametrizations of C.

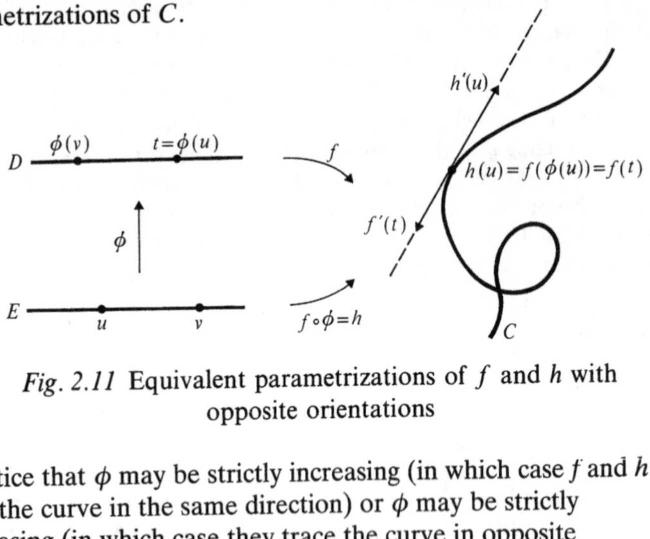

Fig. 2.11 Equivalent parametrizations of f and h with
opposite orientations

Notice that ϕ may be strictly increasing (in which case f and h
trace the curve in the same direction) or ϕ may be strictly
decreasing (in which case they trace the curve in opposite
directions). In the former case we say that f and h are *properly
equivalent* or *equivalent with the same orientation* and in the latter
case that they are *equivalent with opposite orientations*. See
Fig. 2.11.

2.6.10 *Example.*   The functions g and h considered in Example 2.5.8 were
given by $g(t) = (\cos t, \sin t)$, $t \in \mathbb{R}$, and $h(u) = (\sin 2u, \cos 2u)$, $u \in \mathbb{R}$. They are
both C^1 parametrizations of the unit circle in $\mathbb{R}^2$. Let $\phi : \mathbb{R} \to \mathbb{R}$ be defined
by $\phi(u) = \frac{1}{2}\pi - 2u$, $u \in \mathbb{R}$. Then $h = g \circ \phi$ and $\phi'(u) < 0$ for all $u \in \mathbb{R}$. Hence g
and h are equivalent parametrizations with opposite orientation – they trace
the circle in opposite directions.

In order to relate the properties of equivalent parametrizations we need to know how their derivatives are related. The required result is a Chain Rule of the following type.

2.6.11 *Theorem.* (Chain Rule). Let $\phi : E \subseteq \mathbb{R} \to \mathbb{R}$ and $f : D \subseteq \mathbb{R} \to \mathbb{R}^n$ be differentiable functions and let $\phi(E) \subseteq D$. Then $f \circ \phi : E \subseteq \mathbb{R} \to \mathbb{R}^n$ is differentiable, and for each $u \in E$,

2.6.12 $$(f \circ \phi)'(u) = (f'(\phi(u)))\phi'(u).$$

Remark.   We have written the scalar $\phi'(u)$ on the right of the vector $f'(\phi(u))$ in 2.6.12 in order to relate this theorem firstly to the elementary Chain Rule (Theorem 1.6.11) and secondly to generalizations in subsequent chapters.

Proof.   For each $i = 1, \ldots, n$ the coordinate functions of $f \circ \phi$ and f satisfy

2.6.13 $$(f \circ \phi)_i(u) = f_i(\phi(u)), \qquad \text{for all } u \in E.$$

We are given that ϕ is differentiable at $u \in E$, and that f (and therefore f_i) is differentiable at $\phi(u)$. Expression 2.6.13 and the elementary Chain Rule 1.6.11 together imply that $(f \circ \phi)_i$ is differentiable at u and that

$$(f \circ \phi)_i'(u) = f_i'(\phi(u))\phi'(u).$$

The required conclusion follows from Theorem 2.5.7.

2.6.14 *Example.* Let $\phi : \mathbb{R} \to \mathbb{R}$ and $f : \mathbb{R} \to \mathbb{R}^3$ be defined by $\phi(u) = \sin u$, $u \in \mathbb{R}$ and $f(t) = (2t, t^3, 1)$, $t \in \mathbb{R}$. Then $(f \circ \phi)(u) = (2 \sin u, \sin^3 u, 1)$, $u \in \mathbb{R}$. We have by way of illustration of the Chain Rule,

$$f'(\phi(u)) = (2, 3\sin^2 u, 0), \quad \phi'(u) = \cos u, \text{ and}$$
$$(f \circ \phi)'(u) = (2\cos u, 3\sin^2 u \cos u, 0), \qquad u \in \mathbb{R}.$$

The relationship $h = f \circ \phi$ will occur frequently when we compare the properties of two parametrizations f and h of a curve, as in the following theorem.

2.6.15 *Theorem.* Let $f : D \subseteq \mathbb{R} \to \mathbb{R}^n$ and $h : E \subseteq D \to \mathbb{R}^n$ be equivalent C^1 parametrizations of a given curve C in $\mathbb{R}^n$ related by a differentiable function $\phi : E \to D$ with the properties given in Definition 2.6.9. Then, whenever $\phi(u) = t$,

[i] *the tangent vector to f at t, if non-zero, is in the same (opposite) direction as the tangent vector to h at u when f and h have the same (opposite) orientations;*

[ii] *the tangent lines to f at t and to h at u are the same if either of them exists*

Proof. Immediate from the Chain Rule 2.6.11. The theorem is illustrated in Fig. 2.11, where f and h have opposite orientations, the tangent vectors being in opposite directions.

2.6.16 *Example.* The functions

$$g(t) = (\cos t, \sin t) \qquad \text{and} \qquad h(u) = (\sin 2u, \cos 2u)$$

considered in Example 2.6.10 are C^1 parametrizations of the unit circle with opposite orientations. For all $u \in \mathbb{R}$, $h(u) = g(\phi(u))$, where $\phi(u) = \frac{1}{2}\pi - 2u$, and the tangent vectors

$$h'(u) = (2\cos 2u, -2\sin 2u) \quad \text{and} \quad g'(\phi(u)) = (-\cos 2u, \sin 2u)$$

are in opposite directions. The tangent lines to h at u and to g at $\phi(u)$ are the same.

2.6.17 *Example.* Consider the functions

$$f(t) = (\sin t, \sin 2t), \, t \in] - \pi, \pi[$$

and

$$h(u) = (\sin u, \sin 2u), \, u \in]0, 2\pi[$$

of Example 2.6.8 which parametrize the 'figure 8' curve. If there were a differentiable function $\phi :]0, 2\pi[\to] - \pi, \pi[$ such that $h = f \circ \phi$ then, since $(0,0) = h(\pi) = f(\phi(\pi))$, we would have to have $\phi(\pi) = 0$. Furthermore by Theorem 2.6.15 (ii) the tangent line to h at π would be the same as the tangent line to f at $\phi(\pi) = 0$. But $h'(\pi) = (-1, 2)$ and $f'(0) = (1, 2)$, hence the tangent lines are different (see Fig. 2.10). Therefore no such ϕ exists.

So far, both in this section and the previous one, we have considered tangent lines to functions which parametrize curves. Intuitively, however, provided a curve C does not have a 'corner' at $\mathbf{q} \in C$ and provided C does not cross itself at $\mathbf{q}$, the tangent line to C at $\mathbf{q}$ exists and is independent of parametrization. We can exclude the possibility of corners by requiring that the parametrization should have non-zero derivative (see Example 2.5.17). We can exclude the possibility of crossovers by firstly requiring the parametrization to be 1–1 (see Example 2.6.7) and secondly requiring that its domain should be a compact interval $[a, b]$ (compare Example 2.6.8.).

2.6.18 *Definition.* *A function $f : D \subseteq \mathbb{R} \to \mathbb{R}^n$ is said to be* smooth *if it is C^1 (continuously differentiable) and if $f'(t) \neq 0$ for all $t \in D$.*

2.6.19 *Example.* There is no smooth parametrization of the curve $y = |x|$ in $\mathbb{R}^2$. See Fig. 2.8(ii).

2.6.20 *Definition.* *A curve C in $\mathbb{R}^n$ is a (smooth)* simple arc *if C has a 1–1 (smooth) C^1 parametrization of the form $f : [a, b] \subseteq \mathbb{R} \to \mathbb{R}^n$. The points $f(a)$ and $f(b)$ are then called the* end points *of the arc. The function f is called a* simple parametrization *of C.*

2.6.21 *Example.* Let $g : [a, b] \subseteq \mathbb{R} \to \mathbb{R}$ be a C^1 function. The graph G of g is a smooth simple arc in $\mathbb{R}^2$ parametrized by the smooth, 1–1 C^1 function $f : [a, b] \subseteq \mathbb{R} \to \mathbb{R}^2$ defined by

2.6.22 $$f(t) = (t, g(t)), \qquad t \in [a, b].$$

The end points of G are $(a, g(a))$ and $(b, g(b))$.

2.6.23 *Theorem.* *Let C be a smooth simple arc in $\mathbb{R}^n$ simply parametrized by a smooth 1–1 function $f : [a, b] \subseteq \mathbb{R} \to \mathbb{R}^n$. Then any smooth parametrization $h : [c, d] \subseteq \mathbb{R} \to \mathbb{R}^n$ of C is 1–1 and is equivalent to f.*

Proof. The function $\phi = f^{-1} \circ h$ is well defined, maps $[c, d]$ onto $[a, b]$ and satisfies $h = f \circ \phi$. See Fig. 2.12. Choose any $p \in [c, d]$.

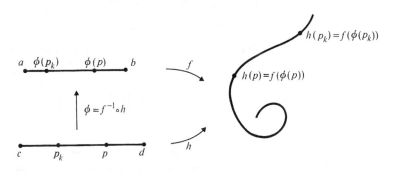

Fig. 2.12

(i) We begin by showing that ϕ is continuous at p. If it were not then there would be a sequence $p_k \to p$ in $[c, d]$ such that $\phi(p_k) \nrightarrow \phi(p)$ in $[a, b]$. Since $[a, b]$ is compact we may assume (by taking a subsequence if necessary – see Theorem 1.6.5) that

2.6.24 $\phi(p_k) \to r$ in $[a, b]$ and $r \neq \phi(p)$.

But f is 1–1 and continuous, and $f \circ \phi = h$. Therefore, applying f to 2.6.24,

$$h(p_k) \to f(r) \text{ in } C \text{ and } f(r) \neq h(p).$$

This contradicts the continuity of h at p.

(ii) We next consider the differentiability of ϕ at p. Consider a sequence

2.6.25 $p_k \to p$ in $[c, d]$ with $p_k \neq p$ for all $k \in \mathbb{N}$.

We need to know that, for large k, $\phi(p_k) \neq \phi(p)$. This follows from the smoothness of h by considering the limit

$$\lim_{k \to \infty} \frac{f(\phi(p_k)) - f(\phi(p))}{p_k - p} = \lim_{k \to \infty} \frac{h(p_k) - h(p)}{p_k - p} \neq 0$$

Now, since f is smooth, we assume without loss of generality that $f_1'(\phi(p)) \neq 0$. Consider the identity

2.6.26 $\dfrac{h_1(p_k) - h_1(p)}{p_k - p} = \dfrac{f_1(\phi(p_k)) - f_1(\phi(p))}{\phi(p_k) - \phi(p)} \dfrac{\phi(p_k) - \phi(p)}{p_k - p}.$

Since ϕ is continuous at p, letting k tend to infinity in 2.6.26 establishes that ϕ is differentiable at p.

(iii) The above argument shows that ϕ is differentiable throughout $[a, b]$. Since $h = f \circ \phi$ the Chain Rule implies that

$$h'(u) = f'(\phi(u))\phi'(u), \qquad u \in [c, d].$$

Since h is smooth, it follows that $\phi'(u) \neq 0$ for all $u \in [c, d]$. Hence (by Rolle's Theorem) the functions ϕ and $h = f \circ \phi$ are 1–1. Finally, since ϕ is continuous and 1–1, ϕ is either strictly increasing (in which case ϕ' is positive) or strictly decreasing (in which case ϕ' is negative).

2.6.27 Corollary.   *Let* **q** *be a point on a smooth simple arc* C *in* $\mathbb{R}^n$. *Then all smooth parametrizations of* C *associate the same tangent line with* **q**.

Remark. In view of this corollary we can talk about the tangent line to a smooth simple arc C at a point $\mathbf{q}$ on C.

Proof. Theorem 2.6.23 implies that two smooth parametrizations of C are equivalent and so the result follows from Theorem 2.6.15.

2.6.28 Example. Consider again the smooth simple arc G which is the graph of a C^1 function $g : [a, b] \subseteq \mathbb{R} \to \mathbb{R}$ (see Example 2.6.21). From 2.6.22 the tangent line to G at $(t, g(t))$ is in direction $(1, g'(t))$.

If G is smoothly parametrized by $(x(u), y(u))$, then the tangent line at $(x(u), y(u))$ has direction $(x'(u), y'(u))$. Hence

2.6.29
$$g'(x(u)) = \frac{y'(u)}{x'(u)}.$$

Informally, 2.6.29 expresses the fact that if a graph $y = g(x)$ is smoothly parametrized by $(x(u), y(u))$ then

$$\frac{dy}{dx} = \frac{dy/du}{dx/du}.$$

Since, by Theorem 2.6.23, any two smooth parametrizations of a smooth simple arc are equivalent, the smooth parametrizations of such an arc divide into two classes according to their orientation.

2.6.30 Definition. *Let C be a smooth simple arc in $\mathbb{R}^n$ parametrized by a smooth function $f : [a, b] \subseteq \mathbb{R} \to \mathbb{R}^n$. The pair $\{C, [f]\}$, where $[f]$ is the set of all smooth parametrizations of C which are properly equivalent to f, is called an* oriented smooth simple arc.

To every smooth simple arc C there correspond precisely two oriented simple arcs. If $\mathbf{a}$ and $\mathbf{b}$ are the end points of C, the two possible orientations can be described as being from $\mathbf{a}$ to $\mathbf{b}$ and from $\mathbf{b}$ to $\mathbf{a}$ respectively. See Fig. 2.13.

We have seen how Definition 2.6.9 establishes an equivalence relation on the set of all parametrizations of a curve in $\mathbb{R}^n$. In any equivalence class either all parametrizations are smooth or none is smooth. In general there may be a large number of equivalence classes of smooth parametrizations. For example, a parametrization of a circle that traces the circle k times cannot be equivalent to one that traces it l times, where $k \neq l$. However, when the curve is a *smooth simple arc,* there are just two classes of properly equivalent smooth parametrizations. These parametrizations are 1–1. They

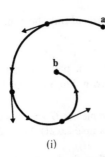 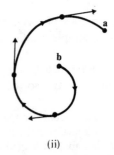

(i)   (ii)

Fig. 2.13 Smooth simple arc with two possible orientations
(i) Orientation from **a** to **b**
(ii) Orientation from **b** to **a**

correspond to the two orientations of the curve from one end point
to the other.

A simple arc that is not smooth also has two possible
orientations. Correspondingly we might expect just two classes of
properly equivalent 1–1 parametrizations of a (not necessarily
smooth) simple arc. However, the equivalence relation suggested by
Definition 2.6.9 is not the appropriate tool for the classifications of
1–1 parametrizations of a simple arc into just two classes.

2.6.31 *Example.* Let C be the simple arc $y = x^2$, $x \in [-1, 1]$. (The curve C
happens to be smooth, but this is not relevant to this example.) Then no
two of the following four 1–1 parametrizations of C are properly
equivalent:

$$f(t) = (t, t^2); \quad g(t) = (-t, t^2); \quad h(t) = (t^3, t^6); \quad k(t) = (-t^3, t^6)$$

for all $t \in [-1, 1]$.

Clearly f and g are not properly equivalent since they trace C in opposite
directions. Suppose next that $h = f \circ \phi$. Then $\phi(u) = u^3$, and so $\phi'(0) = 0$.
Thus by Definition 2.6.9 f and h are not equivalent. The remaining cases
are treated similarly.

The appropriate equivalence relation that reflects the two
possible orientations of a simple arc is obtained by dropping the
differentiability conditions on ϕ in Definition 2.6.9 and replacing
them by continuity and monotonicity. The following theorem sums
up the situation.

2.6.32 *Theorem.* *Let* $f : [a, b] \subseteq \mathbb{R} \rightarrow \mathbb{R}^n$ *and* $h : [c, d] \subseteq \mathbb{R} \rightarrow \mathbb{R}^n$ *be*

1–1 *parametrizations of a simple arc C in* $\mathbb{R}^n$. *Then there exists a unique function* ϕ *from* $[a, b]$ *onto* $[c, d]$ *such that* $h = f \circ \phi$. *Moreover,* ϕ *is continuous and strictly monotonic. In particular,*

  [i] *if* ϕ *is strictly increasing then* $f(a) = h(c)$ *and* $f(b) = h(d)$, *and*

  [ii] *if* ϕ *is strictly decreasing then* $f(a) = h(d)$ *and* $f(b) = h(c)$.

Proof. By the proof of Theorem 2.6.23 the function $\phi = f^{-1} \circ h$ is 1–1, continuous and strictly monotonic.

2.6.33 *Example*. The functions

$f : [-1, 1] \subseteq \mathbb{R} \to \mathbb{R}^2$ defined by $f(t) = (t, \sqrt{(1 - t^2)})$, $\quad t \in [-1, 1]$,

$g : [0, \pi] \subseteq \mathbb{R} \to \mathbb{R}^2$ defined by $g(t) = (\cos t, \sin t)$, $\quad t \in [0, \pi]$,

$h : [-1, 0] \subseteq \mathbb{R} \to \mathbb{R}^2$ defined by $h(t) = (\cos \pi t, -\sin \pi t)$, $\quad t \in [-1, 0]$

are all 1–1 parametrizations of the simple arc $C = \{(x, y) \in \mathbb{R}^2 \mid x^2 + y^2 = 1,$ $y \geqslant 0\}$, the upper half of the unit circle centred at the origin. The end points of C are $(-1, 1)$ and $(1, 1)$. The parametrizations f and h run from $(-1, 0)$ to $(1, 0)$ while g runs from $(1, 0)$ to $(-1, 0)$.

  We find that $h = f \circ \phi$ where $\phi(u) = \cos (\pi u)$, $u \in [-1, 0]$, which is strictly increasing. On the other hand $g = f \circ \psi$, where $\psi(\omega) = \cos \omega$, $\omega \in [0, \pi]$, which is strictly decreasing.

  Notice that g and h are C^1 parametrizations of C but f is not C^1.

Finally in this section we consider how the above material can be adjusted to cope with parametrizations which fail to be C^1 at a finite number of points.

2.6.34 *Example*. Consider the function $f : [-1, 1] \subseteq \mathbb{R} \to \mathbb{R}^2$ defined by

$$f(t) = (t, |t|), \qquad t \in [-1, 1].$$

whose image is shown in Fig. 2.8(ii). The function f is not C^1 because f is not differentiable at 0. However, we may regard f as being composed of two C^1 pieces $f_1 : [-1, 0] \subseteq \mathbb{R} \to \mathbb{R}^2$ and $f_2 : [0, 1] \subseteq \mathbb{R} \to \mathbb{R}^2$, where $f_1(t) = f(t) = (t, -t)$ when $t \in [-1, 0]$ and $f_2(t) = f(t) = (t, t)$ when $t \in [0, 1]$.

Example 2.6.34 illustrates a piecewise C^1 path. The general definition is as follows.

2.6.35 *Definition*. *A function* $f : [a, b] \subseteq \mathbb{R} \to \mathbb{R}^n$ *is said to be* piecewise C^1 (piecewise smooth) *if there is a subdivision* $a = p_0 < p_1 < \cdots < p_r = b$ *of* $[a, b]$ *such that for each* $k = 1, \ldots, r$ *the function* f *restricted to the sub-interval* $[p_{k-1}, p_k]$ *is* C^1 (smooth).

Notice that the piecewise C^1 function f considered in Definition 2.6.35 may not be differentiable at p_k for $k = 1, \ldots, r-1$. However, the fact that f is differentiable on $[p_{k-1}, p_k]$ and on $[p_k, p_{k+1}]$ requires that both the left-hand and right-hand derivatives of f at p_k must exist.

2.6.36 *Example.* The function $f : [-1, 1] \subseteq \mathbb{R} \rightarrow \mathbb{R}^2$ considered in Example 2.6.34 is piecewise C^1 (consider the partition $-1 < 0 < 1$ of $[-1, 1]$). It is also piecewise smooth. The function f is not differentiable at 0. However, the left-hand derivative of f at 0 is $(1, -1)$ and the right-hand derivative of f at 0 is $(1, 1)$.

The image curve C of the function f is sketched in Fig. 2.8(ii). Compare this parametrization of C with the parametrization $g : [-1, 1] \subseteq \mathbb{R} \rightarrow \mathbb{R}^2$ of C given by

$$g(u) = (u^3, |u^3|), \qquad u \in [-1, 1].$$

The function g is a C^1 parametrization of C.

Furthermore f and g have a weaker form of proper equivalence, since if $\phi : [-1, 1] \rightarrow [-1, 1]$ is defined by

$$\phi(t) = t^{1/3}, \qquad t \in [-1, 1]$$

then ϕ is strictly increasing and differentiable except at 0 and $f = g \circ \phi$. Notice that $g = f \circ \phi^{-1}$, where ϕ^{-1} is strictly increasing. Incidentally, ϕ^{-1} is differentiable.

2.6.37 *Remark.* The process adopted in Example 2.6.36 may be thought of as transforming a piecewise C^1 parametrization to give a 'properly equivalent' C^1 parametrization of the same curve in the weaker sense described above. In fact, given any piecewise C^1 parametrization of a curve there is a C^1 parametrization which is 'properly equivalent' to it in this sense. Consequently a curve which has a 1–1 piecewise C^1 parametrization is automatically a simple arc (Definition 2.6.20).

Exercises 2.6

1. Verify that the following subsets C of $\mathbb{R}^2$ are curves in $\mathbb{R}^2$ according to Definition 2.6.2. In each case find a C^1 parametrization of C and sketch the curve.
    (a) The y-axis, $x = 0$;
    (b) part of the y-axis, $x = 0$, $-1 \leqslant y \leqslant 2$;
    (c) the parabola $y + 1 = (x - 2)^2$;
    (d) the circle $(x - 1)^2 + (y - 2)^2 = 4$;
    (e) the ellipse $4x^2 + y^2 = 1$;
    (f) the subset C defined by $x = |y|$;

(g) the subset C defined by $x = |y|, 0 \leqslant x \leqslant 1$;

(h) the subset C defined by $y^2 = x^3, 0 \leqslant x \leqslant 1$;

(i) the subset C parametrized by

$$f(t) = (\sin 2t \cos t, \sin 2t \sin t), \qquad 0 \leqslant t \leqslant 2\pi.$$

Answers: (a) $(0, t), t \in \mathbb{R}$, alternatively $(0, t^3), t \in \mathbb{R}$. For each example there are many possibilities; (b) $(0, t), -1 \leqslant t \leqslant 2$; (c) $(t, (t-2)^2 - 1)$, alternatively $(t + 2, t^2 - 1)$; (d) $(1 + 2\cos t, 2 + 2\sin t)$; (e) $(\frac{1}{2}\cos t, \sin t)$; (f) $(t^2, t|t|)$; (g) $(t^2, t|t|), -1 \leqslant t \leqslant 1$; (h) $(t^2, t^3), -1 \leqslant t \leqslant 1$; (i)$C^1$ parametrization as given. The curve has four 'petals' meeting at the origin.

2. Which of the curves in Exercise 1 can be smoothly parametrized?

Answer: all except (f), (g), (h).

Which of the curves in Exercise 1 are simple arcs? Which are smooth simple arcs?

Answer: (b), (g), (h) are simple arcs; of these, only (b) is smooth.

3. For each of the curves C of Exercise 1 find, where appropriate, the equation of the tangent line to C at the point $\mathbf{q} = (a, b)$ on C. Indicate points on C where no tangent line exists.

(*Hint*: using an appropriate parametrization, apply the method of Section 2.5.)

Answers: (a) $x = 0$,

(c) $y - b = (2a - 4)(x - a)$, where $b = (a - 2)^2 + 1$;

(e) $(x, y) = (\frac{1}{2}\cos \alpha, \sin \alpha) + s(-\frac{1}{2}\sin \alpha, \cos \alpha), s \in \mathbb{R}$, where $a = \frac{1}{2}\cos \alpha$, $b = \sin \alpha$. For example, the tangent line to C at $(\frac{1}{2}, 0)$ is $x = \frac{1}{2}$, and the tangent line to C at $(\frac{1}{4}, -\frac{1}{2}\sqrt{3})$ is $y + \frac{1}{2}\sqrt{3} = (2/\sqrt{3})(x - \frac{1}{4})$. No tangent lines to C at $(0, 0)$ in examples (f) – (i) inclusive.)

4. Indicate on your sketches of the curves C of Exercise 1 the orientation assigned to C by your choice of C^1 parametrization $f : D \subseteq \mathbb{R} \to \mathbb{R}^n$. (Use arrows to indicate the direction in which C is traced.)

Obtain in each case an equivalent parametrization $h = f \circ \phi$ where $\phi(u) = -u$, and verify that the orientation assigned to C by h is opposite to that assigned to C by f.

5. Sketch the simple arc C in $\mathbb{R}^2$ defined by

$$x = y^2, \qquad x \leqslant 4.$$

(a) Verify that the following functions define 1–1 smooth parametrizations of C:

(i) $f : [-3, 1] \subseteq \mathbb{R} \to \mathbb{R}^2$ defined by

$$f(t) = (t^2 + 2t + 1, t + 1), \qquad t \in [-3, 1]$$

(ii) $h:[0, 1] \subseteq \mathbb{R} \to \mathbb{R}^2$ defined by

$$h(u) = (16u^2 - 16u + 4, 2 - 4u), \qquad u \in [0, 1].$$

Show that f and h are equivalent parametrizations of C by constructing a suitable differentiable function ϕ from $[0, 1]$ on to $[-3, 1]$ such that $h = f \circ \phi$. Are f and h *properly* equivalent or not?

Find the equation of the tangent line to C at the point $(\frac{1}{4}, -\frac{1}{2})$ on C (expressing your answer in the form $ax + by = c$), (1) from the parametrization f, (2) from the parametrization h.

(b) Now consider the function $g : [-1, 1] \subseteq \mathbb{R} \to \mathbb{R}^2$ defined by

$$g(w) = (4w^6, 2w^3), \qquad w \in [-1, 1].$$

Show that g is a C^1 parametrization of C. Is g 1–1? Smooth? Are f and g equivalent parametrizations of C (in the sense of Definition 2.6.9)? Find functions $\phi : [-1, 1] \to [-3, 1]$ and $\psi : [-3, 1] \to [-1, 1]$ such that $g = f \circ \phi$ and $f = g \circ \psi$. Is ϕ differentiable? Is ψ differentiable?

Answer: $\phi(w) = 2w^3 - 1$, $\psi(t) = (\frac{1}{2}(t + 1))^{1/3}$, not differentiable at $t = -1$.

6. Sketch the simple arc C parametrized by the (not smooth) C^1 function $f : [-1, 1] \subseteq \mathbb{R} \to \mathbb{R}^2$ defined by $f(t) = (t^3, t^5)$. By finding a suitable smooth C^1 re-parametrization of C, show that C is a smooth simple arc.

Answer: $h(u) = (u, u^{5/3})$, $u \in [-1, 1]$.

7. Sketch the curve C in $\mathbb{R}^2$ smoothly parametrized by

$$g(t) = ((2\cos t - 1)\cos t, (2\cos t - 1)\sin t), t \in [0, 2\pi].$$

Show that $(0, 0) = g(\pi/3) = g(5\pi/3)$, but $g'(\pi/3) \neq g'(5\pi/3)$. Interpret this on your sketch.

Using the method of Examples 2.6.7 and 2.6.8, obtain 1–1 C^1 parametrizations f and h of C such that (a) for $p \in \mathbb{R}$ such that $f(p) = (0, 0)$, $f'(p) = g'(\pi/3)$, (b) for $q \in \mathbb{R}$ such that $h(q) = (0, 0)$, $h'(q) = g'(5\pi/3)$.

8. Prove that the equivalence of C^1 parametrizations given in Definition 2.6.9 is an equivalence relation on the class of all C^1 parametrizations of curves in $\mathbb{R}^n$.

Hint: Prove symmetry by applying Theorem 1.6.7.

9. Given that $\phi(u) = \sin u$, $u \in E \subseteq \mathbb{R}$, and $f(t) = (t^2, \sqrt{t})$, $t \geq 0$, find a suitable interval E such that $f \circ \phi$ is defined and differentiable on E. Apply the Chain Rule 2.6.11 to calculate $(f \circ \phi)'(u)$, $u \in E$.

10. Apply the Chain Rule 2.6.11 to calculate $h'(u)$, where
    (a) $h(u) = (\exp(\sin u), 2(\sin u)^2)$, $u \in \mathbb{R}$;
    (b) $h(u) = (2\cos\sqrt{u}, \sin\sqrt{u})$, $u > 0$.

11. Sketch the curve C parametrized by $f : \mathbb{R} \to \mathbb{R}^2$, where
$f(t) = (t - \sin t, 1 - \cos t), t \in \mathbb{R}$ (The curve C is called a *cycloid*. A point fixed on the rim of a unit circle traces C when the circle rolls along the x-axis.) Prove that C is not the graph of a differentiable function.

Hint: consider the points $t = 2k\pi$, $k \in \mathbb{N}$.

12. Let C be a curve smoothly parametrized by a C^1 function $f(t) = (f_1(t), f_2(t))$, $t \in D$, where $f'_1(t) > 0$ for all $t \in D$. Prove that C is the graph of a C^1 function $g : D_1 \subseteq \mathbb{R} \to \mathbb{R}$, where $D_1 = f_1(D)$.

(*Hint*: apply Theorem 1.6.7. The required function is the composite function $f_2 \circ f_1^{-1}$.)

Deduce that the part of the cycloid (Exercise 11) which is the image of f on the open interval $]0, 2\pi[$ is the graph of a differentiable function.

13. The following exercise illustrates Example 2.6.28. Sketch the simple arc G which is the graph of the C^1 function

$$g(t) = \sqrt{(1 - t^2)}, \qquad t \in [-\tfrac{1}{2}, \tfrac{1}{2}].$$

Indicate the tangent lines to G at $(t, g(t))$ for the cases $t = -\tfrac{1}{2}, t = 0$ and $t = \tfrac{1}{2}$.

Verify that G is smoothly parametrized by

$$(x(\theta), y(\theta)) = (\cos \theta, \sin \theta), \qquad \theta \in [\pi/3, 2\pi/3],$$

and that

$$g'(x(\theta)) = \frac{y'(\theta)}{x'(\theta)} = -\cot \theta.$$

2.7 Path length and length of simple arcs

As a particle moves about in space its position can be regarded as a function of time. Choosing coordinate systems for space and time enables us to describe the motion of the particle over a finite time interval by a function $f : [a, b] \subseteq \mathbb{R} \to \mathbb{R}^3$, where the particle has position vector $f(t)$ at time t units. In most applications there will be no instantaneous jumps in the motion and so the corresponding function will be continuous. This leads us to the following definition.

2.7.1 Definition. *A continuous function $f : [a, b] \subseteq \mathbb{R} \to \mathbb{R}^n$ is called a* path *in $\mathbb{R}^n$ from $f(a)$ to $f(b)$. The path is* differentiable, *C^1 or* smooth *if the function f is respectively differentiable, C^1 or smooth.*

The image of a C^1 path is a curve and the image of a 1–1 smooth path is a smooth simple arc.

When a C^1 path f describes the motion of a particle in $\mathbb{R}^n$, the distance travelled by the particle in the time interval between t and $t + \Delta t$, for small positive Δt, is approximately $\|f(t + \Delta t) - f(t)\|$. This distance, by definition of the derivative, is approximately $\|f'(t)\|\Delta t$. Accordingly, the length of the path is defined as follows.

2.7.2 Definition.   Let $f : [a, b] \subseteq \mathbb{R} \to \mathbb{R}^n$ be a C^1 path in $\mathbb{R}^n$. The length *of* f *is defined to be*

$$l(f) = \int_a^b \|f'(t)\|\,\mathrm{d}t.$$

2.7.3 Example.   The function $g : [-r, r] \subseteq \mathbb{R} \to \mathbb{R}^2$ defined by

$$g(u) = (u, \sqrt{(r^2 - u^2)}), \qquad u \in [-r, r]$$

is a path in $\mathbb{R}^2$ whose image is a semi-circle of radius $r > 0$. However, g is not a C^1 path (why?).

We can reparametrize the semi-circle by defining a C^1 path $f : [0, \pi] \subseteq \mathbb{R} \to \mathbb{R}^2$ by

$$f(t) = (-r\cos t, r\sin t), \qquad t \in [0, \pi].$$

We have $f'(t) = (r\sin t, r\cos t)$, and so $\|f'(t)\| = r$. Hence

$$l(f) = \int_0^\pi r\,\mathrm{d}t = \pi r.$$

2.7.4 Example.   The function $f : [0, 2\pi] \subseteq \mathbb{R} \to \mathbb{R}^2$ defined by

$$f(t) = (\cos 2t, \sin 2t), \qquad t \in [0, 2\pi]$$

is a C^1 path parametrizing the unit circle, and

$$\|f'(t)\| = 2\|(-\sin 2t, \cos 2t)\| = 2.$$

Hence

$$l(f) = \int_0^{2\pi} 2\,\mathrm{d}t = 4\pi.$$

As expected, $l(f)$ is twice the circumference of the unit circle which $f(t)$ traverses twice.

2.7.5 Example.   The 1–1 C^1 path $f : [-1, 1] \subseteq \mathbb{R} \to \mathbb{R}^2$ defined by

$$f(t) = (t^2, t^3), \qquad t \in [-1, 1]$$

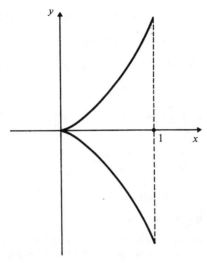

Fig. 2.14 Image of $f(t) = (t^2, t^3)$, $t \in [-1, 1]$

has image curve lying in the semi-cubical parabola $y^2 = x^3$ (see Fig. 2.14). We have $f'(t) = (2t, 3t^2)$, and so

$$l(f) = \int_{-1}^{1} \sqrt{(4t^2 + 9t^4)}\,\mathrm{d}t = 2 \int_{0}^{1} t\sqrt{4 + 9t^2}\,\mathrm{d}t.$$

The substitution $u(t) = 4 + 9t^2$ leads to $l(f) = \frac{2}{27}(13\sqrt{13} - 8)$.

We would expect that if two C^1 paths trace the points of a curve in the same order, then their lengths would be the same. Similarly, we would expect that the length of a C^1 path is the same as the length of the C^1 path which traces the points of the curve in the reverse order. These impressions are confirmed by the following theorem.

2.7.6 Theorem. Let $f : [a, b] \subseteq \mathbb{R} \to \mathbb{R}^n$ and $h : [c, d] \subseteq \mathbb{R} \to \mathbb{R}^n$ be two equivalent C^1 paths parametrizing a curve C in $\mathbb{R}^n$. Then $l(f) = l(h)$.

Proof. Since f and h are equivalent, there exists a $1-1$ differentiable function ϕ mapping $[c, d]$ onto $[a, b]$ such that $h = f \circ \phi$. Furthermore, either $\phi'(u) > 0$ for all $u \in [c, d]$, in which case $a = \inf \phi = \phi(c)$ and $b = \sup \phi = \phi(d)$, or $\phi'(u) < 0$ for all $u \in [c, d]$, in which case $b = \phi(c)$ and $a = \phi(d)$.

It follows from the Chain Rule (Theorem 2.6.11) that

$$\|h'(u)\| = \|f'(\phi(u))\|\,|\phi'(u)|, \qquad u \in [c, d].$$

Hence, by the substitution rule (treating the above alternatives separately),

$$l(h) = \int_c^d \|h'(u)\|\,\mathrm{d}u = \int_c^d \|f'(\phi(u))\|\,|\phi'(u)|\mathrm{d}u$$

$$= \int_a^b \|f'(t)\|\,\mathrm{d}t = l(f).$$

2.7.7 Example.   The function $h : [-2\pi, 2\pi] \subseteq \mathbb{R} \to \mathbb{R}^2$ defined by

$$h(u) = (\cos u,\ -\sin u), \qquad u \in [-2\pi, 2\pi],$$

is a C^1 path parametrizing the unit circle. It is equivalent to the C^1 path f defined in Example 2.7.4 (consider $\phi(u) = \pi - \frac{1}{2}u$). As expected

$$l(h) = \int_{-2\pi}^{2\pi} \|h'(u)\|\,\mathrm{d}u = 4\pi = l(f).$$

In the previous example, $l(h)$ is twice the circumference of the unit circle because h traces the circle twice. In general the length of a C^1 path cannot be judged by considering its image curve C. However, if C is a smooth simple arc then by Theorem 2.6.23 all smooth parametrizations of C are 1–1. In this case we expect the following result.

2.7.8 Corollary.   *All smooth parametrizations of a smooth simple arc in $\mathbb{R}^n$ have the same length.*

Proof.   Theorem 2.6.23 implies that all smooth parametrizations of a smooth simple arc are equivalent. By Theorem 2.7.6, they therefore have the same length.

2.7.9 Definition.   *The* length $l(C)$ *of a smooth simple arc C in $\mathbb{R}^n$ is the length of any smooth parametrization of C.*

2.7.10 Example.   Let G be the graph of the C^1 function $g : [a, b] \subseteq \mathbb{R} \to \mathbb{R}$ considered in Example 2.6.28. Suppose that G is smoothly parametrized by $(x(u), y(u))$, $u \in [c, d]$. Then

$$l(G) = \int_a^b \sqrt{(1 + g'(x)^2)}\,\mathrm{d}x = \int_c^d \sqrt{(x'(u)^2 + y'(u)^2)}\,\mathrm{d}u.$$

The natural parametrization of a curve (the one which would be chosen by a small creature living on the curve!) is one in which the parameter s measures the distance along the curve from some fixed base point p. Such a parametrization $h : E \subseteq \mathbb{R} \to \mathbb{R}^n$, if it exists, is readily identified since it requires that for all a and b in E the distance traversed from $s = a$ to $s = b$ is given by

$$\int_a^b \| h'(s) \| \, \mathrm{d}s = b - a.$$

An equivalent requirement is that $\| h'(s) \| = 1$ for all $s \in E$ (Exercise 2.7.9). This motivates the following definition.

2.7.11 Definition.   *A C^1 parametrization $h : E \subseteq \mathbb{R} \to \mathbb{R}^n$ of a curve C in $\mathbb{R}^n$ is a* path-length parametrization *of C if $\| h'(s) \| = 1$ for all $s \in E$.*
Clearly such a parametrization is smooth.

2.7.12 Example.   The function $h : \mathbb{R} \to \mathbb{R}^2$ defined by

$$h(s) = \left(r\cos\frac{s}{r},\ r\sin\frac{s}{r} \right), \qquad s \in \mathbb{R},$$

is a path-length parametrization of the circle in $\mathbb{R}^2$ with centre $(0, 0)$ and radius r. The length of the path between $s = a$ and $s = b$ is $b - a$.

We shall now show that general smooth parametrizations can be studied by considering equivalent path-length parametrizations.

2.7.13 Definition.   *Let $f : D \subseteq \mathbb{R} \to \mathbb{R}^n$ be a smooth parametrization of a curve C in $\mathbb{R}^n$. Choose a base point $p \in D$. The* path-length *function based at p is the function $\lambda : D \subseteq \mathbb{R} \to \mathbb{R}$ defined by*

2.7.14 $$\lambda(t) = \int_p^t \| f'(u) \| \, \mathrm{d}u, \qquad t \in D.$$

Thinking of f as representing the motion of a particle in $\mathbb{R}^n$ over a time interval D, the number $s = \lambda(t)$ indicates the distance travelled by the particle in the time interval between p and t (with the convention that if $t < p$ then the distance is negative). We find from 2.7.14 that the speed of f at t is given by

2.7.15 $$\lambda'(t) = \| f'(t) \| > 0, \qquad t \in D.$$

This derivative clearly does not depend upon the base point chosen.

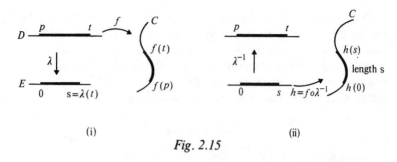

Fig. 2.15

Definition 2.7.13 is illustrated in Fig. 2.15(i). Since λ is continuous, the set $E = \lambda(D)$ is an interval in $\mathbb{R}$. Furthermore, by 2.7.15, λ is a 1–1 function whose inverse function $\lambda^{-1} : E \to D$ has positive derivative throughout E. Comparison of Fig. 2.15(i) and Fig. 2.11 suggests that we can use s as a parameter in a parametrization of C that is properly equivalent to f.

2.7.16 Theorem. *For any smooth parametrization* $f : D \subseteq \mathbb{R} \to \mathbb{R}^n$ *of a curve C in $\mathbb{R}^n$ there is a path-length parametrization* $h : E \subseteq \mathbb{R} \to \mathbb{R}^n$ *of C which is properly equivalent to f.*

Proof. The proof is illustrated in Fig. 2.15(ii). Chose a base point $p \in D$ and let $\lambda : D \subseteq \mathbb{R} \to \mathbb{R}$ be the path-length function based at p given by 2.7.14. Let $E = \lambda(D)$ and define $h = f \circ \lambda^{-1} : E \subseteq \mathbb{R} \to \mathbb{R}^n$. By 2.7.15, h is properly equivalent to f. Also, since $f = h \circ \lambda$, it follows from the Chain Rule and 2.7.15 that

$$\|f'(t)\| = \|h'(\lambda(t))\| \, |\lambda'(t)| = \|h'(\lambda(t))\| \, \|f'(t)\|.$$

Therefore $\|h'(s)\| = 1$ for all $s \in E$.

2.7.17 Example. The image of the continuously differentiable function $f : \mathbb{R} \to \mathbb{R}^3$ defined by

$$f(t) = (at, b \cos \omega t, b \sin \omega t), \qquad t \in \mathbb{R},$$

where $b > 0$ and $\omega > 0$, is a helix wound around a cylinder of radius b whose axis is the x-axis in $\mathbb{R}^3$. The helix corresponding to $a = b = \omega = 1$ is sketched in Fig. 2.2. Now

$$f'(t) = (a, -b\omega \sin \omega t, b\omega \cos \omega t), \qquad t \in \mathbb{R},$$

and

$$\|f'(t)\| = (a^2 + b^2 \omega^2)^{1/2}, \qquad t \in \mathbb{R}.$$

Hence f is smooth, since $a^2 + b^2\omega^2 \neq 0$. Choose $0 \in \mathbb{R}$ as base point. The path-length function $\lambda : \mathbb{R} \to \mathbb{R}$ is defined by

$$\lambda(t) = (a^2 + b^2\omega^2)^{1/2} t, \qquad t \in \mathbb{R}.$$

For example, the image of any closed interval of length $2\pi/\omega$ is a single twist of the helix, and hence the length of a single twist is $2\pi(a^2 + b^2\omega^2)^{1/2}\omega$.

The inverse function $\lambda^{-1} : \mathbb{R} \to \mathbb{R}$ is given by

$$\lambda^{-1}(s) = (a^2 + b^2\omega^2)^{-1/2} s, \qquad s \in \mathbb{R}.$$

Let $c = (a^2 + b^2\omega^2)^{1/2}$. A path-length parametrization properly equivalent to f is given by

$$h(s) = f(\lambda^{-1}(s)) = \left[\frac{as}{c}, b\cos\frac{\omega s}{c}, b\sin\frac{\omega s}{c} \right], \qquad s \in \mathbb{R}.$$

As expected (from Theorem 2.7.16), for all $s \in \mathbb{R}$,

$$\|h'(s)\| = \left\| \left(\frac{a}{c}, -\frac{b\omega}{c}\sin\frac{\omega s}{c}, \frac{b\omega}{c}\cos\frac{\omega s}{c} \right) \right\| = \left[\frac{a^2 + b^2\omega^2}{c^2} \right]^{1/2} = 1.$$

2.7.18 Example.   We saw in Example 2.6.34 that the piecewise C^1 path

$$f(t) = (t, |t|), \qquad t \in [-1, 1]$$

can be regarded as being composed of two C^1 pieces $f_1 : [-1, 0] \subseteq \mathbb{R} \to \mathbb{R}^2$ and $f_2 : [0, 1] \subseteq \mathbb{R} \to \mathbb{R}^2$, where $f_1(t) = (t, -t)$ when $t \in [-1, 0]$ and $f_2(t) = (t, t)$ when $t \in [0, 1]$.

We define the *length* of the path f by

$$l(f) = l(f_1) + l(f_2) = \int_{-1}^{0} \|f_1'(t)\|\, dt + \int_{0}^{1} \|f_2'(t)\|\, dt$$

$$= \int_{-1}^{1} \|f'(t)\|\, dt,$$

where in the last integral the integrand is undefined at $t = 0$. Since $\|f'(t)\| = \sqrt{2}$, $t \neq 0$, it follows that $l(f) = 2\sqrt{2}$.

Finally in this section we introduce two important path constructions.

2.7.19 Definition.   *Let $f : [a, b] \subseteq \mathbb{R} \to \mathbb{R}^n$ be a path. The* inverse path *is defined to be the function $f^- : [a, b] \subseteq \mathbb{R} \to \mathbb{R}^n$ where*

$$f^-(t) = f(a + b - t), \qquad t \in [a, b].$$

Clearly f^- is a path. It traces the image of f from $f(b)$ to $f(a)$ by reversing the action of f. Moreover f^- is C^1 (smooth) if and only if f is C^1 (smooth). The paths f and f^- have the same length.

The second construction is based on the observation that if $f:[a, b]\subseteq\mathbb{R}\rightarrow\mathbb{R}^n$ and $g:[c, d]\subseteq\mathbb{R}\rightarrow\mathbb{R}^n$ are paths in $\mathbb{R}^n$ such that the end point $f(b)$ of f is the initial point $g(c)$ of g, then there is a composite path 'f followed by g' that runs along f from $f(a)$ to $f(b)$ and then runs along g from $f(b)=g(c)$ to $g(d)$. The following definition formalizes this idea.

2.7.20 Definition.   *Let* $f:[a, b]\subseteq\mathbb{R}\rightarrow\mathbb{R}^n$ *and* $g:[c, d]\subseteq\mathbb{R}\rightarrow\mathbb{R}^n$ *be paths in* $\mathbb{R}^n$, *where* $f(b)=g(c)$. *The* product path $h=fg:[a, b+d-c]\subseteq\mathbb{R}\rightarrow\mathbb{R}^n$ *is defined by*

$$h(t)=\begin{cases}f(t) & \textit{when} & a\leqslant t\leqslant b\\ g(c+(t-b)) & \textit{when} & b\leqslant t\leqslant b+d-c.\end{cases}$$

Clearly if f and g are C^1 (smooth) paths then the product path is piecewise C^1 (smooth).

Exercises 2.7

1.   Calculate the lengths of the following smooth simple arcs in $\mathbb{R}^3$.
    (a) The circular helix parametrized by

$$f(t)=(t, \cos t, \sin t), \qquad t\in[a, b];$$

    (b) the curve parametrized by

$$f(t)=(e^t\cos t, e^t\sin t, e^t) \qquad t\in[0, k].$$

Answers: (a) $\sqrt{2}(b-a)$;   (b) $\sqrt{3}(e^k-1)$.

2.   Sketch the following smooth simple arcs in $\mathbb{R}^2$ and calculate their length.
    (a) The curve parametrized by $f(t)=(e^t\cos t, e^t\sin t), t\in[1, 2]$;
    (b) the graph (catenary) $y=\cosh x=\frac{1}{2}(e^x+e^{-x})$ between $x=-1$ and $x=1$;
    (c) the portion of the parabola $y^2=16x$ which lies between the lines $x=0$ and $x=4$.

Answers: (a) $\sqrt{2}(e^2-e)$;   (b) $e-e^{-1}$;   (c) $2\sqrt{2}+2\ln(\sqrt{2}+1)$.

3.   Which of the following two paths in $\mathbb{R}^3$ from $(0, 0, 0)$ to $(1, 1, 1)$ is the longer? (a) $f(t)=(t, t, t^2), 0\leqslant t\leqslant 1$; (b) $g(t)=(t, t^2, t^2), 0\leqslant t\leqslant 1$.

4.   The semi-ellipse E with equation

$$\frac{x^2}{a^2}+\frac{y^2}{b^2}=1, \qquad y\geqslant 0,$$

is smoothly parametrized by $f(t) = (a \cos t, b \sin t)$, $0 \leqslant t \leqslant \pi$. Prove
that if $a \geqslant b > 0$ then the length of E is given by

$$l(E) = 2a \int_0^{\pi/2} \sqrt{(1 - k^2 \cos^2 t)}\,dt,$$

where $k^2 = (a^2 - b^2)/a^2$. The integral is called an *elliptic integral*. There
exist tables of elliptic integrals from which its value can be obtained for
various values of k.

5. (a) Sketch the simple arc C

$$x^{2/3} + y^{2/3} = 1, \qquad 0 \leqslant y \leqslant 1.$$

Verify that C is parametrized by the 1–1 C^1 path

$$f(t) = (\cos^3 t, \sin^3 t), \qquad 0 \leqslant t \leqslant \pi.$$

Show that C is not a smooth simple arc. (Consider the point $f(\tfrac{1}{2}\pi)$.)
(b) Call a simple arc C in $\mathbb{R}^n$ piecewise smooth if it has a piecewise
smooth parametrization. Prove that all piecewise smooth parametrizations of C have the same length. Hence frame a definition of the
length $l(C)$ of C.
(c) Find the length of the simple arc of part (a) of this exercise.

Answer: 3.

6. Calculate the length of an arch of the cycloid $f(t) = (t - \sin t, 1 - \cos t)$,
$t \in [0, 2\pi]$.

Answer: 8.

7. A curve C in $\mathbb{R}^n$ is a (smooth) *simple closed curve* if C has a (smooth) C^1
parametrization of the form $f : [a, b] \subseteq \mathbb{R} \to \mathbb{R}^n$ such that $f(a) = f(b)$ and
f is 1–1 on the half-open interval $[a, b[$. The length of C is then
unambiguously defined (Section 6.2) by

$$l(C) = l(f) = \int_a^b \|f'(t)\|\,dt.$$

Sketch the following simple closed curves and find their length.
(a) The circle $x^2 + y^2 = 9$ (parametrize by $f(t) = (3\cos t, 3\sin t)$,
$0 \leqslant t \leqslant 2\pi$);
(b) the *cardioid* (a heart-shaped curve) parametrized by

$$f(t) = ((1 + \cos t)\cos t, (1 + \cos t)\sin t), \qquad 0 \leqslant t \leqslant 2\pi.$$

(c) the curve (circle!) parametrized by

$$f(t) = (2 - 2\cos t, 2\sin t), \qquad 0 \leqslant t \leqslant 2\pi.$$

(d)   the curve (circle!) parametrized by

$$f(t) = (\sin t \cos t, \sin^2 t) \qquad 0 \leqslant t \leqslant \pi.$$

(e)   the four-pointed star (*astroid*) $x^{2/3} + y^{2/3} = 1$.

Answers: (a) 6;   (b) 8;   (c) 4;   (d) π;   (e) 6.

8.   Let $f : [a, b] \subseteq \mathbb{R} \to \mathbb{R}^2$ be a C^1 path in $\mathbb{R}^2$ defined by

$$f(t) = (r(t) \cos t, r(t) \sin t), \qquad t \in [a, b],$$

where $r(t) \geqslant 0$ for all $t \in [a, b]$.
Prove that the length of f is

$$l(f) = \int_a^b \sqrt{[r(t)^2 + r'(t)^2]} \, dt.$$

Hence find the lengths of the following curves of Exercise 7:
(a)   the circle $r(t) = 3$;
(b)   the cardioid $r(t) = 1 + \cos t$;
(c)   the circle $r(t) = \sin t$.

(*Note*: the path f is said to be in *polar coordinate* form. The point $f(t)$ lies on the circle centre the origin, radius $r(t)$, at an angle t from the x-axis.)

9.   Let $h : E \subseteq \mathbb{R} \to \mathbb{R}^n$ be a C^1 function. Prove that if for all a, b in E

$$\int_a^b \|h'(s)\| \, ds = b - a,$$

then $\|h'(s)\| = 1$ for all $s \in E$.

(*Hint*: fix a, and treat b as variable. Differentiate with respect to b.)

10.   Find path-length parametrizations h that are properly equivalent to the given smooth parametrizations f of the following curves.
(a)   The unit circle $f(t) = (\cos 2t, \sin 2t)$, $t \in \mathbb{R}$;
(b)   the smooth simple arc in $\mathbb{R}^3$ parametrized by

$$f(t) = (e^t \cos t, e^t \sin t, e^t), \qquad t \in [0, \pi].$$

Answers: with base point 0
(a) $\lambda(t) = 2t$, $h(s) = f(\lambda^{-1}(s)) = (\cos s, \sin s)$;
(b) $\lambda(t) = 3(e^t - 1)$, $\lambda^{-1}(s) = \ln(1 + (s/\sqrt{3}))$, $s \in [0, \sqrt{3}(e^\pi - 1)]$,
$h = f \circ \lambda^{-1}$.

11.   Let $f : [0, 1] \subseteq \mathbb{R} \to \mathbb{R}^2$ and $g : [-1, 1] \subseteq \mathbb{R} \to \mathbb{R}^2$ be the C^1 paths defined by

$$f(t) = (-t, t), \qquad t \in [0, 1],$$
$$g(t) = (t, t^2), \qquad t \in [-1, 1].$$

Verify that $f(1) = g(-1)$. Construct the inverse path f^- and the product path fg. Describe by means of a sketch how they trace their images.

2.8 Differential geometry

In this section we shall show how the twisting and turning of a curve in $\mathbb{R}^n$ can be expressed in terms of the derivatives of a path-length parametrization. We shall then use Theorem 2.7.16 to extend the results to more general parametrizations.

Let $h : E \subseteq \mathbb{R} \to \mathbb{R}^n$ be a twice-differentiable path-length parametrization of the curve $C = h(E)$. Since $\|h'(s)\| = 1$ for all $s \in E$, the vector $h''(s)$ indicates the way in which the unit tangent vector $h'(s)$ is changing direction near s. The larger the value of $\|h''(s)\|$, the greater is the 'curving' of C at s.

By Theorem 2.5.14 $h''(s)$ is orthogonal to $h'(s)$.

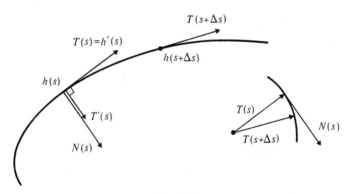

Fig. 2.16

2.8.1 Definition. *Let $h : E \subseteq \mathbb{R} \to \mathbb{R}^n$ be a twice-differentiable path-length parametrization of a curve $h(E)$ in $\mathbb{R}^n$.*

[i] The unit tangent vector $h'(s)$ at $s \in E$ will also be denoted by $T(s)$.

[ii] The non-negative number

$$\kappa(s) = \|T'(s)\| = \|h''(s)\|, \qquad s \in E,$$

is called the curvature *of h at s.*

[iii] If $\kappa(s) \neq 0$, then the unit vector $N(s)$ such that

$$T'(s) = \kappa(s)N(s)$$

is called the principal normal *to h at s.* (*If* $\kappa(s) = 0$, *the principal normal is not defined at s.*)

[iv] *The plane in* ℝ^n *through h(s) parallel to the subspace spanned by T(s) and N(s) is called the* osculating plane *of h at s.*

Since $\kappa(s) > 0$ in (iii), the principal normal $N(s)$ and the vector $T'(s)$ point in the same direction. In fact, as shown in Fig. 2.16, that direction is towards the concave side of the curve.

The osculating plane is illustrated in Fig. 2.17(i). Note that the vectors $T(s)$ and $N(s)$ are orthogonal for all $s \in E$. This follows from the earlier remark that $h'(s)$ and $h''(s)$ are orthogonal.

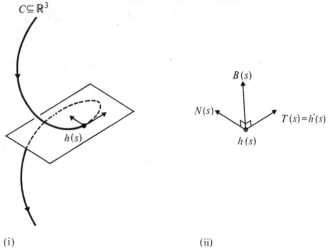

(i)   (ii)

Fig. 2.17 (i) Osculating plane at $h(s)$;
            (ii) Unit tangent, principal normal and binormal at $h(s)$

2.8.2 *Example.* For the path-length parametrization of the helix given in Example 2.7.17,

$$T(s) = h'(s) = \frac{1}{c}\left(a, -b\omega\sin\frac{\omega s}{c}, b\omega\cos\frac{\omega s}{c}\right),$$

and

$$h''(s) = \frac{b\omega^2}{c^2}\left(0, -\cos\frac{\omega s}{c}, -\sin\frac{\omega s}{c}\right), \qquad s \in \mathbb{R}.$$

Hence the curvature $\kappa(s)$ is $b\omega^2/c^2$ and the principal normal is

$$N(s) = \left(0, -\cos\frac{\omega s}{c}, -\sin\frac{\omega s}{c}\right).$$

The subspace spanned by the orthogonal unit vectors $T(s)$ and $N(s)$ in $\mathbb{R}^3$ is the plane

$$b\omega x + a\left(\sin\frac{\omega s}{c}\right)y - a\left(\cos\frac{\omega s}{c}\right)z = 0.$$

The osculating plane to h at s is the plane parallel to this through $h(s)$, given by

2.8.3 $$b\omega x + a\left(\sin\frac{\omega s}{c}\right)y - a\left(\cos\frac{\omega s}{c}\right)z = \frac{ab\omega s}{c}.$$

Suppose now that $a = 0$. The image of h is then a circle of radius b in the y, z-plane. The curvature $b\omega^2/c^2$ is $1/b$, reflecting the fact that the curvature of a circle becomes greater as the radius becomes smaller. From 2.8.3 the osculating plane in this special case is $x = 0$, the y, z plane itself.

Notice that Definition 2.8.1 essentially defines three functions $T : E \subseteq \mathbb{R} \to \mathbb{R}^n$, $N : E \subseteq \mathbb{R} \to \mathbb{R}^n$ and $\kappa : E \subseteq \mathbb{R} \to \mathbb{R}$ associated with the path-length parametrization h. When there is any risk of confusion we shall denote these three functions by T_h, N_h and κ_h.

2.8.4 Theorem. *Let $h : E \subseteq \mathbb{R} \to \mathbb{R}^n$ be a twice differentiable path-length parametrization. Then $\kappa(s) = 0$ for all $s \in E$ if and only if $h(E)$ is a subset of a straight line in $\mathbb{R}^n$.*

Proof. Exercise.

The following theorem is an example of how the functions κ, T and N can be used to study the geometry of curves in $\mathbb{R}^3$. Theorem 2.5.14 implies that the vector $T'(s) = \kappa(s)N(s)$ is orthogonal to $T(s)$ and similarly that $N(s)$ is orthogonal to $N'(s)$. It is therefore possible that $N'(s)$ and $T(s)$ are in the same direction.

2.8.5 Theorem. *Let $h : E \subseteq \mathbb{R} \to \mathbb{R}^3$ be a three times differentiable path-length parametrization of a curve C in $\mathbb{R}^3$ with $\kappa(s) \neq 0$ for all $s \in E$. The curve C lies in a plane in $\mathbb{R}^3$ if and only if $N'(s)$ is a multiple of $T(s)$ for all $s \in E$.*

Proof. Differentiating the identity $N(s) \cdot T(s) = 0$ gives

$$(N'(s) \cdot T(s)) + (N(s) \cdot T'(s)) = 0.$$

Hence

2.8.6 $$N'(s) \cdot T(s) = -\kappa(s), \qquad s \in E.$$

Suppose that $N'(s)$ is a multiple of $T(s)$. It follows from 2.8.6 that

2.8.7 $N'(s) = -\kappa(s)T(s),$ $s \in E.$

Choose any $q \in E$. We shall prove that C lies in the osculating plane of h at q. It is enough to show that for all $s \in E$ the vector $h(s) - h(q)$ lies in that plane. This will follow if we can show that, for all $s \in E$,

2.8.8 $(h(s) - h(q)) \cdot (T(q) \times N(q)) = 0.$

Put $B(s) = T(s) \times N(s)$. Then it follows from $T' = \kappa N$ and (2.8.7) $N' = -\kappa T$ that $B'(s) = \mathbf{0}$. So $B(s) = \text{constant} = B(q)$. But $T(s) \cdot B(s) = 0$, and therefore

2.8.9 $T(s) \cdot (T(q) \times N(q)) = 0$ for all $s \in E.$

Integrating 2.8.9, we obtain

$$h(s) \cdot (T(q) \times N(q)) = \text{constant} = h(q) \cdot (T(q) \times N(q)),$$

and 2.8.8 is proved.

The proof of the converse proposition, that if C lies in a plane then $N'(s)$ is a multiple of $T(s)$ for all $s \in D$ (in fact that $N'(s) = -\kappa(s)T(s)$), is left to the reader.

2.8.10 *Example.* Returning to the path-length parametrization of the helix considered in Example 2.8.2 we find that

$$N'(s) = \frac{\omega}{c}\left(0, \sin\frac{\omega s}{c}, -\cos\frac{\omega s}{c}\right), \qquad s \in E.$$

Hence $N'(s)$ is a multiple of $T(s)$ if and only if $a = 0$ (when the helix becomes a circle). In this case $N'(s) = (-1/b)T(s) = -\kappa(s)T(s)$ as asserted in the proof of Theorem 2.8.5.

It follows from Theorem 2.8.5 and in particular from 2.8.7 that the norm of the vector $N'(s) + \kappa(s)T(s)$ indicates whether or not the curve has any torsion – that is to say whether or not the curve twists out of its osculating planes.

2.8.11 *Theorem.* *With the hypothesis of Theorem 2.8.5 the vector* $N'(s) + \kappa(s)T(s)$ *is orthogonal to the osculating plane to h at s.*

Proof. It is enough to prove that $N'(s) + \kappa(s)T(s)$ is orthogonal to $T(s)$ and to $N(s)$. We have

$$(N'(s) + \kappa(s)T(s)) \cdot T(s) = (N'(s) \cdot T(s)) + \kappa(s) = 0 \qquad \text{by 2.8.6,}$$

$(N'(s) + \kappa(s)T(s)) \cdot N(s) = (N'(s) \cdot N(s)) + 0 = 0$ by Theorem 2.5.14,

and the proof is complete.

2.8.12 Definition.   *Suppose that the function* $h : E \subseteq \mathbb{R} \to \mathbb{R}^3$ *is a three times differentiable path-length parametrization of a curve in* $\mathbb{R}^3$ *with* $\kappa(s) \neq 0$ *for all* $s \in E$.

[i] *For each* $s \in E$ *the unit vector*

$$B(s) = T(s) \times N(s)$$

is called the binormal *to h at s* (see Fig. 2.17(ii)).

[ii] *By Theorem* 2.8.11 *there exists a number* $\tau(s)$ *such that*

2.8.13 $N'(s) + \kappa(s)T(s) = \tau(s)B(s).$

The number $\tau(s)$ *is called the* torsion *of h at s.*

2.8.14 Theorem (J.A. Serret, F-J. Frénet). *In the notation of Definition* 2.8.12, *the three vector-valued functions T, N, B on E and the two real-valued functions* κ *and* τ *on E satisfy the following formulae:*

[i] $T' = \kappa N$
[ii] $N' = -\kappa T + \tau B$
[iii] $B' = -\tau N.$

Proof.   (i) and (ii) are matters of definition. Differentiating $B = T \times N$ we obtain

$$B' = (T' \times N) + (T \times N') = T \times (-\kappa T + \tau B) = \tau(T \times B) = -\tau N.$$

Notice that it is a consequence of (ii) and of Theorem 2.8.5 that $f(E)$ lies in a plane if and only if $\tau(s) = 0$ for all $s \in E$. Hence, from (iii), $h(E)$ is planar if and only if $B'(s) = 0$ for all $s \in E$. Therefore $h(E)$ is planar if and only if the binormal $B(s)$ is constant. See also Exercise 2.8.7.

It will be seen from the relation $B' = -\tau N$ and Fig. 2.17 that the torsion $\tau(s)$ provides a measure of the rate at which the osculating plane twists about the tangent line at $h(s)$, the twist being clockwise in the direction of orientation when $\tau(s)$ is positive and counterclockwise when $\tau(s)$ is negative.

The formula $B' = -\tau N$ can be used to *define* torsion, and 2.8.13 can then be proved as a consequence. See Exercise 2.8.10. In general it is easier to compute the torsion by considering B' rather than N'.

2.8.15 *Example.* Consider again the parametrization of the helix given in Example 2.7.17. It follows from Example 2.8.2 that for all $s\in\mathbb{R}$

$$B(s) = T(s) \times N(s) = \left(\frac{b\omega}{c}, \frac{a}{c}\sin\frac{\omega s}{c}, -\frac{a}{c}\cos\frac{\omega s}{c}\right).$$

Hence

$$B'(s) = -\frac{a\omega}{c^2}\left(0, -\cos\frac{\omega s}{c}, -\sin\frac{\omega s}{c}\right) = -\tau(s)N(s).$$

Therefore $\tau(s) = a\omega/c^2 = a\omega/(a^2 + b^2\omega^2)$. Notice that, since $\omega \neq 0$, $\tau(s) = 0$ if and only if $a = 0$ (when the helix reduces to a circle in the y, z-plane).

In the helix example we found that the curvature and torsion were both constant. This is not generally true as we shall find in the exercises.

The following theorem shows that a curve in $\mathbb{R}^3$ is essentially characterized by its curvature and torsion.

2.8.16 *Theorem.* *Let* $g : E \subseteq \mathbb{R} \to \mathbb{R}^3$ *and* $h : E \subseteq \mathbb{R} \to \mathbb{R}^3$ *be two three times differentiable path-length parametrizations of curves* C_g *and* C_h *in* $\mathbb{R}^3$. *If* $\kappa_g(s) = \kappa_h(s) \neq 0$ *and* $\tau_g(s) = \tau_h(s)$ *for all* $s\in E$, *then the two curves are identical, except possibly in their positions in space.*

Proof. Pick any point $p\in E$. Hold C_g fixed and move C_h rigidly in $\mathbb{R}^3$ until $T_h(p) = T_g(p)$, $N_h(p) = N_g(p)$, $B_h(p) = B_g(p)$. Consider the function $\phi : E \subseteq \mathbb{R} \to \mathbb{R}$ defined by

$$\phi = T_g \cdot T_h + N_g \cdot N_h + B_g \cdot B_h$$

The Serret–Frénet formulae (Theorem 2.8.14) imply that $\phi' = 0$. Hence ϕ is a constant function and therefore $\phi(s) = \phi(p) = 3$ for all $s\in E$. By the Cauchy–Schwarz inequality, for each $s\in E$, $T_g(s)\cdot T_h(s) \leqslant \|T_g(s)\|\,\|T_h(s)\| = 1$, with equality if and only if $T_g(s) = T_h(s)$. A similar statement applies to the terms involving N and B. It follows that

$$T_g(s)\cdot T_h(s) = N_g(s)\cdot N_h(s) = B_g(s)\cdot B_h(s) = 1, \qquad s\in E.$$

Hence $T_g = T_h$, that is $g'(s) = h'(s)$ for all $s\in E$, and so $g(s) = h(s) + \mathbf{k}$ for some $\mathbf{k}\in\mathbb{R}^n$ and all $s\in E$. That is, C_h is a translation of C_g, and the proof is complete.

It is often convenient to study the differential geometry of a curve relative to a smooth parametrization which may not be a path-length parametrization. Suppose that $f : D \subseteq \mathbb{R} \to \mathbb{R}^n$ is a twice differentiable smooth parametrization of a curve $C = f(D)$ in $\mathbb{R}^n$, that a base point is chosen in D and that $\lambda : D \to E$ is the associated path-length function. We therefore have (By Theorem 2.7.16) a twice differentiable path-length parametrization $h = f \circ \lambda^{-1} : E \subseteq \mathbb{R} \to \mathbb{R}^n$ of the curve C. Provided $\kappa_h(s) > 0$ for all $s \in E$, the function h has its associated 'Frénet apparatus' T_h, N_h, κ_h and (if $n = 3$ and provided f is three times differentiable) B_h and τ_h.

2.8.17 Theorem. *With the above notation, and denoting $\lambda'(t) = \| f'(t) \|$ by $v(t)$, we have, for $t \in D$*

[i] $f'(t) = v(t)T_h(\lambda(t))$,
[ii] $f''(t) = v'(t)T_h(\lambda(t)) + (v(t))^2\kappa_h(\lambda(t))N_h(\lambda(t))$,   *if* $\kappa_h(\lambda(t)) > 0$,
[iii] $f''(t) = v'(t)T_h(\lambda(t))$, *if* $\kappa_h(\lambda(t)) = 0$.

Proof. Exercises.

Theorem 2.8.17 motivates the following definition of the Frénet apparatus of the general smooth parametrization f.

2.8.18 Definition. *In the notation used in Theorem 2.8.17 define*

$$T_f(t) = T_h(\lambda(t)), \qquad N_f(t) = N_h(\lambda(t)), \qquad \kappa_f(t) = \kappa_h(\lambda(t))$$

and, provided $n = 3$,

$$B_f(t) = B_h(\lambda(t)) \qquad and \qquad \tau_f(t) = \tau_h(\lambda(t)).$$

The fact that this definition does not depend upon the base point chosen (and hence on λ) follows from Theorem 2.8.17 by expressing $T_f(t)$ and $\kappa_f(t)N_f(t)$ in terms of $v(t)$ and the derivatives of $f(t)$.

From Theorem 2.8.17 we now obtain expressions for the Frénet apparatus associated with a general smooth parametrization f of a curve C.

2.8.19 Theorem. *Let $f : D \subseteq \mathbb{R} \to \mathbb{R}^n$ be a twice-differentiable smooth parametrization of a curve $C = f(D)$ in $\mathbb{R}^n$, and let $v(t) = \| f'(t) \|$, $t \in D$. Then the unit tangent $T_f(t)$, the curvature $\kappa_f(t) \geqslant 0$, and if $\kappa_f(t) > 0$, the principal normal $N_f(t)$ to f at $t \in D$ are given by the relations*

2.8.20 $$f'(t) = v(t)T_f(t)$$

2.8.21 $$f''(t) = v'(t)T_f(t) + (v(t))^2\kappa_f(t)N_f(t).$$

Furthermore, for the case $n = 3$, *if* $\kappa_f(t) \neq 0$, *the binomial* $B_f(t)$ *and torsion* $\tau_f(t)$ *to* f *at* t *are given by*

2.8.22 $B_f(t) = T_f(t) \times N_f(t)$

2.8.23 $B_f'(t) = -v(t)\tau_f(t)N_f(t).$

Proof. Immediate from Theorem 2.8.17.

The reader will have noticed that the differential geometry of a smooth curve C has been developed relative to a particular smooth parametrization of C and not relative to the curve C itself. In general it is not possible to associate a unique Frénet apparatus with a point on a smooth curve. For example, there are four different unit tangent vectors at the point $(0, 0)$ on the figure 8 curve of Fig. 2.9, where the curve crosses itself. However, the differential geometry of an oriented smooth simple arc is independent of the choice of smooth parametrization. For such oriented curves C it is appropriate to refer to the unit tangent vector, the curvature and (if it exists) the principal normal of C at a point $\mathbf{q} \in C$. These quantities can be calculated by applying Theorem 2.8.19 to any smooth parametrization that assigns the desired orientation to the curve. Expression 2.8.21 is awkward to use since it involves the derivative of $v(t)$. The calculation of $v'(t)$ can usually be avoided. (See the following example and Exercises 2.8.1 and 2.8.4.)

2.8.24 Example.   Find the unit tangent, the curvature and the principal normal at the point $\mathbf{q} = (-\tfrac{6}{5}, \tfrac{4}{5})$ on the semi-ellipse

$$\tfrac{1}{4}x^2 + y^2 = 1, \qquad y \geqslant 0,$$

oriented from $(-2, 0)$ to $(2, 0)$. See Fig. 2.18.

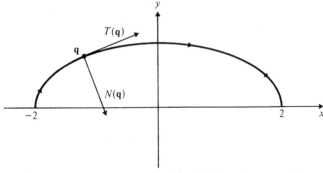

Fig. 2.18 Oriented semi-ellipse $\tfrac{1}{4}x^2 + y^2 = 1$, $y \geqslant 0$

The required orientation is afforded by the parametrization

$$f(t) = (-2\cos t, \sin t), \qquad t \in [0, \pi].$$

Then $\mathbf{q} = (-\frac{6}{5}, \frac{4}{5}) = f(a)$, where $a \in [0, \pi]$ is defined by $\cos a = \frac{3}{5}$, $\sin a = \frac{4}{5}$. We have

$$f'(t) = (2\sin t, \cos t) \quad \text{and} \quad f''(t) = (2\cos t, -\sin t).$$

Hence

$$f'(a) = (\tfrac{8}{5}, \tfrac{3}{5}) \qquad \text{and} \qquad f''(a) = (\tfrac{6}{5}, -\tfrac{4}{5}).$$

By 2.8.20, the unit tangent at $\mathbf{q}$ is the unit vector in the direction of $f'(a)$:

$$T(\mathbf{q}) = \frac{1}{\sqrt{73}}(8, 3).$$

The principal normal $N(\mathbf{q})$ is orthogonal to $T(\mathbf{q})$ and points towards the concave side of the curve at $\mathbf{q}$. Therefore (see Fig. 2.18).

$$N(\mathbf{q}) = \frac{1}{\sqrt{73}}(3, -8).$$

Finally, from 2.8.21, the curvature at $\mathbf{q}$ is given by

$$\kappa(\mathbf{q}) = \frac{f''(a) \cdot N(\mathbf{q})}{\|f'(a)\|^2} = \frac{250}{73^{3/2}}.$$

Expression 2.8.21 splits $f''(t)$ into two orthogonal components — one along the tangent $T_f(t)$ and the other along the principal normal $N_f(t)$. This is significant in the application of these results to dynamics, which we undertake in the next section.

We end this section with some hints on the evaluation of the quantities T, N, B, κ and τ which are associated with a three times differentiable parametrization $f : D \subseteq \mathbb{R} \to \mathbb{R}^3$ of a curve in $\mathbb{R}^3$. If f happens to be a path-length parametrization the process is quite straightforward. All the information is readily obtained from the derivatives f', f'' and f'''. Firstly, the unit tangent T is given by

$$T(s) = f'(s), \qquad s \in D.$$

Next, assuming that $\kappa(s) \neq 0$, the principal normal N and curvature κ are obtained from the relation

$$f''(s) = T'(s) = \kappa(s)N(s), \qquad s \in D$$

where the direction of $N(s)$ is chosen such that $\kappa(s) > 0$. So

$$\kappa(s) = \|f''(s)\| \qquad \text{and} \qquad N(s) = \frac{f''(s)}{\|f''(s)\|}.$$

The binormal $B(s) = T(s) \times N(s)$ can now be calculated. Finally, since

$$f'''(s) = \kappa'(s)N(s) + \kappa(s)N'(s)$$
$$= \kappa'(s)N(s) + \kappa(s)(-\kappa(s)T(s) + \tau(s)B(s))$$

the torsion is given by

$$\tau(s) = \frac{f'''(s) \cdot B(s)}{\kappa(s)} = \frac{f'''(s) \cdot (f'(s) \times f''(s))}{\|f''(s)\|^2}.$$

The situation when f is not a path-length parametrization is a little more complicated. Assuming that $\|f'(t)\| \neq 0$, the unit tangent is obtained from 2.8.20 by

2.8.25 $$T(t) = \frac{f'(t)}{\|f'(t)\|}, \qquad t \in D,$$

where for convenience we drop the suffix in the expression T_f. Theoretically it is now possible to calculate $\kappa(t)$ and $N(t)$ from 2.8.21, but this would require a preliminary evaluation of $v'(t)$. To avoid this, we *first calculate the binormal* $B(t)$. We again assume non-zero curvature. By 2.8.20 and 2.8.21

2.8.26

$$f'(t) \times f''(t) = (v(t))^3 \kappa(t)(T(t) \times N(t)) = (v(t))^3 \kappa(t)B(t).$$

Hence $B(t)$ is the unit vector in the direction of the non-zero vector $f'(t) \times f''(t)$. So

2.8.27 $$B(t) = \frac{f'(t) \times f''(t)}{\|f'(t) \times f''(t)\|}.$$

Taking norms in 2.8.26 leads to a formula for the curvature $\kappa(t)$ (see Exercise 2.8.1).

Now we can find the principal normal $N(t)$ quite easily. Since $N(t) = B(t) \times T(t)$ (see Fig. 2.17(ii)), and $N(t)$ is a unit vector, it follows from 2.8.27 and 2.8.25 that

2.8.28 $$N(t) = \frac{(f'(t) \times f''(t)) \times f'(t)}{\|(f'(t) \times f''(t)) \times f'(t)\|}.$$

The torsion $\tau(t)$ depends on the value of $f'''(t)$. A tedious differentiation of 2.8.21 and application of Theorem 2.8.14 leads to an expression for $f'''(t)$ in terms of the vectors $T(t)$, $N(t)$ and $B(t)$, the curvature $\kappa(t)$ and the torsion $\tau(t)$. By taking the dot product of $f'''(t)$ with $B(t)$ one obtains a formula for τ in terms of f', f'' and f'''. See Exercise 2.8.4.

Exercises 2.8

1. Let C be a simple arc smoothly parametrized by $f : D \subseteq \mathbb{R} \to \mathbb{R}^3$. Show that at the point $\mathbf{q} \in C$, where $\mathbf{q} = f(a)$, $a \in D$, the curvature is given by

$$\kappa(\mathbf{q}) = \frac{\|f'(a) \times f''(a)\|}{\|f'(a)\|^3}.$$

Hint: apply 2.8.20 and 2.8.21.

2. Let the plane curve C be the graph $y = g(x)$ of a C^2 function $g : D \subseteq \mathbb{R} \to \mathbb{R}$. Prove that the curvature at $\mathbf{q} = (a, b) \in C$ is

$$\kappa(\mathbf{q}) = \frac{|g''(a)|}{(1 + (g'(a))^2)^{3/2}}.$$

Hint: regard C as a space curve parametrized by $f(t) = (t, g(t), 0)$ and apply Exercise 1.

3. Sketch the semi-ellipse C defined by

$$\frac{x^2}{9} + \frac{y^2}{4} = 1, \qquad x \leqslant 0.$$

Assuming a 'clockwise' orientation of C (from $(0, -2)$ to $(0, 2)$), find the unit tangent, the principal normal and the curvature to C at $(-3/\sqrt{2}, 2/\sqrt{2}) \in C$.

4. Let C be an oriented simple arc in $\mathbb{R}^3$ smoothly parametrized by the C^3 function $f : D \subseteq \mathbb{R} \to \mathbb{R}^3$. Prove that at the point $\mathbf{q} \in C$ where $\mathbf{q} = f(a)$, $a \in D$, the unit vectors $T(\mathbf{q})$, $B(\mathbf{q})$, and $N(\mathbf{q})$ have the directions of $f'(a)$, $f'(a) \times f''(a)$ and $(f'(a) \times f''(a)) \times f'(a)$ respectively.

   Show also that the torsion is given by

$$\tau(\mathbf{q}) = \frac{(f'(a) \times f''(a)) \cdot f'''(a)}{\|f'(a) \times f''(a)\|^2}.$$

5. Let C be the curve (*twisted cubic*) in $\mathbb{R}^3$ with orientation determined by the parametrization

$$f(t) = (2t, t^2, t^3/3), \qquad t \in \mathbb{R}.$$

   Find T, N, B, κ, τ and the equation of its osculating plane at $(2, 1, \frac{1}{3}) \in C$.

Hint: Apply Exercise 4.

Answer: $\frac{1}{3}(2, 2, 1)$, $\frac{1}{3}(-2, 1, 2)$, $\frac{1}{3}(1, -2, 2)$, $\frac{2}{9}, \frac{2}{9}$, $x - 2y + 2z = \frac{2}{3}$.

6. Calculate the (constant) curvature and torsion of the following circular helices.
   (a) the right-handed helix

   $$f(t) = (2t, \cos t, \sin t), \qquad t \in \mathbb{R};$$

   (b) the left-handed helix

   $$f(t) = (-2t, \cos t, \sin t), \qquad t \in \mathbb{R}.$$

   Answer: (a) $\kappa = \tfrac{1}{5}$, $\tau = \tfrac{2}{5}$;   (b) $\kappa = \tfrac{1}{5}$, $\tau = -\tfrac{2}{5}$.

7. Prove that if all osculating planes of a curve are parallel, then the curve is planar.

8. Show that the curve parametrized by

   $$f(t) = \left(t, \frac{1+t}{t}, \frac{1-t^2}{t}\right), \qquad t > 0$$

   is planar.

9. Find the most general function g such that the curve parametrized by

   $$f(t) = (a\cos t, a\sin t, g(t)), \qquad t \in \mathbb{R}$$

   lies in a plane.

   Hint: the binormal must be constant. Calculate $f'(t) \times f''(t)$ and apply Exercise 4.

   Answer: $g(t) = b\cos t + c\sin t + d$.

10. Let $h : E \subseteq \mathbb{R} \to \mathbb{R}^3$ be a three times differentiable path-length parametrization of a curve in $\mathbb{R}^3$. It is immediate from Definition 2.8.12(i) that for each $s \in E$

    $$B(s) \cdot T(s) = 0 \quad \text{and} \quad B(s) \cdot B(s) = 1.$$

    By differentiating these expressions, show that $B'(s)$ is a multiple of $N(s)$. One can now *define* the torsion of h at s by $B'(s) = -\tau(s)N(s)$. Hence deduce the expression 2.8.13 for $N'(s)$.

    Hint: differentiate $N(s) = T(s) \times B(s)$.

11. Prove that a circle of radius r has constant curvature $\kappa = 1/r$.

12. Prove that a curve in $\mathbb{R}^2$ with non-zero curvature is characterized by its curvature.

    Hint: adapt the proof of Theorem 2.8.16 by suppressing reference to binormals. The relation $N' = -\kappa T$ is needed in the proof. This can be established directly in $\mathbb{R}^2$ by differentiating $N \cdot T = 0$ and $N \cdot N = 1$.

13. Let C and C^* be curves in $\mathbb{R}^2$ parametrized respectively by
$f:D \subseteq \mathbb{R} \to \mathbb{R}^2$ and $f^*:D \subseteq \mathbb{R} \to \mathbb{R}^2$, where $D = [-1, 1]$ and

$$f(t) = (t, t^3) \qquad \text{and} \qquad f^*(t) = (t, |t^3|)$$

for all $t \in D$. Using Exercise 1, show that for each $t \in D$ the curvature
of C at $f(t) \in C$ is equal to the curvature of C^* at $f^*(t) \in C^*$. Sketch
the curves C and C^* and observe that they are not identical in shape.
Why does this not contradict Exercise 12?

2.9   Particle dynamics

In this section we consider the motion of a point or 'particle' in $\mathbb{R}^n$
over a 'time interval' $D \subseteq \mathbb{R}$. Such a motion is described by a function
$f : D \subseteq \mathbb{R} \to \mathbb{R}^n$, where $f(t)$ specifies the position of the particle at
time $t \in D$. The route traced by the particle lies on the curve
$C = f(D)$. We shall assume that the function f is at least twice
differentiable. The most important cases for practical applications
are $n = 2$ and $n = 3$. We then have the machinery to obtain, relative
to appropriate coordinate systems for space and time, a description
of the motion of a physical particle in a plane ($n = 2$) or in space
($n = 3$).

The integral 2.7.14

$$s = \lambda(t) = \int_p^t \| f'(u) \| \, du$$

provides a formula for the distance travelled in $\mathbb{R}^n$ by the particle in
the interval between $p \in D$ and $t \in D$. The derivative $f'(t)$ is called the
velocity of f (or the velocity of the particle) at t. It is a tangent
vector to f at t, which indicates the instantaneous magnitude and
direction of the motion at t. By 2.7.15, $ds/dt = \| f'(t) \|$, and so $\| f'(t) \|$
measures at t the rate of increase of the distance s. Appropriately,
$\| f'(t) \|$ is called the *scalar velocity* or the *speed* of f at t.

The second derivative $f''(t)$ is called the *acceleration* of f at t.
Since $f'' = df'/dt$, the acceleration at t measures the rate of change
of the velocity at t. We emphasize that this is a vectorial measure.
The vector

$$f''(t) = \lim_{h \to 0} (f'(t + h) - f'(t))/h$$

is related to the rates of change of speed and of direction of velocity
at t. See Fig. 2.19 and the following examples.

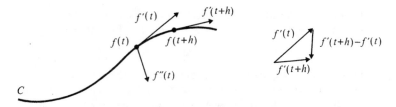

Fig. 2.19 Position $f(t)$, velocity $f'(t)$, acceleration $f''(t)$

2.9.1 Example. Consider a particle moving in $\mathbb{R}^2$ such that its position at $t \in \mathbb{R}$ is $f(t) = (t^2, t^3) \in \mathbb{R}^2$.

The velocity at t is $f'(t) = (2t, 3t^2)$ and the acceleration at t is $f''(t) = (2, 6t)$. The particle traces the semi-cubical parabola $y^2 = x^3$ (see Fig. 2.14 and Example 2.7.5). The speed at t is $\|f'(t)\| = \sqrt{(4t^2 + 9t^4)}$.

The reader is invited to sketch the progress of the particle and to note the relationship between the velocity and the acceleration for various $t \in \mathbb{R}$. In particular, at $t = 0$ the particle is momentarily at rest and the acceleration is $(2, 0)$. Since at $t = 0$ the velocity changes from a 'westerly' to an 'easterly' direction, we would expect the acceleration to be 'easterly', and this is confirmed by our calculation.

2.9.2 Example. A particle moves on a helix H in $\mathbb{R}^3$ such that its position at $t \in \mathbb{R}$ *is*

$$f(t) = (at, b\cos\omega t, b\sin\omega t),$$

where $b > 0$, and $\omega > 0$. (See Example 2.7.17. The case $a = 0$ corresponds to a circular motion in the y, z plane.) At $t \in \mathbb{R}$ the particle has velocity

$$f'(t) = (a, -b\omega\sin\omega t, b\omega\cos\omega t)$$

and speed

$$\|f'(t)\| = (a^2 + b^2\omega^2)^{1/2} = c.$$

It therefore moves at the uniform (constant) speed c. Its velocity is not constant, however, for its direction of motion changes continually. The acceleration at $t \in \mathbb{R}$ is

$$f''(t) = -b\omega^2(0, \cos\omega t, \sin\omega t).$$

Therefore the acceleration is of constant magnitude $b\omega^2$ for all $t \in \mathbb{R}$. Since

$$f''(t) = -\omega^2(at, b\cos\omega t, b\sin\omega t) + \omega^2 at(1, 0, 0)$$
$$= -\omega^2 f(t) + \omega^2 at(1, 0, 0),$$

the acceleration $f''(t)$ lies in the plane spanned by $f(t)$ and the x-axis.

Moreover,

$$f''(t) \cdot f'(t) = 0 \quad \text{and} \quad f''(t) \cdot (1, 0, 0) = 0,$$

and so the direction of the acceleration at t is at right angles to both the velocity at t and the x-axis.

The motion of a particle whose position is given by a twice differentiable smooth function $f : D \subseteq \mathbb{R} \to \mathbb{R}^n$ is related to the differential geometry of the function f as follows. From Theorem 2.8.19 we have

2.9.3 $\quad f'(t) = v(t) T(t)$
2.9.4 $\quad f''(t) = v'(t) T(t) + (v(t))^2 \kappa(t) N(t) \qquad \text{if } \kappa(t) > 0$
2.9.5 $\quad f''(t) = v'(t) T(t) \qquad\qquad\qquad \text{if } \kappa(t) = 0.$

where T, κ and N are the unit tangent, curvature and principal normal associated with f, and $v(t)$ is the speed of the particle at $t \in D$.

2.9.6 Remarks. [i] Expression 2.9.4 resolves the acceleration into two orthogonal components, one along the tangent and the other along the principal normal.

[ii] The acceleration is in the tangential direction at a point if and only if the curvature is zero there.

[iii] If a particle moves at constant speed (that is, $v'(t) = 0$) and if the acceleration is non-zero, then the curvature is non-zero and the acceleration is in the direction of the principal normal.

[iv] The normal component of acceleration depends on both the speed and the curvature. If $\kappa(t) \neq 0$ then the magnitude of the normal component, $(v(t))^2 \kappa(t)$, is often given in the form $(v(t))^2/\rho(t)$ where $\rho(t) = 1/\kappa(t)$. The number $\rho(t)$ is called the *radius of curvature* of f at t.

2.9.7 Example. Continuing Example 2.9.1, $f(t) = (t^2, t^3)$, $f'(t) = (2t, 3t^2)$ and $f''(t) = (2, 6t)$; we have

2.9.8 $\quad v(t) = \sqrt{(4t^2 + 9t^4)} \quad \text{and} \quad T(t) = \frac{1}{v(t)}(2t, 3t^2), \qquad t \neq 0.$

The tangential component of acceleration has magnitude

$$v'(t) = \frac{4 + 18t^2}{\sqrt{(4 + 9t^2)}}.$$

The principal normal $N(t)$ is orthogonal to $T(t)$ and points towards the concave side of the particle's trajectory, so

$$N(t) = \frac{1}{\sqrt{(4 + 9t^2)}}(-3|t|, 2), \qquad t \neq 0.$$

Hence by 2.9.4 the normal component of acceleration has magnitude

$$\frac{v^2(t)}{\rho(t)} = \frac{6|t|}{\sqrt{(4 + 9t^2)}}, \qquad t \neq 0.$$

Note that the point $t = 0$ must be excluded from stage 2.9.8 of our calculation, since the function f is not smooth at $t = 0$. Of course the acceleration at $t = 0$ is $f''(0) = (2, 0)$.

The motion of a particle in $\mathbb{R}^2$ is conveniently considered in terms of polar functions which we now described. We take the opportunity of introducing a notation which is common in the literature of planar particle dynamics.

A point P in $\mathbb{R}^2$ is uniquely specified by its position vector $\mathbf{r} = (x, y)$, $x \in \mathbb{R}$, $y \in \mathbb{R}$. Here x and y are called the *Cartesian coordinates* of P. Alternatively the point P is specified by a pair of numbers r, θ, where $r = \|\mathbf{r}\|$ and

2.9.9 $x = r\cos\theta, \qquad y = r\sin\theta.$

For example, if $\mathbf{r} = (1, -1)$, then $r = \sqrt{2}$ and we may choose $\theta = -\pi/4$. The choice of r in 2.9.9 is unique, but if $\theta = \theta_1$ is appropriate, then so is $\theta = \theta_1 + 2k\pi$ for any integer k. The numbers r, θ satisfying 2.9.9 are called *polar coordinates* of the point P.

Suppose now that a particle moves in $\mathbb{R}^2$ over a time interval $D \subseteq \mathbb{R}$. Then its position is given in polar coordinates by a rule

2.9.10 $\mathbf{r}(t) = (x(t), y(t)), \qquad t \in D,$

where

2.9.11 $x(t) = r(t)\cos\theta(t), \qquad y(t) = r(t)\sin\theta(t), \qquad t \in D.$

In 2.9.11, x, y, r, and θ are regarded as real-valued functions on $D \subseteq \mathbb{R}$. The functions r, θ are called *polar functions* corresponding to the Cartesian functions x, y. According to the notation in 2.9.10, we regard $\mathbf{r}$ as the vector-valued function on D that describes the motion of our particle. In the general case of motion in $\mathbb{R}^n$ we used the symbol f instead of $\mathbf{r}$. We now temporarily abandon this notation. We shall also change the notation for velocity and acceleration. Whereas f' and f'' were used to denote the derivatives

$\mathrm{d}f/\mathrm{d}t$ and $\mathrm{d}^2 f/\mathrm{d}t^2$, we now employ the more usual 'dot' notation and denote the derivatives $\mathrm{d}\mathbf{r}/\mathrm{d}t$ and $\mathrm{d}^2\mathbf{r}/\mathrm{d}t^2$ by $\dot{\mathbf{r}}$ and $\ddot{\mathbf{r}}$ respectively. The derivatives of other functions are expressed similarly.

The formulae 2.9.3 and 2.9.4 show how the velocity and the acceleration of a particle can be resolved into tangential and normal components. For planar motion given by polar functions there is another useful resolution—into radial and transverse components— which we now describe.

Suppose then that a particle moves in $\mathbb{R}^2$ according to a rule 2.9.10, 2.9.11, that is,

2.9.12 $\mathbf{r}(t) = r(t)(\cos\theta(t), \sin\theta(t))$, $t \in D$,

where $r : D \subseteq \mathbb{R} \to \mathbb{R}$ and $\theta : D \subseteq \mathbb{R} \to \mathbb{R}$ are twice-differentiable functions. Corresponding to this motion, we define two 'unit vector' functions $\hat{\mathbf{r}} : D \subseteq \mathbb{R} \to \mathbb{R}^2$ and $\hat{\mathbf{s}} : D \subseteq \mathbb{R} \to \mathbb{R}^2$ by

$$\hat{\mathbf{r}}(t) = (\cos\theta(t), \sin\theta(t))$$

2.9.13

$$\hat{\mathbf{s}}(t) = (-\sin\theta(t), \cos\theta(t)),   t \in D.$$

Notice that $\hat{\mathbf{r}}(t)$ is the vector of unit length in the direction of $\mathbf{r}(t)$ and that $\hat{\mathbf{s}}(t)$ is a unit vector at right angles to $\mathbf{r}(t)$. (Calculate the dot product $\hat{\mathbf{r}}(t) \cdot \hat{\mathbf{s}}(t)$). See Fig. 2.20.

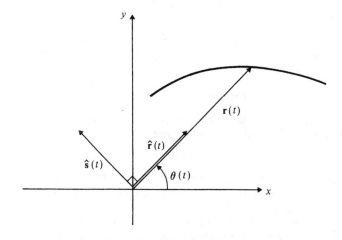

Fig. 2.20

Differentiating 2.9.13, we obtain, at $t \in D$,

2.9.14
$$\frac{d\hat{\mathbf{r}}}{dt}(t) = \dot{\theta}(t)(-\sin\theta(t), \cos\theta(t)) = \dot{\theta}(t)\hat{\mathbf{s}}(t)$$
$$\frac{d\hat{\mathbf{s}}}{dt}(t) = -\dot{\theta}(t)(\cos\theta(t), \sin\theta(t)) = -\dot{\theta}(t)\hat{\mathbf{r}}(t).$$

Let us now calculate the velocity and the acceleration of the particle at $t \in D$. From 2.9.12 and 2.9.13,

2.9.15
$$\mathbf{r}(t) = r(t)\hat{\mathbf{r}}(t), \qquad t \in D.$$

Differentiating 2.9.15 and applying 2.9.14, we obtain a formula for the velocity function $\dot{\mathbf{r}} : D \subseteq \mathbb{R} \to \mathbb{R}^2$:

2.9.16
$$\dot{\mathbf{r}} = \dot{r}\hat{\mathbf{r}} + r\dot{\theta}\hat{\mathbf{s}}.$$

A further differentiation, and applications of 2.9.14 give a formula for the acceleration function $\ddot{\mathbf{r}} : D \subseteq \mathbb{R} \to \mathbb{R}^2$:

2.9.17
$$\ddot{\mathbf{r}} = (\ddot{r} - r\dot{\theta}^2)\hat{\mathbf{r}} + (2\dot{r}\dot{\theta} + r\ddot{\theta})\hat{\mathbf{s}}.$$

The expressions 2.9.16 and 2.9.17 provide resolutions of the velocity and the acceleration at $t \in D$ into the *radial component* in the direction of $\hat{\mathbf{r}}(t)$ and the *transverse component* in the direction of $\hat{\mathbf{s}}(t)$. The transverse component is at right angles to the radial component.

From 2.9.16 we also find that the speed of the particle at $t \in D$ is

2.9.18
$$v(t) = \|\dot{\mathbf{r}}(t)\| = \sqrt{[(\dot{r}(t))^2 + (r(t)\dot{\theta}(t))^2]}.$$

2.9.19 *Example.* If a particle moves in $\mathbb{R}^2$ on a circle C of radius a and centre at the origin then, for all t, $r(t) = a$ and $\dot{r}(t) = 0$. Hence its velocity and acceleration functions are

$$\dot{\mathbf{r}} = a\dot{\theta}\hat{\mathbf{s}}, \qquad \ddot{\mathbf{r}} = -a\dot{\theta}^2\hat{\mathbf{r}} + a\ddot{\theta}\hat{\mathbf{s}}.$$

If, in addition, the particle moves at constant speed, then $\dot{\theta} = \omega$, a constant (which is called the *angular speed* of the particle). In this case

2.9.20
$$\ddot{\mathbf{r}} = -a\omega^2\hat{\mathbf{r}} = -\omega^2\mathbf{r}.$$

This means that the acceleration has constant magnitude $a\omega^2$ and its direction is at all times towards the origin. The transverse component of acceleration is zero.

As an illustration, a particle moving in $\mathbb{R}^2$ according to the rule $\mathbf{r} = (a\cos\omega t, a\sin\omega t)$, $t \in \mathbb{R}$ traces the circle C at constant angular speed ω. In contrast, the rule $\mathbf{r} = (a\cos(t^2), a\sin(t^2))$, $t \in \mathbb{R}$, also defines a motion on the circle *C*, but now the speed and the angular speed are not constant, and there is a non-trivial transverse component of acceleration. In both cases the radial component of acceleration is non-trivial.

2.9.21 Definition. *A particle moving in $\mathbb{R}^2$ is said to experience a* central acceleration *if the transverse component of acceleration is zero, that is, if its acceleration is a multiple of* $\mathbf{r}(t)$.

A central acceleration may be pictured as drawn from the particle towards or away from the origin.

2.9.22 Theorem. *If a twice-differentiable function $\mathbf{r} : D \subseteq \mathbb{R} \to \mathbb{R}^2$ describes a motion in $\mathbb{R}^2$ under a central acceleration, then there is a constant h such that*

$$(r^2\dot\theta)(t) = h \qquad \text{for all } t \in D.$$

Proof. By 2.9.17, if the function $\ddot{\mathbf{r}}$ specifies a central acceleration, then $2\dot r\dot\theta + r\ddot\theta$ is the zero function on *D*. Since

$$r(2\dot r\dot\theta + r\ddot\theta) = \frac{\mathrm{d}}{\mathrm{d}t}(r^2\dot\theta),$$

the theorem follows.

2.9.23 Example. *Planetary motion. Central orbits.* In studying the motion of a planet around the sun it is usual to regard the sun and the planet as particles (points) in space with the sun fixed at an origin. Considering the enormous distances involved, this is a reasonable simplification. As the planet moves in space, the sun exerts on it a gravitational force of attraction according to the Inverse Square Law of gravitation. The law states that two particles in space attract each other with a force which is proportional to the product of their masses and inversely proportional to the square of the distance between them. All other forces on the planet are negligible in comparison. Consequently the planet moves under a central acceleration. Its orbit is called a central orbit. It can be shown that central orbits are planar orbits (Exercise 2.9.6).

From the Inverse Square Law, Isaac Newton (1642–1727) deduced the three celebrated laws of planetary motion discovered experimentally by Johannes Kepler (1571–1630) (from a detailed study of the astronomer Tycho Brahe's observations of the planet Mars).

Kepler's First Law states that planets move in elliptical orbits with the sun at a focus of the ellipse. For a proof of this law from the Inverse Square Law see Smith & Smith, *Mechanics*, Chapter 9.

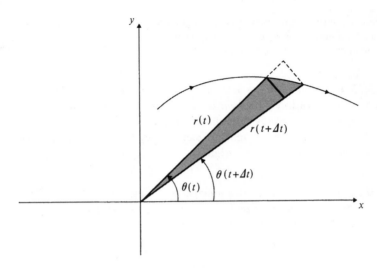

Fig. 2.21

Kepler's Second Law states that a planet moves in its orbit in such a manner that a line drawn from the origin (sun) to the planet sweeps over equal areas in equal times. This is a general property enjoyed by particles moving under a central acceleration. We can prove this as follows. Consider a particle moving in $\mathbb{R}^2$ over a time interval D, and let its position be specified by the function $\mathbf{r} : D \subseteq \mathbb{R} \to \mathbb{R}^2$. With reference to Fig. 2.21, the area swept out by a line joining the origin and the particle in an interval $[t, t + \Delta t] \subseteq D$, where Δt is small, lies between the circle sector areas $\frac{1}{2}r^2(t)(\theta(t + \Delta t) - \theta(t))$ and $\frac{1}{2}r^2(t + \Delta t)(\theta(t + \Delta t) - \theta(t))$. Dividing by Δt and letting $\Delta t \to 0$, we deduce that the rate at which area is swept out at $t \in D$ is $|\frac{1}{2}(r^2\dot\theta)(t)|$. If the particle moves under a central acceleration, then, by Theorem 2.9.22, this rate is constant.

Exercises 2.9

1. Sketch the trajectory of a particle whose motion in $\mathbb{R}^2$ is described by $f(t) = (t^2, t^3)$, $t \in \mathbb{R}$. (See Example 2.9.1). Indicate the velocity and the acceleration at $t = 0$, $t = 1$ and $t = -1$.

2. Find (i) the velocity, and (ii) the tangential and normal components of acceleration at $t = 1$ of a particle whose motion in $\mathbb{R}^3$ is described by
    (a) $f(t) = (a\cos \pi t, a\sin \pi t, bt)$, $a > 0, b \neq 0$;
    (b) $f(t) = (t^2, \frac{2}{3}t^3, t)$.

Answer: (b) tangential acceleration 4, normal acceleration 2.

3. Prove that if a particle moves at constant speed, then its acceleration is orthogonal to its velocity. (This is illustrated in Example 2.9.2.)

4. Find (i) the velocity $\dot{\mathbf{r}}(t)$, (ii) the speed $\|\dot{\mathbf{r}}(t)\|$, (iii) the acceleration $\ddot{\mathbf{r}}(t)$, (iv) the radial and transverse components of velocity and acceleration at t, of a particle whose motion in $\mathbb{R}^2$ is described by
   (a) $\mathbf{r}(t) = (a\cos(t^2), a\sin(t^2))$, $\qquad\qquad t \in \mathbb{R}$;
   (b) $\mathbf{r}(t) = (a\cos \omega t, b\sin \omega t)$, $\qquad\qquad t \in \mathbb{R}$, ω constant;
   (c) $\mathbf{r}(t) = (a(1 + \cos t)\cos t, a(1 + \cos t)\sin t)$, $\quad t \in \mathbb{R}$.
   (In Example 4(c) the particle moves on the cardioid $r(t) = a(1 + \cos t)$.)

5. A particle moves with constant speed V on the cardioid $r(t) = a(1 + \cos t)$, $t \in \mathbb{R}$. Prove that $\theta(t) = (V/2a)\sec\frac{1}{2}t$, and that the radial component of acceleration is constant.

Hint: describe the motion in the form 2.9.12.

6. If the function $\mathbf{r} : D \subseteq \mathbb{R} \to \mathbb{R}^3$ describes a central orbit, then there exists a function $F : D \subseteq \mathbb{R} \to \mathbb{R}$ such that
$$\ddot{\mathbf{r}}(t) = F(t)\mathbf{r}(t), \qquad\qquad t \in D.$$

Deduce that the central orbit is planar as follows:
   (a) Prove that $\ddot{\mathbf{r}}(t) \times \mathbf{r}(t) = \mathbf{0}, t \in D$.
   (b) Show that there exists a constant vector $\mathbf{c} \in \mathbb{R}^3$ such that
$$\dot{\mathbf{r}}(t) \times \mathbf{r}(t) = \mathbf{c}, t \in D.$$

   (c) Show that the dot product $\mathbf{c} \cdot \mathbf{r}(t) = 0, t \in D$.
   (d) Conclude that the motion is planar. (The case $\mathbf{c} = \mathbf{0}$ needs separate consideration. In this case the motion is linear.)

2.10   Differentiability and linear approximation

We now look at the differentiability of a function $f : D \subseteq \mathbb{R} \to \mathbb{R}^n$ at a point p in an interval D from a different point of view. This new approach will lead naturally to a definition of the differentiability of a function from a subset of $\mathbb{R}^m$ into $\mathbb{R}^n$.

We are concerned with the way in which the value of a function f changes as we move away from p (see Fig. 2.22(i)). Accordingly we consider the 'difference' function $\delta_{f,p} : D_p \subseteq \mathbb{R} \to \mathbb{R}^n$, where $D_p = \{h \in \mathbb{R} \mid p + h \in D\}$ and

2.10.1 $\qquad \delta_{f,p}(h) = f(p + h) - f(p)$, $\qquad\qquad h \in D_p$.

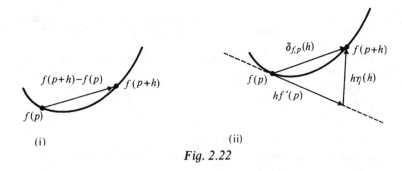

Fig. 2.22

2.10.2 Example.   Let $D = \,]0, 3[$ and let $f : D \subseteq \mathbb{R} \to \mathbb{R}$ be given by $f(x) = 1/x(x - 3)$. Then $D_1 = \,]-1, 2[$ and for $h \in D_1$,

$$\delta_{f,1}(h) = \frac{1}{(h + 1)(h - 2)} + \tfrac{1}{2}.$$

Suppose that the function $f : D \subseteq \mathbb{R} \to \mathbb{R}^n$ is differentiable at $p \in D$. Then

$$\lim_{h \to p} \frac{f(p + h) - f(p)}{h} = f'(p).$$

It follows that for each $h \in D_p$, $h \neq 0$, there exists a vector $\eta(h) \in \mathbb{R}^n$ such that

$$\frac{f(p + h) - f(p)}{h} = f'(p) + \eta(h) \qquad \text{and} \qquad \lim_{h \to 0} \eta(h) = \mathbf{0}.$$

Therefore, by defining $\eta(0) = \mathbf{0}$, we have a function $\eta : D_p \subseteq \mathbb{R} \to \mathbb{R}^n$ such that

2.10.3 $\delta_{f,p}(h) = hf'(p) + h\eta(h),$

and

2.10.4 $\displaystyle\lim_{h \to 0} \eta(h) = \mathbf{0}.$

2.10.5 Example.   Define $f : \mathbb{R} \to \mathbb{R}^2$ by $f(t) = (t^3 + 1, t)$. For fixed $p \in \mathbb{R}$ we have $f'(p) = (3p^2, 1)$ and for all $h \in \mathbb{R}$,

$$\delta_{f,p}(h) = ((p + h)^3 + 1, p + h) - (p^3 + 1, p) = (3p^2 h + 3ph^2 + h^3, h)$$
$$= h(3p^2, 1) + h(3ph + h^2, 0).$$

In this case $\eta(h) = (3ph + h^2, 0)$, $h \in \mathbb{R}$.

2.10.6 *Example*. Define $f : \mathbb{R} \to \mathbb{R}^2$ by $f(t) = (1, 2) + t(1, 1)$. The image of f is a straight line through $(1, 2)$. At $p \in \mathbb{R}$ we have

$$\delta_{f,p}(h) = (1 + p + h, 2 + p + h) - (1 + p, 2 + p) = h(1, 1) + h(0, 0).$$

In this case $\eta(h) = (0, 0)$ for all h.

It is worth studying 2.10.3 very carefully. The left-hand side, which is just the difference between two values that f takes, is split up on the right as the sum of two vectors. The first of these, $hf'(p)$, is a tangent vector to f at p. The second vector, $h\eta(h)$, is essentially the error we make in moving along this tangent rather than moving along the curve. This decomposition of $\delta_{f,p}$ is illustrated in Fig. 2.22(ii).

Expression 2.10.3 can be developed further by defining a *linear* function $L_{f,p} : \mathbb{R} \to \mathbb{R}^n$ by

2.10.7 $$L_{f,p}(t) = tf'(p), \qquad t \in \mathbb{R}.$$

We then have, from 2.10.3,

2.10.8 $$\delta_{f,p}(h) = L_{f,p}(h) + h\eta(h), \qquad h \in D_p.$$

$L_{f,p}$ is called the *linear approximation* to f at p or the *differential* of f at p.

2.10.9 *Example*. Consider again the function $f : \mathbb{R} \to \mathbb{R}^2$, defined by $f(t) = (t^3 + 1, t)$. See Fig. 2.23. Then $f(0) = (1, 0)$ and $f(-1) = (0, -1)$, and we

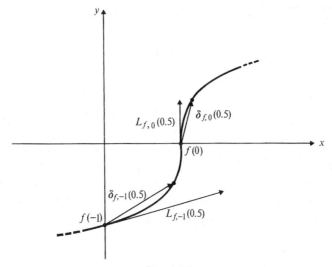

Fig. 2.23

have

$$\delta_{f,0}(h) = (h^3, h), \qquad L_{f,0}(h) = h f'(0) = h(0, 1),$$
$$\eta(h) = (h^2, 0),$$

and

$$\delta_{f,-1}(h) = (3h - 3h^2 + h^3, h), \ L_{f,-1}(h) = h f'(-1) = h(3, 1),$$
$$\eta(h) = (-3h + h^2, 0).$$

2.10.10 *Example.* The function $f : \mathbb{R} \to \mathbb{R}^2$ defined by $f(t) = (t|t|, t^2)$ has the image sketched in Fig. 2.9(ii). The function is differentiable at 0, where its image has a 'corner', and expression 2.10.8 takes the form

$$\delta_{f,0}(h) = h(0, 0) + h(|h|, h).$$

The differential $L_{f,0}$ is the zero function in this case. This is to be expected, since a non-zero vector $L_{f,0}(h)$ would produce a tangent line to f at 0, and no such line exists.

The following theorem shows that the existence of a linear function L satisfying 2.10.8, together with the requirement that $\lim_{h \to 0} \eta(h) = \mathbf{0}$, is also a sufficient condition for the differentiability of f at p. We remind the reader that throughout this chapter the domain D of a function $f : D \subseteq \mathbb{R} \to \mathbb{R}^n$ is assumed to be an interval.

2.10.11 *Theorem.* *A function* $f : D \subseteq \mathbb{R} \to \mathbb{R}^n$ *is differentiable at* $p \in D$ *if and only if there exists a linear function* $L : \mathbb{R} \to \mathbb{R}^n$ *and a function* $\eta : D_p \subseteq \mathbb{R} \to \mathbb{R}^n$ *such that*
[i] $\delta_{f,p}(h) = L(h) + h\eta(h),$ *for all* $h \in D_p,$
and

[ii] $\lim_{h \to 0} \eta(h) = \mathbf{0}.$
If such a linear function exists then it is unique and equal to $L_{f,p}.$

Proof. We have already shown that if f is differentiable, then such a linear function and such an η exist satisfying conditions (i) and (ii).
Conversely suppose we can find a linear function $L : \mathbb{R} \to \mathbb{R}^n$ and a function $\eta : D_p \to \mathbb{R}^n$ satisfying (i) and (ii). Since L is linear,

2.10.12 $$L(t) = t L(1), \qquad t \in \mathbb{R}.$$

By condition (i),

$$f(p + h) - f(p) = h L(1) + h\eta(h), \qquad \text{for all } h \in D_p.$$

Hence for $h \neq 0$,

$$\frac{f(p+h) - f(p)}{h} = L(1) + \eta(h).$$

Using condition (ii) we can see that

$$\lim_{h \to 0} \frac{f(p+h) - f(p)}{h} = L(1).$$

The function f is therefore differentiable at p and $f'(p) = L(1)$.

Finally, if such a linear function exists, then by the above argument f is differentiable and $L(1) = f'(p)$. Hence $L(t) = t f'(p)$, $t \in \mathbb{R}$, and therefore $L = L_{f,p}$.

The following definition puts the relationship between $\delta_{f,p}$ and $L_{f,p}$ into a more general setting.

2.10.13 Definition. *Let $g : N \subseteq \mathbb{R} \to \mathbb{R}^n$ and $g^* : M \subseteq \mathbb{R} \to \mathbb{R}^n$ be two functions on intervals N and M where $0 \in N$ and $N \subseteq M$. We say that g^* closely approximates g near 0 if*

$$\lim_{h \to 0} \frac{g(h) - g^*(h)}{h} = \mathbf{0}, \qquad h \in N.$$

2.10.14 Example. Let $g :]-1, \infty[\subseteq \mathbb{R} \to \mathbb{R}^2$ and $g^* : \mathbb{R} \to \mathbb{R}^2$ be defined by $g(t) = (\ln(1+t), t^2)$ and $g^*(t) = (t + t^2, t^3)$. Then g^* is a close approximation to g near 0.

We now have the following restatement of Theorem 2.10.11.

2.10.15 Corollary. *A function $f : D \subseteq \mathbb{R} \to \mathbb{R}^n$ is differentiable at $p \in D$ if and only if the difference function $\delta_{f,p} : D_p \subseteq \mathbb{R} \to \mathbb{R}^n$ can be closely approximated by a linear function near 0. If such a close linear approximation exists then it is unique and is $L_{f,p}$.*

Proof. If $\delta_{f,p}$ is closely approximated by a linear function $L : \mathbb{R} \to \mathbb{R}^n$ near 0, then define a function $\eta : D_p \subseteq \mathbb{R} \to \mathbb{R}^n$ by

$$\eta(h) = \begin{cases} \dfrac{\delta_{f,p}(h) - L(h)}{h} & \text{when } h \in D_p \backslash \{0\} \\ \mathbf{0} & \text{when } h = 0. \end{cases}$$

Then $\lim_{h\to 0} \eta(h) = 0$ and $\delta_{f,p}(h) = L(h) + h\eta(h)$. Hence, by Theorem 2.10.11, f is differentiable and $L = L_{f,p}$.

The converse is left as an exercise.

Our intention in this section has been to link the definition of the derivative of $f : D \subseteq \mathbb{R} \to \mathbb{R}^n$ given in Section 2.5 to the definition of the derivative of $f : D \subseteq \mathbb{R}^m \to \mathbb{R}^n$ given in later chapters. To do this we need to make a slight adjustment to the conditions for differentiability given in Theorem 2.10.11. For future reference this change is incorporated in the following new definition of differentiability.

2.10.16 Definition.   *The function* $f : D \subseteq \mathbb{R} \to \mathbb{R}^n$ *is* differentiable *at* $p \in D$ *if there is a linear function* $L_{f,p} : \mathbb{R} \to \mathbb{R}^n$(*called the* differential *of* f *at* p) *and a function* $\eta : D_p \subseteq \mathbb{R} \to \mathbb{R}^n$ *such that*

[i] $f(p + h) - f(p) = L_{f,p}(h) + |h|\eta(h)$, *for all* $h \in D_p$.
and

[ii] $\lim_{h \to 0} \eta(h) = 0$.

The function η of Definition 2.10.16 is obtained from that of Theorem 2.10.11 by changing the sign of $\eta(h)$ for negative h. This change does not alter the limit in (ii).

We now illustrate the new approach to differentiability contained in the above definition by rephrasing the Chain Rule in a form which leads to a natural generalization in later chapters.

2.10.17 Definition.   *If the function* $f : D \subseteq \mathbb{R} \to \mathbb{R}^n$ *is differentiable at* $p \in D$. *then the* $n \times 1$ *matrix* $J_{f,p}$ *which represents the linear transformation* $L_{f,p} : \mathbb{R} \to \mathbb{R}^n$ *with respect to standard bases is called the* Jacobian *of* f *at* p.

Since $L_{f,p}(1) = f'(p)$ (see 2.10.7), the Jacobian of f at p is the $n \times 1$ matrix

2.10.18 $$J_{f,p} = \begin{bmatrix} f'_1(p) \\ \vdots \\ f'_n(p) \end{bmatrix}.$$

2.10.19 Theorem. (Chain Rule)   *Let* $\phi : E \subseteq \mathbb{R} \to \mathbb{R}$ *and* $f : D \subseteq \mathbb{R} \to \mathbb{R}^n$ *be differentiable functions and let* $\phi(E) \subseteq D$. *Then*

$f \circ \phi : E \subseteq \mathbb{R} \to \mathbb{R}^n$ *is differentiable, and*

2.10.20 $\qquad L_{f \circ \phi, p} = L_{f, \phi(p)} \circ L_{\phi, p}, \qquad p \in E$

with the corresponding Jacobian relation

2.10.21 $\qquad J_{f \circ \phi, p} = J_{f, \phi(p)} J_{\phi, p}, \qquad p \in E.$

Proof. The fact that $f \circ \phi$ is differentiable is part of Theorem 2.6.11. Expression 2.10.21 follows immediately from 2.6.12. (This gives further justification for writing the scalar on the right in that expression.)

Expression 2.10.20 follows directly from 2.10.21. Alternatively it can be proved by noting that, for all $t \in \mathbb{R}$,

$$L_{f \circ \phi, p}(t) = t(f \circ \phi)'(p) = tf'(\phi(p))\phi'(p) = L_{f, \phi(p)}(t\phi'(p))$$
$$= L_{f, \phi(p)}(L_{\phi, p}(t)),$$

the identities following from 2.10.7.

2.10.22 Example. As in Example 2.6.14, let $\phi : \mathbb{R} \to \mathbb{R}$ and $f : \mathbb{R} \to \mathbb{R}^3$ be defined by $\phi(t) = \sin t$, $t \in \mathbb{R}$ and $f(u) = (2u, u^3, 1)$, $u \in \mathbb{R}$. Then, for any $p \in \mathbb{R}$,

$$L_{f \circ \phi, p}(t) = t(f \circ \phi)'(p) = t(2\cos p, 3\sin^2 p \cos p, 0),$$
$$L_{f, \phi(p)}(u) = uf'(\phi(p)) = u(2, 3\sin^2 p, 0),$$
$$L_{\phi, p}(t) = t\phi'(p) = t\cos p,$$

and the expression 2.10.20 is satisfied.

The related Jacobians are as follows:

$$J_{f \circ \phi, p} = \begin{bmatrix} 2\cos p \\ 3\sin^2 p \cos p \\ 0 \end{bmatrix}, \quad J_{f, \phi(p)} = \begin{bmatrix} 2 \\ 3\sin^2 p \\ 0 \end{bmatrix}, \quad J_{\phi, p} = [\cos p].$$

Clearly 2.10.21 is satisfied.

Exercises 2.10

1. Calculate the difference function $\delta_{f, p}$ and the differential $L_{f, p}$ (i) at $p = 0$, (ii) at $p = 1$, (iii) at $p = -1$ for the following functions $f : \mathbb{R} \to \mathbb{R}^2$, and in each case obtain the relation of Theorem 2.10.11.
   (a) $f(t) = (t, t^2)$, $\qquad t \in \mathbb{R}$;
   (b) $f(t) = (t, t^3)$, $\qquad t \in \mathbb{R}$;
   (c) $f(t) = (t^2, t^3)$, $\qquad t \in \mathbb{R}$.

2. Prove that the function $f : \mathbb{R} \to \mathbb{R}^2$ defined by $f(t) = (|t|, t)$ is not

differentiable at 0, by calculating $\delta_{f,0}(h)$ and showing that the condition for differentiability in Theorem 2.10.11 fails.

3. Define $g : \mathbb{R} \to \mathbb{R}^2$ and $g^* : \mathbb{R} \to \mathbb{R}^2$ by

$$g(t) = (e^t, \sin t), \qquad g^*(t) = (1 + t - t^2, t), \qquad t \in \mathbb{R}.$$

Prove that g^* closely approximates g near 0.

4. Verify the expression $L_{f \circ \phi, p} = L_{f, \phi(p)} \circ L_{\phi, p}$, $p \in \mathbb{R}$ for the following functions $f : \mathbb{R} \to \mathbb{R}^3$, $\phi : \mathbb{R} \to \mathbb{R}$:
    (a) $f(t) = (e^t, t, 1), \qquad \phi(t) = t^2$;
    (b) $f(t) = (0, -t, t^2 + 1), \qquad \phi(t) = \sin t$.
    Verify also the expression 2.10.21 for the corresponding Jacobians.

Real-valued functions of $\mathbb{R}^m$

3.1   Introduction. Level sets and graphs

We now come to the study of what are often called real-valued functions of many variables. Such functions $f : S \subseteq \mathbb{R}^m \to \mathbb{R}$, for various dimensions m, frequently occur in physical problems.

3.1.1 *Example.*   In the study of changing weather conditions the temperature is measured at various points in space at different times. By choosing frames of reference for time and space we can define an underlying 'temperature function' $\theta : \mathbb{R}^4 \to \mathbb{R}$ by the rule that $\theta(x, y, z, t)$ is the temperature in degrees centigrade at the point in space with position vector (x, y, z) at time t units.

3.1.2 *Example.*   The volume of a cylinder is given by the formula $\pi r^2 h$ in terms of its radius r and its height h. The underlying volume function $V : D \subseteq \mathbb{R}^2 \to \mathbb{R}$ is defined by $V(r, h) = \pi r^2 h$, where D is the first quadrant $\{(r, h) \in \mathbb{R}^2 \mid r > 0, h > 0\}$ in $\mathbb{R}^2$.

A function $f : S \subseteq \mathbb{R}^m \to \mathbb{R}$ is sometimes called a *scalar field* on S. It is a rule associating a real number $f(\mathbf{p})$ with each point $\mathbf{p} \in S$. For any real number c, the set of all points in S at which the function takes the value c will be called a level set of the function. More formally, the *level set corresponding to c* is

3.1.3 $$f^{-1}(c) = \{\mathbf{x} \in S \mid f(\mathbf{x}) = c\}.$$

3.1.4 *Example.*   The surface of the earth corresponds to a subset S of $\mathbb{R}^3$ with respect to a frame of reference in space. The pressure at each point at a certain time, the mean annual temperature at each point, the height of each point above sea level all lead to scalar fields on S. The level sets of these scalar fields correspond to isobars, isotherms and contours respectively.

3.1.5 *Example.*   Consider the function defined in Example 3.1.2. The level sets corresponding to $c \leqslant 0$ are all empty. For $c > 0$ the level sets are graphs $h = c/\pi r^2$ in the (r, h)-plane. See Fig. 3.1(i).

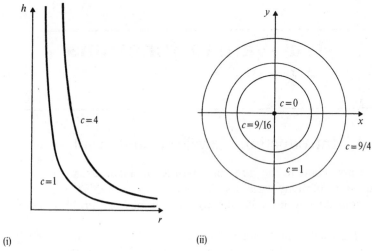

(i)   (ii)

Fig. 3.1 (i) Level sets of $f(r, h) = \pi r^2 h$;
(ii) Level sets of $f(x, y) = x^2 + y^2$

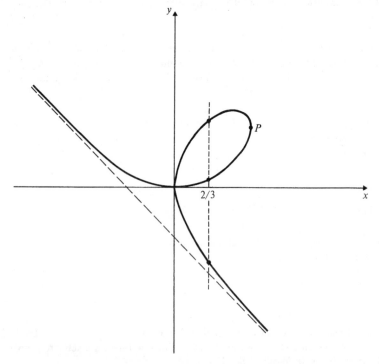

Fig. 3.2 Folium of Descartes $x^3 + y^3 - 3xy = 0$

3.1.6 *Example*.   The level sets of the function $f : \mathbb{R}^2 \to \mathbb{R}$ defined by $f(x, y) = x^2 + y^2$ are concentric circles centre the origin (corresponding to $c > 0$), the origin itself (corresponding to $c = 0$), and the empty set (corresponding to $c < 0$). See Fig. 3.1(ii).

3.1.7 *Example*.   Many classical curves are defined as level sets of functions $f : \mathbb{R}^2 \to \mathbb{R}$. The function defined by $f(x, y) = x^3 + y^3 - 3xy$ has the folium of Descartes (see Fig. 3.2) as the level set corresponding to 0.

The equation, $x^3 + y^3 = 3xy$, is said to define y *implicitly* in terms of x, although there are difficulties at $(0, 0)$ and at P where the curve folds back. Note that to a given value of x there might correspond one, two or three values of y. For example, when $x = \frac{2}{3}$ the equation is satisfied when $y = \frac{4}{3}$ and $y = (-2 \pm \sqrt{6})/3$. We will study such implicit relationships in greater detail in the next chapter. The curve itself was suggested by Descartes in 1638 as a test for Fermat's procedure for finding the tangents to curves.

3.1.8 *Example*.   Consider a function $f : \mathbb{R}^2 \to \mathbb{R}$ defined by

$$f(x, y) = \begin{cases} -|y| & \text{if } |y| \le |x|, \\ -|x| & \text{if } |x| \le |y|. \end{cases}$$

The level set corresponding to 0 is the union of the x and y axes. The level set corresponding to -2 (and indeed to all negative values of c) consists of four components (the case $c = -2$ is shown in Fig. 3.3).

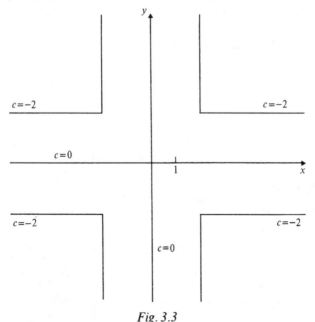

Fig. 3.3

In the previous chapter we found it helpful to picture functions $f : \mathbb{R} \to \mathbb{R}^n$, $n > 2$, by drawing their images as subsets of $\mathbb{R}^n$. There was usually no need to consider the graph. In the case of a function $f : D \subseteq \mathbb{R}^m \to \mathbb{R}$, however, the image of f is squashed up within $\mathbb{R}$ and the subtleties of the function are lost. In this case, as is usual in particular when $m = 1$, we can picture the function by considering its graph.

3.1.9 Definition.   *The graph of $f : S \subseteq \mathbb{R}^m \to \mathbb{R}$ is the subset of $\mathbb{R}^{m+1}$ given by*

$$\{(x_1, \ldots, x_m, x_{m+1}) \in \mathbb{R}^{m+1} \mid x_{m+1} = f(x_1, \ldots, x_m)\}.$$

Equivalently the graph is $\{(\mathbf{x}, f(\mathbf{x})) \in \mathbb{R}^{m+1} \mid \mathbf{x} \in S\}$. The expression $x_{m+1} = f(\mathbf{x})$ is called the equation *of the graph.*

In sketching the graph (when $m = 2$) knowledge of the level sets of f is useful since the intersection of the plane $z = c$ and the graph $z = f(x, y)$ is a copy in the plane $z = c$ of the level set $f(x, y) = c$.

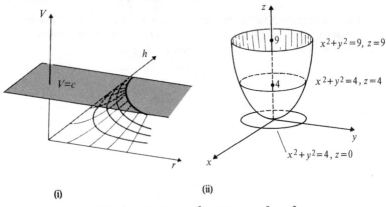

Fig. 3.4 (i) $V = \pi r^2 h$; (ii) $z = x^2 + y^2$

3.1.10 Example.   The graph of the function $f(r, h) = \pi r^2 h$ has equation $V = \pi r^2 h$ in an r, h, V coordinate system. The plane $V = c$ intersects the graph at points at which $\pi r^2 h = c$. This set lies above the level set corresponding to c in the (r, h)-plane. See Fig. 3.4 (i).

3.1.11 Example.   The graph of the function $f(x, y) = x^2 + y^2$ has equation $z = x^2 + y^2$. For $c > 0$ the plane $z = c$ intersects the graph in a circle lying above the level set $x^2 + y^2 = c$ in the (x, y)-plane. The graph is a parabolic bowl with circular horizontal cross-sections. See Fig. 3.4(ii).

3.1.12 *Example*.   The graph of the function defined in Example 3.1.8 is shown in Fig. 3.5. It is worth looking at this graph carefully to see how the level sets illustrated in Fig. 3.3 are related to it.

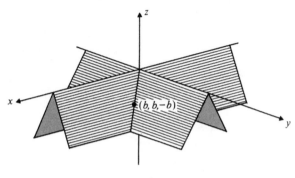

Fig. 3.5

3.1.13 *Example*.   The graph of the linear function $f : \mathbb{R}^2 \to \mathbb{R}$ given by $f(x, y) = 2x - y$ is the subset S of points $(x, y, z) \in \mathbb{R}^3$ such that

$$z = 2x - y.$$

This is the equation of the graph of f. The set S is the two-dimensional subspace of $\mathbb{R}^3$ consisting of all points $(x, y, 2x - y)$. It is the plane in $\mathbb{R}^3$ through $(0, 0, 0)$ containing $(1, 0, 2)$ and $(0, 1, -1)$.

In general, the graph of any linear function $f : \mathbb{R}^2 \to \mathbb{R}$ is a plane in $\mathbb{R}^3$ through the origin.

If the domain of a function is $\mathbb{R}^3$, then we can still draw the level sets even though we can no longer sketch the graph.

3.1.14 *Example*.   Consider the linear function $f : \mathbb{R}^3 \to \mathbb{R}$ given by

$$f(x, y, z) = x + z.$$

The level set corresponding to 0 is the kernel of f. It is the plane $x + z = 0$ through $(0, 0, 0)$. See Fig. 3.6. The other level sets are planes $x + z = c$ parallel to ker f.

Exercises 3.1

1.   Sketch the level sets $f(x, y) = c$ of the function $f : \mathbb{R}^2 \to \mathbb{R}$ defined by
   (a) $f(x, y) = x^2 - y^2$ (consider separately the case $c > 0$, $c = 0$, $c < 0$),
   (b) $f(x, y) = (x - 1)^2 + (y + 2)^2$,
   (c) $f(x, y) = x^2 - y$,
   (d) $f(x, y) = x + 3y$,
   (e) $f(x, y) = xy$.

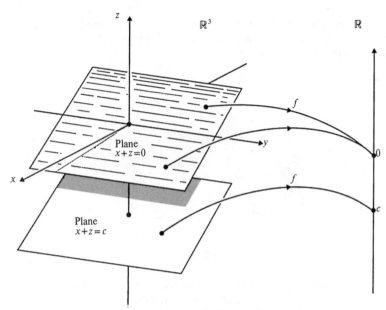

Fig. 3.6 Level sets of $f(x, y, z) = x + z$

2. Attempt a sketch (in $\mathbb{R}^3$) of the graphs of the functions f of Exercise 1.

3. Let $f : \mathbb{R}^m \to \mathbb{R}$ be a non-trivial linear function. Prove that (a) the level set of f corresponding to 0 is ker f, an $(m-1)$-dimensional subspace of $\mathbb{R}^m$, (b) if $f(\mathbf{p}) = c$, then the level set corresponding to c is the subset $\mathbf{p} + \ker f \subseteq \mathbb{R}^m$, the translation of ker f through $\mathbf{p}$.

Hint: for part (b): if $f(\mathbf{p}) = c$ and $f(\mathbf{q}) = c$, then $0 = f(\mathbf{q}) - f(\mathbf{p}) = f(\mathbf{q} - \mathbf{p})$.

4. Prove that the graph of a linear function $f : \mathbb{R}^m \to \mathbb{R}$ is an m-dimensional subspace of $\mathbb{R}^{m+1}$.

Hint: show that the graph of f is the set spanned by the vectors $(\mathbf{e}_i, f(\mathbf{e}_i)) \in \mathbb{R}^{m+1}$, $i = 1, \ldots, m$.

3.2   Continuity and limits

The definition of the continuity of a function $f : S \subseteq \mathbb{R}^m \to \mathbb{R}$ will present little difficulty if the work on sequences, continuity and limits in the last chapter has been understood. We need only make minor changes to Definition 2.3.1 as follows.

3.2.1 Definition.   *The function* $f : S \subseteq \mathbb{R}^m \to \mathbb{R}$ *is said to be con-*

tinuous *at* $\mathbf{p} \in S$ *if* $f(\mathbf{a}_k) \to f(\mathbf{p})$ *whenever* $\mathbf{a}_k \to \mathbf{p}$, *where the sequence* $(\mathbf{a}_k)$ *lies in* S. *The function is continuous* (*on its domain* S) *if it is continuous at every point of* S.

3.2.2 Theorem. *A linear function* $f : \mathbb{R}^m \to \mathbb{R}$ *is continuous.*

Proof. Let $\mathbf{e}_1, \ldots, \mathbf{e}_m$ be the usual basis of $\mathbb{R}^m$. By the linearity of f and the Cauchy–Schwarz inequality, for any $\mathbf{x} \in \mathbb{R}^m$,

$$|f(\mathbf{x})|^2 = |x_1 f(\mathbf{e}_1) + \cdots + x_m f(\mathbf{e}_m)|^2$$
$$\leqslant (x_1^2 + \cdots + x_m^2)(f(\mathbf{e}_1)^2 + \cdots + f(\mathbf{e}_m)^2).$$

Therefore, for all $\mathbf{x} \in D$

$$|f(\mathbf{x})| \leqslant M\|\mathbf{x}\|, \qquad \text{where } M = (f(\mathbf{e}_1)^2 + \cdots + f(\mathbf{e}_m)^2)^{1/2}.$$

It follows that for any $\mathbf{p} \in \mathbb{R}^m$ and any sequence $(\mathbf{a}_k)$ in $\mathbb{R}^m$

$$|f(\mathbf{a}_k) - f(\mathbf{p})| = |f(\mathbf{a}_k - \mathbf{p})| \leqslant M\|\mathbf{a}_k - \mathbf{p}\|.$$

Hence $f(\mathbf{a}_k) \to f(\mathbf{p})$ whenever $\mathbf{a}_k \to \mathbf{p}$. This establishes that f is continuous at each $\mathbf{p} \in \mathbb{R}^m$.

3.2.3 *Theorem*
[i] *All constant functions on* $S \subseteq \mathbb{R}^m$ *are continuous.*

[ii] *If* f *and* g *are real-valued functions on* $S \subseteq \mathbb{R}^m$ *and if they are both continuous at* $\mathbf{p} \in S$, *then so are* $f + g$ *and* fg, *and* (*provided* $g(\mathbf{p}) \neq 0$) *so is* f/g.

[iii] *If* f *is a real-valued function on* $S \subseteq \mathbb{R}^m$ *which is continuous at* $\mathbf{p} \in S$ *and if the function* $\phi : \mathbb{R} \to \mathbb{R}$ *is continuous at* $f(\mathbf{p})$, *then the composition* $\phi \circ f : S \subset \mathbb{R}^m \to \mathbb{R}$ *is continuous at* $\mathbf{p}$.

Proof. Exercise.

3.2.4 *Example*. The function $f : \mathbb{R}^2 \to \mathbb{R}$ defined by

$$f(x_1, x_2) = \begin{cases} x_1 + 2x_2 & \text{when } (x_1, x_2) \neq (0, 0), \\ 1 & \text{when } (x_1, x_2) = (0, 0), \end{cases}$$

is continuous everywhere except at the origin. To see this we relate f to the linear function $g : \mathbb{R}^2 \to \mathbb{R}$ given by $g(x_1, x_2) = x_1 + 2x_2$. Let $\mathbf{p} \in \mathbb{R}^2$, where $\mathbf{p} \neq (0, 0)$. If $\mathbf{a}_k \to \mathbf{p}$, then for large enough $k \in \mathbb{N}$, $\mathbf{a}_k \neq \mathbf{0}$. Hence for large enough k, $f(\mathbf{a}_k) = g(\mathbf{a}_k)$. But g is continuous at $\mathbf{p}$ (Theorem 3.2.2) and therefore

$$\lim_{k \to \infty} f(\mathbf{a}_k) = \lim_{k \to \infty} g(\mathbf{a}_k) = g(\mathbf{p}) = f(\mathbf{p}).$$

This establishes that f is continuous at $\mathbf{p} \neq (0, 0)$. To show that f is not continuous at $\mathbf{0}$ it is enough to find one sequence which does not satisfy the conditions of Definition 3.2.1. Consider the sequence $\mathbf{b}_k = (1/k, 1/k)$, $k \in \mathbb{N}$. Then $\mathbf{b}_k \to \mathbf{0}$ but $f(\mathbf{b}_k) \nrightarrow f(\mathbf{0})$ since $f(\mathbf{b}_k) = 3/k$, whereas $f(\mathbf{0}) = 1$.

The graph of the above example is the plane $x_3 = x_1 + 2x_2$ in $\mathbb{R}^3$ with the point $(0, 0, 0)$ replaced by $(0, 0, 1)$. In general if there is such a jump or tear in the graph then the function has discontinuities.

3.2.5 *Example.* Let $f : \mathbb{R}^2 \to \mathbb{R}$ be defined by

$$f(x_1, x_2) = \begin{cases} -\tfrac{1}{2}x_2 + 2 & \text{when } x_1 \geqslant 0 \text{ and } x_2 \geqslant 0, \\ 2 & \text{otherwise.} \end{cases}$$

The discontinuities of f are revealed by sketching its graph (Fig. 3.7). The function is not continuous at $(0, 1)$, for example. It is enough to find just one sequence $(\mathbf{b}_k)$ in $\mathbb{R}^2$ such that $\mathbf{b}_k \to (0, 1)$ but $f(\mathbf{b}_k) \nrightarrow f(0, 1) = \tfrac{3}{2}$. Consideration of the graph suggests the choice $\mathbf{b}_k = (-1/k, 1)$, $k \in \mathbb{N}$. Then $f(\mathbf{b}_k) = 2$ for all k and we have the required counter example. A similar argument can be used to show that f is discontinuous at all points of the form $(0, x_2)$, $x_2 > 0$. The function is however continuous at every other point, as we now show. Let $\mathbf{a}_k = (a_{1k}, a_{2k})$.

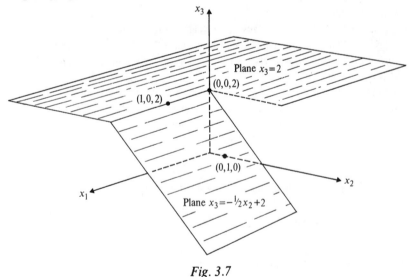

Fig. 3.7

    [i] If $\mathbf{a}_k \to \mathbf{0}$ then $a_{2k} \to 0$ and it is clear from the definition that $f(\mathbf{a}_k) \to 2 = f(\mathbf{0})$. So f is continuous at $\mathbf{0}$.

    [ii] If $\mathbf{a}_k \to \mathbf{p}$ where $p_1 > 0$ and $p_2 > 0$, then for large enough $k \in \mathbb{N}$, $a_{1k} > 0$

and $a_{2k} > 0$. Hence the function f on $\mathbf{a}_k$ is, for large k, the same as the function g given by $g(x_1, x_2) = -\frac{1}{2}x_2 + 2$. Now g, being the sum of a linear function and a constant function, is continuous at $\mathbf{p}$. Therefore

$$\lim_{k \to \infty} f(\mathbf{a}_k) = \lim_{k \to \infty} g(\mathbf{a}_k) = g(\mathbf{p}) = f(\mathbf{p}).$$

So f is continuous at $\mathbf{p}$.

[iii] Similarly if either $p_1 < 0$ or $p_2 < 0$ (or both) then the function f is the same as the constant function $h(x_1, x_2) = 2$ near $\mathbf{p}$. Again we have that f is continuous at $\mathbf{p}$.

[iv] The remaining cases where $\mathbf{p} = (p_1, 0)$, $p_1 > 0$, will be considered later in this section. See Example 3.2.24.

The domains which are particularly appropriate for the calculus of real valued functions on $\mathbb{R}^m$ are the open subsets of $\mathbb{R}^m$. They are generalizations of open intervals $]a, b[$ in $\mathbb{R}$. Roughly speaking they are subsets of $\mathbb{R}^m$ which 'entirely surround' each of their points. In order to make this precise we need two preparatory definitions.

3.2.6 Definition. *Given $\varepsilon > 0$ and $\mathbf{p} \in \mathbb{R}^m$, the set $\{\mathbf{x} \in \mathbb{R}^m \mid \|\mathbf{x} - \mathbf{p}\| < \varepsilon\}$ is called the ε-neighbourhood of $\mathbf{p}$ and will be denoted by $N(\mathbf{p}, \varepsilon)$.*

3.2.7 Example. See Fig. 3.8.

[i] In $\mathbb{R}$, with $p = 2$, an ε-neighbourhood of p is an *open interval*

$$\{t \in \mathbb{R} \mid |t - 2| < \varepsilon\} =]2 - \varepsilon, 2 + \varepsilon[.$$

[ii] In $\mathbb{R}^2$, with $\mathbf{p} = (-1, 1)$, an ε-neighbourhood of $\mathbf{p}$ is an *open disc* centre $\mathbf{p}$,

$$\{(x_1, x_2) \in \mathbb{R}^2 \mid \sqrt{[(x_1 + 1)^2 + (x_2 - 1)^2]} < \varepsilon\}.$$

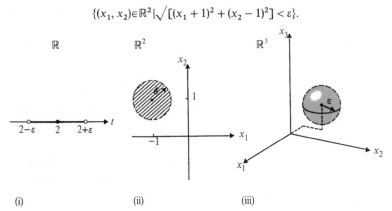

Fig. 3.8 (i) open interval in $\mathbb{R}$; (ii) open disc in $\mathbb{R}^2$; (iii) open ball in $\mathbb{R}^3$

[iii] In $\mathbb{R}^3$, with $\mathbf{p} = (-1, 1, 2)$, an ε-neighbourhood of $\mathbf{p}$ is an *open ball* centre $\mathbf{p}$,

$$\{(x_1, x_2, x_3) \in \mathbb{R}^3 \,|\, \sqrt{[(x_1 + 1)^2 + (x_2 - 1)^2 + (x_3 - 2)^2]} < \varepsilon\}.$$

3.2.8 Definition. *Let S be a subset of $\mathbb{R}^m$ and let $\mathbf{p} \in S$. We say that $\mathbf{p}$ is an* interior point *of S if there is a real number $\varepsilon > 0$ such that $N(\mathbf{p}, \varepsilon) \subseteq S$. The set of all interior points of S is denoted by* Int S.

3.2.9 Example.   **[i]** Let $A = \,]1, 2[\subseteq \mathbb{R}$ and let $p = 1.8$. Then p is an interior point of A since $N(p, 0.1) \subseteq A$. See Fig. 3.9(i). One could show similarly that every point of A is an interior point of A.
   [ii] Let $B = \{(x, y) \in \mathbb{R}^2 \,|\, 1 < x < 2, y = 0\} \subseteq \mathbb{R}^2$ and let $\mathbf{p} = (1.8, 0)$. Then $\mathbf{p}$ is not an interior point of B since for all $\varepsilon > 0$, $(1.8, \frac{1}{2}\varepsilon) \in N(\mathbf{p}, \varepsilon)$ but $(1.8, \frac{1}{2}\varepsilon) \notin B$. See Fig. 3.9(ii).

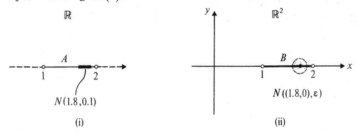

Fig. 3.9 (i) $A \subseteq \mathbb{R}$ has interior points; (ii) $B \subseteq \mathbb{R}^2$ has no interior points

3.2.10 Definition.   *A subset D of $\mathbb{R}^m$ is* open *in $\mathbb{R}^m$ if every point of D is an interior point of D.*

In Example 3.2.9(i) the set A is open in $\mathbb{R}$, and indeed one can use a similar argument to show that all open intervals are open in $\mathbb{R}$. In Example 3.2.9(ii) the set B is not open in $\mathbb{R}^2$ since the point $(1.8, 0)$ is not an interior point.

3.2.11 Example.   For any $\mathbf{p} \in \mathbb{R}^m$ and any $\varepsilon > 0$, the ε-neighbourhood $N(\mathbf{p}, \varepsilon)$ is an open subset of $\mathbb{R}^m$. For if $\mathbf{q} \in N(\mathbf{p}, \varepsilon)$ and if $\|\mathbf{q} - \mathbf{p}\| = \gamma < \varepsilon$, then $N(\mathbf{q}, \varepsilon - \gamma) \subseteq N(\mathbf{p}, \varepsilon)$ (use the triangle inequality and the identity $\mathbf{x} - \mathbf{p} = \mathbf{x} - \mathbf{q} + \mathbf{q} - \mathbf{p}$).

3.2.12 Example.   The rectangle $R = \{(x, y) \in \mathbb{R}^2 \,|\, 0 < x < 1, 0 < y < 3\}$ is an open subset of $\mathbb{R}^2$ since for any point $\mathbf{p} \in R$ there exists an $\varepsilon > 0$ such that $N(\mathbf{p}, \varepsilon) \subseteq R$. See Fig. 3.10.

Notice that in Example 3.2.12 for each $\mathbf{p} \in R$ there is a sequence

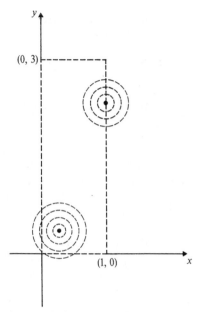

Fig. 3.10 Cluster points of open rectangle

$(\mathbf{a}_k)$ such that (i) $\mathbf{a}_k \in R$ for all $k \in \mathbb{N}$, (ii) $\mathbf{a}_k \neq \mathbf{p}$ for all $k \in \mathbb{N}$, and (iii) $\mathbf{a}_k \to \mathbf{p}$. There are also other points $\mathbf{p}$ of the plane for which such a sequence exists. Each point $\mathbf{p}$ on the edge of the rectangle is so 'close' to R that every neighbourhood $N(\mathbf{p}, 1/k), k \in \mathbb{N}$, intersects R. By choosing $\mathbf{a}_k \in N(\mathbf{p}, 1/k) \cap R$, we can generate a sequence satisfying the above three properties. See Fig. 3.10. We are led to the following definition.

3.2.13 Definition. *Let S be a subset of $\mathbb{R}^m$. A point $\mathbf{c}$ is called a* cluster point (*or a* limit point) *of S if there is a sequence of points $(\mathbf{x}_k)$ in S such that $\mathbf{x}_k \neq \mathbf{c}$ for all $k \in \mathbb{N}$, but $\mathbf{x}_k \to \mathbf{c}$.*

3.2.14 Example. **[i]** The cluster points of the set A of Example 3.2.9(i) form the closed interval $[1, 2]$. In general the set of cluster points of any interval in $\mathbb{R}$ is the union of the interval and its end points. (Compare Definition 2.4.1.)

[ii] In Example 3.2.9(ii) the cluster points of B are the points of B together with the points $(1, 0)$ and $(2, 0)$. Every point of B is also a cluster point of $\mathbb{R}^2 \backslash B$, the complement of B in $\mathbb{R}^2$.

[iii] The cluster points of $N(\mathbf{p}, \varepsilon)$ in $\mathbb{R}^m$ form the *closed ball*

$$B(\mathbf{p}, \varepsilon) = \{\mathbf{x} \in \mathbb{R}^m \mid \|\mathbf{x} - \mathbf{p}\| \leq \varepsilon\}.$$

[iv] The cluster points of the rectangle R in Example 3.2.12 form the
closed rectangle $[(x, y) \in \mathbb{R}^2 | 0 \leqslant x \leqslant 1, 0 \leqslant y \leqslant 3 \}$.

3.2.15 Definition.   *A subset S of* $\mathbb{R}^m$ *is* closed *in* $\mathbb{R}^m$ *if it contains all
its cluster points.*

3.2.16 Example.   A closed interval in $\mathbb{R}$, a closed ball in $\mathbb{R}^m$, a closed
rectangle in $\mathbb{R}^2$ are all closed sets in their respective Euclidean spaces.

Notice that there are sets which are neither open nor closed – in
this sense the terms are unfortunate. In Example 3.2.9**(ii)** for instance,
the set B is not open in $\mathbb{R}^2$ since $(1.8, 0)$ is not an interior point, and
it is not closed in $\mathbb{R}^2$ since it does not contain its cluster point $(1, 0)$.
We do however have the following important result.

3.2.17 Theorem.   *A subset D of* $\mathbb{R}^m$ *is open if and only if its complement in* $\mathbb{R}^m$ *is closed.*

Proof.   Exercise.

We are now in a position to adapt Definition 2.4.5 and the
theorems which follow to our present situation.

3.2.18 Definition.   *Given a cluster point* $\mathbf{p}$ *of* $S \subseteq \mathbb{R}^m$, *a point q in* $\mathbb{R}$
and a function $f : S \subseteq \mathbb{R}^m \to \mathbb{R}$, *we write* $\lim_{\mathbf{x} \to \mathbf{p}} f(\mathbf{x}) = q$ *if* $f(\mathbf{a}_k) \to q$
whenever $\mathbf{a}_k \to \mathbf{p}$ *where the sequence* $(\mathbf{a}_k)$ *lies in S and* $\mathbf{a}_k \neq \mathbf{p}$ *for all*
$k \in \mathbb{N}$.

3.2.19 Example.   **[i]** In the case of the function of Example 3.2.4,
$\lim_{\mathbf{x} \to \mathbf{0}} f(\mathbf{x}) = 0$.
    [ii] In the case of the function of Example 3.2.5, $\lim_{\mathbf{x} \to \mathbf{0}} f(\mathbf{x}) = 2$.

3.2.20 Theorem.   *The function* $f : S \subseteq \mathbb{R}^m \to \mathbb{R}$ *is continuous at a
cluster point* $\mathbf{p} \in S$ *if and only if* $\lim_{\mathbf{x} \to \mathbf{p}} f(\mathbf{x}) = f(\mathbf{p})$.

Proof.   Exercise (compare Theorem 2.4.9).

3.2.21 Theorem.   *Given a function* $f : S \subseteq \mathbb{R}^m \to \mathbb{R}$ *and* $\mathbf{p}$ *a cluster
point of S, then* $\lim_{\mathbf{x} \to \mathbf{p}} f(\mathbf{x}) = q$ *if and only if to each* $\varepsilon > 0$ *(however
small) there corresponds a real number* $\delta > 0$ *such that* $|f(\mathbf{x}) - q| < \varepsilon$
whenever $\mathbf{x} \in S$ *and* $0 < \|\mathbf{x} - \mathbf{p}\| < \delta$.

Proof.   Exercise (compare Theorem 2.4.11).

We note again that the limiting process is not concerned with the value that f takes at $\mathbf{p}$, nor even whether f is defined at $\mathbf{p}$.

We can combine Theorems 3.2.20 and 3.2.21 to provide a manipulative test for continuity.

3.2.22 Theorem. *The function $f : S \subseteq \mathbb{R}^m \to \mathbb{R}$ is continuous at $\mathbf{p} \in S$ if and only if to each $\varepsilon > 0$ there corresponds $\delta > 0$ such that $f(\mathbf{x}) \in N(f(\mathbf{p}), \varepsilon)$ whenever $\mathbf{x} \in S$ and $\mathbf{x} \in N(p, \delta)$.*

Proof. Exercise (compare Corollary 2.4.16).

Theorem 3.2.22 is useful for proving the continuity of a function f at a point $\mathbf{p}$ in its domain. There is some risk of becoming lost in the calculations, and it is as well to establish a systematic procedure. We recommend the following when the domain of f is an open set D.

1. Consider a δ-neighbourhood $N(\mathbf{p}, \delta)$ of $\mathbf{p}$ lying in D. Locate the image $f(N(\mathbf{p}, \delta))$ as a subset of $\mathbb{R}$.
2. Consider the possible values of ε such that $f(N(\mathbf{p}, \delta))$ is a subset of the ε-neighbourhood of $f(\mathbf{p})$; that is, find values of ε such that

3.2.23 $$f(N(\mathbf{p}, \delta)) \subseteq N(f(\mathbf{p}), \varepsilon).$$

3. Finally, prove continuity by showing that to each choice of ε, however small, there corresponds a δ (which will depend on ε) for which the inclusion 3.2.23 is true.

3.2.24 Example. Let us apply the above procedure to prove the continuity of the function $f : \mathbb{R}^2 \to \mathbb{R}$ of Example 3.2.5 at the point $\mathbf{p} = (1, 0)$. We repeat the definition

$$f(x_1, x_2) = \begin{cases} -\frac{1}{2}x_2 + 2 & \text{when } x_1 \geq 0 \text{ and } x_2 \geq 0, \\ 2 & \text{otherwise.} \end{cases}$$

We have $f(\mathbf{p}) = f(1, 0) = 2$. (The point $(\mathbf{p}, f(\mathbf{p}))$ on the graph of f marked as $(1, 0, 2)$ in Fig. 3.7.)

1. Consider a small $\delta > 0$. Now $N(\mathbf{p}, \delta)$ is a disc in $\mathbb{R}^2$ centred at $\mathbf{p}$ (Fig. 3.11). We wish to locate its image under f as a subset of $\mathbb{R}$.

Note that $N(\mathbf{p}, \delta)$ is a subset of the open square $S(\mathbf{p}, \delta)$ centred at $\mathbf{p}$ of edge 2δ. It is clear from the definition of f that

$$f(N(\mathbf{p}, \delta)) \subseteq f(S(\mathbf{p}, \delta)) = \{t \in \mathbb{R} | 2 - \tfrac{1}{2}\delta < t < 2\}.$$

2. We deduce that

$$f(N(\mathbf{p}, \delta)) \subseteq N(2, \tfrac{1}{2}\delta),$$

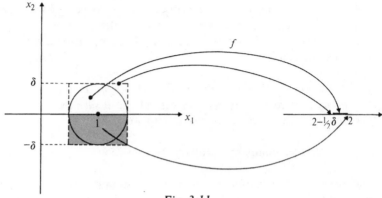

Fig. 3.11

and hence that

$$f(N(\mathbf{p}, \delta) \subseteq N(f(\mathbf{p}), \varepsilon) = N(2, \varepsilon)$$

whenever $\varepsilon \geqslant \frac{1}{2}\delta$.

3. We conclude that to each $\varepsilon > 0$ there corresponds a δ such that condition 3.2.23 is satisfied. We need only choose δ such that $\delta \leqslant 2\varepsilon$. This proves that f is continuous at $\mathbf{p}$.

The following theorem provides a useful test for cluster points. We use the notation $\hat{N}(\mathbf{c}, \varepsilon) = N(\mathbf{c}, \varepsilon) \backslash \{\mathbf{c}\}$ for the ε-neighbourhood of $\mathbf{c}$ 'punctured' at its centre.

3.2.25 Theorem.   *A point $\mathbf{c} \in \mathbb{R}^m$ is a cluster point of $S \subseteq \mathbb{R}^m$ if and only if $\hat{N}(\mathbf{c}, \varepsilon)$ meets S for every $\varepsilon > 0$.*

Proof.   Immediate from Definition 3.2.13.

3.2.26 Definition.   *Let S be a subset of $\mathbb{R}^m$. The* closure *of S (in $\mathbb{R}^m$) is the set S together with all its cluster points. It is denoted by $\bar{S}$.*

3.2.27 Example.   The closure of the ε-neighbourhood $N(\mathbf{p}, \varepsilon)$ of $\mathbf{p} \in \mathbb{R}^m$ is the closed ball $B(\mathbf{p}, \varepsilon)$.

A set $S \subseteq \mathbb{R}^m$ is closed if and only if $S = \bar{S}$.

3.2.28 Theorem.   *Let S be a subset of $\mathbb{R}^m$. Then $\mathbf{p} \in \bar{S}$ if and only if $N(\mathbf{p}, \varepsilon)$ meets S for every $\varepsilon > 0$.*

Proof.   From Definition 3.2.26, $\mathbf{p} \in \bar{S}$ if and only if *either $\mathbf{p} \in S$ or $\mathbf{p}$*

is a cluster point of S, in which case for every $\varepsilon > 0$, $\hat{N}(\mathbf{p}, \varepsilon)$ meets S. The result follows.

As an application of Theorem 3.2 28 we prove

3.2.29 Theorem. *Let D, S be subsets of $\mathbb{R}^m$. Then*
[i] $\bar{D}$ *is closed in $\mathbb{R}^m$;* **[ii]** *if $D \subseteq S$, then $\bar{D} \subseteq \bar{S}$;* **[iii]** *if S is closed in $\mathbb{R}^m$ and D is open in $\mathbb{R}^m$ then $S \backslash D$ is closed in $\mathbb{R}^m$;* **[iv]** *S is closed in $\mathbb{R}^m$ if and only if $\mathbb{R}^m \backslash S$ is open in $\mathbb{R}^m$.*

Proof. [*i*] It follows from Theorem 3.2.28 that if an *open* set contains a cluster point of D then it contains a point of D. Let $T = \bar{D}$. For any point $\mathbf{p}$ in $\bar{T}$, the set $N(\mathbf{p}, \varepsilon)$ meets T for all $\varepsilon > 0$. Hence $N(\mathbf{p}, \varepsilon)$ meets D for all $\varepsilon > 0$ and therefore $\mathbf{p} \in T$. So $\bar{T} \subseteq T$. Since also $T \subseteq \bar{T}$, it follows that $\bar{T} = T$.
  [*ii*] By Theorem 3.2.28, $\mathbf{p} \in \bar{D}$ implies that $\mathbf{p} \in \bar{S}$, where $D \subseteq S$.
  [*iii*], [*iv*] Exercises.

3.2.30 Definition. **[i]** *A subset $S \subseteq \mathbb{R}^m$ is* bounded *if it is contained in a closed ball $B(\mathbf{p}, k)$ for some $\mathbf{p} \in \mathbb{R}^m$, $k > 0$.* **[ii]** *A subset $S \subseteq \mathbb{R}^m$ is* compact *(in $\mathbb{R}^m$) if it is closed in $\mathbb{R}^m$ and is bounded.*

3.2.31 Example. The closed ball $B(\mathbf{p}, k)$ $\mathbf{p} \in \mathbb{R}^m$, is compact. The real line $\mathbb{R}$ is closed but is not compact.

The following theorems provide two useful criteria for compactness in $\mathbb{R}^m$. We omit the proofs.

3.2.32 Theorem. *A set $K \subseteq \mathbb{R}^m$ is compact if and only if every sequence in K has a subsequence which converges to a point in K.*

For our second criterion we require the concept of an open cover. Given a set $K \subseteq \mathbb{R}^m$, a collection $\mathscr{C} = \{V_\lambda \mid \lambda \in \Lambda\}$ of subsets of $\mathbb{R}^m$ is called a *cover* of K if $K \subseteq \cup_\lambda V_\lambda$. The cover is said to be an *open cover* of K if each of the sets V_λ is open in $\mathbb{R}^m$. Furthermore, if there are a finite number of sets $V_{\lambda_1}, \ldots, V_{\lambda_r}$ in $\mathscr{C}$ such that $K \subseteq V_{\lambda_1} \cup \cdots \cup V_{\lambda_r}$, then $\mathscr{C}$ is said to contain the *finite subcover* $\mathscr{S} = \{V_{\lambda_1}, \ldots, V_{\lambda_r}\}$ of K.

3.2.33 Theorem. *A set K in $\mathbb{R}^m$ is compact if and only if every open cover of K contains a finite subcover of K.*

Sometimes the condition of Theorem 3.2.33 is taken as the definition of compactness (for example in the general theory of topological spaces) and then 3.2.30[ii] and 3.2.32 appear as consequences.

Exercises 3.2

1. Which of the following subsets of $\mathbb{R}^2$ are (i) open, (ii) closed, (iii) both open and closed, (iv) neither open nor closed, (v) compact?
    (a) $1 < x_1 < 2$ and $-1 < x_2 < 2$;
    (b) $1 < x_1 < 2$ and $-1 \leqslant x_2 < 2$;
    (c) $x_1^2 + x_2^2 > 0$;
    (d) $x_1^2 + \frac{1}{2}x_2^2 \leqslant 1$;
    (e) $\|\mathbf{x} - (1, 3)\| < 1$;
    (f) $x_1 > x_2$;
    (g) $\mathbb{R}^2$
    In each case determine the cluster points of the set.

Answers: (a), (c), (e), (f), (g) open;   (d), (g) closed;   (d) compact.

2. Prove that the function of Example 3.2.5 is (a) continuous at (2, 0); (b) discontinuous at (0, 2). (In (a) apply the procedure of Example 3.2.24. In (b) construct a sequence demonstrating the failure of Definition 3.2.1.)

3. Apply the method of Example 3.2.24 to prove that the function $f : \mathbb{R}^2 \rightarrow \mathbb{R}$ defined by $f(x, y) = x^2 + y^2$ is continuous at $\mathbf{p} = (1, 2)$.

Hint: Show that, provided $0 < \delta < 1$, $f(N(p, \delta)) \subseteq N(5, 8\delta)$. Hence for a given small $\varepsilon > 0$, we can choose $\delta \leqslant \varepsilon/8$ in Theorem 3.2.22.

Note: The method applies to an arbitrary point $\mathbf{p} \in \mathbb{R}^2$, and so f is continuous on $\mathbb{R}^2$.

4. Prove Theorem 3.2.3.

5. Assuming the continuity of the exponential function, prove that $g : \mathbb{R}^2 \rightarrow \mathbb{R}$ defined by $g(x, y) = \exp(x^2 + y^2)$ is continuous on $\mathbb{R}^2$. (Apply Exercise 3 and Theorem 3.2.3(iii).)

6. The function $f : \mathbb{R}^2 \rightarrow \mathbb{R}$ given by

$$f(x_1, x_2) = \begin{cases} \dfrac{x_1 + x_2}{\sqrt{(x_1^2 + x_2^2)}} & \text{when } (x_1, x_2) \neq (0, 0), \\ 0 & \text{when } (x_1, x_2) = (0, 0) \end{cases}$$

    is discontinuous at the origin.

Hint: consider the sequence $\mathbf{a}_k = \frac{1}{k}(\cos \alpha, \sin \alpha)$, $k \in \mathbb{N}$, α fixed.

7. Prove that the function $f : \mathbb{R}^2 \to \mathbb{R}$ given by

$$f(x, y) = \begin{cases} \dfrac{2xy}{x^2 + y^2} & \text{when } (x, y) \neq (0, 0), \\ 0 & \text{when } (x, y) = (0, 0) \end{cases}$$

is discontinuous at the origin.

8. Prove that the functions $f : \mathbb{R}^2 \to \mathbb{R}$ and $g : \mathbb{R}^2 \to \mathbb{R}$ given by

$$f(x, y) = \begin{cases} \dfrac{x^2 y}{x^2 + y^2} & \text{when } (x, y) \neq (0, 0) \\ 0 & \text{when } (x, y) = (0, 0), \end{cases}$$

$$g(x, y) = \begin{cases} \dfrac{2xy}{\sqrt{(x^2 + y^2)}} & \text{when } (x, y) \neq (0, 0) \\ 0 & \text{when } (x, y) = (0, 0) \end{cases}$$

are everywhere continuous.

Hint: to prove continuity at $\mathbf{p} \neq \mathbf{0}$ apply Theorem 3.2.3; to prove continuity at $\mathbf{p} = \mathbf{0}$, put $(x, y) = (r\cos\theta, r\sin\theta)$, $r \geq 0$.

9. Let $f : S \subseteq \mathbb{R}^m \to \mathbb{R}$ be continuous. Prove that the level sets of f are closed subsets in S.

10. Let S and T be subsets of $\mathbb{R}^m$. Prove that $\overline{S \cup T} = \bar{S} \cup \bar{T}$, and $\overline{S \cap T} \subseteq \bar{S} \cap \bar{T}$. Show that strict inclusion is possible.

Hint: In $\mathbb{R}$ consider the open intervals $S = \,]0, 1[$ and $T = \,]1, 2[$.

3.3   Linear approximation and differentiability

Our aim in this section is to formulate a definition of the differentiability of a function $f : D \subseteq \mathbb{R}^m \to \mathbb{R}$ at a point in its domain. When $m = 1$ the new definition must agree with that given (in Section 2.5) of the differentiability of function $f : D \subseteq \mathbb{R} \to \mathbb{R}^n$ when $n = 1$. To avoid becoming tangled in special cases *we consider only those functions whose domain D is an open subset of $\mathbb{R}^m$*.

It is more appropriate to take the approach of Section 2.10 than that of Section 2.5 (see Exercise 3.3.1) and accordingly we introduce the relevant ideas.

3.3.1 Definition.   *For an open set D in $\mathbb{R}^m$ with $\mathbf{p} \in D$, let* $D_\mathbf{p} = \{\mathbf{h} \in \mathbb{R}^m \mid \mathbf{p} + \mathbf{h} \in D\}$.

Notice that $D_\mathbf{p}$, which is merely the translation of D through $-\mathbf{p}$, contains $\mathbf{0}$ and is itself an open set. See Fig. 3.12(i).

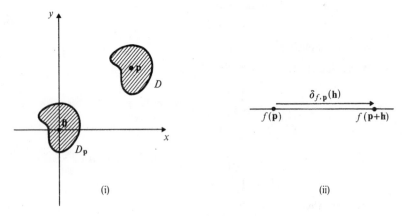

Fig. 3.12

3.3.2 *Example*.   Let $D = \{(x, y)|x \in]1, 3[, y \in]2, 6[\}$. Then

$$D_{(2, 5)} = \{(x, y)|x \in]-1, 1[, y \in]-3, 1[\}.$$

3.3.3 *Definition*.   *Consider a function* $f : D \subseteq \mathbb{R}^m \to \mathbb{R}$, *where D is an open subset of* $\mathbb{R}^m$, *and let* $\mathbf{p} \in D$. *The* difference function $\delta_{f,\mathbf{p}} : D_\mathbf{p} \subset \mathbb{R}^m \to \mathbb{R}$ *is defined by*

$$\delta_{f, \mathbf{p}}(\mathbf{h}) = f(\mathbf{p} + \mathbf{h}) - f(\mathbf{p}), \qquad \mathbf{h} \in D_\mathbf{p}.$$

The difference function expresses the change in $f(\mathbf{x})$ as $\mathbf{x}$ moves away from $\mathbf{p}$. See Fig. 3.12(ii).

3.3.4 *Example*.   If $f : \mathbb{R}^2 \to \mathbb{R}$ is defined by $f(x_1, x_2) = x_1^2 + x_2^2$, then

$$\delta_{f, \mathbf{p}}(\mathbf{h}) = 2h_1 p_1 + 2h_2 p_2 + h_1^2 + h_2^2.$$

3.3.5 *Definition*.   *Let* $g : N \subseteq \mathbb{R}^m \to \mathbb{R}$ *and* $g^* : M \subseteq \mathbb{R}^m \to \mathbb{R}$ *be two functions defined on open domains N and M such that* $\mathbf{0} \in N \cap M$. *We say that g and g^** closely approximate *each other near* $\mathbf{0} \in \mathbb{R}^m$ *if*

$$g(\mathbf{0}) = g^*(\mathbf{0}) \quad and \quad \lim_{\mathbf{h} \to \mathbf{0}} \frac{g(\mathbf{h}) - g^*(\mathbf{h})}{\|\mathbf{h}\|} = 0.$$

Equivalently, g and g^* closely approximate each other near $\mathbf{0}$ *if there* *exists a function* $\eta : N \cap M \subseteq \mathbb{R}^m \to \mathbb{R}$ *such that*

[i] $g(\mathbf{h}) = g^*(\mathbf{h}) + \|\mathbf{h}\|\eta(\mathbf{h})$ *for all* $\mathbf{h} \in N \cap M$ *and*

[ii] $\lim_{\mathbf{h} \to 0} \eta(\mathbf{h}) = 0.$

The definition expresses the order of the difference between g and g^* near $\mathbf{0}$.

3.3.6 Definition. *A function* $f : D \subseteq \mathbb{R}^m \to \mathbb{R}$ *which is defined on an open subset D of $\mathbb{R}^m$ is* differentiable *at* $\mathbf{p} \in D$ *if the difference function* $\delta_{f,\mathbf{p}} : D_\mathbf{p} \subseteq \mathbb{R}^m \to \mathbb{R}$ *can be closely approximated by a linear function near* $\mathbf{0}$.

Equivalently, f is differentiable at $\mathbf{p}$ if there exists a linear function $L : \mathbb{R}^m \to \mathbb{R}$ *and a function* $\eta : D_\mathbf{p} \subseteq \mathbb{R}^m \to \mathbb{R}$ *such that*

3.3.7 $f(\mathbf{p} + \mathbf{h}) - f(\mathbf{p}) = L(\mathbf{h}) + \|\mathbf{h}\|\eta(\mathbf{h}), \qquad \mathbf{h} \in D_\mathbf{p}$

and

3.3.8 $$\lim_{\mathbf{h} \to 0} \eta(\mathbf{h}) = 0.$$

Clearly when $m = 1$ this definition is in accordance with Definition 2.10.16 as applied to the case $n = 1$, since if $h \in \mathbb{R}$, then $\|h\| = |h|$.

3.3.9 Example. A linear function $L : \mathbb{R}^m \to \mathbb{R}$ is differentiable everywhere.

3.3.10 Example. Let $f(x_1, x_2) = x_1^2 + x_2^2$, as in Example 3.3.4. For each $\mathbf{p} \in \mathbb{R}^2$ the condition 3.3.7 is satisfied if we take $L(\mathbf{h}) = 2p_1 h_1 + 2p_2 h_2$ and $\eta(\mathbf{h}) = \|\mathbf{h}\| = \sqrt{(h_1^2 + h_2^2)}$. Since $\lim_{\mathbf{h} \to 0} \|\mathbf{h}\| = 0$, the function f is differentiable at $\mathbf{p}$.

3.3.11 Theorem. *A function* $f : D \subseteq \mathbb{R}^m \to \mathbb{R}$ *which is differentiable at* $\mathbf{p} \in D$ *is also continuous there.*

Proof. Let $(\mathbf{a}_k)$ be a sequence in D such that $\mathbf{a}_k \to \mathbf{p}$. For each $k \in \mathbb{N}$, let $\mathbf{h}_k = \mathbf{a}_k - \mathbf{p}$. If f is differentiable at $\mathbf{p}$, then there exists a linear function L and a function η satisfying 3.3.7 and 3.3.8. Hence

3.3.12 $f(\mathbf{a}_k) = f(\mathbf{p}) + L(\mathbf{h}_k) + \|\mathbf{h}_k\|\eta(\mathbf{h}_k), \qquad k \in \mathbb{N}.$

But $\mathbf{h}_k \to \mathbf{0}$ and so (by Theorem 3.2.2) $L(\mathbf{h}_k) \to 0$ and (from 3.3.8) $\|\mathbf{h}_k\|\eta(\mathbf{h}_k) \to 0$. It follows from 3.3.12 that $f(\mathbf{a}_k) \to f(\mathbf{p})$ and the proof is complete.

We shall describe some continuous functions which are not differentiable later in this section.

3.3.13 Theorem.   Let $f : D \subseteq \mathbb{R}^m \to \mathbb{R}$ be differentiable at $\mathbf{p} \in D$. Then there is only one close linear approximation to $\delta_{f,\mathbf{p}}$ near $\mathbf{0}$.

Proof.   Let $\mathbf{e}_1, \ldots, \mathbf{e}_m$ be the usual basis of $\mathbb{R}^m$, let $L : \mathbb{R}^m \to \mathbb{R}$ be a close linear approximation to $\delta_{f,\mathbf{p}}$ near $\mathbf{0}$ and let $\eta : D_\mathbf{p} \to \mathbb{R}$ be a function satisfying 3.3.7 and 3.3.8. Since f is differentiable at $\mathbf{p}$ we have for each j, $1 \leqslant j \leqslant m$,

[i] $f(\mathbf{p} + t\mathbf{e}_j) - f(\mathbf{p}) = L(t\mathbf{e}_j) + \|t\mathbf{e}_j\|\eta(t\mathbf{e}_j), \qquad t\mathbf{e}_j \in D_\mathbf{p}$

[ii] $\lim\limits_{t \to 0} \eta(t\mathbf{e}_j) = 0.$

It follows that

$$f(\mathbf{p} + t\mathbf{e}_j) - f(\mathbf{p}) = tL(\mathbf{e}_j) + |t|\eta(t\mathbf{e}_j),$$

and therefore, for $t \neq 0$,

$$\frac{f(\mathbf{p} + t\mathbf{e}_j) - f(\mathbf{p})}{t} = L(\mathbf{e}_j) + \frac{|t|}{t}\eta(t\mathbf{e}_j).$$

Using (ii) above, we obtain

3.3.14 $$\lim_{t \to 0} \frac{f(\mathbf{p} + t\mathbf{e}_j) - f(\mathbf{p})}{t} = L(\mathbf{e}_j).$$

Now if $L^* : \mathbb{R}^m \to \mathbb{R}$ were another close linear approximation to f near $\mathbf{p}$, the above argument would also show that

$$\lim_{t \to 0} \frac{f(\mathbf{p} + t\mathbf{e}_j) - f(\mathbf{p})}{t} = L^*(\mathbf{e}_j).$$

Hence $L(\mathbf{e}_j) = L^*(\mathbf{e}_j)$ for each j, $1 \leqslant j \leqslant m$. But a linear function on $\mathbb{R}^m$ is uniquely determined by its effect on $\mathbf{e}_1, \ldots, \mathbf{e}_m$. We have therefore proved that $L = L^*$.

3.3.15 Definition.   If $f : D \subseteq \mathbb{R}^m \to \mathbb{R}$ is differentiable at $\mathbf{p} \in D$, then the (unique) close linear approximation to $\delta_{f,\mathbf{p}}$ at $\mathbf{0}$ is denoted by $L_{f,\mathbf{p}}$ and is called the differential of f at $\mathbf{p}$.

The differentiability of $f : D \subseteq \mathbb{R}^m \to \mathbb{R}$ at $\mathbf{p} \in D$ implies that there exists a function $\eta : D_\mathbf{p} \subseteq \mathbb{R}^m \to \mathbb{R}$ and a linear function

$L_{f,\mathbf{p}} : \mathbb{R}^m \to \mathbb{R}^m$ such that for all $\mathbf{h} \in D_{\mathbf{p}}$,

$$f(\mathbf{p} + \mathbf{h}) - f(\mathbf{p}) = L_{f,\mathbf{p}}(h_1\mathbf{e}_1 + \cdots + h_m\mathbf{e}_m) + \|\mathbf{h}\|\eta(\mathbf{h})$$
$$= h_1 L_{f,\mathbf{p}}(\mathbf{e}_1) + \cdots + h_m L_{f,\mathbf{p}}(\mathbf{e}_m) + \|\mathbf{h}\|\eta(\mathbf{h}),$$

where $\lim_{\mathbf{h} \to \mathbf{0}} \eta(\mathbf{h}) = 0$.

It is important to know how to find $L_{f,\mathbf{p}}(\mathbf{e}_j)$ for each $j = 1, \ldots, m$. We have seen in the proof of Theorem 3.3.13 (see in particular 3.3.14) that

3.3.16 $L_{f,\mathbf{p}}(\mathbf{e}_j) = \lim\limits_{t \to 0} \dfrac{f(\mathbf{p} + t\mathbf{e}_j) - f(\mathbf{p})}{t}, \qquad j = 1, \ldots, m.$

The limit on the right-hand side of 3.3.16 may exist even if the function f is not differentiable at $\mathbf{p}$. If it does exist, then it is called the *partial derivative* of f with respect to the jth coordinate x_j and is denoted by

3.3.17

$$\frac{\partial f}{\partial x_j}(\mathbf{p}) = \lim_{t \to 0} \frac{f(\mathbf{p} + t\mathbf{e}_j) - f(\mathbf{p})}{t}$$

$$= \lim_{t \to 0} \frac{f(p_1, \ldots, p_{j-1}, p_j + t, p_{j+1}, \ldots, p_m) - f(p_1, \ldots, p_m)}{t}$$

The techniques of elementary calculus are all that is required to evaluate partial derivatives. In order to determine $(\partial f/\partial x_j)(\mathbf{p})$ we simply have to differentiate $f(p_1, \ldots, p_{j-1}, x_j, p_{j+1}, \ldots, p_m)$ with respect to x_j and evaluate the derivative at $x_j = p_j$.

3.3.18 Example. Let $f : \mathbb{R}^3 \to \mathbb{R}$ be defined by

$$f(x_1, x_2, x_3) = x_1^2 x_2 + 2x_3.$$

Then

$$\frac{\partial f}{\partial x_1}(\mathbf{p}) = 2p_1 p_2, \qquad \frac{\partial f}{\partial x_2}(\mathbf{p}) = p_1^2, \qquad \frac{\partial f}{\partial x_3}(\mathbf{p}) = 2.$$

Provided partial derivatives of $f : D \subseteq \mathbb{R}^m \to \mathbb{R}$ exist at every point $\mathbf{p}$ of the (open) domain D we can regard $\partial f/\partial x_1, \partial f/\partial x_2, \ldots$ themselves as functions from D into $\mathbb{R}$. We find $(\partial f/\partial x_j)(\mathbf{x})$ by differentiating the expression for $f(x_1, \ldots, x_m)$ with respect to x_j, regarding the other variables as constants. In Example 3.3.18 we have $(\partial f/\partial x_1)(\mathbf{x}) = 2x_1 x_2$, for example.

3.3.19 Remark. The notation is adjusted to suit the problem. For example, in Example 3.1.2 we considered the relationship, $V(r, h) = \pi r^2 h$, between the volume V, the radius r and the height h of the cylinder. We find that $(\partial V/\partial r)(r, h) = 2\pi rh$ and that $(\partial V/\partial h)(r, h) = \pi r^2$.

Our notation for partial derivatives, which is the one most commonly used, is meaningful only with reference to the notation used for the vectors in the domain. For example, if a function $f : \mathbb{R}^3 \to \mathbb{R}$ is defined by its action on vectors (x_1, x_2, x_3) then it makes no sense to speak of $\partial f/\partial y$. An alternative notation is to let $f_j(\mathbf{p})$ stand for the partial derivative of f at $\mathbf{p}$ with respect to the jth coordinate. Thus $f_2(x_1, x_2, x_3)$ stands for $(\partial f/\partial x_2)(x_1, x_2, x_3)$ and $f_2(x, y, z)$ stands for $(\partial f/\partial y)(x, y, z)$. This notation has much to recommend it, but we refrain from using it to avoid confusion with our notation (adopted in Chapter 2) for coordinate functions. Coordinate functions and partial derivatives play a very important role in vector calculus. There are many theorems which involve both at the same time and they must be clearly distinguished in the notation.

Now that we have a notation for the limit in 3.3.16 we can state the following important corollary of Theorem 3.3.13.

3.3.20 Theorem.   Let $f : D \subseteq \mathbb{R}^m \to \mathbb{R}$ be differentiable at $\mathbf{p} \in \mathbb{R}^m$. Then
[i] the partial derivative $(\partial f/\partial x_j)(\mathbf{p})$ exists and is equal to $L_{f,\mathbf{p}}(\mathbf{e}_j)$ for each $1 \leqslant j \leqslant m$,
[ii] for all $\mathbf{h} \in \mathbb{R}^m$,

$$L_{f,\mathbf{p}}(\mathbf{h}) = h_1 \frac{\partial f}{\partial x_1}(\mathbf{p}) + \cdots + h_m \frac{\partial f}{\partial x_m}(\mathbf{p}),$$

[iii] the $1 \times m$ matrix representing $L_{f,\mathbf{p}}$ (with respect to the usual bases) is

$$J_{f,\mathbf{p}} = \left[\frac{\partial f}{\partial x_1}(\mathbf{p}) \quad \cdots \quad \frac{\partial f}{\partial x_m}(\mathbf{p}) \right].$$

Proof.   Exercise. The matrix representing $L_{f,\mathbf{p}}$ is $[L_{f,\mathbf{p}}(\mathbf{e}_1) \quad \cdots \quad L_{f,\mathbf{p}}(\mathbf{e}_m)]$.

3.3.21 Definition.   The above matrix $J_{f,\mathbf{p}}$ representing the differential $L_{f,\mathbf{p}}$ is called the Jacobian matrix of f at $\mathbf{p}$.

We can also rephrase the definition of differentiability in terms of partial derivatives.

3.3.22 Theorem. *The function $f : D \subseteq \mathbb{R}^m \to \mathbb{R}$ is differentiable at $\mathbf{p} \in D$ if and only if*
  [i] *all the partial derivatives of f exist at $\mathbf{p}$, and*
  [ii] *there exists a function $\eta : D_{\mathbf{p}} \subseteq \mathbb{R}^m \to \mathbb{R}$ such that for all $\mathbf{h} \in D_{\mathbf{p}}$*

3.3.23 $\quad f(\mathbf{p} + \mathbf{h}) - f(\mathbf{p}) = h_1 \dfrac{\partial f}{\partial x_1}(\mathbf{p}) + \cdots + h_m \dfrac{\partial f}{\partial x_m}(\mathbf{p}) + \|\mathbf{h}\| \, \eta(\mathbf{h}),$

where

3.3.24 $$\lim_{\mathbf{h} \to \mathbf{0}} \eta(\mathbf{h}) = 0.$$

Proof. Exercise.

3.3.25 Example. Consider again (see Examples 3.3.4 and 3.3.10) the function $f : \mathbb{R}^2 \to \mathbb{R}$ given by

$$f(x_1, x_2) = x_1^2 + x_2^2.$$

We have seen that

$$f(\mathbf{x} + \mathbf{h}) - f(\mathbf{x}) = 2x_1 h_1 + 2x_2 h_2 + \|\mathbf{h}\| \, \|\mathbf{h}\|.$$

Since $(\partial f / \partial x_1)(\mathbf{x}) = 2x_1$ and $(\partial f / \partial x_2)(\mathbf{x}) = 2x_2$, the Jacobian matrix of f at $\mathbf{x}$ is $J_{f, \mathbf{x}} = [2x_1 \quad 2x_2]$. Hence $L_{f, \mathbf{x}}(\mathbf{h}) = 2x_1 h_1 + 2x_2 h_2$. By taking $\eta(\mathbf{h}) = \|\mathbf{h}\|$ we have explained the presence of the various components in the above expression.

Theorem 3.3.20 tells us that if any one of the partial derivatives of a function $f : D \subseteq \mathbb{R}^n \to \mathbb{R}$ does not exist at $\mathbf{p} \in D$, then f is not differentiable at $\mathbf{p}$.

3.3.26 Example. Let $f : \mathbb{R}^m \to \mathbb{R}$ be the function defined by $f(\mathbf{x}) = \|\mathbf{x}\|$. The partial derivative $(\partial f / \partial x_1)(\mathbf{0})$, if it exists, is found by considering

$$\frac{f(\mathbf{0} + t\mathbf{e}_1) - f(\mathbf{0})}{t} = \frac{|t|}{t}.$$

But $\lim_{t \to 0} |t|/t$ does not exist, and so f is not differentiable at $\mathbf{0}$.

The converse of Theorem 3.3.20(i) is not true: the existence of the partial derivatives of a function $f : D \subseteq \mathbb{R}^m \to \mathbb{R}$ at $\mathbf{p} \in D$ is not enough to guarantee that f is differentiable at $\mathbf{p}$. We show this in the following example by using Theorem 3.3.22.

3.3.27 Example. Refer again to the function f defined in Example 3.1.8.

Since $f(t, 0) = f(0, t) = f(0, 0)$ for all $t \in \mathbb{R}$, the partial derivatives of f both exist at $\mathbf{0}$ and are both 0. However, f is not differentiable at $\mathbf{0}$, for if it were, then the close linear approximation L to $\delta_{f, 0}$ at $\mathbf{0}$ would have to be given by

$$L(h_1, h_2) = 0h_1 + 0h_2, \qquad (h_1, h_2) \in \mathbb{R}^2.$$

The corresponding function $\eta : \mathbb{R}^2 \to \mathbb{R}$ would have to satisfy (compare Theorem 3.3.22)

$$f(h_1, h_2) - f(0, 0) = 0h_1 + 0h_2 + \sqrt{(h_1^2 + h_2^2)}\eta(h_1, h_2), \text{ for all } (h_1, h_2) \in \mathbb{R}^2.$$

Hence, for $(h_1, h_2) \neq (0, 0)$,

$$\eta(h_1, h_2) = \frac{f(h_1, h_2)}{\sqrt{(h_1^2 + h_2^2)}}.$$

In particular,

$$f(h_1, h_2) = -|h| \qquad \text{when } h_1 = h_2 = h,$$

and so

$$\eta(h, h) = \frac{-|h|}{|h|\sqrt{2}} = -\frac{1}{\sqrt{2}}, \qquad \text{when } h \neq 0.$$

Hence $\lim_{\mathbf{h} \to \mathbf{0}} \eta(\mathbf{h}) \neq 0$. This establishes that f is not differentiable at $\mathbf{0}$.

3.3.28 *Example*.   In Example 3.3.18 we considered a function $f : \mathbb{R}^3 \to \mathbb{R}$ defined by $f(x_1, x_2, x_3) = x_1^2 x_2 + 2x_3$. The partial derivatives of f are given by

$$\frac{\partial f}{\partial x_1}(\mathbf{x}) = 2x_1 x_2, \qquad \frac{\partial f}{\partial x_2}(\mathbf{x}) = x_1^2, \qquad \frac{\partial f}{\partial x_3}(\mathbf{x}) = 2.$$

To determine whether f is differentiable at $\mathbf{x} \in \mathbb{R}^3$ we must ask whether the linear function $L : \mathbb{R}^3 \to \mathbb{R}$ given by

$$L(\mathbf{h}) = 2x_1 x_2 h_1 + x_1^2 h_2 + 2h_3, \qquad \mathbf{h} \in \mathbb{R}^3$$

is a close linear approximation to $\delta_{f, \mathbf{x}}$ near $\mathbf{0}$.
Now

$$f(\mathbf{x} + \mathbf{h}) - f(\mathbf{x}) = 2x_1 x_2 h_1 + x_1^2 h_2 + 2h_3 + 2x_1 h_1 h_2 + h_1^2 x_2 + h_1^2 h_2.$$

and so

$$\delta_{f, \mathbf{x}}(\mathbf{h}) = L(\mathbf{h}) + \|\mathbf{h}\|\eta(\mathbf{h}),$$

where

$$\eta(\mathbf{h}) = \begin{cases} \dfrac{2x_1 h_1 h_2 + h_1^2 x_2 + h_1^2 h_2}{\|\mathbf{h}\|} & \text{if } \mathbf{h} \neq \mathbf{0} \\[2mm] 0 & \text{if } \mathbf{h} = \mathbf{0}. \end{cases}$$

It is not difficult to see (since $|h_i| \leq \|\mathbf{h}\|$ for $i = 1, 2, 3$) that $\lim_{\mathbf{h} \to \mathbf{0}} \eta(\mathbf{h}) = 0$. Therefore by Theorem 3.3.22 f is differentiable at $\mathbf{x}$. Notice that different points $\mathbf{x}$ lead to different functions η, but in each case our conclusion is valid.

In the last example partial differentiation leads to three *continuous* functions $\partial f / \partial x_1$, $\partial f / \partial x_2$ and $\partial f / \partial x_3$ on $\mathbb{R}^3$. This is a significant property which gives us a partial converse to Theorem 3.3.20(i).

3.3.29 Theorem. *Let $f : D \subseteq \mathbb{R}^m \to \mathbb{R}$ be a function whose partial derivatives exist throughout a neighbourhood $N(\mathbf{p}, \varepsilon)$ of $\mathbf{p} \in D$. If $\partial f / \partial x_j : N(\mathbf{p}, \varepsilon) \subseteq \mathbb{R}^m \to \mathbb{R}$ is continuous at $\mathbf{p}$ for each $j = 1, \ldots, m$, then f is differentiable at $\mathbf{p}$.*

Proof. Define a function $\eta : D_{\mathbf{p}} \subseteq \mathbb{R}^m \to \mathbb{R}$ to satisfy

3.3.30 $$f(\mathbf{p} + \mathbf{h}) - f(\mathbf{p}) = \frac{\partial f}{\partial x_1}(\mathbf{p})h_1 + \cdots + \frac{\partial f}{\partial x_m}(\mathbf{p})h_m + \|\mathbf{h}\| \eta(\mathbf{h}).$$

We prove the theorem by showing that under its hypothesis $\lim_{\mathbf{h} \to \mathbf{0}} \eta(\mathbf{h}) = 0$. It is clearly enough to confine ourselves to values of $\mathbf{h}$ such that $0 < \|\mathbf{h}\| < \varepsilon$. The left-hand side of 3.3.30 can be expressed as a 'telescopic sum' in which the inside terms cancel pairwise, leaving unpaired the first and last terms only, thus:

$$f(\mathbf{p} + \mathbf{h}) - f(\mathbf{p}) =$$
$$f(p_1 + h_1, p_2 + h_2, \ldots, p_m + h_m) - f(p_1, p_2 + h_2, \ldots, p_m + h_m)$$
$$+ f(p_1, p_2 + h_2, p_3 + h_3, \ldots, p_m + h_m) - f(p_1, p_2, p_3 + h_3, + \ldots + p_m + h_m)$$
$$+ \ldots$$
$$+ f(p_1, p_2, \ldots, p_{m-1}, p_m + h_m) - f(p_1, p_2, \ldots, p_{m-1}, p_m).$$

In trying to find an alternative form for the jth line of this telescopic sum, there are two cases to consider. Remember that $\|\mathbf{h}\| < \varepsilon$.

[i] Suppose that $h_j \neq 0$. The two points of $\mathbb{R}^m$ involved in the jth line lie in $N(\mathbf{p}, \varepsilon)$, as does the segment joining them. On that segment f is effectively a function of its jth coordinate alone. This function of x_j satisfies the conditions of the Mean-Value Theorem of elementary calculus on the closed interval with end points p_j and $p_j + h_j$. So we can replace the jth line by

3.3.31 $$\frac{\partial f}{\partial x_j}(\mathbf{q}_j)h_j$$

where for some $0 < \theta_j < 1$

$$\mathbf{q}_j = (p_1, \ldots, p_{j-1}, p_j + \theta_j h_j, p_{j+1} + h_{j+1}, \ldots, p_m + h_m).$$

[ii] If $h_j = 0$, then the jth line is 0, and so is equal to the expression given in 3.3.31 for any choice of $0 < \theta_j < 1$.

In either case (with $\mathbf{q}_j$ appropriately chosen in the above manner for $j = 1, \ldots, m$) we have

3.3.32 $$f(\mathbf{p} + \mathbf{h}) - f(\mathbf{p}) = \frac{\partial f}{\partial x_1}(\mathbf{q}_1)h_1 + \cdots + \frac{\partial f}{\partial x_m}(\mathbf{q}_m)h_m.$$

From 3.3.30 and 3.3.32

$$\|\mathbf{h}\|\eta(\mathbf{h}) = \sum_{j=1}^{m} \left(\frac{\partial f}{\partial x_j}(\mathbf{q}_j) - \frac{\partial f}{\partial x_i}(\mathbf{p}) \right) h_j.$$

Now for each j, $|h_j| \leqslant \|\mathbf{h}\|$, so we can apply the triangle inequality to obtain

$$\|\mathbf{h}\| \, |\eta(\mathbf{h})| \leqslant \|\mathbf{h}\| \sum_{j=1}^{m} \left| \frac{\partial f}{\partial x_j}(\mathbf{q}_j) - \frac{\partial f}{\partial x_j}(\mathbf{p}) \right|.$$

Since $\|\mathbf{h}\| \neq 0$, we obtain

3.3.33 $$|\eta(\mathbf{h})| \leqslant \sum_{j=1}^{m} \left| \frac{\partial f}{\partial x_j}(\mathbf{q}_j) - \frac{\partial f}{\partial x_j}(\mathbf{p}) \right|.$$

For each j, however, $\lim_{h_j \to 0} \mathbf{q}_j = \mathbf{p}$ (by 3.3.31). Therefore, if $\partial f / \partial x_j$ is continuous at $\mathbf{p}$,

$$\lim_{h_j \to 0} \left| \frac{\partial f}{\partial x_j}(\mathbf{q}_j) - \frac{\partial f}{\partial x_j}(\mathbf{p}) \right| = 0.$$

Hence, from 3.3.33, $\lim_{\mathbf{h} \to 0} \eta(\mathbf{h}) = 0$ and the proof is complete.

3.3.34 Definition.   *A function $f : D \subseteq \mathbb{R}^m \to \mathbb{R}$, whose partial derivatives (i) exist throughout a neighbourhood of $\mathbf{p} \in D$ and (ii) are continuous at $\mathbf{p}$, is said to be* continuously differentiable at $\mathbf{p}$.

3.3.35 Definition.   *A function $f : D \subseteq \mathbb{R}^m \to \mathbb{R}$ is said to be* continuously differentiable, *or a C^1 function, if it is continuously differentiable at each point $\mathbf{p} \in D$.*

Theorem 3.3.29 tells us that if a function $f : D \subseteq \mathbb{R}^m \to \mathbb{R}$ is continuously differentiable at a point in D, then it is also differentiable there. The following example shows that there are functions which are differentiable at a point but which are not continuously differentiable there.

3.3.36 *Example.* Consider the function $g : \mathbb{R} \to \mathbb{R}$ defined by

$$g(x) = \begin{cases} x^2 \sin(1/x), & \text{for } x \neq 0, \\ 0 & \text{for } x = 0. \end{cases}$$

We find (Exercise 3.3.9) that

$$g'(x) = \begin{cases} 2x \sin(1/x) - \cos(1/x), & \text{for } x \neq 0, \\ 0, & \text{for } x = 0. \end{cases}$$

It will be seen that g is continuously differentiable at all $x \neq 0$, and that g is differentiable at 0, but g' is not continuous there.

To fill out the picture further, Exercise 3.3.10 describes a function which is differentiable at a point but is not differentiable anywhere else, and Exercises 3.3.7, 8, 11 describe functions which are continuously differentiable everywhere except at a single point, where they are not differentiable.

Exercises 3.3

1. Let $f : \mathbb{R}^2 \to \mathbb{R}$ be defind by $f(x_1, x_2) = x_1 + x_2$. Prove that, for any $\mathbf{p} \in \mathbb{R}^2$

$$\lim_{\mathbf{h} \to 0} \frac{f(\mathbf{p} + \mathbf{h}) - f(\mathbf{p})}{\|\mathbf{h}\|}$$

    does not exist. This illustrates the difficulty of attempting to define the differentiability of functions $f : D \subseteq \mathbb{R}^m \to \mathbb{R}, m \geqslant 2$, by generalizing Definition 2.5.1.

2. Prove that a linear function $L : \mathbb{R}^m \to \mathbb{R}$ is differentiable everywhere, and that it is equal to its own differential at all points in $\mathbb{R}^m$.

3. Prove from Definition 3.3.6 that the function $f : \mathbb{R}^2 \to \mathbb{R}$ defined by $f(x_1, x_2) = x_1^3 + x_2^3$, is differentiable everywhere. Find the linear function $L_{f, \mathbf{p}}$ that closely approximates $\delta_{f, \mathbf{p}}$ near $\mathbf{0}$.

Answer: $L_{f, \mathbf{p}}(\mathbf{h}) = 3p_1^2 h_1 + 3p_2^2 h_2$.

4. In Exercise 3 check your result against Theorem 3.3.22. Also write down the Jacobian matrix $J_{f, \mathbf{p}}$.

5. (a) Given that $f(v, s) = v^2/2s, s \neq 0$, calculate $(\partial f / \partial v)(v, s)$ and $(\partial f / \partial s)(v, s)$.

(b) Find $\partial f/\partial x$ and $\partial f/\partial y$ where $f(x, y)$ is equal to (i) $\sqrt{(x^2 + y^2)}$;
(ii) $(x - y)/(x + y)$, $x + y \neq 0$.
(c) Given that $f(x, y, z) = x^2y + y^2z + z^2x$, show that

$$\left(\frac{\partial f}{\partial x} + \frac{\partial f}{\partial y} + \frac{\partial f}{\partial z}\right)(x, y, z) = (x + y + z)^2$$

and that

$$x\frac{\partial f}{\partial x} + y\frac{\partial f}{\partial y} + z\frac{\partial f}{\partial z} = 3f.$$

6. (a) Given that $r(x, y) = \sqrt{(x^2 + y^2)}$, find $(\partial r/\partial x)(x, y)$.
(b) Given that $x(r, \theta) = r\cos\theta$, find $(\partial x/\partial r)(r, \theta)$.
Interpret these results geometrically with reference to Cartesian coordinates x, y and polar coordinates r, θ of points in $\mathbb{R}^2$.

7. Consider the function $f : \mathbb{R}^2 \to \mathbb{R}$ given by

$$f(x, y) = \begin{cases} \dfrac{2xy}{x^2 + y^2} & \text{when } (x, y) \neq (0, 0), \\ 0 & \text{when } (x, y) = (0, 0). \end{cases}$$

(a) Prove, using Definition 3.3.17, that $(\partial f/\partial x)(0, 0) = (\partial f/\partial y)(0, 0) = 0$. In Exercise 3.2.7 we saw that f is discontinuous at $(0, 0)$. Thus a function can have a discontinuity at a point where all its partial derivatives exist.
(b) Show that if $(x, y) \neq (0, 0)$ then

$$\frac{\partial f}{\partial x}(x, y) = \frac{2y(y^2 - x^2)}{(x^2 + y^2)^2},$$

and calculate $(\partial f/\partial y)(x, y)$, $(x, y) \neq (0, 0)$.
(c) Prove that the functions $\partial f/\partial x : \mathbb{R}^2 \to \mathbb{R}$ and $\partial f/\partial y : \mathbb{R}^2 \to \mathbb{R}$ are discontinuous at the origin.

Hint: consider the sequence $(x_k, y_k) = (1/k)(\cos\alpha, \sin\alpha)$, α fixed, as $k \to \infty$.

Observe that the discontinuity of at least one of $\partial f/\partial x$ and $\partial f/\partial y$ at the origin follows from the discontinuity of f. For if both were continuous, then, by Theorems 3.3.29 and 3.3.11, f would have to be continuous at the origin.
(d) Prove by the method of Example 3.3.27 that f is not differentiable at the origin.

8. We have seen (Exercise 3.2.8) that the functions $f : \mathbb{R}^2 \to \mathbb{R}$ and $g : \mathbb{R}^2 \to \mathbb{R}$ given by

$$f(x, y) = \begin{cases} \dfrac{x^2y}{x^2 + y^2} & \text{when } (x, y) \neq (0, 0) \\ 0 & \text{when } (x, y) = (0, 0) \end{cases}$$

and

$$g(x, y) = \begin{cases} \dfrac{2xy}{\sqrt{(x^2 + y^2)}} & \text{when } (x, y) \neq (0, 0) \\ 0 & \text{when } (x, y) = (0, 0) \end{cases}$$

are continuous on $\mathbb{R}^2$.

Prove for each of f and g that the partial derivative (a) exist on $\mathbb{R}^2$, (b) take the value 0 at $(0, 0)$, and (c) are discontinuous at the origin and continuous elsewhere. Show also that neither f nor g is differentiable at $(0, 0)$.

Hint: if f were differentiable at $(0, 0)$, formula 3.3.23 would give

$$f(\mathbf{h}) = \|\mathbf{h}\| \, \eta(\mathbf{h}), \qquad \text{where } \lim_{\mathbf{h} \to \mathbf{0}} \eta(\mathbf{h}) = 0.$$

Prove this false by considering the sequence $(\mathbf{h}_k)$, where $\mathbf{h}_k = (1/k, 1/k)$, $k \in \mathbb{N}$. How does the sequence $(f(\mathbf{h}_k)/\|\mathbf{h}_k\|)$ behave?

9. Complete Example 3.3.36.

Hint: calculate $g'(x)$, $x \neq 0$, by standard techniques. Using the definition of the derivative, calculate

$$g'(0) = \lim_{h \to 0} (g(h) - g(0))/h.$$

Finally prove that g' is not continuous at 0.

10. Consider the function $f : \mathbb{R} \to \mathbb{R}$ defined by

$$f(x) = \begin{cases} x^2 & \text{when } x \text{ is rational,} \\ 0 & \text{when } x \text{ is irrational.} \end{cases}$$

Prove that f is differentiable at 0, but not differentiable anywhere else.

Hint: for the last part show that f is discontinuous at $x \neq 0$ and apply Theorem 3.3.11.

11. Prove that the function $f : \mathbb{R} \to \mathbb{R}$ defined by

$$f(x) = \begin{cases} x \sin(1/x) & \text{when } x \neq 0, \\ 0 & \text{when } x = 0 \end{cases}$$

is continuously differentiable everywhere except at 0, where it is not differentiable. Is it continuous at 0?

12. A function $f : D \subseteq \mathbb{R}^m \to \mathbb{R}$ may or may not have the following properties at a point $\mathbf{p} \in D$: (a) continuity, (b) existence of partial derivatives, (c) continuity of partial derivatives (if they exist), (d) differentiability. List the combinations that can arise and try to find examples in the text or the exercises that illustrate them. Fill in the missing possibilities, if any.

13. Given that $f : D \subseteq \mathbb{R}^m \to \mathbb{R}$ and $g : D \subseteq \mathbb{R}^m \to \mathbb{R}$ are differentiable at $\mathbf{p} \in D$, prove that the functions $f + g$ and fg are differentiable at $\mathbf{p}$.

Show also that the corresponding Jacobian matrices are

$$J_{f+g, \mathbf{p}} = J_{f, \mathbf{p}} + J_{g, \mathbf{p}}$$

and

$$J_{fg, \mathbf{p}} = f(\mathbf{p}) J_{g, \mathbf{p}} + J_{f, \mathbf{p}} g(\mathbf{p}).$$

Hint for the case fg: consider

$$
\begin{aligned}
(fg)(\mathbf{p} + \mathbf{h}) - (fg)(\mathbf{p}) &= f(\mathbf{p} + \mathbf{h})g(\mathbf{p} + \mathbf{h}) - f(\mathbf{p})g(\mathbf{p}) \\
&= (f(\mathbf{p} + \mathbf{h}) - f(\mathbf{p}))g(\mathbf{p} + \mathbf{h}) \\
&\quad + f(\mathbf{p})(g(\mathbf{p} + \mathbf{h}) - g(\mathbf{p})),
\end{aligned}
$$

and apply Definition 3.3.6 and Theorem 3.3.11.

14. Verify the Jacobian matrix formulae of Exercise 13 for the differentiable functions $f : \mathbb{R}^2 \to \mathbb{R}$ and $g : \mathbb{R}^2 \to \mathbb{R}$ given by $f(x_1, x_2) = x_1^2 + x_2^2$, $g(x_1, x_2) = \sin(x_1 x_2)$.

Hint: use Theorem 3.3.20(iii). Note that from elementary calculus

$$\frac{\partial(fg)}{\partial x_i} = f\frac{\partial g}{\partial x_i} + \frac{\partial f}{\partial x_i}g.$$

3.4   Differentiability and tangent planes

The differentiability of a function $\phi : \mathbb{R} \to \mathbb{R}$ is equivalent to the existence of a tangent line to the graph of ϕ at every point. In this section we consider the corresponding property for a function $f : D \subseteq \mathbb{R}^2 \to \mathbb{R}$ and then give a generalization to functions defined on higher-dimensional domains.

Suppose that $f : D \subseteq \mathbb{R}^2 \to \mathbb{R}$ is differentiable at $\mathbf{p} \in D$ with differential $L_{f, \mathbf{p}}$. Then there is a function $\eta : D_\mathbf{p} \subseteq \mathbb{R}^2 \to \mathbb{R}$ such that for all $\mathbf{x} \in D$,

3.4.1 $$f(\mathbf{x}) = f(\mathbf{p}) + L_{f, \mathbf{p}}(\mathbf{x} - \mathbf{p}) + \|\mathbf{x} - \mathbf{p}\| \eta(\mathbf{x} - \mathbf{p}),$$

$$\text{where } \lim_{\mathbf{x} \to \mathbf{p}} \eta(\mathbf{x} - \mathbf{p}) = 0.$$

See 3.3.7, replacing $\mathbf{p} + \mathbf{h}$ by $\mathbf{x}$.

The graph of f is the set G in $\mathbb{R}^3$ with equation $z = f(\mathbf{x})$. Consideration of 3.4.1 suggests that we compare G with the set T in $\mathbb{R}^3$ with equation

3.4.2 $$z = f(\mathbf{p}) + L_{f, \mathbf{p}}(\mathbf{x} - \mathbf{p}).$$

3.4.3 Lemma. *With the above notation, the set T defined by 3.4.2 is a plane in $\mathbb{R}^3$ through $(\mathbf{p}, f(\mathbf{p}))$ parallel to the graph S of $L_{f,\mathbf{p}}$.*

Proof. We know from Example 3.1.13 that S, the graph of $L_{f,\mathbf{p}}$, is a plane in $\mathbb{R}^3$. Now

$$T = \{(\mathbf{x}, f(\mathbf{p}) + L_{f,\mathbf{p}}(\mathbf{x} - \mathbf{p})) \in \mathbb{R}^3 \mid \mathbf{x} \in \mathbb{R}^2\}$$
$$= (\mathbf{p}, f(\mathbf{p})) + \{(\mathbf{x} - \mathbf{p}, L_{f,\mathbf{p}}(\mathbf{x} - \mathbf{p})) \in \mathbb{R}^3 \mid \mathbf{x} \in \mathbb{R}^2\}.$$

Hence $T = (\mathbf{p}, f(\mathbf{p})) + S$ as required.

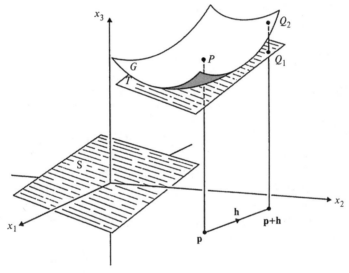

Fig. 3.13

The situation is illustrated in Fig. 3.13, where P is the point $(\mathbf{p}, f(\mathbf{p}))$.

If we move away from $\mathbf{p}$ to $\mathbf{p} + \mathbf{h}$ in $\mathbb{R}^2$, then the corresponding point Q_1 on the *plane T* has third coordinate

3.4.4 $$f(\mathbf{p}) + L_{f,\mathbf{p}}(\mathbf{h}).$$

On the other hand, the corresponding point Q_2 on the *graph G* has third coordinate

$$f(\mathbf{p} + \mathbf{h}) = f(\mathbf{p}) + L_{f,\mathbf{p}}(\mathbf{h}) + \|\mathbf{h}\|\,\eta(\mathbf{h}).$$

The vector $\overrightarrow{Q_1 Q_2}$ indicates the error we make in assuming that the change in the value of f is linear as we move away from $\mathbf{p}$. The differentiability of f at $\mathbf{p}$ tells us that

$$\overrightarrow{Q_1Q_2} = (\mathbf{0}, \|\mathbf{h}\|\,\eta(\mathbf{h}))$$

and so this error is 'small' in the sense that

3.4.5
$$\lim_{\mathbf{h} \to \mathbf{0}} \frac{\|\overrightarrow{Q_1Q_2}\|}{\|\mathbf{h}\|} = 0.$$

3.4.6 Definition.   Let $G \subseteq \mathbb{R}^3$ be the graph of a differentiable function $f : D \subseteq \mathbb{R}^2 \to \mathbb{R}$. For any point $\mathbf{p} \in D$, the plane in $\mathbb{R}^3$ with equation

$$z = f(\mathbf{p}) + L_{f,\mathbf{p}}(\mathbf{x} - \mathbf{p}), \qquad \mathbf{x} = (x, y) \in \mathbb{R}^2,$$

is called the tangent plane to G at $(\mathbf{p}, f(\mathbf{p}))$.

The tangent plane to G at $(\mathbf{p}, f(\mathbf{p}))$ is the only plane which provides a close fit to G near $(\mathbf{p}, f(\mathbf{p}))$ in the sense that 3.4.5 is satisfied.

Conversely, for an arbitrary function f, if there is a plane such that the corresponding points Q_1 and Q_2 as in Fig. 3.13 satisfy condition 3.4.5, then f is differentiable at $\mathbf{p}$. The linear approximation $L_{f,\mathbf{p}}$ may be deduced from the equation of the plane by referring to 3.4.2.

By Theorem 3.3.20(ii) the equation of the tangent plane to the graph of f at $(a, b, f(a, b))$ is

3.4.7
$$z = f(a, b) + (x - a)\frac{\partial f}{\partial x}(a, b) + (y - b)\frac{\partial f}{\partial y}(a, b)$$

3.4.8 Example.   The function $f(x, y) = x^2 + y^2$ has graph with equation $z = x^2 + y^2$. There is a sketch of the graph in Fig. 3.4(ii). We have

$$\frac{\partial f}{\partial x}(x, y) = 2x \quad \text{and} \quad \frac{\partial f}{\partial y}(x, y) = 2y.$$

The equation of the tangent plane to the graph at $(a, b, a^2 + b^2)$ is, by 3.4.7,

$$z = (a^2 + b^2) + (x - a)2a + (y - b)2b,$$

that is,

$$z = 2ax + 2by - (a^2 + b^2).$$

In particular the tangent plane at $(0, 0, 0)$ is $z = 0$, as we would expect on glancing at Fig. 3.14.

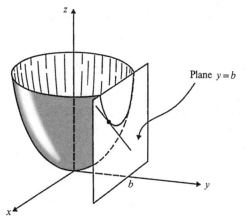

Fig. 3.14 The graph $z = x^2 + y^2$

The partial derivatives have a geometrical interpretation. In the above example we calculate $(\partial f/\partial x)(a, b)$ by holding y constant at b. This means that we limit ourselves to the parabola $z = x^2 + b^2$ which is the cross section of the graph of f sliced by the plane $y = b$ (see Fig. 3.14). The equation $(\partial f/\partial x)(x, b) = 2x$ is a statement about the slope of this parabola at the point $(x, b, x^2 + b^2)$.

A sketch of the graph of a function can reveal points at which the partial derivatives do not exist. Consider for example the function f defined in Example 3.1.8. Its graph is sketched in Fig. 3.5. The fact that there are corners at the appropriate places tells us that $\partial f/\partial x$ does not exist at all points $(0, b)$, (b, b) and $(-b, b)$ in the domain of f, except at $(0, 0)$, where $(\partial f/\partial x)(0, 0) = 0$. A similar remark applies to $\partial f/\partial y$. (Note that $\partial f/\partial y$ exists at $(0, b)$ but not at $(a, 0)$, $a \neq 0$.)

According to Theorem 3.3.20, a function is not differentiable at points where a partial derivative does not exist. We have seen (Example 3.3.27) that the function f is not differentiable at $(0, 0)$, and this together with the above considerations shows that f is not differentiable anywhere along the lines $x = 0$, $y = 0$, $x = y$ and $x = -y$. Intuitively, the graph of f is too 'sharp' at such points for a tangent plane to fit closely to it.

The idea of the tangent plane to a graph in $\mathbb{R}^3$ is generalized as follows.

3.4.9 Definition. *Let $G \subseteq \mathbb{R}^m$ be the graph of a differentiable function*

$f : D \subseteq \mathbb{R}^{m-1} \to \mathbb{R}$. *For any point* $\mathbf{p} \in D$, *the set* $T \in \mathbb{R}^m$ *with equation*

$$x_m = f(\mathbf{p}) + L_{f,\mathbf{p}}(\mathbf{x} - \mathbf{p}), \qquad \mathbf{x} = (x_1, \ldots, x_{m-1}) \in \mathbb{R}^{m-1},$$

is called the tangent space *to G at* $(\mathbf{p}, f(\mathbf{p}))$.

The tangent space T is the translation through $(\mathbf{p}, f(\mathbf{p}))$ of the $(m-1)$-dimensional subspace of $\mathbb{R}^m$ which forms the graph of $L_{f,\mathbf{p}} : \mathbb{R}^{m-1} \to \mathbb{R}$. The tangent space has equation

3.4.10 $x_m = f(\mathbf{p}) + (x_1 - p_1)\dfrac{\partial f}{\partial x_1}(\mathbf{p}) + \cdots + (x_{m-1} - p_{m-1})\dfrac{\partial f}{\partial x_{m-1}}(\mathbf{p})$.

3.4.11 *Example.* The differentiable function $f : \mathbb{R}^3 \to \mathbb{R}$ defined by

$$f(x_1, x_2, x_3) = x_1^2 + x_2^2 + x_3^2$$

has graph $x_4 = x_1^2 + x_2^2 + x_3^2$ in $\mathbb{R}^4$. The tangent space to the graph at

$$(p_1, p_2, p_3, p_1^2 + p_2^2 + p_3^2)$$

is

$$\begin{aligned} x_4 &= p_1^2 + p_2^2 + p_3^2 + 2p_1(x_1 - p_1) + 2p_2(x_2 - p_2) + 2p_3(x_3 - p_3) \\ &= p_1 x_1 + p_2 x_2 + p_3 x_3 - (p_1^2 + p_2^2 + p_3^2). \end{aligned}$$

Exercises 3.4

1. Define $f : \mathbb{R}^2 \to \mathbb{R}$ by $f(x, y) = x^2 - y^2$. Show that the equation of the tangent plane to the graph G of f at $(a, b, a^2 - b^2)$ is

$$z = (a^2 - b^2) + 2a(x - a) - 2b(y - b).$$

Sketch the graph G of f (Exercise 3.1.1(i)); its equation is $z = x^2 - y^2$. Fit the tangent planes for various choices of (a, b).

2. Sketch the graphs $z = f(x, y)$ and tangent planes at $(a, b, f(a, b))$ for various choices of (a, b) when
    (a) $f(x, y) = x^2 - y$,
    (b) $f(x, y) = x + 3y$,
    (c) $f(x, y) = xy$.
    Consider in particular the cases $(a, b) = (0, 0)$ and $(a, b) = (1, 1)$.

3. In your sketches for Exercise 2 interpret geometrically the partial derivatives $(\partial f / \partial x)(a, b)$ as discussed in the text.

4. Given a function $f : D \subseteq \mathbb{R} \to \mathbb{R}$, the equation $z = f(x)$ can be interpreted as the equation of the graph G of f in the x, z plane. Show that the equation

$$z = f(a) + (x - a)\frac{df}{dx}(a)$$

is the equation of the tangent line to G at $(a, f(a))$. Compare with Equation 3.4.7.

5. Consider the function $f : \mathbb{R}^2 \to \mathbb{R}$ of Exercise 3.3.8 given by

$$f(x, y) = \begin{cases} \dfrac{x^2 y}{x^2 + y^2} & \text{when } (x, y) \neq (0, 0) \\ 0 & \text{when } (x, y) = (0, 0). \end{cases}$$

Show that for any constant α the straight line through the origin parametrized by

$$(t\cos\alpha, t\sin\alpha, t\cos^2\alpha\sin\alpha), \qquad t \in \mathbb{R},$$

lies entirely in the graph G of f with equation $z = f(x, y)$. Hence a three-dimensional model of G can be constructed using thin straight rods (straws, for example). The 'wavy' nature of G near the origin suggests that no tangent plane can be fitted to G at the origin and hence that f is not differentiable at $(0, 0)$.

Does a similar analysis apply to the function g of Exercise 3.3.8?

3.5   Error estimation

In this section we discuss briefly a practical application of the linear approximation of differentiable functions to the estimation of how errors in measurements affect calculations based on those measurements.

Suppose that a physical quantity y is related to measurable variables $x_1, \ldots, x_m$ by the formula $y = f(x_1, \ldots, x_m)$. In an experiment to determine y, the recorded values of the variables form a vector $\mathbf{x} = (x_1, \ldots, x_m)$. These readings are however inaccurate, and the true values form the vector

$$\mathbf{x} + \Delta\mathbf{x} = (x_1 + \Delta x_1, \ldots, x_m + \Delta x_m).$$

The experimenter calculates the value of y as $f(\mathbf{x})$, whereas the true value is $f(\mathbf{x} + \Delta\mathbf{x})$. The resulting error Δy in the value of y is given by

$$\Delta y = f(\mathbf{x} + \Delta\mathbf{x}) - f(\mathbf{x}).$$

Suppose that f is differentiable at $\mathbf{x}$. Then, by 3.3.23, we obtain the approximation

3.5.1 $$\Delta y \approx \frac{\partial f}{\partial x_1}(\mathbf{x})\Delta x_1 + \cdots + \frac{\partial f}{\partial x_m}(\mathbf{x})\Delta x_m,$$

where the symbol $\approx$ indicates that the difference between the two sides is of the form $\|\Delta\mathbf{x}\|\eta(\Delta\mathbf{x})$, where $\eta(\Delta\mathbf{x}) \to 0$ as $\Delta\mathbf{x} \to \mathbf{0}$.

In a practical situation the values of $(\partial f/\partial x_1)(\mathbf{x}), \ldots, (\partial f/\partial x_m)(\mathbf{x})$ will be known, as will upper bounds for $|\Delta x_1|, \ldots, |\Delta x_m|$. We can therefore find an approximate upper bound for Δy by choosing $\Delta x_1, \ldots, \Delta x_m$ appropriately in 3.5.1.

Using classical (Leibniz) terminology, if $\Delta x_1, \ldots, \Delta x_m$ are the (infinitesimal) 'differentials' $\mathrm{d}x_1, \ldots, \mathrm{d}x_m$ respectively, then the approximation in 3.5.1 gives us an equation involving the (infinitesimal) 'differential' $\mathrm{d}y$, where

3.5.2
$$\mathrm{d}y = \frac{\partial f}{\partial x_1}\mathrm{d}x_1 + \cdots + \frac{\partial f}{\partial x_m}\mathrm{d}x_m.$$

This useful expression can be made respectable if $\mathrm{d}y, \mathrm{d}x_1, \ldots, \mathrm{d}x_m$ are defined as differentials in our sense (as linear approximations), but for the purpose of this section it is best to regard 3.5.2 as a mnemonic.

3.5.3 *Example.*   The (constant) acceleration a metres/(second)^2 of a vehicle is calculated from the formula

3.5.4
$$a = \frac{v^2}{2s}$$

where v metres/second is the speed of the vehicle when it has covered a distance s metres from the start of its journey. The speed is observed as $30.00\,\mathrm{m\,s^{-1}}$ at a point $300.0\,\mathrm{m}$ from the start. Given that the speed measurement is accurate to within $\pm 1.00\,\mathrm{m\,s^{-1}}$ and the distance measurement is accurate to within $\pm 3.0\,\mathrm{m}$, estimate the accuracy of the calculation of acceleration.

Solution.   Formula 3.5.4 defines a differentiable function $f : D \subseteq \mathbb{R}^2 \to \mathbb{R}$ given by

3.5.5
$$a = f(v, s) = \frac{v^2}{2s}$$

where $D = \{(v, s) \in \mathbb{R}^2 \mid s \neq 0\}$. By 3.5.1

3.5.6
$$\Delta a \approx \frac{v}{s}\Delta v - \frac{v^2}{2s^2}\Delta s.$$

Hence, when $v = 30$ and $s = 300$,

3.5.7
$$\Delta a \approx \frac{1}{10}\Delta v - \frac{1}{200}\Delta s.$$

Now the bounds of accuracy are $|\Delta v| \leqslant 1$ and $|\Delta s| \leqslant 3$. The right-hand side of 3.5.6 attains its maximum when $\Delta v = 1$ and $\Delta s = -3$ and its minimum when $\Delta v = -1$ and $\Delta s = 3$. Hence we have the approximation

3.5.8 $$-0.115 \leqslant \Delta a \leqslant 0.115.$$

A direct calculation from the formula

$$\Delta a = f(v + \Delta v, s + \Delta s) - f(v, s)$$

at the point $(v, s) = (30, 300)$ leads to the true bounds for the error in the acceleration

3.5.9 $$-0.112 \leqslant \Delta a \leqslant 0.118.$$

Exercises 3.5

1. A rectangular box has a square base. An edge of the base is measured to be 10 cm to an accuracy of 0.1 cm. The height is measured to be 20 cm to an accuracy of 0.1 cm. Using the formula volume = base area × height, the volume V of the box is calculated to be $100 \times 20 = 2000 \, \text{cm}^3$. Estimate the accuracy of the calculation by using Formula 3.5.1.

Answer: $|\Delta V| \leqslant 50$.

Compare with the accurate bounds for the error.

Answer: $-49.601 \leqslant \Delta V \leqslant 50.401$.

2. If the error in a measurement $x \neq 0$ is Δx, the *relative error* is defined to be $\rho(x) = \Delta x / x$. For example, given a measurement 10 with error 0.2, the relative error is 0.02 or 2%. Prove that (a) $\rho(xy) \approx \rho(x) + \rho(y)$, (b) $\rho(x/y) \approx \rho(x) - \rho(y)$.

3. The radius r of a cylindrical tube is found from the formula $\pi r^2 l = V$. Given that the relative errors in the measurements of l and V are 0.2% and 0.3% respectively, estimate the maximum relative error in the radius calculation.

Hint: prove that $2\rho(r) \approx \rho(V) - \rho(l)$ in the notation of Exercise 2.

Answer: $|\rho(r)| \leqslant 0.25\%$.

3.6 The Chain Rule. Rate of change along a path

The reader will be familiar with the rules for differentiating the product and quotient of two differentiable real-valued functions of $\mathbb{R}$.

Namely

3.6.1   if $F(t) = x(t)y(t)$, then $F'(t) = x'(t)y(t) + x(t)y'(t)$, and

3.6.2   if $F(t) = x(t)/y(t)$, then $F'(t) = (x'(t)y(t) - x(t)y'(t))/y^2(t)$.

More generally, given m real-valued functions $x_1(t), \ldots, x_m(t)$ and given a function $f : \mathbb{R}^m \to \mathbb{R}$, one could ask about the differentiability of the function $F : \mathbb{R} \to \mathbb{R}$ defined by

3.6.3 $$F(t) = f(x_1(t), \ldots, x_m(t)).$$

If we define $g : \mathbb{R} \to \mathbb{R}^m$ by $g(t) = (x_1(t), \ldots, x_m(t))$, then we can interpret 3.6.3 as defining F as the composition $f \circ g$ of the two functions f and g.

In particular, if g is a path in $\mathbb{R}^m$, then $g(t)$ moves along the image of the path, and the derivative of $f \circ g$, if it exists, defines the rate of change of the scalar field f along the path g. For example, if the image of the path lies in a level set of f, then the rate of change along the path is 0.

The following theorem shows how the derivative of $f \circ g$ is related to those of f and g.

3.6.4 Theorem. *Chain Rule. Let $g : E \subseteq \mathbb{R} \to \mathbb{R}^m$ be defined on an open interval E and let $f : D \subseteq \mathbb{R}^m \to \mathbb{R}$ be defined on an open set D such that $g(E) \subseteq D$. Define $F : E \subseteq \mathbb{R} \to \mathbb{R}$ to be the composite function given by*

$$F(t) = (f \circ g)(t) = f(g(t)), \qquad t \in E.$$

Suppose that g is differentiable at $a \in E$ and that f is differentiable at $g(a) \in D$. Then F is differentiable at a and its differential is given by

3.6.5 $$L_{F,a} = L_{f,g(a)} \circ L_{g,a}.$$

The theorem will be proved in a more general situation in Section 4.4. Notice however that

$$\delta_{f \circ g, a}(h) = (f \circ g)(a+h) - (f \circ g)(a) = f(g(a+h)) - f(g(a))$$
$$= \delta_{f,g(a)}(g(a+h) - g(a)) = \delta_{f,g(a)}(\delta_{g,a}(h)).$$

Hence

3.6.6 $$\delta_{f \circ g, a} = \delta_{f,g(a)} \circ \delta_{g,a}.$$

The identity 3.6.5 will follow from the fact (Lemma 4.4.4) that the composition of the close linear approximations near 0 to $\delta_{f,g(a)}$ and $\delta_{g,a}$ is a close linear approximation near 0 to their composition.

3.6.7 Corollary. *Let* $g : E \subseteq \mathbb{R} \to \mathbb{R}^m$ *be differentiable at* $a \in E$, *let* $g(E) \subseteq D$ *and let* $f : D \subseteq \mathbb{R}^m \to \mathbb{R}$ *be differentiable at* $g(a)$. *Then*

3.6.8 $$J_{f \circ g, a} = J_{f, g(a)} J_{g, a}.$$

Alternatively, putting $g(t) = (x_1(t), \ldots, x_m(t))$ *and* $F(t) = f(g(t)) = f(x_1(t), \ldots, x_m(t))$,

3.6.9 $$\frac{dF}{dt} = \frac{\partial f}{\partial x_1} \frac{dx_1}{dt} + \cdots + \frac{\partial f}{\partial x_m} \frac{dx_m}{dt},$$

where dF/dt, $dx_1/dt, \ldots, dx_m/dt$ *are evaluated at* a *and* $\partial f/\partial x_1, \ldots, \partial f/\partial x_m$ *are evaluated at* $g(a)$.

Proof. The identity 3.6.5 between differentials leads to the identity 3.6.8 between representing matrices (Jacobians).
If we put $g(t) = (x_1(t), \ldots, x_m(t))$ and let $g(a) = \mathbf{p}$, then by Theorem 3.3.20 and 2.10.18,

$J_{F, a}$ is the 1×1 matrix $[F'(a)]$,

$J_{f, g(a)}$ is the $1 \times m$ matrix $\left[\dfrac{\partial f}{\partial x_1}(\mathbf{p}) \quad \cdots \quad \dfrac{\partial f}{\partial x_m}(\mathbf{p}) \right]$

$J_{g, a}$ is the $m \times 1$ matrix $\begin{bmatrix} g_1'(a) \\ \vdots \\ g_m'(a) \end{bmatrix} = \begin{bmatrix} \dfrac{dx_1}{dt}(a) \\ \vdots \\ \dfrac{dx_m}{dt}(a) \end{bmatrix}$

Expression 3.6.9 now follows from 3.6.8.

The reader should notice the relationship between 3.6.9 and the formal expression 3.5.2 used in error estimation.

3.6.10 Example. Consider $f : D \subseteq \mathbb{R}^2 \to \mathbb{R}$ defined by $f(x, y) = x/y$, where $D = \{(x, y) \subseteq \mathbb{R}^2 | y \neq 0\}$. Then $(\partial f/\partial x)(x, y) = 1/y$ and $(\partial f/\partial y)(x, y) = -x/y^2$. Application of 3.6.9 now gives the rule for differentiating the quotient $x(t)/y(t)$ contained in 3.6.2. Similarly, consideration of $f(x, y) = xy$ leads to 3.6.1.

3.6.11 Example. If $F(t) = x(t)\sin(y(t)z(t))$ then, by 3.6.9,

$$F'(t) = x'(t)\sin(y(t)z(t)) + y'(t)x(t)z(t)\cos(y(t)z(t))$$
$$+ z'(t)x(t)y(t)\cos(y(t)z(t)).$$

3.6.12 Example.   Consider the function $g :]-\pi/2, \pi/2[\to \mathbb{R}^2$ defined by

$$g(t) = (\tan t, \cos^2 t), \qquad t \in]-\pi/2, \pi/2[.$$

The image of g is known as the Witch of Agnesi (first studied by
Fermat and later by Maria Agnesi). See Fig. 3.15.

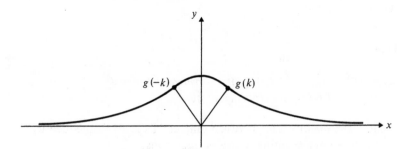

Fig. 3.15 $g(t) = (\tan t, \cos^2 t), \ -\pi/2 < t < \pi/2$

We can find the points at which the curve is closest to the origin as
follows. Define a function $f : \mathbb{R}^2 \to \mathbb{R}$ by

$$f(x, y) = (x^2 + y^2)^{1/2} = \|(x, y)\|.$$

The composite function $F = f \circ g$ has derivative (by 3.6.9)

$$F'(t) = x(x^2 + y^2)^{-1/2} \sec^2 t - 2y(x^2 + y^2)^{-1/2} \sin t \cos t,$$

where $x = \tan t$ and $y = \cos^2 t$. It follows that

$$F'(t) = (x^2 + y^2)^{-1/2} \sin t (\sec^3 t - 2\cos^3 t).$$

In particular $F'(t) = 0$ when $t = 0$ and at the two values k and $-k$ in
$]-\pi/2, \pi/2[$ at which $\cos^2 k = 2^{-1/3}$. It will be found that $F(0) = 1$, and
$F(\pm k) = 0.943$. Therefore $g(k)$ and $g(-k)$ are the required points on the
curve.

Exercises 3.6

1.   Given that $f(x, y) = \tan^{-1}(y/x), x \neq 0$ and $g(t) = (x(t), y(t))$, where
     $x(t) = \cos t, y(t) = \sin t$, apply the Chain Rule in the form 3.6.9 to find
     $F'(a)$, where $F = f \circ g$. Interpret your result geometrically.

Answer: $F'(a) = 1$ for all a.

     Obtain the Jacobian matrix expression 3.6.8.

2.   (a) Using the Chain Rule in the form 3.6.9 in each of the following
     cases, calculate $(f \circ g)'(t)$:
     (i) $f(x_1, x_2) = x_1 x_2, g(t) = (\sin t, \cos t)$;

(ii) $f(x, y) = x^2y + y^3$, $g(t) = (t + 1, e^t)$.

(b) If $f(x_1, x_2, x_3) = x_1^2x_2 + x_2^2x_3 + x_3^2x_1$ and $g(t) = (ta, tb, tc)$, where a, b, c are constants, apply the Chain Rule 3.6.9 to calculate $(f \circ g)'(t)$. Show also that

$$a\frac{\partial f}{\partial x_1}(a, b, c) + b\frac{\partial f}{\partial x_2}(a, b, c) + c\frac{\partial f}{\partial x_3}(a, b, c) = 3f(a, b, c).$$

Answer: $(f \circ g)'(t) = 3t^3(a^2b + b^2c + c^2a)$.

3. A function $f : D \subseteq \mathbb{R}^m \to \mathbb{R}$ is said to be *homogeneous* of degree k if, whenever $\mathbf{x} \in D$ and $t\mathbf{x} \in D$, $t > 0$, $f(t\mathbf{x}) = t^k f(\mathbf{x})$. For example, linear functions are homogeneous of degree 1, the function f of Exercise 2(b) is homogeneous of degree 3 and the functions given by

$$f(x, y, z) = \frac{z}{\sqrt{(x^2 + y^2)}}, \qquad x^2 + y^2 \neq 0,$$

and

$$f(x, y, z) = \frac{1}{\sqrt{(x^2 + y^2 + z^2)}}, \qquad x^2 + y^2 + z^2 \neq 0,$$

are homogeneous of degrees 0 and -1 respectively.

Prove *Euler's Theorem*, that a differentiable homogeneous function $f : D \subseteq \mathbb{R}^m \to \mathbb{R}$ of degree k satisfies the identity

$$p_1\frac{\partial f}{\partial x_1}(\mathbf{p}) + \cdots + p_m\frac{\partial f}{\partial x_m}(\mathbf{p}) = kf(\mathbf{p}), \qquad \mathbf{p} \in D.$$

Hint: use the method of Exercise 2(b).

4. Let $f : \mathbb{R}^2 \to \mathbb{R}$ be defined by

$$f(x, y) = \begin{cases} \dfrac{x^2y}{x^2 + y^2} & \text{when } (x, y) \neq (0, 0) \\ 0 & \text{when } (x, y) = (0, 0). \end{cases}$$

In Exercise 3.3.8 we saw that f is not differentiable at $(0, 0)$ but that its partial derivatives exist there. Check that $(\partial f/\partial x)(0, 0) = (\partial f/\partial y)(0, 0) = 0$.

Now define $g : \mathbb{R} \to \mathbb{R}^2$ by $g(t) = (x(t), y(t))$, where $x(t) = t$, $y(t) = t$, $t \in \mathbb{R}$. Given that $F = f \circ g$,

(a) prove that F is differentiable at $(0, 0)$ and show that $F'(0) = 1$.

(b) verify that

$$\frac{\partial f}{\partial x}(0, 0)\frac{dx}{dt}(0) + \frac{\partial f}{\partial y}(0, 0)\frac{dy}{dt}(0) = 0.$$

This example illustrates the breakdown of the Chain Rule 3.6.9 at a point where the function f is not differentiable.

5. The temperature in a region of space is given by the formula

$$f(x, y, z) = kx^2(y - z),$$

where k is a positive constant. An insect flies so that its position at time t is $g(t) = (t, t, 2t)$. Find the rate of change of temperature along its path.

Hint: calculate $-T'(t)$, where $T = f \circ g$. This problem is solved most simply by determining $T(t)$ and then differentiating. Use it alternatively as an exercise on the Chain Rule.

3.7   Directional derivatives

We recall that the partial derivatives of $f : D \subseteq \mathbb{R}^m \to \mathbb{R}$ at $\mathbf{p} \in D$ are given by the formula 3.3.17

3.7.1 $$\frac{\partial f}{\partial x_j}(\mathbf{p}) = \lim_{t \to 0} \frac{f(\mathbf{p} + t\mathbf{e}_j) - f(\mathbf{p})}{t}, \qquad j = 1, \ldots, m.$$

For each $t \in \mathbb{R}$ the vector $\mathbf{p} + t\mathbf{e}_j$ lies on the straight line through $\mathbf{p}$ in the direction $\mathbf{e}_j$. We may regard the partial derivative as the rate of change of $f(\mathbf{x})$ as $\mathbf{x}$ moves through $\mathbf{p}$ along this line (the case $m = 2$ is illustrated in Fig. 3.16(i)).

The right-hand side of 3.7.1 is the derivative of $f(\mathbf{p} + t\mathbf{e}_j)$ with respect to t at $t = 0$. We have therefore an alternative expression for

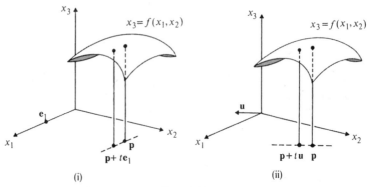

Fig. 3.16

the partial derivative

$$\frac{\partial f}{\partial x_j}(\mathbf{p}) = \frac{\mathrm{d}}{\mathrm{d}t} f(\mathbf{p} + t\mathbf{e}_j)\bigg|_{t=0}$$

From this point of view there is no reason why we should restrict ourselves to lines in directions $\mathbf{e}_1, \ldots, \mathbf{e}_m$. For any unit vector $\mathbf{u} \in \mathbb{R}^m$, the straight line through $\mathbf{p}$ in direction $\mathbf{u}$ consists of the vectors $\mathbf{p} + t\mathbf{u}$, $t \in \mathbb{R}$. Accordingly we have the following generalization of 3.7.1 (see Fig. 3.16(ii)).

3.7.2 Definition.   *Given a function* $f : D \subseteq \mathbb{R}^m \to \mathbb{R}$, *a point* $\mathbf{p} \in D$, *and a unit vector* $\mathbf{u} \in \mathbb{R}^m$, *the* directional derivative *of* f *at* $\mathbf{p}$ *in direction* $\mathbf{u}$ *is defined by*

$$\frac{\partial f}{\partial \mathbf{u}}(\mathbf{p}) = \lim_{t \to 0} \frac{f(\mathbf{p} + t\mathbf{u}) - f(\mathbf{p})}{t} = \frac{\mathrm{d}}{\mathrm{d}t} f(\mathbf{p} + t\mathbf{u})\bigg|_{t=0},$$

provided this limit exists.

Notice that the partial derivatives are directional derivatives and that

$$(\partial f/\partial x_j)(\mathbf{p}) = (\partial f/\partial \mathbf{e}_j)(\mathbf{p}), \qquad j = 1, \ldots, m.$$

The following example shows that the existence of all the partial derivatives is not enough to guarantee the existence of other directional derivatives.

3.7.3 Example.   Consider again (see Example 3.1.8) the function $f : \mathbb{R}^2 \to \mathbb{R}$ defined by

$$f(x, y) = \begin{cases} -|y| & \text{if } |y| \leqslant |x|, \\ -|x| & \text{if } |x| \leqslant |y|. \end{cases}$$

For any $t \in \mathbb{R}$ and any unit vector $\mathbf{u} \in \mathbb{R}^2$, $f(\mathbf{0} + t\mathbf{u}) = |t| f(\mathbf{u})$. Now $f(\mathbf{u})$ is constant and therefore $f(\mathbf{0} + t\mathbf{u})$ is differentiable as a function of t if and only if $f(\mathbf{u}) = 0$. This in turn is true if and only if $\mathbf{u}$ is $(\pm 1, 0)$ or $(0, \pm 1)$. The only directional derivatives at $\mathbf{0}$ which exist therefore are $(\partial f/\partial \mathbf{e}_1)(\mathbf{0}), (\partial f/\partial(-\mathbf{e}_1))(\mathbf{0}), (\partial f/\partial \mathbf{e}_2)(\mathbf{0}), (\partial f/\partial(-\mathbf{e}_2))(\mathbf{0})$. Each of these directional derivatives is 0. The sketch of the graph of f (Fig. 3.5) suggests the same result.

The following theorem shows however that if f is differentiable at $\mathbf{p} \in D$, then each directional derivative $(\partial f/\partial \mathbf{u})(\mathbf{p})$ exists and is a value of the differential. The corollary which follows it shows how the directional derivative is linearly related to the partial derivatives.

3.7.4 Theorem.   *Let* $f : D \subseteq \mathbb{R}^m \to \mathbb{R}$ *be differentiable at* $\mathbf{p} \in D$. *Then* $(\partial f / \partial \mathbf{u})(\mathbf{p})$ *exists for every unit vector* $\mathbf{u} \in \mathbb{R}^m$.
    Furthermore

[i] $\dfrac{\partial f}{\partial \mathbf{u}}(\mathbf{p}) = L_{f,\mathbf{p}}(\mathbf{u}),$

and, in the notation of Section 3.5,

[ii] $f(\mathbf{p} + t\mathbf{u}) \approx f(\mathbf{p}) + t\dfrac{\partial f}{\partial \mathbf{u}}(\mathbf{p}).$

Proof.   Since f is differentiable at $\mathbf{p}$ and since $L_{f,\mathbf{p}}$ is linear, we have

3.7.5 $f(\mathbf{p} + t\mathbf{u}) - f(\mathbf{p}) = t L_{f,\mathbf{p}}(\mathbf{u}) + |t|\eta(t\mathbf{u}),$ $t\mathbf{u} \in D_{\mathbf{p}}$

where $\lim_{t \to 0} \eta(t\mathbf{u}) = 0$. *Hence*

$$\frac{f(\mathbf{p} + t\mathbf{u}) - f(\mathbf{p})}{t} = L_{f,\mathbf{p}}(\mathbf{u}) + \frac{|t|}{t}\eta(t\mathbf{u}) \qquad t \neq 0, t\mathbf{u} \in D_{\mathbf{p}}.$$

Taking the limit as t tends to 0 on both sides of this expression establishes (i). Replacing $L_{f,\mathbf{p}}(\mathbf{u})$ by $(\partial f / \partial \mathbf{u})(\mathbf{p})$ in 3.7.5 leads to (ii).

3.7.6 Corollary.   *Let* $f : D \subseteq \mathbb{R}^m \to \mathbb{R}$ *be differentiable at* $\mathbf{p} \in D$. *Then for every unit vector* $\mathbf{u} = (u_1, \ldots, u_m) \in \mathbb{R}^m$,

$$\frac{\partial f}{\partial \mathbf{u}}(\mathbf{p}) = u_1 \frac{\partial f}{\partial x_1}(\mathbf{p}) + \cdots + u_m \frac{\partial f}{\partial x_m}(\mathbf{p}).$$

Proof.   The result follows from Theorem 3.7.4(i) by replacing $\mathbf{h}$ by $\mathbf{u}$ in Theorem 3.3.20(ii).

3.7.7 Example.   Let $f : D \subseteq \mathbb{R}^2 \to \mathbb{R}$ be defined by

$$f(x, y) = \sqrt{(9 - x^2 - y^2)}, \qquad D = \{(x, y) \in \mathbb{R}^2 \mid x^2 + y^2 < 9\},$$

and let $\mathbf{p} = (2, 1)$. The graph of f is an inverted hemispherical bowl. The partial derivatives of f are

$$\frac{\partial f}{\partial x}(x, y) = \frac{-x}{\sqrt{(9 - x^2 - y^2)}}, \qquad \frac{\partial f}{\partial y}(x, y) = \frac{-y}{\sqrt{(9 - x^2 - y^2)}}.$$

    In particular, $(\partial f / \partial x)(\mathbf{p}) = -1$ and $(\partial f / \partial y)(\mathbf{p}) = -\frac{1}{2}$. The function is differentiable at $\mathbf{p}$ since (Theorem 3.3.29) the partial derivatives are continuous there. It follows from Corollary 3.7.6 that for any unit vector $\mathbf{u} \in \mathbb{R}^2$,

3.7.8
$$\frac{\partial f}{\partial \mathbf{u}}(\mathbf{p}) = -u_1 - \frac{1}{2}u_2.$$

When u_1 and u_2 are both positive $(\partial f/\partial \mathbf{u})(\mathbf{p})$ is negative and so $f(\mathbf{x})$ decreases as we move from $\mathbf{p}$ in the direction $\mathbf{u}$.

The directional derivative in 3.7.8 is 0 when $\mathbf{u}$ is a unit vector such that $u_2 = -2u_1$. This is true if and only if $\mathbf{u} = \pm (1/\sqrt{5}, -2/\sqrt{5})$.

3.7.9 *Example*. Let $f : \mathbb{R}^2 \to \mathbb{R}$ be defined by $f(x_1, x_2) = x_1^2 - x_2^2$, and let $\mathbf{p} = (-2, 1)$. By the above argument, for any unit vector $\mathbf{u} \in \mathbb{R}^2$, $(\partial f/\partial \mathbf{u})(\mathbf{p}) = -4u_1 - 2u_2$. Thus for example when $\mathbf{u} = (2/\sqrt{5}, 1/\sqrt{5})$ we have $(\partial f/\partial \mathbf{u})(\mathbf{p}) = -2/\sqrt{5}$. Hence the estimated value of $f(\mathbf{p} + t\mathbf{u})$ using Theorem 3.7.4(ii) is $3 - 2t/\sqrt{5}$. The correct value is $3 - 2t\sqrt{5} + \frac{3}{5}t^2$.

Exercises 3.7

1. Find the directional derivative $(\partial f/\partial \mathbf{u})(\mathbf{p})$ for $\mathbf{u} = (\frac{3}{5}, \frac{4}{5})$ and $\mathbf{p} = (0, -1)$, where

$$\text{(a) } f(x, y) = \sqrt{(x^2 + y^2)}, \quad \text{(b) } f(x) = \frac{x - y}{x + y}, \quad (x + y) \neq 0.$$

Hint: apply Corollary 3.7.6.

2. The roof of an exhibition hall is constructed in the shape of a hemisphere rising from ground level. A workman is standing at a point on the roof at a height above ground level which is half the maximum height of the roof. When he faces due east the vertical axis through the centre of the building is directly behind him.

   Calculate the slope of his ascent or descent if he moves on the roof (a) in an easterly direction; (b) in a westerly direction; (c) in a south-easterly direction.

   In what direction must be set off in order to 'traverse' in a horizontal direction?

   Hint: consider the surface of the roof as the graph of the function f, where $f(x, y) = \sqrt{(1 - x^2 - y^2)}$, $x^2 + y^2 \leq 1$, and choose x and y axes in directions south and east. The workman is then located at the point $(\mathbf{p}, f(\mathbf{p})) = (0, \frac{1}{2}\sqrt{3}, \frac{1}{2})$. To solve part (a) calculate $(\partial f/\partial \mathbf{u})(\mathbf{p})$ where $\mathbf{u} = (0, 1)$.

   Answers: (a) descent at $60°$ to horizontal; (b) ascent at $60°$; (c) descent at $\tan^{-1}\sqrt{\frac{3}{2}}$. Horizontal motion due north or south.

3. Continuing Exercise 3.6.5, when the insect is at the point $(1, 1, 2)$ how should he change his course to proceed at constant temperature?

Hint: find unit vector $\mathbf{u} \in \mathbb{R}^3$ such that $(\partial f/\partial \mathbf{u})(1, 1, 2) = 0$.

Answer: any direction $\mathbf{u} = (u_1, u_2, u_3)$ such that $-u_1 + u_2 - u_3 = 0$ will do.

4. Let $f : \mathbb{R}^2 \to \mathbb{R}$ be defined by

$$f(x, y) = \begin{cases} \dfrac{x^2 y}{x^2 + y^2} & \text{when } (x, y) \neq (0, 0) \\ 0 & \text{when } (x, y) = (0, 0). \end{cases}$$

Prove from Definition 3.7.2 that, given any unit vector $\mathbf{u} = (u_1, u_2) \in \mathbb{R}^2$, $(\partial f/\partial \mathbf{u})(\mathbf{0})$ exists, and

$$\frac{\partial f}{\partial \mathbf{u}}(\mathbf{0}) = \frac{u_1^2 u_2}{u_1^2 + u_2^2}.$$

In Exercise 3.6.4 we observed that $(\partial f/\partial x)(\mathbf{0}) = (\partial f/\partial y)(\mathbf{0}) = 0$. Hence the formula for $\partial f/\partial \mathbf{u}$ of Corollary 3.7.6 breaks down at $\mathbf{p} = \mathbf{0}$. Note that f is not differentiable at $\mathbf{0}$ (Exercise 3.3.8) although all its directional derivatives exist.

Are similar conclusions true for the function $g : \mathbb{R}^2 \to \mathbb{R}$ of Exercise 3.3.8?

3.8   The gradient. Smooth surfaces

We begin this section by solving the following problems concerning the directional derivative of a function $f : D \subseteq \mathbb{R}^m \to \mathbb{R}$ that is differentiable at $\mathbf{p} \in D$.

3.8.1   *Find all unit vectors $\mathbf{u} \in \mathbb{R}^m$ such that $(\partial f/\partial \mathbf{u})(\mathbf{p}) = 0$.*

3.8.2   *Find all unit vectors $\mathbf{u} \in \mathbb{R}^m$ such that $(\partial f/\partial \mathbf{u})(\mathbf{p})$ takes maximum value.*

The solutions of these two problems follow easily from Theorem 3.7.4 with the help of the following very important concept.

3.8.3 Definition.   *Let $f : D \subseteq \mathbb{R}^m \to \mathbb{R}$ be a function that is differentiable at $\mathbf{p} \in D$. The gradient of f at $\mathbf{p}$ is defined to be the vector*

$$(\operatorname{grad} f)(\mathbf{p}) = \left(\frac{\partial f}{\partial x_1}(\mathbf{p}), \dots, \frac{\partial f}{\partial x_m}(\mathbf{p}) \right) \in \mathbb{R}^m.$$

When f is differentiable throughout D, then this expression defines a (vector-valued) function $\operatorname{grad} f : D \subseteq \mathbb{R}^m \to \mathbb{R}^m$.

3.8.4 Example. Define $f : \mathbb{R}^2 \to \mathbb{R}$ by $f(x, y) = x^2 - y^2$. Then
$$(\operatorname{grad} f)(x, y) = (2x, -2y).$$

3.8.5 Example. Define $F : \mathbb{R}^3 \to \mathbb{R}$ by $F(x, y, z) = x^2 + y^2 + z^2$. Then
$$(\operatorname{grad} F)(x, y, z) = (2x, 2y, 2z).$$

We shall consider vector-valued functions in greater detail in later chapters. For the moment we continue our study of the function $\operatorname{grad} f$ by explaining its connection with differentials, directional derivatives and the Chain Rule.

3.8.6 Theorem. *Let $f : D \subseteq \mathbb{R}^m \to \mathbb{R}$ be differentiable at $\mathbf{p} \in D$. Then*
[i] *the differential $L_{f, \mathbf{p}}$ is related to the gradient through the scalar product*
$$L_{f, \mathbf{p}}(\mathbf{h}) = \mathbf{h} \cdot (\operatorname{grad} f)(\mathbf{p}). \qquad \mathbf{h} \in \mathbb{R}^m;$$

[ii] *for a given unit vector $\mathbf{u} \in \mathbb{R}^m$, the directional derivative $(\partial f / \partial \mathbf{u})(\mathbf{p})$ is expressible as the scalar product*
$$\frac{\partial f}{\partial \mathbf{u}}(\mathbf{p}) = \mathbf{u} \cdot (\operatorname{grad} f)(\mathbf{p}).$$

Proof. (i) Theorem 3.3.20(ii); (ii) Corollary 3.7.6.

3.8.7 Theorem (*Chain Rule*). *Let $g : E \subseteq \mathbb{R} \to \mathbb{R}^m$ be defined on an open interval E, let $f : D \subseteq \mathbb{R}^m \to \mathbb{R}$ be defined on an open set D such that $g(E) \subseteq D$, and let $F = f \circ g$.*

Suppose that g is differentiable at $a \in E$ and that f is differentiable at $g(a) \in D$. Then F is differentiable at a and
$$F'(a) = (f \circ g)'(a) = (\operatorname{grad} f)(g(a)) \cdot g'(a).$$

Proof. Put $g(t) = (x_1(t), \dots, x_m(t))$ and apply 3.6.9.

There is a striking similarity between this form of the Chain Rule and the Chain Rule of elementary calculus (see 1.6.11).

We can now solve the problems raised in 3.8.1 and 3.8.2. The solution is trivial when $(\operatorname{grad} f)(\mathbf{p}) = \mathbf{0}$, for then $(\partial f / \partial \mathbf{u})(\mathbf{p}) = 0$ for all unit vectors $\mathbf{u} \in \mathbb{R}^m$. The non-trivial case is dealt with in the following theorem.

3.8.8 Theorem. *Let the function $f : D \subseteq \mathbb{R}^m \to \mathbb{R}$ be differentiable at*

$\mathbf{p} \in D$ and let $(\operatorname{grad} f)(\mathbf{p}) \neq \mathbf{0}$. *Then*

[i] *the directional derivative* $(\partial f / \partial \mathbf{u})(\mathbf{p})$ *takes maximum value if and only if* $\mathbf{u}$ *is the unit vector in the direction* $(\operatorname{grad} f)(\mathbf{p})$. *The maximum value is* $\|(\operatorname{grad} f)(\mathbf{p})\|$;

[ii] $(\partial f / \partial \mathbf{u})(\mathbf{p}) = 0$ *if and only if* $\mathbf{u}$ *is a unit vector orthogonal to* $(\operatorname{grad} f)(\mathbf{p})$.

Proof. Immediate from Theorem 3.8.6(ii).

Theorem 3.8.8 tells us that $(\operatorname{grad} f)(\mathbf{p})$ is in the direction of maximum rate of change and is orthogonal to the directions of zero rate of change of f at $\mathbf{p}$. The gradient takes on even more significance if we consider 3.8.1 and 3.8.2 from a geometrical point of view and in particular examine how $(\operatorname{grad} f)(\mathbf{p})$ is related to the level set of f which contains $\mathbf{p}$. Let the level set be $S = \{\mathbf{x} \in \mathbb{R}^m \mid f(\mathbf{x}) = f(\mathbf{p})\}$. Since f is constant in S, it is reasonable to expect that $(\partial f / \partial \mathbf{u})(\mathbf{p})$ will be zero when $\mathbf{u}$ is 'tangential' to S at $\mathbf{p}$. It would follow that $(\partial f / \partial \mathbf{u})(\mathbf{p})$ takes maximum value when $\mathbf{u}$ is a unit vector 'normal' to the level set S through $\mathbf{p}$. This in turn suggests that $(\operatorname{grad} f)(\mathbf{p})$ is 'normal' to the level set S.

In attempting to apply these ideas to general functions $f : D \subseteq \mathbb{R}^m \rightarrow \mathbb{R}$ we meet the difficulty that we do not have a definition of tangent lines and normals to arbitrary sets. We avoid this problem by using $(\operatorname{grad} f)(\mathbf{p})$ to *define* the concepts (Definition 3.8.18). But first we consider some illustrative examples and introduce the important notion of a surface in $\mathbb{R}^m$.

3.8.9 *Example.*   Consider again the function $f : \mathbb{R}^2 \rightarrow \mathbb{R}$ given by $f(x, y) = x^2 - y^2$. The level set S through $\mathbf{p} = (-2, 1)$ is the hyperbola

$$x^2 - y^2 = f(-2, 1) = 3.$$

We know from elementary analytic geometry that the tangent to the hyperbola at $(-2, 1)$ has slope -2. The unit vector $\mathbf{u} = (-1/\sqrt{5}, 2/\sqrt{5})$ is in this direction. It is straightforward to check that

$$\frac{\partial f}{\partial \mathbf{u}}(-2, 1) = -\frac{1}{\sqrt{5}} \frac{\partial f}{\partial x}(-2, 1) + \frac{2}{\sqrt{5}} \frac{\partial f}{\partial y}(-2, 1) = 0.$$

Similarly we know that the normal to S through $(-2, 1)$ has slope $\frac{1}{2}$. The vector $(\operatorname{grad} f)(-2, 1) = (-4, -2)$ is in the direction of the normal (see Fig. 3.17(i)). It points in the direction of increasing values of $f(x, y)$.

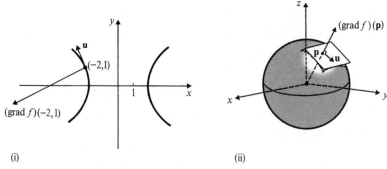

Fig. 3.17 (i) Hyperbola $f(x, y) = x^2 - y^2 = 3$
(ii) Sphere $f(x, y, z) = x^2 + y^2 + z^2 = 9$

3.8.10 Example. The level sets of the function $f : \mathbb{R}^3 \to \mathbb{R}$ given by $f(x, y, z) = x^2 + y^2 + z^2$ are concentric spheres centre **0**. In particular the level set S through $\mathbf{p} = (1, 2, 2)$ is the sphere with equation

$$x^2 + y^2 + z^2 = f(1, 2, 2) = 9.$$

The unit vector $\mathbf{u} = (0, 1/\sqrt{2}, -1/\sqrt{2})$ is orthogonal to the radius vector through $\mathbf{p}$ (see Fig. 3.17(ii)). Hence $\mathbf{u}$ is the direction of a tangent line to the sphere at $\mathbf{p}$. It is straightforward to check that

$$\frac{\partial f}{\partial \mathbf{u}}(1, 2, 2) = 0 \frac{\partial f}{\partial x}(1, 2, 2) + \frac{1}{\sqrt{2}} \frac{\partial f}{\partial y}(1, 2, 2) - \frac{1}{\sqrt{2}} \frac{\partial f}{\partial z}(1, 2, 2) = 0.$$

Finally, $(\text{grad } f)(1, 2, 2) = (2, 4, 4)$, a vector which is a scalar multiple of the radius vector through $\mathbf{p}$. It is therefore in the direction of the normal to the sphere at $\mathbf{p}$.

3.8.11 Definition. *A set $S \subseteq \mathbb{R}^m$ is called a* smooth surface *if S is a level set of a C^1 function $f : D \subseteq \mathbb{R}^m \to \mathbb{R}$ such that $(\text{grad } f)(\mathbf{x}) \neq \mathbf{0}$ for all $\mathbf{x} \in S$.*

3.8.12 Example. The sphere $x^2 + y^2 + z^2 = 9$ is a smooth surface in $\mathbb{R}^3$. It is the level set corresponding to the value 9 of the C^1 function $f : \mathbb{R}^3 \to \mathbb{R}$ given by $f(x, y, z) = x^2 + y^2 + z^2$, and $(\text{grad } f)(x, y, z) \neq (0, 0, 0)$ at points on the sphere.

Similarly, the hyperbola $x^2 - y^2 = 3$ is a smooth surface in $\mathbb{R}^2$. It is a level set of the C^1 function $f : \mathbb{R}^2 \to \mathbb{R}$ given by $f(x, y) = x^2 - y^2$. We observe that $(\text{grad } f)(x, y) \neq (0, 0)$ at points on the hyperbola.

Notice that the branch of the hyperbola containing the point $(-2, 1)$ is also defined as the level set corresponding to 0 of the differentiable function

$g(x, y) = x + \sqrt{(y^2 + 3)}$, and that $(\text{grad } g)(x, y) \neq (0, 0)$ everywhere. At a point (x, y) on the hyperbola the vectors $(\text{grad } f)(x, y)$ and $(\text{grad } g)(x, y)$ must both be in the direction of the normal to the hyperbola (Example 3.8.9) and so $(\text{grad } g)(x, y)$ is a scalar multiple of $(\text{grad } f.)(x, y)$. For example, $(\text{grad } g)(-2, 1) = (1, 1/2)$, and $(\text{grad } f)(-2, 1) = (-4, -2)$.

The condition in Definition 3.8.11 that f should be continuously differentiable is there to ensure that the normal lines to S (which we shall define) change direction continuously over S. The requirement that the gradient shall be non-zero is there to exclude 'non-smooth' points at which it is not possible to fit a tangent plane, as the following example illustrates.

3.8.13 Example.   The function $f : \mathbb{R}^3 \to \mathbb{R}$ defined by $f(x, y, z) = x^2 + y^2 - z^2$ has as level set corresponding to 0 the right circular ⟨ one $z = \pm \sqrt{(x^2 + y^2)}$. See Fig. 3.18(i). At the apex, $(\text{grad } f)(\mathbf{0}) = \mathbf{0}$. Such points must be avoided to allow sensible definitions of normal and tangent plane to a surface at a point on it.

Note that the level sets $x^2 + y^2 - z^2 = c \neq 0$ are smooth surfaces. If $c = a^2 > 0$ the level set is a hyperboloid of one sheet (Fig. 3.18(ii)), and if $c = -a^2 < 0$ it is a hyperboloid of two sheets (Fig. 3.18(iii)).

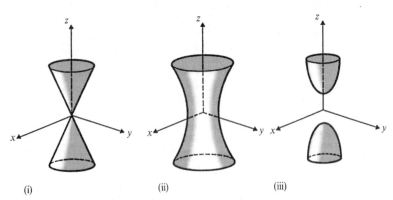

(i)   (ii)   (iii)

Fig. 3.18 (i) Cone $x^2 + y^2 - z^2 = 0$; (ii) Hyperboloid $x^2 + y^2 - z^2 = a^2$; (iii) Hyperboloid $x^2 + y^2 - z^2 = -a^2$

In view of the above comments we have the following definition.

3.8.14 Definition.   *A function $f : D \subseteq \mathbb{R}^m \to \mathbb{R}$ is said to be* smooth *if it is continuously differentiable and if* $(\text{grad } f)(\mathbf{x}) \neq \mathbf{0}$ *for all* $\mathbf{x} \in D$.

We can now say that a smooth surface is a level set of a smooth function (Exercise 3.8.10). An important class of smooth surfaces is considered in the following theorem.

3.8.15 Theorem.   *The graph of a C^1 function $f : D \subseteq \mathbb{R}^{m-1} \to \mathbb{R}$ is a smooth surface in $\mathbb{R}^m$.*

Proof.   Consider the function $F : D \times \mathbb{R} \subseteq \mathbb{R}^m \to \mathbb{R}$ defined by the rule

3.8.16 $F(\mathbf{x}, x_m) = f(\mathbf{x}) - x_m, \qquad \mathbf{x} \in D, x_m \in \mathbb{R}.$

The graph G of f is the level set of F corresponding to the value 0, that is

$$G = \{(\mathbf{x}, x_m) \in \mathbb{R}^m \mid f(\mathbf{x}) - x_m = 0, \mathbf{x} \in D\}.$$

To prove that G is a smooth surface in $\mathbb{R}^m$ we show that (i) F is a C^1 function and (ii) $(\operatorname{grad} F)(\mathbf{x}, x_m) \neq (\mathbf{0}, 0)$ for all $\mathbf{x} \in D, x_m \in \mathbb{R}$.

[i] It follows from 3.8.16 that for each $i = 1, \ldots, m-1$ and each $\mathbf{x} \in D, x_m \in \mathbb{R}$

$$\frac{\partial F}{\partial x_i}(\mathbf{x}, x_m) = \frac{\partial f}{\partial x_i}(\mathbf{x}) \qquad \text{and} \qquad \frac{\partial F}{\partial x_m}(\mathbf{x}, x_m) = -1.$$

Therefore the partial derivatives of F are continuous and so F is a C^1 function.

[ii] For any $\mathbf{x} \in D, x_m \in \mathbb{R}$

$$(\operatorname{grad} F)(\mathbf{x}, x_m) = ((\operatorname{grad} f)(\mathbf{x}), -1) \neq (\mathbf{0}, 0).$$

This completes the proof.

One of the most important theorems in this book, the Implicit Function Theorem (Theorem 4.8.1), establishes a local converse to Theorem 3.8.15. It shows that corresponding to any point in a smooth surface, there is a neighbourhood within which the surface is the graph of a C^1 function. That is to say, every smooth surface is 'locally' graph-like.

We are now in a position to define the normal and the tangent space to a smooth surface at a point on it. Before we do that, however, we prove the following theorem by way of justification.

3.8.17 Theorem.   *Let S be a surface which is a level set of a smooth function $f : D \subseteq \mathbb{R}^m \to \mathbb{R}$ and let $\mathbf{q}$ be a point in S. Then for any* *differentiable path* $\alpha : E \subseteq \mathbb{R} \to \mathbb{R}^m$ *such that* $\alpha(E) \subseteq S$ *and any* $a \in E$ *such that* $\alpha(a) = \mathbf{q}$, *the tangent vector* $\alpha'(a)$ *is orthogonal to the vector* $(\text{grad } f)(\mathbf{q})$.

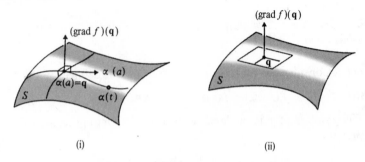

Fig. 3.19

Proof. The theorem is illustrated in Fig. 3.19(i). Since $\alpha(E) \subseteq S$ we have

$$f(\alpha(t)) = f(\mathbf{q}) \qquad \text{for all } t \in E.$$

We obtain the derivative at a by using the Chain Rule in the form 3.8.7:

$$(\text{grad } f)(\alpha(a)) \cdot \alpha'(a) = 0.$$

This is the required result.

Theorem 3.8.17 tells us that $(\text{grad } f)(\mathbf{q})$ is orthogonal to every curve in S through $\mathbf{q}$. We are led to the following definitions.

3.8.18 Definition.   *Let the surface S in $\mathbb{R}^m$ be a level set of a smooth function* $f : D \subseteq \mathbb{R}^m \to \mathbb{R}$. *Then, for any* $\mathbf{q} \in S$,
   [i] *the straight line through* $\mathbf{q}$ *in direction* $(\text{grad } f)(\mathbf{q})$ *is called the* normal line *to the surface S at* $\mathbf{q}$,
   [ii] *the set of all vectors* $\mathbf{x}$ *such that* $(\mathbf{x} - \mathbf{q}) \cdot (\text{grad } f)(\mathbf{q}) = 0$ *is called the* tangent space *to S at* $\mathbf{q}$ (tangent line *when $m = 2$*, tangent plane *when $m = 3$*).

The tangent space to S at $\mathbf{q}$ is the translation of an $(m - 1)$-dimensional subspace of $\mathbb{R}^m$ through $\mathbf{q}$. The situation is illustrated in Fig. 3.19(ii).

3.8.19 Example.   The ellipsoid $x^2 + 2y^2 + 3z^2 = 6$ is a smooth surface since it is a level set of the smooth function $f: \mathbb{R}^3 \backslash \{\mathbf{0}\} \to \mathbb{R}$ defined by

$$f(x, y, z) = x^2 + 2y^2 + 3z^2.$$

Consider the point $\mathbf{q} = (1, 1, 1)$ on the ellipsoid. Since $(\text{grad } f)(1, 1, 1) = (2, 4, 6)$,

[i] the normal line to the surface at $\mathbf{q}$ is the straight line

$$(x, y, z) = (1, 1, 1) + t(2, 4, 6);$$

[ii] the tangent plane at $\mathbf{q}$ has equation

$$((x, y, z) - (1, 1, 1)) \cdot (2, 4, 6) = 0,$$

namely,

$$2x + 4y + 6z = 12.$$

The above definition of a tangent space to a smooth surface S is given in terms of a smooth function which has the surface as a level set. In general there are many different smooth functions with this property. We shall show in Exercise 4.8.9 (once we have proved the Implicit Function Theorem) that they all determine the same tangent space to S. We can however prove this result now in the particular case when S is a graph of a C^1 function. (We know from Theorem 3.8.15 that such a graph is a smooth surface.) The following theorem shows that if G is the graph of a C^1 function $f: D \subseteq \mathbb{R}^{m-1} \to \mathbb{R}$ and if $F: D \times \mathbb{R} \subseteq \mathbb{R}^m \to \mathbb{R}$ is any smooth function having G as a level set, then the equation of the tangent space to G at $(\mathbf{p}, f(\mathbf{p}))$, as calculated by applying the above definition to F, is

$$x_m = f(\mathbf{p}) + L_{f, \mathbf{p}}(\mathbf{x} - \mathbf{p}), \qquad \mathbf{x} \in \mathbb{R}^{m-1}, \mathbf{p} \in D.$$

Since this equation depends on f and $\mathbf{p}$ alone, it follows that all appropriate smooth functions F determine the same tangent space. We also observe that this equation is the equation of the tangent space to the graph G of f as given in Section 3.4. The two definitions of tangent spaces to G (Definitions 3.4.9 and 3.8.18(ii)) are therefore consistent.

3.8.20 Theorem. *Let the surface $G \subseteq \mathbb{R}^m$ be a level set of a smooth function $F: D \times \mathbb{R} \subseteq \mathbb{R}^m \to \mathbb{R}$ and also the graph of a C^1 function $f: D \subseteq \mathbb{R}^{m-1} \to \mathbb{R}$. Then in the sense of Definition 3.8.18(ii), the equation of the tangent space to G at $(\mathbf{p}, f(\mathbf{p})) \in G$ is*

$$x_m = f(\mathbf{p}) + (x_1 - p_1)\frac{\partial f}{\partial x_1}(\mathbf{p}) + \cdots + (x_{m-1} - p_{m-1})\frac{\partial f}{\partial x_{m-1}}(\mathbf{p}).$$

Proof. Since G is both a level set of F and the graph of f, there exists $k \in \mathbb{R}$ such that

$$F(\mathbf{x}, f(\mathbf{x})) = k \qquad \text{for all } \mathbf{x} \in D.$$

For each $j = 1, \ldots, m - 1$, taking partial derivatives of this expression with respect to x_j, we obtain (by the Chain Rule)

$$\frac{\partial F}{\partial x_j}(\mathbf{p}, f(\mathbf{p})) + \frac{\partial F}{\partial x_m}(\mathbf{p}, f(\mathbf{p}))\frac{\partial f}{\partial x_j}(\mathbf{p}) = 0.$$

Hence

$$(\text{grad } F)(\mathbf{p}, f(\mathbf{p})) = -\frac{\partial F}{\partial x_m}(\mathbf{p}, f(\mathbf{p}))\left(\frac{\partial f}{\partial x_1}(\mathbf{p}), \ldots, \frac{\partial f}{\partial x_{m-1}}(\mathbf{p}), -1\right).$$

Therefore, since F is smooth, $(\partial F/\partial x_m)(\mathbf{p}, f(\mathbf{p})) \neq 0$ and from 3.8.18(ii) the equation of the tangent space to G at $(\mathbf{p}, f(\mathbf{p}))$ is

$$((\mathbf{x}, x_m) - (\mathbf{p}, f(\mathbf{p})))\cdot\left(\frac{\partial f}{\partial x_1}(\mathbf{p}), \ldots, \frac{\partial f}{\partial x_{m-1}}(\mathbf{p}), -1\right) = 0.$$

The result follows.

The level sets of smooth functions which are considered in this section are defined *implicitly*. For example, the equation of the sphere $x^2 + y^2 + z^2 = 9$ is not an explicit formula for points (x, y, z) on it. For some purposes, in particular when we wish to integrate over the sphere, this situation is unsatisfactory. In Chapter 8, with integration theory in mind, we shall define surfaces in $\mathbb{R}^3$ explicitly in terms of parametrizations.

We close this section by proving a theorem which generalizes the fundamental result that if the derivative of a real-valued function $f: E \subseteq \mathbb{R} \to \mathbb{R}$ is zero on the interval E, then f is a constant function. We need a preparatory definition.

3.8.21 Definition. *An open subset D of $\mathbb{R}^m$ is* path connected *if, given any two vectors $\mathbf{a}$ and $\mathbf{b}$ in D, there is a path $\alpha:[0, 1] \subseteq \mathbb{R} \to \mathbb{R}^m$ in D such that $\alpha(0) = \mathbf{a}$ and $\alpha(1) = \mathbf{b}$.*

We state without proof that if such a path α exists then there is a differentiable path with the same property.

3.8.22 Theorem. *Let D be an open path-connected subset of $\mathbb{R}^m$ and let $f: D \subseteq \mathbb{R}^m \to \mathbb{R}$ be a differentiable function such that $(\text{grad } f)(\mathbf{x}) = 0$ for all $\mathbf{x} \in D$. Then f is a constant function.*

Proof. It is enough to prove that, for any two vectors **a** and **b** in D, $f(\mathbf{a}) = f(\mathbf{b})$. Let $\alpha : [0, 1] \subseteq \mathbb{R} \to \mathbb{R}^m$ be a differentiable path in D such that $\alpha(0) = \mathbf{a}$ and $\alpha(1) = \mathbf{b}$. Then $f \circ \alpha : [0, 1] \subseteq \mathbb{R} \to \mathbb{R}^m$ has derivative given by

$$(f \circ \alpha)'(t) = (\operatorname{grad} f)(\alpha(t)) \cdot \alpha'(t) = 0.$$

Hence $f \circ \alpha$ is a constant function (Exercise 1.6.3) and therefore $(f \circ \alpha)(0) = (f \circ \alpha)(1)$. It follows that $f(\mathbf{a}) = f(\mathbf{b})$.

Exercises 3.8

1. Find $(\operatorname{grad} f)(x, y)$ when

    (a) $f(x, y) = \sqrt{(x^2 + y^2)}$, (b) $f(x) = \dfrac{x - y}{x + y}$, $x + y \neq 0$.

    Hence find $(\partial f / \partial \mathbf{u})(\mathbf{p})$ when $\mathbf{u} = (\frac{3}{5}, \frac{4}{5})$, $\mathbf{p} = (0, -1)$. Compare with Exercise 3.7.1.

2. For the following functions $f : \mathbb{R}^2 \to \mathbb{R}$ and $g : \mathbb{R} \to \mathbb{R}^2$ calulate $(f \circ g)'(t)$ from the formula $(f \circ g)'(t) = (\operatorname{grad} f)(g(t)) \cdot g'(t)$.
    (a) $f(x_1, x_2) = x_1 x_2$, $g(t) = (\sin t, \cos t)$;
    (b) $f(x, y) = x^2 y + y^3$, $g(t) = (t + 1, e^t)$.
    Compare with Exercise 3.6.2.

3. Continuing Exercise 3.6.5, when the insect is at the point $(1, 1, 2)$ and the temperature in space is given by $f(x, y, z) = kx^2(y - z)$, what course should he take to warm up as quickly as possible?

    Hint: find **u** such that $(\partial f / \partial \mathbf{u})(1, 1, 2)$ is a maximum from Theorem 3.8.8(i).

    Check: by Theorem 3.8.8(ii) the direction of maximum rate of increase of temperature at $(1, 1, 2)$ must be orthogonal to any direction of constant temperature at $(1, 1, 2)$ as calculated in Exercise 3.7.3.

4. Let $f : \mathbb{R}^2 \to \mathbb{R}$ be the function defined by $f(x, y) = x^2 - y^2$. Sketch the level set

    $$S = \{(x, y) \in \mathbb{R}^2 \mid f(x, y) = -1\}.$$

    On your sketch draw at points **p** in S the vector $(\operatorname{grad} f)(\mathbf{p})$, where
    (a) $\mathbf{p} = (0, 1)$, (b) $\mathbf{p} = (1, \sqrt{2})$, (c) $p = (1, -\sqrt{2})$. Observe that in each case $(\operatorname{grad} f)(\mathbf{p})$ is normal to the level set of f through **p**, and points in the direction of increasing values of f.

5. Repeat Exercise 4 for various level sets of the functions of Exercise 3.1.1.

6. Sketch the hyperboloid of two sheets $x^2 - y^2 - z^2 = 1$. Show that it is a smooth surface according to Definition 3.8.11. Find (i) the normal at **q**, (ii) the tangent plane at **q** to the hyperboloid when (a) $\mathbf{q} = (1, 0, 0)$;

(b) $\mathbf{q} = (-1, 0, 0)$; (c) $\mathbf{q} = (2, 0, \sqrt{3})$; (d) $\mathbf{q} = (\sqrt{3}, 1, -1)$. Sketch the normals and tangent planes.

Answer to part (c): the normal to the hyperboloid at $(2, 0, \sqrt{3})$ is the line $(x, y, z) = (2, 0, \sqrt{3}) + t(4, 0, -2\sqrt{3})$; the tangent plane at $(2, 0, \sqrt{3})$ has equation $2x - \sqrt{3}z = 1$.

7. Prove that a plane in $\mathbb{R}^3$ is a smooth surface.

8. Let $F : \mathbb{R}^3 \backslash \{0\} \to \mathbb{R}$ be defined by

$$F(\mathbf{r}) = \frac{1}{r}, \qquad \mathbf{r} \neq \mathbf{0}$$

where $\mathbf{r} = (x, y, z)$ and $r = \|\mathbf{r}\| = \sqrt{(x^2 + y^2 + z^2)}$. Prove that

$$(\text{grad } F)(\mathbf{r}) = -\frac{1}{r^3}\mathbf{r}.$$

Show that $(\text{grad } F)(a, b, c)$ is normal to the sphere $r^2 = a^2 + b^2 + c^2$ at the point (a, b, c) and points in the direction of increasing values of F. Find the equation of the tangent plane to this sphere at (a, b, c).

Answer: $ax + by + cz = a^2 + b^2 + c^2$.

9. Let $F : \mathbb{R}^3 \to \mathbb{R}$ be defined by $F(x, y, z) = y^3 - z^3$. Prove that

$$(\text{grad } F)(x, 0, 0) = (0, 0, 0), \qquad x \in \mathbb{R}.$$

Is the level set

$$S = \{(x, y, z) \in \mathbb{R}^3 | F(x, y, z) = 0\}$$

a smooth surface in $\mathbb{R}^3$?

Hint: note that Definition 3.8.11 does not apply to the function F. This does *not* imply that S is not a smooth surface in $\mathbb{R}^3$. Prove that $y^3 - z^3 = 0$ if and only if $y - z = 0$. Hence show that S is a smooth surface in $\mathbb{R}^3$.

10. Let $S \subseteq \mathbb{R}^m$ be a smooth surface (Definition 3.8.11). Show that there exists an open subset E in $\mathbb{R}^m$ such that (a) f is smooth on E and (b) S is a level set of the function $f : E \subseteq \mathbb{R}^m \to \mathbb{R}$. (This result allows us to define a smooth surface as a level set of a smooth function.)

Hint: starting from Definition 3.8.11, prove that the set $K = \{\mathbf{x} \in D \mid (\text{grad} f)(\mathbf{x}) = \mathbf{0}\}$ is closed. Then f is smooth on the open domain $E = D \backslash K$, and $S \subseteq E$.

11. Prove that the cone $S = \{(x, y, z) \in \mathbb{R}^3 \mid x^2 + y^2 - z^2 = 0\}$ (see Example 3.8.13) is not a smooth surface.

Hint: suppose that there exists a smooth function $f : E \subseteq \mathbb{R}^3 \to \mathbb{R}$ such that

$$S = \{(x, y, z) \in E \mid f(x, y, z) = 0\}.$$

Consider the three differentiable paths α, β, γ in S given by

$$\alpha(t) = (t, 0, t), \qquad \beta(t) = (0, t, t), \qquad \gamma(t) = (t, t, \sqrt{2}t),$$
$$t \in [-1, 1].$$

Then $\alpha(0) = \beta(0) = \gamma(0) = \mathbf{0}$, the apex of the cone. Prove that the tangent vectors $\alpha'(0), \beta'(0), \gamma'(0)$ are linearly independent. Argue by contradiction from Exercise 10.

12. Let $f : \mathbb{R}^3 \to \mathbb{R}$ be defined by $f(x, y, z) = x^2 - y^2, (x, y, z) \in \mathbb{R}^3$. Show that f is not a smooth function on the level set $S = \{(x, y, z) \in \mathbb{R}^3 \mid f(x, y, z) = 0\}$. Sketch S.

3.9 The Mean-Value Theorem

The Mean-Value Theorem of elementary calculus can be extended as follows.

3.9.1 Theorem. *Let the function $f : D \subseteq \mathbb{R}^m \to \mathbb{R}$ be differentiable and let* $\mathbf{p}$ *lie in D. Suppose that for some non-zero vector $\mathbf{h} \in \mathbb{R}^m$ the line segment joining $\mathbf{p}$ and $\mathbf{p} + \mathbf{h}$ is a subset of D. Then there exists a point $\mathbf{c} = \mathbf{p} + \theta \mathbf{h}, 0 < \theta < 1$, in that line segment such that*

$$f(\mathbf{p} + \mathbf{h}) - f(\mathbf{p}) = h_1 \frac{\partial f}{\partial x_1}(\mathbf{c}) + \cdots + h_m \frac{\partial f}{\partial x_m}(\mathbf{c}).$$

3.9.2 Corollary. *With the notation and hypothesis of Theorem 3.9.1*

$$f(\mathbf{p} + \mathbf{h}) - f(\mathbf{p}) = \|\mathbf{h}\| \frac{\partial f}{\partial \mathbf{u}}(\mathbf{c})$$

where $\mathbf{u} = \|\mathbf{h}\|^{-1}\mathbf{h}$ is the unit vector in the direction of $\mathbf{h}$.

Proof of Theorem 3.9.1. Since we are working on the line segment joining $\mathbf{p}$ and $\mathbf{p} + \mathbf{h}$ we can reduce our problem to one concerning a function $F : [0, 1] \subseteq \mathbb{R} \to \mathbb{R}$ by defining

3.9.3 $$F(t) = f(\mathbf{p} + t\mathbf{h}), \qquad t \in [0, 1].$$

Notice that F is the composition of f with the differentiable function $g : [0, 1] \subseteq \mathbb{R} \to \mathbb{R}^m$ defined by $g(t) = \mathbf{p} + t\mathbf{h}, t \in [0, 1]$.

The Mean-Value Theorem applies to $F = f \circ g$ since F is continuous on $[0, 1]$ and differentiable on $]0, 1[$. Therefore there exists $0 < \theta < 1$ such that

3.9.4 $$f(\mathbf{p} + \mathbf{h}) - f(\mathbf{p}) = F(1) - F(0) = F'(\theta).$$

By the Chain Rule 3.6.9

3.9.5 $$F'(\theta) = h_1 \frac{\partial f}{\partial x_1}(\mathbf{p} + \theta\mathbf{h}) + \cdots + h_m \frac{\partial f}{\partial x_m}(\mathbf{p} + \theta\mathbf{h}).$$

The result follows from 3.9.4 and 3.9.5 by taking $\mathbf{c} = \mathbf{p} + \theta\mathbf{h}$.

Proof of Corollary 3.9.2.   By Corollary 3.7.6

3.9.6 $$\frac{\partial f}{\partial \mathbf{u}}(\mathbf{c}) = u_1 \frac{\partial f}{\partial x_1}(\mathbf{c}) + \cdots + u_m \frac{\partial f}{\partial x_m}(\mathbf{c})$$

and therefore, multiplying 3.9.6 by $\|\mathbf{h}\|$

3.9.7 $$\|\mathbf{h}\| \frac{\delta f}{\delta \mathbf{u}}(\mathbf{c}) = h_1 \frac{\partial f}{\partial x_1}(\mathbf{c}) + \cdots + h_m \frac{\partial f}{\partial x_m}(\mathbf{c}).$$

Combining 3.9.7 and Theorem 3.9.1 gives the required result.

3.9.8 Example.   We now take a second look at our estimation of the acceleration in Example 3.5.3. The relationship is $a = f(v, s) = v^2/2s$ and the observed values are $(v, s) = (30, 300)$ with possible errors given by $|\Delta v| \le 1$ and $|\Delta s| \le 3$. We know from the direct solution 3.5.9 that if the true acceleration is $a_t\, m s^{-2}$, then $-0.112 \le a_t - a \le 0.118$. We cannot hope to improve on that inequality since all values in that range could occur, given the error bounds on v and s.

The Mean-Value Theorem tells us that

3.9.9 $$a_t - a = \Delta v \frac{\partial f}{\partial v}(v + \theta\Delta v, s + \theta\Delta s) + \Delta s \frac{\partial f}{\partial s}(v + \theta\Delta v, s + \theta\Delta s)$$

for some $0 < \theta < 1$.
Hence

$$a_t - a = \Delta v \frac{v + \theta\, \Delta v}{s + \theta\, \Delta s} - \Delta s \frac{1}{2}\left(\frac{v + \theta\, \Delta v}{s + \theta\, \Delta s}\right)^2.$$

We can now find an upper bound for $|a_t - a|$ by taking $\Delta v = 1$, $\Delta s = -3$ and $\theta = 1$, giving

$$|a_t - a| \le \frac{31}{297} + \frac{3}{2}\left(\frac{31}{297}\right)^2 = 0.121.$$

We therefore have $-0.121 \leqslant a_t - a \leqslant 0.121$. As expected this range contains the achievable range given above.

It is interesting to note that the approximation 3.5.8 which was obtained by neglecting a term $\|\Delta\mathbf{x}\|\eta(\Delta\mathbf{x})$ can also be obtained from 3.9.9 by taking $\theta = 0$. We saw in Example 3.5.3 that this leads to $-0.115 \leqslant a_t - a \leqslant 0.115$, which does not contain the achievable range as a subset.

Of course in this simple example a direct calculation is most appropriate. But the example illustrates how the Mean-Value Theorem provides a correct upper bound for the error, which a crude application of differentiability at (30, 300) did not do.

Exercises 3.9

1. Let $f : \mathbb{R}^2 \to \mathbb{R}$ be defined by $f(x_1, x_2) = x_1^2 x_2$. Obtain the expression for $f(\mathbf{p} + \mathbf{h}) - f(\mathbf{p})$ of Theorem 3.9.1 for the case $\mathbf{p} = (10, 20)$, $\mathbf{h} = (0.1, 0.1)$.

2. Prove that, if $f(x, y) = e^x \cos y$, then there exists $0 < \theta < 1$ such that

    $$f(x, y) = 1 + xe^{\theta x}\cos \theta y - ye^{\theta x}\sin \theta y.$$

    Deduce that provided $|x|$ and $|y|$ are 'small', the polynomial $1 + x$ is a reasonable approximation to $f(x, y)$.

    Hint: In Theorem 3.9.1 take $\mathbf{p} = \mathbf{0}$ and $\mathbf{h} = (x, y)$.

3. By the method of Exercise 2 obtain a polynomial of degree 1 in $h_1 = (x + 1)$ and $h_2 = (y - 1)$ which is a reasonable approximation to $\sqrt{(1 + x + 4y)}$ when x is close to -1 and y is close to 1.

    Hint: In Theorem 3.9.1 take $\mathbf{p} = (-1, 1)$ and $\mathbf{p} + \mathbf{h} = (x, y)$ and consider small values of $h_1 = x + 1$ and $h_2 = y - 1$.

    Answer: $2 + \frac{1}{4}h_1 + h_2$.

4. Consider again the rectangular box of Exercise 3.5.1. Estimate the accuracy of the calculation for the volume V by applying the Mean-Value Theorem in the manner of Example 3.9.8. (Use the calculations of Exercise 1.)

3.10 Higher-order derivatives

In the next section we propose to prove Taylor's Theorem which generalizes the Mean-Value Theorem 3.9.1 and Corollary 3.9.2. The theorem is conveniently stated in terms of higher-order directional derivatives. However, for the purpose of calculation there is a useful alternative version that involves higher-order partial derivatives. As a preliminary to Taylor's Theorem we will consider higher-order derivatives in this section.

3.10.1 Definition. *The* nth-*order directional derivative of the function* $f : D \subseteq \mathbb{R}^m \to \mathbb{R}$ *in the direction of a unit vector* $\mathbf{u} \in \mathbb{R}^m$ *is the function* $\partial^n f / \partial \mathbf{u}^n : D \subseteq \mathbb{R}^m \to \mathbb{R}$ *defined inductively by the rule*

$$\frac{\partial^{r+1} f}{\partial \mathbf{u}^{r+1}} = \frac{\partial}{\partial \mathbf{u}} \left(\frac{\partial^r f}{\partial \mathbf{u}^r} \right), \qquad r = 1, \ldots, n-1,$$

whenever these derivatives exist.

For the case $m = 1$, with $\mathbf{u}$ the unit vector in the positive direction in $\mathbb{R}$, the directional derivatives are simply the ordinary derivatives $f', f'', \ldots$.

Recall from Section 3.7 that for arbitrary m the directional derivative $\delta f / \delta \mathbf{e}_j$ in the direction $\mathbf{e}_j$ in $\mathbb{R}^m$ is by definition equal to the partial derivative $\partial f / \partial x_j$. Similarly the higher-order derivative $\partial^n f / \partial \mathbf{e}_j^n$ is denoted by $\partial^n f / \partial x_j^n$. This is in agreement with the following general definition of higher-order partial derivatives.

3.10.2 Definition. *The partial derivatives of the partial derivatives of a function* $f : D \subseteq \mathbb{R}^m \to \mathbb{R}$, *provided they exist, are called* second-order partial derivatives. *The second-order partial derivative*

$$\frac{\partial}{\partial x_j} \left(\frac{\partial f}{\partial x_i} \right)$$

is denoted by

$$\frac{\partial^2 f}{\partial x_j \partial x_i}$$

if $i \neq j$ *and by*

$$\frac{\partial^2 f}{\partial x_i^2}$$

if $i = j$. *The notation for third-order partial derivatives is*

$$\frac{\partial^3 f}{\partial x_k \partial x_j \partial x_i} = \frac{\partial}{\partial x_k} \left(\frac{\partial^2 f}{\partial x_j \partial x_i} \right).$$

Similar notation applies to fourth and higher orders.

3.10.3 Example. Define $f : \mathbb{R}^3 \to \mathbb{R}$ by

$$f(x_1, x_2, x_3) = x_1^3 x_2 - x_1 x_3^2, \qquad \mathbf{x} \in \mathbb{R}^3.$$

Then $\partial f / \partial x_1 (\mathbf{x}) = 3x_1^2 x_2 - x_3^2$ and $\partial f / \partial x_2 (\mathbf{x}) = x_1^3$. Hence

$$\frac{\partial^2 f}{\partial x_1^2}(\mathbf{x}) = 6x_1 x_2, \qquad \frac{\partial^2 f}{\partial x_2 \partial x_1}(\mathbf{x}) = 3x_1^2 \quad \text{and} \quad \frac{\partial^2 f}{\partial x_1 \partial x_2}(\mathbf{x}) = 3x_1^2.$$

Finally,

$$\frac{\partial^3 f}{\partial x_2 \partial x_1^2}(\mathbf{x}) = 6x_1, \qquad \frac{\partial^3 f}{\partial x_1 \partial x_2 \partial x_1}(\mathbf{x}) = 6x_1 \quad \text{and} \quad \frac{\partial^3 f}{\partial x_1^2 \partial x_2}(\mathbf{x}) = 6x_1.$$

It is no coincidence that the two mixed second-order partial derivatives and the three third-order partial derivatives are equal, as the following theorem shows.

3.10.4 Theorem. *Let $f : D \subseteq \mathbb{R}^2 \to \mathbb{R}$ be a function such that*
[i] *$\partial f / \partial x_1$, $\partial f / \partial x_2$ and $\partial^2 f / \partial x_2 \partial x_1$ exist throughout D, and*
[ii] *$\partial^2 f / \partial x_2 \partial x_1$ is continuous at $\mathbf{p} \in D$.*
Then $\partial^2 f / \partial x_1 \partial x_2$ exists at $\mathbf{p}$ and

$$\frac{\partial^2 f}{\partial x_1 \partial x_2}(\mathbf{p}) = \frac{\partial^2 f}{\partial x_2 \partial x_1}(\mathbf{p}).$$

Proof. We need to express the relevant partial derivatives in terms of limits.

$$\frac{\partial f}{\partial x_1}(\mathbf{x}) = \lim_{h_1 \to 0} \frac{f(x_1 + h_1, x_2) - f(x_1, x_2)}{h_1},$$

$$\frac{\partial^2 f}{\partial x_2 \partial x_1}(\mathbf{p}) = \lim_{h_2 \to 0} \frac{1}{h_2} \left(\lim_{h_1 \to 0} \frac{f(p_1 + h_1, p_2 + h_2) - f(p_1, p_2 + h_2)}{h_1} \right.$$
$$\left. - \lim_{h_1 \to 0} \frac{f(p_1 + h_1, p_2) - f(p_1, p_2)}{h_1} \right).$$

Therefore

$$\frac{\partial^2 f}{\partial x_2 \partial x_1}(\mathbf{p}) = \lim_{h_2 \to 0} \lim_{h_1 \to 0} \frac{F(h_1, h_2)}{h_1 h_2},$$

where

3.10.5
$$F(h_1, h_2) = f(p_1 + h_1, p_2 + h_2) - f(p_1, p_2 + h_2)$$
$$- f(p_1 + h_1, p_2) + f(p_1, p_2).$$

Similarly, we find that

$$\frac{\partial^2 f}{\partial x_1 \partial x_2}(\mathbf{p}) = \lim_{h_1 \to 0} \lim_{h_2 \to 0} \frac{F(h_1, h_2)}{h_1 h_2},$$

provided that this limit exists.

We intend to show that under the hypothesis of the theorem

3.10.6
$$\lim_{h_1 \to 0} \lim_{h_2 \to 0} \frac{F(h_1, h_2)}{h_1 h_2} = \frac{\partial^2 f}{\partial x_2 \partial x_1}(\mathbf{p}).$$

The result will follow.

For fixed h_2, let $g(x_1) = f(x_1, p_2 + h_2) - f(x_1, p_2)$. The domain of g is an open subset of $\mathbb{R}$ and, since $\partial f / \partial x_1$ exists, g is differentiable. The Mean-Value Theorem implies then that

$$F(h_1, h_2) = g(p_1 + h_1) - g(p_1) \qquad \text{(by 3.10.5)}$$
$$= h_1 g'(p_1 + \phi h_1) \qquad \text{for some } 0 < \phi < 1$$
$$= h_1 \left(\frac{\partial f}{\partial x_1}(p_1 + \phi h_1, p_2 + h_2) - \frac{\partial f}{\partial x_1}(p_1 + \phi h_1, p_2) \right).$$

A second application of the Mean-Value Theorem is possible since $(\partial / \partial x_2)(\partial f / \partial x_1)$ exists. Therefore

$$F(h_1, h_2) = h_1 h_2 \frac{\partial^2 f}{\partial x_2 \partial x_1}(p_1 + \phi h_1, p_2 + \theta h_2), \qquad \text{for some } 0 < \theta < 1.$$

By the continuity of $\partial^2 f / \partial x_2 \partial x_1$ at $\mathbf{p}$,

$$F(h_1, h_2) = h_1 h_2 \left(\frac{\partial^2 f}{\partial x_2 \partial x_1}(\mathbf{p}) + \eta(\mathbf{h}) \right), \qquad \text{where } \lim_{\mathbf{h} \to 0} \eta(\mathbf{h}) = 0.$$

Expression 3.10.6 follows.

3.10.7 Corollary. *Let* $f : D \subseteq \mathbb{R}^m \to \mathbb{R}$ *be a function such that for particular i and j,*

[i] $\dfrac{\partial f}{\partial x_i}, \dfrac{\partial f}{\partial x_j}, \dfrac{\partial^2 f}{\partial x_j \partial x_i}$ *exist throughout D, and*

[ii] $\dfrac{\partial^2 f}{\partial x_j \partial x_i}$ *is continuous at* $\mathbf{p} \in D$.

Then $\partial^2 f / \partial x_i \partial x_j$ *exists at* $\mathbf{p}$ *and*

$$\frac{\partial^2 f}{\partial x_i \partial x_j}(\mathbf{p}) = \frac{\partial^2 f}{\partial x_j \partial x_i}(\mathbf{p}).$$

Proof. Fix x_r at p_r except when $r = i$ or j. Then f can be regarded as acting on $\mathbb{R}^2$ and Theorem 3.10.4 applies.

3.10.8 *Example.* Consider again the function $f(x_1, x_2, x_3) = x_1^3 x_2 - x_1 x_3^2$ of Example 3.10.3. Clearly all partial derivatives are continuous. It is no surprise, in view of Theorem 3.10.4, that

$$\frac{\partial^2 f}{\partial x_2 \partial x_1}(\mathbf{x}) = \frac{\partial^2 f}{\partial x_1 \partial x_2}(\mathbf{x}) \quad \text{and} \quad \frac{\partial^3 f}{\partial x_2 \partial x_1^2}(\mathbf{x}) = \frac{\partial^3 f}{\partial x_1 \partial x_2 \partial x_1}(\mathbf{x}) = \frac{\partial^3 f}{\partial x_1^2 \partial x_2}(\mathbf{x}).$$

We now consider how directional derivatives can be expressed in terms of partial derivatives. For first-order derivatives the relation is given by Corollary 3.7.6, which states that if $f : D \subseteq \mathbb{R}^m \to \mathbb{R}$ is differentiable, then the derivative in the direction of the unit vector $\mathbf{u} = (u_1, \ldots, u_m)$ is given by

3.10.9 $$\frac{\partial f}{\partial \mathbf{u}} = u_1 \frac{\partial f}{\partial x_1} + \cdots + u_m \frac{\partial f}{\partial x_m} = \sum_i u_i \frac{\partial f}{\partial x_i}$$

We can think of the process of forming the function $\partial f / \partial \mathbf{u} : D \subseteq \mathbb{R}^m \to \mathbb{R}$ from the differentiable function $f : D \subseteq \mathbb{R}^m \to \mathbb{R}$ as a transformation of f by the operator $\partial / \partial \mathbf{u}$. In these terms 3.10.9 gives an identity between operators

3.10.10 $$\frac{\partial}{\partial \mathbf{u}} = u_1 \frac{\partial}{\partial x_1} + \cdots + u_m \frac{\partial}{\partial x_m},$$

provided we restrict the action of the operators to differentiable functions.

Suppose now that the function f has continuous first-order partial derivatives throughout D. By Theorem 3.3.29, f is differentiable, and so we have the expression 3.10.9 for $\partial f / \partial \mathbf{u}$. Moreover, by 3.10.9, $\partial f / \partial \mathbf{u}$ is continuous since the partial derivatives are continuous.

Suppose in addition that the second-order partial derivatives of f are all continuous throughout D. By Theorem 3.3.29 the first-order partial derivatives are differentiable and hence, by 3.10.9, so is $\partial f / \partial \mathbf{u}$.

Therefore, applying the operator 3.10.10,

3.10.11 $\dfrac{\partial^2 f}{\partial \mathbf{u}^2} = \left(u_1 \dfrac{\partial}{\partial x_1} + \cdots + u_m \dfrac{\partial}{\partial x_m} \right) \dfrac{\partial f}{\partial \mathbf{u}}$

$$= \left(u_1 \frac{\partial}{\partial x_1} + \cdots + u_m \frac{\partial}{\partial x_m} \right) \left(u_1 \frac{\partial f}{\partial x_1} + \cdots + u_m \frac{\partial f}{\partial x_m} \right)$$

$$= \sum_{j,\,i} u_j u_i \frac{\partial^2 f}{\partial x_j \partial x_i}.$$

It follows that $\partial^2 f / \partial \mathbf{u}^2$ is continuous since the second-order derivatives are.

Similarly, if in addition the third-order partial derivatives are all continuous throughout D then, by Theorem 3.3.29, the second-order partial derivatives are differentiable and hence, by 3.10.11, so is $\partial^2 f / \partial \mathbf{u}^2$. Therefore by 3.10.10,

$$\frac{\partial^3 f}{\partial \mathbf{u}^3} = \left(u_1 \frac{\partial}{\partial x_1} + \cdots + u_m \frac{\partial}{\partial x_m} \right) \frac{\partial^2 f}{\partial \mathbf{u}^2}$$

$$= \left(u_1 \frac{\partial}{\partial x_1} + \cdots + u_m \frac{\partial}{\partial x_m} \right) \sum_{j,\,i} u_j u_i \frac{\partial^2 f}{\partial x_j \partial x_i}$$

$$= \sum_{k,\,j,\,i} u_k u_j u_i \frac{\partial^3 f}{x_k x_j x_i}.$$

It follows that $\partial^3 f / \partial \mathbf{u}^3$ is continuous since the third-order derivatives are.

The above process leads to the general result stated in the following theorem.

3.10.12 Definition.   *A function* $f : D \subseteq \mathbb{R}^m \to \mathbb{R}$ *all of whose partial derivatives up to order n are continuous throughout D is said to be n* times continuously differentiable *or a* C^n *function.*

3.10.13 Theorem.   *Let* $f : D \subseteq \mathbb{R}^m \to \mathbb{R}$ *be a* C^n *function, and let* $\mathbf{u}$ *be a unit vector in* $\mathbb{R}^m$. *Then the functions* $f, \partial f / \partial \mathbf{u}, \ldots, \partial^{n-1} f / \partial \mathbf{u}^{n-1}$ *are differentiable,* $\partial^n f / \partial \mathbf{u}^n$ *is continuous and, for* $r = 1, \ldots, n$,

3.10.14 $$\frac{\partial^r f}{\partial \mathbf{u}^r} = \left(u_1 \frac{\partial}{\partial x_1} + \cdots + u_m \frac{\partial}{\partial x_m} \right)^r f$$

$$= \sum_{i_r, \ldots, i_1} u_{i_r} \cdots u_{i_1} \frac{\partial^r f}{\partial x_{i_r} \cdots \partial x_{i_1}}.$$

Proof. Continue the above process.

3.10.15 Remark. The notation

$$\left(u_1 \frac{\partial}{\partial x_1} + \cdots + u_m \frac{\partial}{\partial x_m} \right)^r$$

in 3.10.14 indicates an r-fold application of the operator 3.10.10. In view of the conditions on f, this is meaningful.

3.10.16 Example. In the case of a C^2 function $f : D \subseteq \mathbb{R}^2 \to \mathbb{R}$,

$$\frac{\partial^2 f}{\partial \mathbf{u}^2} = \left(u_1 \frac{\partial}{\partial x_1} + u_2 \frac{\partial}{\partial x_2} \right)^2 f$$

$$= u_1^2 \frac{\partial^2 f}{\partial x_1^2} + u_1 u_2 \frac{\partial^2 f}{\partial x_1 \partial x_2} + u_2 u_1 \frac{\partial^2 f}{\partial x_2 \partial x_1} + u_2^2 \frac{\partial^2 f}{\partial x_2^2}$$

$$= u_1^2 \frac{\partial^2 f}{\partial x_1^2} + 2 u_1 u_2 \frac{\partial^2 f}{\partial x_1 \partial x_2} + u_2^2 \frac{\partial^2 f}{\partial x_2^2}.$$

The last step is obtained by an application of Theorem 3.10.4 to the C^2 function f.
Similarly, if f is a C^3 function, then, by Theorem 3.10.4,

$$\frac{\partial^3 f}{\partial \mathbf{u}^3} = \left(u_1 \frac{\partial}{\partial x_1} + u_2 \frac{\partial}{\partial x_2} \right)^3 f$$

$$= u_1^3 \frac{\partial^3 f}{\partial x_1^3} + 3 u_1^2 u_2 \frac{\partial^3 f}{\partial x_1^2 \partial x_2} + 3 u_1 u_2^2 \frac{\partial^3 f}{\partial x_1 \partial x_2^2} + u_2^3 \frac{\partial^3 f}{\partial x_2^3}.$$

The similarity between these expressions and the expansions of $(a + b)^2$ and $(a + b)^3$ continues through higher orders and makes the expression for $\partial^n f / \partial \mathbf{u}^n$ easy to remember.

Exercises 3.10

1. Compute the value at (x, y) taken by the second-order partial derivatives $\partial^2 f / \partial x^2$, $\partial^2 f / \partial x \partial y$, $\partial^2 f / \partial y \partial x$ and $\partial^2 f / \partial y^2$ for each of the following C^2 functions $f : \mathbb{R}^2 \to \mathbb{R}$. Check in each case that the mixed second-order partial derivatives are equal.

    (a) $f(x, y) = x^3 y$;   (b) $f(x, y) = \sin(x^3 y)$.

    Show also in each case that $\partial^3 f / \partial x \partial y \partial x = \partial^3 f / \partial x^2 \partial y$.

2. For the functions f of Exercise 1 compute, using 3.10.11, $(\partial^2 f / \partial \mathbf{u}^2)(x, y)$, where $\mathbf{u}$ is a unit vector (u_1, u_2). Verify your earlier calculations of $\partial^2 f / \partial x^2$ and $\partial^2 f / \partial y^2$ by choosing $\mathbf{u} = (1, 0)$ and $\mathbf{u} = (0, 1)$.

3. The following partial differential equations are of some importance in classical physics.

    (a) *Laplace's Equation:* $\dfrac{\partial^2 f}{\partial x^2} + \dfrac{\partial^2 f}{\partial y^2} + \dfrac{\partial^2 f}{\partial z^2} = 0;$

    (b) *One-dimensional Wave Equation:* $\dfrac{\partial^2 f}{\partial t^2} - c^2\dfrac{\partial^2 f}{\partial x^2} = 0,$ c constant;

    (c) *One-dimensional Heat Equation:* $\dfrac{\partial f}{\partial t} - k\dfrac{\partial^2 f}{\partial x^2} = 0,$ k constant.

    Verify the following respective solutions of equations (a), (b) and (c):
    (a) $f(x, y, z) = 1/r$, where $r = \sqrt{(x^2 + y^2 + z^2)} \neq 0$;
    (b) $f(x, t) = \sin(x - ct)$, or more generally,
       $f(x, t) = \phi(x - ct)$, where $\phi : \mathbb{R} \to \mathbb{R}$ is any real-valued twice differentiable function on $\mathbb{R}$;
    (c) $f(x, t) = e^{-kt}\sin x$.

4. Consider the function $f : \mathbb{R}^2 \to \mathbb{R}$ defined by

    $$f(x, y) = xy\frac{x^2 - y^2}{x^2 + y^2}, \qquad (x, y) \neq (0, 0), \qquad f(0, 0) = 0.$$

    Prove that

    $$\frac{\partial^2 f}{\partial x \partial y}(x, y) = \frac{\partial^2 f}{\partial y \partial x}(x, y) \qquad \text{when } (x, y) \neq (0, 0).$$

    Show also that

    $$\frac{\partial^2 f}{\partial x \partial y}(0, 0) = 1 \qquad \text{and} \qquad \frac{\partial^2 f}{\partial y \partial x}(0, 0) = -1.$$

    Deduce by Theorem 3.10.4 that neither of the mixed second-order partial derivatives is continuous at $(0, 0)$, and prove this by direct calculation.

    Hint: the first part is a routine calculation. For the second part show that

    $$\frac{\partial f}{\partial y}(x, 0) = \lim_{k \to 0} (f(x, k) - f(x, 0))/k = x$$

    and hence that $(\partial^2 f/\partial x \partial y)(0, 0) = 1$.

3.11   Taylor's Theorem

The method of Section 3.9 can be used to extend Taylor's Theorem of elementary calculus (the General Mean-Value Theorem) to any suitable function $f : D \subseteq \mathbb{R}^m \to \mathbb{R}$ near a point $\mathbf{p} \in D$. We again consider the composition of f with a function g whose image contains the straight line segment in D joining two distinct points $\mathbf{p}$ and $\mathbf{p} + \mathbf{h}$.

Let E be an open interval in $\mathbb{R}$ containing $[0, 1]$ such that $\mathbf{p} + t\mathbf{h} \in D$ for all $t \in E$. Let $g : E \subseteq \mathbb{R} \to \mathbb{R}^m$ be given by $g(t) = \mathbf{p} + t\mathbf{h}$, $t \in E$. Define a function $F : E \subseteq \mathbb{R} \to \mathbb{R}$ by

3.11.1 $\qquad F(t) = (f \circ g)(t) = f(\mathbf{p} + t\mathbf{h}), \qquad t \in E.$

We know from Taylor's Theorem that provided F is n times differentiable on E there exists $0 < \theta < 1$ such that

3.11.2 $\quad \begin{aligned} f(\mathbf{p} + \mathbf{h}) - f(\mathbf{p}) &= F(1) - F(0) \\ &= F'(0) + \frac{F''(0)}{2!} + \cdots + \frac{F^{(n-1)}(0)}{(n-1)!} + \frac{F^{(n)}(\theta)}{n!}. \end{aligned}$

Our problem is to interpret these higher derivatives of F in terms of the given function f.

Suppose that f is differentiable on D. Then Theorem 3.8.7 (the Chain Rule) implies that F is differentiable and

3.11.3 $\qquad F'(t) = (\operatorname{grad} f)(\mathbf{p} + t\mathbf{h}) \cdot \mathbf{h}, \qquad t \in E.$

Hence by Theorem 3.8.6(ii)

3.11.4 $\qquad F'(t) = \|\mathbf{h}\| \frac{\partial f}{\partial \mathbf{u}}(\mathbf{p} + t\mathbf{h})$

where $\mathbf{u} = \|\mathbf{h}\|^{-1}\mathbf{h}$ is the unit vector in the direction of $\mathbf{h}$.

Suppose now in addition that the function $\partial f/\partial \mathbf{u} : D \subseteq \mathbb{R}^m \to \mathbb{R}$ is itself differentiable in D. The same argument can be used to deduce from 3.11.4 that F' is differentiable and that

3.11.5 $\qquad F''(t) = \|\mathbf{h}\|^2 \frac{\partial^2 f}{\partial \mathbf{u}^2}(\mathbf{p} + t\mathbf{h}), \qquad t \in E.$

The process begun in 3.11.4 and 3.11.5 can now be continued. A simple inductive argument shows that if $f, \partial f/\partial \mathbf{u}, \ldots, \partial^{n-1} f/\partial \mathbf{u}^{n-1}$ are differentiable, then F is n times differentiable and

3.11.6 $\qquad F^{(r)}(t) = \|\mathbf{h}\|^r \frac{\partial^r f}{\partial \mathbf{u}^r}(\mathbf{p} + t\mathbf{h}), \qquad t \in E, r = 1, \ldots, n.$

We can now state and prove the following extension of Taylor's Theorem.

3.11.7 *Theorem* (Taylor).   *Let* **p** *and* **p** + **h** *be two distinct points in an open subset D of* $\mathbb{R}^m$ *such that the straight line segment joining* **p** *and* **p** + **h** *lies in D. Let* **u** *be the unit vector in direction* **h**. *Consider a function* $f : D \subseteq \mathbb{R}^m \to \mathbb{R}$ *such that, for some* $n \in \mathbb{N}$, *the functions*

$$f, \frac{\partial f}{\partial \mathbf{u}}, \cdots, \frac{\partial^{n-1} f}{\partial \mathbf{u}^{n-1}}$$

are all differentiable in D.

Then there exists $0 < \theta < 1$ *such that*

3.11.8 $f(\mathbf{p} + \mathbf{h}) = f(\mathbf{p}) + \|\mathbf{h}\| \dfrac{\partial f}{\partial \mathbf{u}}(\mathbf{p}) + \dfrac{\|\mathbf{h}\|^2}{2!} \dfrac{\partial^2 f}{\partial \mathbf{u}^2}(\mathbf{p}) + \cdots$

$$+ \frac{\|\mathbf{h}\|^{n-1}}{(n-1)!} \frac{\partial^{n-1} f}{\partial \mathbf{u}^{n-1}}(\mathbf{p}) + \frac{\|\mathbf{h}\|^n}{n!} \frac{\partial^n f}{\partial \mathbf{u}^n}(\mathbf{p} + \theta \mathbf{h}).$$

Proof.   Since D is open in $\mathbb{R}^m$ we can choose E, an open interval in $\mathbb{R}$, such that $[0, 1] \subseteq E$ and $\mathbf{p} + t\mathbf{h} \in D$ for all $t \in E$. Define $F : E \subseteq \mathbb{R} \to \mathbb{R}$ by the rule $F(t) = f(\mathbf{p} + t\mathbf{h})$, $t \in E$. Then by the above inductive argument and by 3.11.6 the function F is n times differentiable in E. That is, F satisfies the hypothesis of Taylor's Theorem of elementary calculus. Substituting for $F^{(r)}$ from 3.11.6 in 3.11.2, we obtain the required result.

We note in passing that the hypothesis in Theorem 3.11.7 could be weakened. By just considering f on the segment joining **p** and **p** + **h** and by considering $\partial f / \partial \mathbf{u}$ and higher-order directional derivatives from first principles we could prove 3.11.8 by just assuming that f is well behaved on the segment. We have chosen the above approach since it is simpler, fits in with the more applicable form of the theorem (Theorem 3.11.10) and relies on the link between differentiability and linear approximations.

For a fixed vector $\mathbf{p} \in D$, 3.11.8 expresses $f(\mathbf{p} + \mathbf{h})$ as a polynomial in $\|\mathbf{h}\|$ plus a remainder term

3.11.9 $$R_n = \frac{\|\mathbf{h}\|^n}{n!} \frac{\partial^n f}{\partial \mathbf{u}^n}(\mathbf{p} + \theta \mathbf{h}).$$

In many important examples the remainder term can be made as small as we like by taking large enough values of n. To evaluate the polynomial and the remainder we need to be able to calculate the higher-order directional derivatives. Theorem 3.10.13 gives an expression for directional derivatives in terms of partial derivatives.

In order to use that expression we strengthen the hypothesis of Taylor's Theorem and assume n times continuous differentiability of f. It is this version of Taylor's Theorem which is useful for applications.

3.11.10 Theorem. *Let* $\mathbf{p}$ *and* $\mathbf{p} + \mathbf{h}$ *be two distinct points in an open subset D of $\mathbb{R}^m$ such that the straight line segment joining $\mathbf{p}$ and $\mathbf{p} + \mathbf{h}$ lies in D. Let $f : D \subseteq \mathbb{R}^m \to \mathbb{R}$ be a function which, for some $n \in \mathbb{N}$, is n times continuously differentiable on D. Then there exists $0 < \theta < 1$ such that*

3.11.11 $$f(\mathbf{p} + \mathbf{h}) = f(\mathbf{p}) + \sum_{r=1}^{n-1} \frac{1}{r!}\left[\left(h_1\frac{\partial}{\partial x_1} + \cdots + h_m\frac{\partial}{\partial x_m}\right)^r f\right](\mathbf{p}) + R_n$$

where

3.11.12 $$R_n = \frac{1}{n!}\left[\left(h_1\frac{\partial}{\partial x_1} + \cdots + h_m\frac{\partial}{\partial x_m}\right)^n f\right](\mathbf{p} + \theta\mathbf{h}).$$

Proof. Let $\mathbf{u}$ be the unit vector in direction $\mathbf{h}$. By Theorem 3.10.13 the hypothesis of this theorem implies the hypothesis of Theorem 3.11.7. The conclusion of the theorem follows from the conclusion of Theorem 3.11.7 by a second reference to Theorem 3.10.13, bearing in mind that $\|\mathbf{h}\|^r u_i^r = h_i^r$, for $i = 1, \ldots, m$.

We call 3.11.11 the nth *order Taylor expansion of f at* $\mathbf{p}$. The first-order expansion is that obtained from the Mean-Value Theorem (Theorem 3.9.1): $f(\mathbf{p} + \mathbf{h}) = f(\mathbf{p}) + R_1$.

3.11.13 Example. Consider the function $f : \mathbb{R}^2 \to \mathbb{R}$ defined by $f(x_1, x_2) = x_1^2 x_2$ near $\mathbf{p} = (10, 20)$. The partial derivatives are as follows.

$$(\partial f/\partial x_1)(\mathbf{p}) = 2p_1 p_2 = 400; \qquad (\partial f/\partial x_2)(\mathbf{p}) = p_1^2 = 100;$$
$$(\partial^2 f/\partial x_1^2)(\mathbf{p}) = 40; \qquad (\partial^2 f/\partial x_2^2)(\mathbf{p}) = 0$$

and $\qquad (\partial^2 f/\partial x_1 \partial x_2)(\mathbf{p}) = (\partial^2 f/\partial x_2 \partial x_1)(\mathbf{p}) = 2p_1 = 20.$

Taking $h = (0.1, 0.2)$ we obtain
$f(10.1, 20.2) = f(10, 20) + R_1 = 2000 + R_1,$
$f(10.1, 20.2) = 2000 + 400h_1 + 100h_2 + R_2 = 2060 + R_2,$
$f(10.1, 20.2) = 2000 + 400h_1 + 100h_2 + \frac{1}{2}(40h_1^2 + 2 \times 20 \times h_1 h_2 + 0h_2^2) + R_3$
$\qquad = 2060.6 + R_3.$
In fact, $f(10.1, 20.2) = 2060.602$, so that $R_1 = 60.602$, $R_2 = 0.602$ and $R_3 = 0.002$.

This example shows how the Taylor expansion can be used to estimate the value of a function in the vicinity of a point **p**. The importance of Theorem 3.11.10 lies in the expression for the remainder 3.11.12.

We now consider the remainder term R_n in more detail. Remember that the function $f : D \subseteq \mathbb{R}^m \to \mathbb{R}$ is n times continuously differentiable.

From 3.11.12, using the fact that $\|\mathbf{h}\|^n u_i^n = h_i^n$ for $i = 1, \dots, m$, where **u** is the unit vector in the direction of **h**,

$$R_n = \frac{\|\mathbf{h}\|^n}{n!}\left[\left(u_1\frac{\partial}{\partial x_1} + \dots + u_m\frac{\partial}{\partial x_m}\right)^n f\right](\mathbf{p} + \theta\mathbf{h}).$$

In particular the nth-order partial derivatives of f are continuous and so we can express the remainder in the form

3.11.14 $R_n = \dfrac{\|\mathbf{h}\|^n}{n!}\left[\left(u_1\dfrac{\partial}{\partial x_1} + \dots + u_m\dfrac{\partial}{\partial x_m}\right)^n f\right](\mathbf{p}) + \|\mathbf{h}\|^n\eta(\mathbf{h})$

where $\lim_{\mathbf{h} \to 0} \eta(\mathbf{h}) = 0$. The function η depends of course on our choice of n.

From 3.11.11 we obtain

3.11.15

$$f(\mathbf{p} + \mathbf{h}) = f(\mathbf{p}) + \sum_{r=1}^{n}\frac{1}{r!}\left[\left(h_1\frac{\partial}{\partial x_1} + \dots + h_m\frac{\partial}{\partial x_m}\right)^r f\right](\mathbf{p}) + \|\mathbf{h}\|^n\eta(\mathbf{h})$$

where $\lim_{\mathbf{h} \to 0} \eta(\mathbf{h}) = 0$.

3.11.16 Definition. *Let* $f : D \subseteq \mathbb{R}^m \to \mathbb{R}$ *be a* C^n *function and let* $\mathbf{p} \in D$. *Define a function* $T_n : \mathbb{R}^m \to \mathbb{R}$ *by*

$$T_n(\mathbf{h}) = \sum_{r=1}^{n}\frac{1}{r!}\left[\left(h_1\frac{\partial}{\partial x_1} + \dots + h_m\frac{\partial}{\partial x_m}\right)^r f\right](\mathbf{p}), \qquad \mathbf{h} \in \mathbb{R}^m.$$

Then $f(\mathbf{p}) + T_n(\mathbf{h})$ *is called the* **Taylor polynomial** *of degree n of f at* **p**. *It is a polynomial in* $h_1, \dots, h_m$.

Referring to Example 3.11.13 the first two Taylor polynomials of the function $f : \mathbb{R}^2 \to \mathbb{R}$ defined by $f(x_1, x_2) = x_1^2 x_2$ are

$$T_1(\mathbf{h}) = 2000 + 400h_1 + 100h_2,$$
$$T_2(\mathbf{h}) = 2000 + 400h_1 + 100h_2 + 20h_1^2 + 20h_1 h_2.$$

We can now express 3.11.15 in terms of the difference function and the polynomial T_n as follows:

3.11.17 $\delta_{f,\mathbf{p}}(\mathbf{h}) = f(\mathbf{p} + \mathbf{h}) - f(\mathbf{p}) = T_n(\mathbf{h}) + \|\mathbf{h}\|^n \eta(\mathbf{h})$

where $\lim_{\mathbf{h}\to\mathbf{0}} \eta(\mathbf{h}) = 0$.

When $n = 1$, 3.11.17 simply becomes 3.3.7. The polynomial function T_1 is the unique linear function (the differential) $L_{f,\mathbf{p}}$ that closely approximates $\delta_{f,\mathbf{p}}$ near the origin. In this sense T_n is a generalization of the differential. To explore this further we use the following extension of 'close approximation' (compare Definition 3.3.5).

3.11.18 Definition.   *Let $g : N \subseteq \mathbb{R}^m \to \mathbb{R}$ and $g^* : M \subseteq \mathbb{R}^m \to \mathbb{R}$ be two functions, both of whose open domains contain $\mathbf{0}$. We say that g and g^* closely approximate each other to the nth degree near $\mathbf{0}$ if there is a function $\eta : N \cap M \subseteq \mathbb{R}^m \to \mathbb{R}$ such that*

$$g^*(\mathbf{h}) - g(\mathbf{h}) = \|\mathbf{h}\|^n \eta(\mathbf{h}), \qquad \mathbf{h} \in N \cap M$$

and $\lim_{\mathbf{h}\to\mathbf{0}} \eta(\mathbf{h}) = 0$.

The above condition can be written

3.11.19 $g(\mathbf{0}) = g^*(\mathbf{0})$ and $\lim\limits_{\mathbf{h}\to\mathbf{0}} \dfrac{|g^*(\mathbf{h}) - g(\mathbf{h})|}{\|\mathbf{h}\|^n} = 0.$

Notice that if g^* closely approximates g to the nth degree near $\mathbf{0}$, then it also closely approximates g to the rth degree for all $r < n$. So a higher-degree approximation is 'better' than a lower-degree approximation.

We now have the following generalization of Theorem 3.3.13.

3.11.20 Theorem.   *Let $f : D \subseteq \mathbb{R}^m \to \mathbb{R}$ be n times continuously differentiable and let $\mathbf{p} \in D$. Then the polynomial T_n of Definition 3.11.16 is the only nth degree polynomial that closely approximates $\delta_{f,\mathbf{p}}$ to the nth degree near $\mathbf{0}$.*

Proof.   The relation 3.11.15 establishes that T_n is a sufficiently close approximation. The fact that it is the only one follows from the fact that if two nth-degree polynomials are close approximates to the nth-degree near $\mathbf{0}$ then they are equal. See Exercise 3.11.4.

3.11.21 *Example.* Find the best approximation to the function $f(x, y) = e^x \cos y$ by a second degree polynomial near **0**.

The best second-degree approximation is the Taylor polynomial $f(\mathbf{0}) + T_2(\mathbf{x})$. We could calculate the partial derivatives of f up to second-order and solve our problem by substituting in the expression

$$f(\mathbf{0}) + x\frac{\partial f}{\partial x}(\mathbf{0}) + y\frac{\partial f}{\partial y}(\mathbf{0}) + \frac{1}{2!}\left(x^2\frac{\partial^2 f}{\partial x^2}(\mathbf{0}) + 2xy\frac{\partial^2 f}{\partial x\partial y}(\mathbf{0}) + y^2\frac{\partial^2 f}{\partial y^2}(\mathbf{0})\right).$$

Alternatively we could use the known series for e^x and $\cos y$ as follows.

$$e^x = 1 + x + \tfrac{1}{2}x^2 + x^2\alpha(x), \qquad \text{where } \lim_{x \to 0} \alpha(x) = 0,$$

$$\cos y = 1 - \tfrac{1}{2}y^2 + y^2\beta(y), \qquad \text{where } \lim_{y \to 0} \beta(y) = 0.$$

Multiplying these two series together, we have

3.11.22 $\qquad\qquad e^x \cos y = 1 + x + \tfrac{1}{2}x^2 - \tfrac{1}{2}y^2 + R$

where R involves terms of the form $y^2\beta(y)$ and $x^2\alpha(x)$ and others of higher than second order. Hence

$$\lim_{(x, y) \to (0, 0)} R/\|(x, y)\|^2 = 0$$

and, from 3.11.22, $1 + x + \tfrac{1}{2}x^2 - \tfrac{1}{2}y^2$ closely approximates $e^x \cos y$ to the second degree. By Theorem 3.11.20, $1 + x + \tfrac{1}{2}x^2 - \tfrac{1}{2}y^2$ is the required approximation.

3.11.23 *Example.* Obtain a second-degree polynomial suitable for estimating the values of $f(x, y) = \sqrt{(1 + x + 4y)}$ near $(-1, 1)$.

We require the Taylor polynomial of degree 2 of f at $\mathbf{p} = (-1, 1)$. Define (h, k) by the rule $(x, y) = (-1, 1) + (h, k)$.

Method 1. Calculate all partial derivatives of f up to the second-order at $(-1, 1)$. Then from 3.11.15

$$\begin{aligned} f(x, y) &= f((-1, 1) + (h, k)) \\ &= f(-1, 1) + T_2(h, k) + \|(h, k)\|^2\eta(h, k). \end{aligned}$$

The required polynomial is $f(-1, 1) + T_2(h, k)$.

Method 2. As in the previous example we may reduce the problem to an application of Taylor's Theorem for functions defined on $\mathbb{R}$. We have

$$f(x, y) = f(-1 + h, 1 + k) = \sqrt{[1 + (-1 + h) + 4(1 + k)]} = \sqrt{(4 + t)}$$

where $t = h + 4k$. Now by the Binomial Theorem (Taylor's Theorem!)

$$\sqrt{(4 + t)} = 2\left(1 + \frac{t}{4}\right)^{1/2} = 2\left(1 + \frac{t}{8} - \frac{t^2}{128} + t^2\eta(t)\right)$$

where $\lim_{t \to 0} \eta(t) = 0$. Hence

$$f(-1+h, 1+k) = 2\left(1 + \tfrac{1}{8}(h+4k) - \tfrac{1}{128}(h+4k)^2\right) + R$$

where $\lim_{(h,k) \to (0,0)} R/(h^2+k^2) = 0$.

The required Taylor polynomial at $(-1, 1)$ is therefore

3.11.24 $\qquad\qquad 2 + \tfrac{1}{4}h + k - \tfrac{1}{64}h^2 - \tfrac{1}{8}hk - \tfrac{1}{4}k^2.$

By taking $h = -0.1$ and $k = -0.1$ we obtain an estimate of 1.8711 for $f(-1.1, 0.9)$ from this polynomial. The actual value of $f(-1.1, 0.9)$ (correct to four decimal places) is 1.8708.

Warning. The student may be tempted in this example to expand $\sqrt{(1+x+4y)}$ by the Binomial Theorem in the form $\sqrt{(1+u)}$, where $u = x + 4y$. This is inappropriate for this problem because we are interested in values of (x, y) near $(-1, 1)$ which correspond to values of u near 3. The Taylor polynomial of $\sqrt{(1+u)}$ of degree 2 is uninformative near $u = 3$, for here the remainder term is large.

3.11.25 *Example.* We now take a further look at our estimation of the acceleration in Example 3.5.3. The relationship is $a = f(v, s) = v^2/2s$ and the observed values are $(v, s) = (30, 300)$ with possible errors given by $|\Delta v| \leqslant 1$ and $|\Delta s| \leqslant 3$. Letting the true acceleration be $a_t\, \mathrm{m\,s^{-2}}$ we have

[i] the possible range is $1.388 \leqslant a_t \leqslant 1.618$ (see 3.5.9);

[ii] the approximate range given by 3.5.7 (effectively the Taylor polynomial of degree 1 at (30, 300)) is $1.385 \leqslant a_t \leqslant 1.615$,

[iii] the approximate range given by the Mean-Value Theorem (3.9.9) is $-1.379 \leqslant a_t \leqslant 1.621$.

We now look at the estimate provided by the Taylor polynomial of degree 2 at (30, 300). Using the indirect technique developed in the previous examples,

$$a_t = \tfrac{1}{2}(v+\Delta v)^2(s+\Delta s)^{-1} = \frac{1}{2s}(v+\Delta v)^2\left(1 + \frac{\Delta s}{s}\right)^{-1}$$

$$= \frac{1}{2s}(v^2 + 2v\Delta v + \Delta v^2)\left(1 - \frac{\Delta s}{s} + \left(\frac{\Delta s}{s}\right)^2 + \cdots\right)$$

$$= \frac{v^2}{2s} + \frac{v}{s}\Delta v - \frac{v^2}{2s^2}\Delta s + \frac{\Delta v^2}{2s} - \frac{v}{s^2}\Delta v\,\Delta s + \frac{v^2}{2s^3}\Delta s^2 + R.$$

where R involves terms of higher than second order in Δv and Δs.

The Taylor polynomial of degree 2 near $(v, s) = (30, 300)$ is therefore

$$\frac{v^2}{2s} + \frac{v}{s}\Delta v - \frac{v^2}{2s^2}\Delta s + \frac{1}{2s}\Delta v^2 - \frac{v}{s^2}\Delta v\,\Delta s + \frac{v^2}{2s^3}\Delta s^2.$$

This clearly takes its maximum value when $\Delta v = 1$ and $\Delta s = -3$ and its minimum value when $\Delta v = -1$ and $\Delta s = 3$. To three places of decimals the resulting approximation is $1.388 \leqslant a_t \leqslant 1.618$. We have therefore come very close to the achievable range.

It must be stressed that before applying Taylor's Theorem to a function f one should check that the continuity conditions on the partial derivatives of f are satisfied. We took this for granted in the above examples, since well-known functions were involved. Failure of the conditions often manifests itself in the breakdown of calculations. For example, suppose that we are looking for a Taylor polynomial approximating $f(x, y) = \sqrt{(1 + x + 4y)}$ near $(x, y) = (3, -1)$. We begin with the substitution $x = 3 + h$, $y = -1 + k$ and obtain $f(3 + h, 1 + k) = \sqrt{(h + 4k)}$. What next? Putting $t = h + 4k$ fails, for there is no polynomial in t that closely approximates $\sqrt{t}$ near the origin. The trouble can be diagnosed by calculating $(\partial f/\partial x)(x, y)$. It does not exist at $(3, -1)$.

Exercises 3.11

1. For each of the following functions f, obtain the second-degree Taylor polynomial $f(2, 1) + T_2(h, k)$ suitable for estimating the values of $f(2 + h, 1 + k)$ for numerically small values of h, k. For the case $h = 0.2$, $k = -0.1$ compare your estimate with the accurate value.

    (a) $f(x, y) = x^2 y^2$;   (b) $f(x, y) = \ln(1 + x + 2y)$
    (c) $f(x, y) = e^x \sin y$.

Answers: (a) $4 + 4h + 8k + h^2 + 8hk + 4k^2$;
(b) $\ln 5 + \frac{1}{50}(10h + 20k - h^2 - 4hk - 4k^2)$.

    Compare your estimates with those obtained from the Taylor polynomial of degree 3 of f at $(2, 1)$.

2. Find the best approximation to the functions of Exercise 1 by a second-degree polynomial near $(0, 0)$.

Answers: (a) 0;   (b) $x + 2y - \frac{1}{2}(x^2 + 4xy + 4y^2)$;   (c) $y + xy$.

3. Do the following limits exist?

    (a) $\displaystyle \lim_{(x, y) \to (0, 0)} \frac{\sin(xy)}{x}$ (b) $\displaystyle \lim_{(x, y) \to (0, 0)} \frac{\cos(xy) - 1}{x^2 y}$

Hint: consider Taylor expansions with remainder near $(0, 0)$.

Answer: Both limits exist and are 0.

4. Let $f(x_1, \ldots, x_m)$ and $f^*(x_1, \ldots, x_m)$ be two polynomials of degree n. Prove that if f^* closely approximates f to the nth degree near 0 (Definition 3.11.18), then $f = f^*$.

Hint: in the expression $f^*(\mathbf{h}) - f(\mathbf{h}) = \|\mathbf{h}\|^n \eta(\mathbf{h})$, prove that $\eta(\mathbf{h}) = 0$, $\mathbf{h} \in \mathbb{R}^m$.)

5. Prove that close approximation to the nth degree near **0** (Definition 3.11.18) defines an equivalence relation on the set of functions $f : N \subseteq \mathbb{R}^m \to \mathbb{R}$ whose domains are open sets containing **0**.

3.12 Local extrema

Remember that we are considering functions $f : D \subseteq \mathbb{R}^m \to \mathbb{R}$ whose domain D is an open subset of $\mathbb{R}^m$.

3.12.1 Definition. [i] *The function f is said to have a* local maximum *at $\mathbf{p} \in D$ if there is a neighbourhood N of $\mathbf{p}$ such that*

$$f(\mathbf{x}) \leqslant f(\mathbf{p}) \text{ for all } \mathbf{x} \in N.$$

[ii] *The function f is said to have a* local minimum *at $\mathbf{p}$ if there is a neighbourhood N of $\mathbf{p}$ such that*

$$f(\mathbf{x}) \geqslant f(\mathbf{p}) \text{ for all } \mathbf{x} \in N.$$

3.12.2 Definition. *We say that the function f has a* local extremum *at $\mathbf{p} \in D$ if f has either a local maximum or a local minimum at $\mathbf{p}$. In that case $\mathbf{p}$ is called an* extremal point *of f.*

In the case $m = 2$, provided f is continuous, the graph of f will look like a landscape. If $(\mathbf{p}, f(\mathbf{p}))$ is a mountain peak, then f has a local maximum at $\mathbf{p}$. If $(\mathbf{p}, f(\mathbf{p}))$ is at the bottom of a valley, then f has a local minimum at $\mathbf{p}$. Notice however that if $(\mathbf{p}, f(\mathbf{p}))$ is in the middle of a plain, then f has both a local maximum and a local minimum at the point $\mathbf{p}$.

3.12.3 Theorem. *Let $f : D \subseteq \mathbb{R}^m \to \mathbb{R}$ have a local extremum at $\mathbf{p} \in D$ and let $\mathbf{u}$ be a unit vector in $\mathbb{R}^m$ such that $(\partial f / \partial \mathbf{u})(\mathbf{p})$ exists. Then $(\partial f / \partial \mathbf{u})(\mathbf{p}) = 0$.*

Proof. If f has a local maximum at $\mathbf{p}$, then $f(\mathbf{p} + t\mathbf{u}) - f(\mathbf{p}) \leqslant 0$ for all numerically small t. Hence

$$\lim_{t \to 0+} \frac{f(\mathbf{p} + t\mathbf{u}) - f(\mathbf{p})}{t} \leqslant 0 \quad \text{and} \quad \lim_{t \to 0} \frac{f(\mathbf{p} + t\mathbf{u}) - f(\mathbf{p})}{t} \geqslant 0.$$

It follows that $(\partial f/\partial \mathbf{u})(\mathbf{p}) \leqslant 0$ and $(\partial f/\partial \mathbf{u})(\mathbf{p}) \geqslant 0$. Hence $(\partial f/\partial \mathbf{u})(\mathbf{p}) = 0$. A similar argument applies if f has a local minimum at $\mathbf{p}$.

3.12.4 Corollary.   *Let a differentiable function* $f : D \subseteq \mathbb{R}^m \to \mathbb{R}$ *have a local extremum at* $\mathbf{p} \in D$. *Then*

[i] $\dfrac{\partial f}{\partial x_j}(\mathbf{p}) = 0$ *for* $j = 1, \ldots, m$;

[ii] $L_{f,\mathbf{p}} : \mathbb{R}^m \to \mathbb{R}$ *is the zero transformation.*

Proof.   (i) $\dfrac{\partial f}{\partial x_j}(\mathbf{p}) = \dfrac{\partial f}{\partial \mathbf{e}_j}(\mathbf{p}) = 0$ for $i = 1, \ldots, m$, by Theorem 3.12.3.
(ii) Immediate from (i) and Theorem 3.3.20.

For the case $m = 2$ this Corollary tells us that if f has a local extremum at $\mathbf{p} \in D$ then the tangent plane to the graph of f at $(\mathbf{p}, f(\mathbf{p}))$ is horizontal. Our experience of landscapes, however, suggests that there may be points on the graph which do not correspond to local extrema, but where nevertheless the tangent planes are horizontal. (See Fig. 3.20). We are led to the following definition.

3.12.5 Definition.   *Let* $f : D \subseteq \mathbb{R}^m \to \mathbb{R}$ *be a differentiable function. Any point* $\mathbf{p} \in D$ *for which* $L_{f,\mathbf{p}}$ *is the zero transformation is called a* critical point *of* f. *Equivalently (see 3.3.20(ii)) the critical points are those points* $\mathbf{p} \in D$ *for which* $(\partial f/\partial x_j)(\mathbf{p}) = 0$ *for all* $j = 1, \ldots, m$.

In the case $m = 2$, $\mathbf{p}$ is a critical point if and only if the graph of f has a horizontal tangent plane at $(\mathbf{p}, f(\mathbf{p}))$. Corollary 3.12.4 tells us that extremal points of differentiable functions are also critical points.
The following examples explore this relationship further.

3.12.6 Example.   Define $f : \mathbb{R} \to \mathbb{R}$ by the rule $f(x) = |x|$. Then f has a local minimum at 0. But 0 is not a critical point of f since f is not differentiable at 0.

3.12.7 Example.   Define $f : \mathbb{R} \to \mathbb{R}$ by the rule $f(x) = x^3$. The function is strictly increasing and so has no extremal points, but $f'(0) = 0$, and so 0 is a critical point of f.

3.12.8 Example.   Define $f : \mathbb{R}^2 \to \mathbb{R}$ by the rule $f(x, y) = x^2 - y^2$. The graph of f is sketched in Fig. 3.20. Now f is differentiable and $(\partial f/\partial x)(x, y) = 2x$

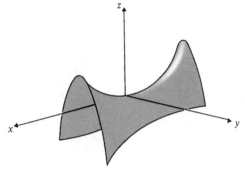

Fig. 3.20 Graph $z = x^2 - y^2$

and $(\partial f / \partial y)(x, y) = -2y$. It follows that f has only one critical point, the origin $(0, 0)$ where it takes the value 0. The tangent plane at $(0, 0)$ is the plane $z = 0$ in $\mathbb{R}^3$. Now $f(h, 0) = h^2 > 0$ for $h \neq 0$ and $f(0, k) = -k^2 < 0$ for $k \neq 0$. Hence in any neighbourhood of $(0, 0)$ the function takes values greater than $f(0, 0)$ and values less than $f(0, 0)$. So $(0, 0)$ is not an extremal point of f.

The graph of f is the surface $z = x^2 - y^2$ in $\mathbb{R}^3$. The horizontal sections (the intersection of the graph with the planes $z = c$) are hyperbolas. The vertical sections (the intersection of the graph with the planes $y = c$ and $x = c$) are parabolas. For this reason the surface is called a hyperbolic paraboloid. The saddle-like shape of this surface gives us the following definition.

3.12.9 Definition. *A point* $\mathbf{p} \in D$ *is called a* saddle point *of a differentiable function* $f : D \subseteq \mathbb{R}^m \to \mathbb{R}$ *if* $\mathbf{p}$ *is a critical point but* f *does not have a local extremum at* $\mathbf{p}$.

Do not be misled by the name—some very peculiar saddles turn up as the graphs of functions near a saddle point.

3.12.10 Example. An amusing example is that of a 'monkey saddle' which allows for the monkey's anatomy (Fig. 3.21 (i)). This is illustrated by the graph of the function $f : \mathbb{R}^2 \to \mathbb{R}$ defined by

$$f(x, y) = x^3 - 3xy^2 = x(x - \sqrt{3}y)(x + \sqrt{3}y).$$

The function is differentiable. The partial derivatives are $(\partial f / \partial x)(x, y) = 3x^2 - 3y^2$ and $(\partial f / \partial y)(x, y) = -6xy$. The only critical point of the function therefore is the origin $(0, 0)$. This point can be identified as a 'monkey', saddle point by marking the regions of $\mathbb{R}^2$ in the neighbourhood of $(0, 0)$ in

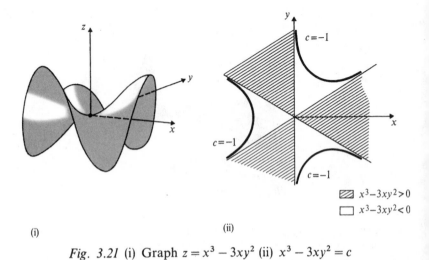

Fig. 3.21 (i) Graph $z = x^3 - 3xy^2$ (ii) $x^3 - 3xy^2 = c$

which f takes positive and negative values. See Fig. 3.21(ii). For example, f takes positive values when $x > 0$ and $x > \sqrt{3}y$ and $x > -\sqrt{3}y$. More detailed information is obtainable by drawing the level sets of f in $\mathbb{R}^2$.

Experience of elementary calculus suggests that it is important to develop a technique for finding and identifying local extrema of differentiable functions. Corollary 3.12.4 tells us that we can narrow the search by first picking out the critical points—those points where the partial derivatives all vanish. We shall use our work on Taylor expansions to establish conditions which will, in many cases, enable us to pick out extremal points from among the critical points and to identify which ones correspond to local maxima and which to local minima.

Let $\mathbf{p}$ be a critical point of a twice continuously differentiable function $f : D \subseteq \mathbb{R}^m \to \mathbb{R}$. Since D is an open set in $\mathbb{R}^m$, there exists $\varepsilon > 0$ such that $N(\mathbf{p}, \varepsilon) \subseteq D$. The straight line segment joining $\mathbf{p}$ and $\mathbf{p} + \mathbf{h}$ lies in D for any vector $\mathbf{h} \in \mathbb{R}^m$ such that $\|\mathbf{h}\| < \varepsilon$. By 3.11.15, for such a vector $\mathbf{h}$, since the first-order partial derivatives of f all vanish at $\mathbf{p}$,

3.12.11
$$f(\mathbf{p} + \mathbf{h}) - f(\mathbf{p}) = \frac{1}{2}\left[\left(h_1 \frac{\partial}{\partial x_1} + \cdots + h_m \frac{\partial}{\partial x_m} \right)^2 f \right](\mathbf{p}) + \|\mathbf{h}\|^2 \eta(\mathbf{h}),$$

where $\lim_{\mathbf{h} \to \mathbf{0}} \eta(\mathbf{h}) = 0$. The second-order partial derivatives are

continuous and therefore, by the symmetry established in
Corollary 3.10.7,

3.12.12 $$\left[\left(h_1\frac{\partial}{\partial x_1}+\cdots+h_m\frac{\partial}{\partial x_m}\right)^2 f\right](\mathbf{p})$$

$$=\left(h_1^2\frac{\partial^2 f}{\partial x_1^2}(\mathbf{p})+\cdots+2h_1h_2\frac{\partial^2 f}{\partial x_1\partial x_2}(\mathbf{p})+\cdots\right).$$

The right-hand side of 3.12.12 is a quadratic polynomial $q(\mathbf{h})$
where q is a quadratic form on $\mathbb{R}^m$ (Section 1.4). In this case
$q(\mathbf{h})=[\mathbf{h}]^t A[\mathbf{h}]$, where A is the symmetric $m\times m$ matrix $[a_{ij}]$ with

3.12.13 $$a_{ij}=\frac{\partial^2 f}{\partial x_i\partial x_j}(\mathbf{p}),\qquad i,j=1,\ldots,m,$$

and $[\mathbf{h}]$ is the column whose entries are the coordinates $h_1,\ldots,h_m$
of $\mathbf{h}$. With this notation 3.12.11 becomes

3.12.14 $f(\mathbf{p}+\mathbf{h})-f(\mathbf{p})=\frac{1}{2}[\mathbf{h}]^t A[\mathbf{h}]+\|\mathbf{h}\|^2\eta(\mathbf{h})$, $\qquad\|\mathbf{h}\|<\varepsilon$.

We also know from Theorem 1.4.6 that all eigenvalues of A are real
and that if $\lambda_{\min}$ is the minimum eigenvalue and $\lambda_{\max}$ is the maximum
eigenvalue, then

3.12.15 $\|\mathbf{h}\|^2\lambda_{\min}\leqslant[\mathbf{h}]^t A[\mathbf{h}]\leqslant\|\mathbf{h}\|^2\lambda_{\max}$, $\qquad\mathbf{h}\in\mathbb{R}^m$.

It follows from 3.12.14 that

3.12.16 $\|\mathbf{h}\|^2(\frac{1}{2}\lambda_{\min}+\eta(\mathbf{h}))\leqslant f(\mathbf{p}+\mathbf{h})-f(\mathbf{p})$
$\leqslant\|\mathbf{h}\|^2(\frac{1}{2}\lambda_{\max}+\eta(\mathbf{h}))$, $\qquad\|\mathbf{h}\|<\varepsilon$.

We can now state and prove our main result. Recall that the
symmetric matrix A is said to be positive definite if $[\mathbf{h}]^t A[\mathbf{h}]>0$ for
all $\mathbf{h}\neq\mathbf{0}$, it is said to be negative definite if $[\mathbf{h}]^t A[\mathbf{h}]<0$ for all $\mathbf{h}\neq\mathbf{0}$
and it is indefinite if $[\mathbf{h}]^t A[\mathbf{h}]$ takes both positive and negative
values. Moreover, the matrix A is positive definite (negative definite)
if and only if all the eigenvalues of A are positive (negative), and it is
indefinite if and only if there are both positive and negative eigenvalues (see Theorem 1.4.12). We can therefore decide the nature of
A if we have an estimation of the largest and smallest eigenvalues
of A.

3.12.17 Theorem. *Suppose that the C^2 function $f:D\subseteq\mathbb{R}^m\to\mathbb{R}$ has a
critical point at $\mathbf{p}\in D$. Define a (symmetric) $m\times m$ matrix $A=[a_{ij}]$ by
the rule given in 3.12.13.*

[i] *If A is positive definite, then f has a local minimum at $\mathbf{p}$.*
[ii] *If A is negative definitive, then f has a local maximum at $\mathbf{p}$.*
[iii] *If A is indefinite, then $\mathbf{p}$ is a saddle point of f.*

Proof. Choose $\varepsilon > 0$ such that for $\|\mathbf{h}\| < \varepsilon$ the straight line segment joining $\mathbf{p}$ and $\mathbf{p} + \mathbf{h}$ is contained in D. We therefore have identity 3.12.11 and its consequences. The results will follow from 3.12.16 where $\lambda_{\min}$ and $\lambda_{\max}$ are respectively the minimum and maximum eigenvalues of A.

[i] If A is positive definite then, by Theorem 1.4.12, $\lambda_{\min} > 0$. Now since $\lim_{\mathbf{h} \to 0} \eta(\mathbf{h}) = 0$ we can find $\delta > 0$ such that

3.12.18 $\qquad\qquad |\eta(\mathbf{h})| < \tfrac{1}{2}\lambda_{\min}$ whenever $\|\mathbf{h}\| < \delta$.

Therefore if $\|\mathbf{h}\| < \min\{\varepsilon, \delta\}$, then from 3.12.16 and 3.12.18

$$0 < \|\mathbf{h}\|^2(\tfrac{1}{2}\lambda_{\min} + \eta(\mathbf{h})) \leqslant f(\mathbf{p} + \mathbf{h}) - f(\mathbf{p}).$$

It follows that f has a local minimum at $\mathbf{p}$.

[ii] If A is negative definite then again by Theorem 1.4.12, $\lambda_{\max} < 0$. We can find $\delta > 0$ such that

$$|\eta(\mathbf{h})| < -\tfrac{1}{2}\lambda_{\max} \qquad\qquad \text{whenever } \|\mathbf{h}\| < \delta.$$

Therefore if $\|\mathbf{h}\| < \min\{\varepsilon, \delta\}$, then from 3.12.16 and 3.12.18

$$0 > \|\mathbf{h}\|^2(\tfrac{1}{2}\lambda_{\max} + \eta(\mathbf{h})) \geqslant f(\mathbf{p} + \mathbf{h}) - f(\mathbf{p}).$$

It follows that f has a local maximum at $\mathbf{p}$.

[iii] If A is indefinite, then (Theorem 1.4.12) it has both positive and negative eigenvalues. In particular $\lambda_{\max} > 0$ and $\lambda_{\min} < 0$. Let $[\mathbf{v}_{\max}]$ and $[\mathbf{v}_{\min}]$ be any unit eigenvectors of A corresponding to $\lambda_{\max}$ and $\lambda_{\min}$ respectively. Then $A[\mathbf{v}_{\max}] = \lambda_{\max}[\mathbf{v}_{\max}]$, and from 3.12.14

$$f(\mathbf{p} + s\mathbf{v}_{\max}) - f(\mathbf{p}) = s^2(\tfrac{1}{2}\lambda_{\max} + \eta(s\mathbf{v}_{\max})).$$

Similarly,

$$f(\mathbf{p} + s\mathbf{v}_{\min}) - f(\mathbf{p}) = s^2(\tfrac{1}{2}\lambda_{\min} + \eta(s\mathbf{v}_{\min})).$$

Therefore, by the argument used in [i] and [ii], for all small enough s

$$f(\mathbf{p} + s\mathbf{v}_{\max}) - f(\mathbf{p}) > 0 \text{ and } f(\mathbf{p} + s\mathbf{v}_{\min}) - f(\mathbf{p}) < 0.$$

Hence $\mathbf{p}$ is a saddle point of f.

Remark. Theorem 3.12.17 generalizes the familiar test for local maxima and local minima of functions $f : \mathbb{R} \to \mathbb{R}$ in terms of the sign of $f''(x)$ (Exercise 3.12.5).

3.12.19 *Example.* The origin $(0, 0)$ is a critical point of the function $f : \mathbb{R}^2 \to \mathbb{R}$ defined by $f(x, y) = x^2 + y^2$. The associated matrix of second-order partial derivatives at $(0, 0)$ is

$$\begin{bmatrix} 2 & 0 \\ 0 & 2 \end{bmatrix}.$$

As expected, this is positive definite, indicating a minimum.

3.12.20 *Example.* The origin $(0, 0)$ is a critical point of the function $f(x, y) = x^2 - y^2$ considered in Example 3.12.8. The associated matrix of second-order partial derivatives at $(0, 0)$ is

$$\begin{bmatrix} 2 & 0 \\ 0 & -2 \end{bmatrix}$$

This is an indefinite matrix, indicating that $(0, 0)$ is a saddle point (see Fig. 3.20).

Theorem 3.12.17 does not give a complete answer to our problem. The theorem tells us nothing if the matrix of second derivatives at the critical point is either positive semi-definite or negative semi-definite. In such cases a direct approach from first principles is usually appropriate.

3.12.21 *Example.* The origin is a critical point of all three functions on $\mathbb{R}^2$,

$$f(x, y) = x^4 + y^4, \qquad g(x, y) = x^4 - y^4 \quad \text{and} \quad k(x, y) = -x^4 - y^4.$$

The matrix of second derivatives at $(0, 0)$ is the zero matrix in all three cases and yet f, g and k have respectively a maximum, a saddle and a minimum at $(0, 0)$.

In order to apply Theorem 3.12.17 we need to be able to recognize the nature of the symmetric matrix A. The most direct method is to take the polynomial $[\mathbf{x}]^t A [\mathbf{x}]$ and by completing squares express it as the sum and difference of squares. Alternatively one can examine the eigenvalues of A.

A useful test for positive definiteness is given by the following set of determinant conditions.

3.12.22 *Theorem.* *Let A be an $m \times m$ symmetric matrix and let A_r be* the $r \times r$ submatrix which is obtained by taking just the first r rows and the first r columns of A. The matrix A is positive definite if and only if the determinants det A_r are positive for $r = 1, \ldots, m$.

Proof.   A proof can be found G. Hadley, *Linear Algebra*. The case $m = 2$ is covered in Exercise 3.12.6.

3.12.23 Example.   The 3×3 symmetric matrix A is positive definite if and only if

$$a_{11} > 0, \quad \det \begin{bmatrix} a_{11} & a_{12} \\ a_{21} & a_{22} \end{bmatrix} > 0 \quad \text{and} \quad \det \begin{bmatrix} a_{11} & a_{12} & a_{13} \\ a_{21} & a_{22} & a_{23} \\ a_{31} & a_{32} & a_{33} \end{bmatrix} > 0.$$

3.12.24 Definition.   *Let $f : D \subseteq \mathbb{R}^m \to \mathbb{R}$ be a C^2 function with a critical point at $\mathbf{p} \in D$. Let $A = (a_{ij})$ be the $m \times m$ symmetric matrix defined by 3.12.13. The determinant of A is called the* discriminant *of f at $\mathbf{p}$ and is denoted by $\Delta_f(\mathbf{p})$.*

In these terms Theorem 3.12.17 gives us information about the nature of a critical point $\mathbf{p}$ of f when $\Delta_f(\mathbf{p}) \neq 0$.

3.12.25 Example.   In the case $m = 2$,

$$\Delta_f(\mathbf{p}) = \frac{\partial^2 f}{\partial x_1^2}(\mathbf{p}) \frac{\partial^2 f}{\partial x_2^2}(\mathbf{p}) - \left(\frac{\partial^2 f}{\partial x_1 \partial x_2}(\mathbf{p}) \right)^2.$$

For $m = 2$ we can state Theorem 3.12.17 in the following form.

3.12.26 Theorem.   *Let $\mathbf{p}$ be a critical point of a C^2 function $f : D \subseteq \mathbb{R}^2 \to \mathbb{R}$. Then*

[i]   *f has a local minimum at $\mathbf{p}$ if $\dfrac{\partial^2 f}{\partial x_1^2}(\mathbf{p}) > 0$ and $\Delta_f(\mathbf{p}) > 0$,*

[ii]   *f has a local maximum at $\mathbf{p}$ if $\dfrac{\partial f}{\partial x_1^2}(\mathbf{p}) < 0$ and $\Delta_f(\mathbf{p}) > 0$,*

[iii]   *$\mathbf{p}$ is a saddle point of f if $\Delta_f(\mathbf{p}) < 0$.*

Proof.   Let A be the 2×2 matrix defined in relation to f at $\mathbf{p}$ by 3.12.13.

[*i*]   By Theorem 3.12.22, if $(\partial^2 f / \partial x_1^2)(\mathbf{p}) > 0$ and $\Delta_f(\mathbf{p}) > 0$ then A is positive definite. It follows from Theorem 3.12.17(i) that f has a local minimum at $\mathbf{p}$.

[*ii*]   Define a function $g : D \subseteq \mathbb{R}^2 \to \mathbb{R}$ by $g(\mathbf{x}) = -f(\mathbf{x})$. Then $\mathbf{p}$ is also a critical point of g. If f satisfies the hypothesis of [ii], then g

satisfies the hypothesis of [i]. Hence g has a local minimum at $\mathbf{p}$ and therefore f has a local maximum at $\mathbf{p}$.

[*iii*] The product of the eigenvalues of A (with multiplicity allowed for) is $\det A = \Delta_f(\mathbf{p})$. Hence if $\Delta_f(\mathbf{p}) < 0$, then A has a positive and a negative eigenvalue. Therefore A is indefinite and Theorem 3.12.17(iii) implies that $\mathbf{p}$ is a saddle point of f.

3.12.27 Example. Consider the function $f : \mathbb{R}^2 \to \mathbb{R}$ defined by

$$f(x_1, x_2) = x_1^3 + x_2^3 - 6x_1 x_2.$$

The first-order partial derivatives are

$$\frac{\partial f}{\partial x_1}(x) = 3x_1^2 - 6x_2 \quad \text{and} \quad \frac{\partial f}{\partial x_2}(x) = 3x_2^2 - 6x_1.$$

The critical points are located at (x_1, x_2) where $3x_1^2 = 6x_2$ and $3x_2^2 = 6x_1$, giving $(0, 0)$ and $(2, 2)$ as the only solutions. The second-order partial derivatives are

$$\frac{\partial^2 f}{\partial x_1^2}(\mathbf{x}) = 6x_1, \quad \frac{\partial^2 f}{\partial x_2^2}\lambda(\mathbf{x}) = 6x_2, \quad \frac{\partial^2 f}{\partial x_1 \partial x_2}(\mathbf{x}) = -6.$$

Hence $\Delta_f(\mathbf{x}) = 36x_1 x_2 - 36$. By Theorem 3.12.26, $(0, 0)$ is a saddle point since $\Delta_f(0, 0) = -36 < 0$. Similarly f has a local minimum at $(2, 2)$ since $\Delta_f(2, 2) = 108 > 0$ and $(\partial^2 f/\partial x_1^2)(2, 2) = 12 > 0$. Since f is differentiable, all its extreme points are critical points and so there are no other local maxima or minima.

Notice that although the value at the local minimum $(2, 2)$ is -8 the function takes values less than -8 since $f(x_1, 0) = x_1^3$.

3.12.28 Example. Consider again the function $f(x, y) = x^3 - 3xy^2$ of Example 3.12.10. We found then that the only critical point is $(0, 0)$. However $\Delta_f(x, y) = -36x^2 - 36y^2$ and in particular $\Delta_f(0, 0) = 0$. In this case Theorem 3.12.26 is uninformative and the direct approach taken earlier is appropriate.

3.12.29 Example. Consider the function $f : \mathbb{R}^3 \to \mathbb{R}$ defined by $f(x, y, z) = xy + 2xz + 2yz$. We find that

$$\frac{\partial f}{\partial x}(x, y, z) = y + 2z, \quad \frac{\partial f}{\partial y}(x, y, z) = x + 2z, \quad \frac{\partial f}{\partial z}(x, y, z) = 2x + 2y$$

and deduce that $(0, 0, 0)$ is the only critical point. Calculation of the 3×3 matrix A of second-order partial derivatives at $(0, 0, 0)$ (see 3.12.13) gives

$$A = \begin{bmatrix} 0 & 1 & 2 \\ 1 & 0 & 2 \\ 2 & 2 & 0 \end{bmatrix}.$$

It will be found that A has two negative and one positive eigenvalues. Hence A is indefinite and by Theorem 3.12.17(iii) $(0, 0, 0)$ is a saddle point of f.

Our example 3.12.29 illustrates a general method. Sometimes one can find a shorter route towards the solution. Here, for example, we need only note that $f(h, k, 0) = hk$ and $f(-h, k, 0) = -hk$ to conclude that f takes both positive and negative values in every neighbourhood of the origin.

Exercises 3.12

1. Determine the nature of the critical points of the following functions $f : \mathbb{R}^3 \to \mathbb{R}$ by the method of Theorem 3.12.17.
    (a) $f(x, y, z) = x^2 + y^2 + z^2 + xy$;
    (b) $f(x, y, z) = xy + xz$.
    Find an alternative short method, if possible.

Answers: (a) local minimum at $(0, 0, 0)$. Complete the square.
    (b) saddle points at $(0, k, -k)$, any $k \in \mathbb{R}$.

2. Determine the nature of the critical points of the following functions $f : \mathbb{R}^2 \to \mathbb{R}$ by the method of Theorem 3.12.26.
    (a) $f(x, y) = x^5 y + xy^5 + xy$;
    (b) $f(x, y) = x^3 + y^2 - 3x$;
    (c) $f(x, y) = \sin x + \sin y + \cos(x + y)$;
    (d) $f(x, y) = x^2 + y^2 + (1/x^2 y^2)$, $xy \neq 0$.

Answers: (a) saddle point at $(0, 0)$;   (b) local minimum -2 at $(1, 0)$,
    saddle point at $(-1, 0)$;   (c) local minimum -3 at $(3\pi/2, 3\pi/2)$, local
    maximum $\frac{3}{2}$ at $(\pi/6, \pi/6)$ and $(5\pi/6, 5\pi/6)$, saddle points at $(\frac{1}{2}\pi, \frac{1}{2}\pi)$,
    $(\frac{1}{2}\pi, -\frac{1}{2}\pi)$, $(3\pi/2, -3\pi/2)$, further critical points by periodicity;
    (d) local minimum 3 at $(\pm 1, \pm 1)$.

3. Determine the nature of the critical points of the following functions $f : \mathbb{R}^2 \to \mathbb{R}$. Where the method of Theorem 3.12.26 fails, proceed either by inspection or by developing the Taylor expansion of f at the critical point to at least third order.
    (a) $f(x, y) = (x + y)^2 + x^4$;
    (b) $f(x, y) = (x + y)e^{-xy}$;
    (c) $f(x, y) = (y - x^2)(y - 2x^2)$.

Answers: (a) local minimum at $(0, 0)$ by inspection;
    (b) $(\pm 1/\sqrt{2}, \pm 1/\sqrt{2})$;   (c) saddle point at $(0, 0)$. Sketch a sign table
    for f near $(0, 0)$ in $\mathbb{R}^2$, indicating that $f(x, y) < 0$ when $x^2 < y < 2x^2$ and
    $f(x, y) > 0$ when $y < x^2$ or $y > 2x^2$.

4. Find the shortest distance from the origin of $\mathbb{R}^3$ to the surface $xyz = 1$.

Hint: find the minimum of $x^2 + y^2 + z^2$, where $z = 1/xy$. Use the calculations of Exercise 2 (d).

5. Apply Theorem 3.12.17 to a C^2 function $f : D \subseteq \mathbb{R} \to \mathbb{R}$ and prove in particular that if $p \in D$ is a critical point of f (so $f'(p) = 0$) and if $f''(p)$ is positive, then f has a local minimum at p.

Note: part (iii) of Theorem 3.12.17 does not apply since a 1×1 matrix has only one eigenvalue.

6. Prove that a 2×2 real symmetric matrix $A = [a_{ij}]$ is positive definite if and only if

$$a_{11} > 0 \quad \text{and} \quad \det \begin{bmatrix} a_{11} & a_{12} \\ a_{21} & a_{22} \end{bmatrix} > 0.$$

Hint: let λ_1, λ_2 be the eigenvalues of A. By Exercise 1.4.2, $\lambda_1 \lambda_2 = \det A$. Now the condition $a_{11} > 0$ means that if $\mathbf{h} = (1, 0)$, then $[\mathbf{h}]^t A [\mathbf{h}] = a_{11} > 0$, and hence that A is not negative definite. Deduce that the two conditions $a_{11} > 0$ and $\det A > 0$ hold if and only if $\lambda_1 > 0$ and $\lambda_2 > 0$.

7. Decide which of the local minima and local maxima of f calculated in Exercises 1, 2, and 3 are absolute minima or maxima of f. (For example, the local minimum of f at $(1, 0)$ in Exercise 2(b) is not the absolute minimum of f.)

4

Vector-valued functions of $\mathbb{R}^m$

4.1   Introduction

The vector-valued functions which we consider in this chapter are those of the form $f : S \subseteq \mathbb{R}^m \to \mathbb{R}^n$, $m \in \mathbb{N}$, $n \in \mathbb{N}$. Sometimes we have to impose restrictions on the domain of f. In particular, the domain of a differentiable function $f : D \subseteq \mathbb{R}^m \to \mathbb{R}^n$ is, by definition, an open subset D of $\mathbb{R}^m$. We have considered the particular cases $m = 1$ and $n = 1$ in Chapters 2 and 3 respectively. Many of the new ideas in this chapter are simply generalizations of those in the earlier ones. We rely heavily upon the results of Chapter 3 by relating properties of vector-valued functions to properties of their real-valued coordinate functions.

As in the case of real-valued functions we are sometimes helped in picturing a function by considering its level sets. The *level set corresponding to* $\mathbf{c} \in \mathbb{R}^n$ of the function $f : S \subseteq \mathbb{R}^m \to \mathbb{R}^n$ is (compare 3.1.3)

4.1.1 $f^{-1}(\mathbf{c}) = \{\mathbf{x} \in S \mid f(\mathbf{x}) = \mathbf{c}\}.$

Although we can only sketch the graph of f if $m + n \leqslant 3$ (and all possible such cases have already been considered) we can draw level sets of f for arbitrary n provided $m \leqslant 3$.

4.1.2 *Example.* Define $f : \mathbb{R}^3 \to \mathbb{R}^2$ by $f(\mathbf{x}) = (x_1^2 + x_2^2 + x_3^2, x_1 + x_2 + x_3)$. The coordinate functions of f are f_i, $i = 1, 2$, defined by $f_1(\mathbf{x}) = x_1^2 + x_2^2 + x_3^2$ and $f_2(\mathbf{x}) = x_1 + x_2 + x_3$. The level set corresponding to $(-1, 1)$ is empty. The level set corresponding to $(1, 1)$ is

$$\{(x_1, x_2, x_3) \in \mathbb{R}^3 \mid x_1^2 + x_2^2 + x_3^2 = 1 \quad \text{and} \quad x_1 + x_2 + x_3 = 1\}.$$

This set is a circle – the intersection of a sphere and a plane in $\mathbb{R}^3$.

4.1.3 *Example.* We have seen how to associate a vector-valued function $\operatorname{grad} f : D \subseteq \mathbb{R}^m \to \mathbb{R}^m$ with a differentiable real-valued function f on D. The coordinate functions of $\operatorname{grad} f$ are the partial derivative functions $\partial f / \partial x_i : D \subseteq \mathbb{R}^m \to \mathbb{R}$, $i = 1, \ldots, m$.

The level set of grad f corresponding to $\mathbf{0}$ is the set of critical points of f (see Definition 3.12.5).

Vector-valued functions $F : D \subseteq \mathbb{R}^m \to \mathbb{R}^m$ whose domain and image are subsets of the same Euclidean space, as in the case of grad f considered in the above example, are particularly significant from a physical point of view. They are called *vector fields*. When m is 2 or 3 we can picture $F(\mathbf{p})$ (provided that it is not $\mathbf{0}$) as an arrow emanating from $\mathbf{p}$.

A curve in D whose tangent at each of its points $\mathbf{p}$ is in the direction of the field vector $F(\mathbf{p})$ at that point is called a *field line of* F. In the case of fluid motion, field lines are also called *stream lines* (see the following examples).

4.1.4 *Example*.   Figure 4.1 illustrates three examples of vector fields and field lines arising in physical situations.

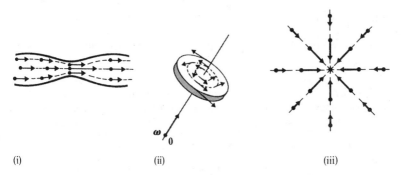

(i)   (ii)   (iii)

Fig. 4.1 (i) Fluid flow; (ii) rotating body; (iii) gravitational field

(a) *A velocity field describing the steady flow of a fluid in a pipe.*   At each point in the fluid an arrow indicates the magnitude and direction of the velocity at that point. The illustration indicates that the fluid flows fastest at the narrowest part of the pipe. The study of hydrodynamics concerns the search for vector functions (fields) $F : D \subseteq \mathbb{R}^3 \to \mathbb{R}^3$ which most accurately describe fluid flow subject to given physical conditions, in the sense that $\mathbf{v} = F(\mathbf{p})$ corresponds to the velocity at the point with position vector $\mathbf{p}$. A speck of dust in the fluid will be carried along a field line.

(b) *The velocity field of a rigid body rotating about a fixed axis with angular speed ω radians/sec.* This field $F : D \subseteq \mathbb{R}^3 \to \mathbb{R}^3$ is given by

$$\mathbf{v} = F(\mathbf{r}) = \omega \times \mathbf{r}$$

where ω represents the angular velocity vector and $\mathbf{v}$ corresponds to the velocity of the point on the body with position vector **r** relative to a right-handed Cartesian coordinate system with origin on the axis of rotation. The field lines are circles in planes orthogonal to the axis with centres on it.

(c) *The gravitational field due to a particle of mass M.* This is described by the function $F : \mathbb{R}^3 \backslash \{\mathbf{0}\} \to \mathbb{R}^3$, where

4.1.5
$$F(\mathbf{r}) = -\frac{GM}{r^3}\mathbf{r}, \qquad \mathbf{r} \neq \mathbf{0}.$$

Here $\mathbf{r} = (x, y, z)$ is measured relative to axes with origin at the particle and $r = \|\mathbf{r}\|$. The vector $F(\mathbf{r})$ indicates the magnitude and direction of the force of attraction exerted by the mass M on a particle of unit mass with position vector **r**. The constant G is called the gravitational constant. Expression 4.1.5 is Newton's celebrated Inverse Square Law. It is so called because the magnitude of the attractive force at **r** is proportional to the inverse of the square of r:

$$\|F(\mathbf{r})\| = \frac{GM}{r^3}\|\mathbf{r}\| = \frac{GM}{r^2}.$$

The field lines, or lines of force, are straight lines through the origin.

Finally we note that we can combine vector-valued functions as in Definitions 1.5.4 and 1.5.6.

4.1.6 *Example.* Let **u** be a unit vector in $\mathbb{R}^m$ and let $g : \mathbb{R}^m \to \mathbb{R}$ be the constant function defined by $g(\mathbf{x}) = \mathbf{u}$ for all $\mathbf{x} \in \mathbb{R}^m$. For any differentiable function $f : D \subseteq \mathbb{R}^m \to \mathbb{R}$ the function $(\text{grad} f)\cdot g : D \subseteq \mathbb{R}^m \to \mathbb{R}$ is defined by

$$((\text{grad } f)\cdot g)(\mathbf{x}) = (\text{grad } f)(\mathbf{x})\cdot g(\mathbf{x}) = (\text{grad } f)(\mathbf{x})\cdot\mathbf{u}, \qquad \mathbf{x} \in D.$$

By Theorem 3.8.6 however, $(\text{grad} f)(\mathbf{x})\cdot\mathbf{u} = (\partial f/\partial \mathbf{u})(\mathbf{x})$. We therefore have that $(\text{grad} f)\cdot g = \partial f/\partial\mathbf{u}$. This relationship is often written $\mathbf{u}\cdot\text{grad} f = \partial f/\partial\mathbf{u}$ or alternatively, in operator form, $\mathbf{u}\cdot\text{grad} = \partial/\partial\mathbf{u}$ (see 3.10.10).

Exercises 4.1

1. Define $f : \mathbb{R}^3 \to \mathbb{R}^2$ by $f(x, y, z) = (x^2 + y^2 - z^2, y)$. Prove that the level set of f corresponding to $(0, c)$ is a hyperbola. Interpret the result geometrically with a sketch.

2. Define $f : \mathbb{R}^2 \to \mathbb{R}$ by $f(x, y) = x^2 + y^2$. Prove that the gradient function $\text{grad } f : \mathbb{R}^2 \to \mathbb{R}^2$ satisfied $(\text{grad } f)(x, y) = (2x, 2y)$. Sketch the gradient (vector) field in $\mathbb{R}^2$ and show that the field lines of $\text{grad} f$ are lines through the origin. In the same sketch draw a selection of level sets of f. Relate your sketch to Theorem 3.8.17 and Definition 3.8.18(i).

3. Repeat Exercise 2 with $f : \mathbb{R}^2 \to \mathbb{R}$ defined by $f(x, y) = x^2 - y^2$. (Consider level sets of f corresponding to $c > 0$, $c = 0$ and $c < 0$.)

4. Sketch the following vector fields $F : \mathbb{R}^2 \to \mathbb{R}^2$. In each case draw some field lines.

(a) $F(x, y) = (1, \frac{1}{2}x^2)$ (b) $F(x, y) = (-y, x)$
(c) $F(x, y) = (\frac{1}{4}x, \frac{1}{4})$.

Answers: Field lines given by (a) $\alpha(t) = (t, \frac{1}{6}t^3 + k)$, k constant;   (b) circles centred at origin;   (c) $\alpha(t) = (t, k + \ln|t|)$, k constant, $t \neq 0$.

5. Sketch the three-dimensional vector field $F : \mathbb{R}^3 \to \mathbb{R}^3$ given by $F(\mathbf{r}) = (1/r)\mathbf{r}$, $\mathbf{r} \neq \mathbf{0}$, where $\mathbf{r} = (x, y, z)$ and $r = \|\mathbf{r}\|$.

Note: $F(\mathbf{r}) = \hat{\mathbf{r}}$ in the notation 2.9.13.

4.2   Continuity and limits

The definition and properties of continuous vector-valued functions on $\mathbb{R}^m$ follow naturally from the work done in Sections 2.3, 2.4 and 3.2. Accordingly we shall merely state a theorem when the proof is a simple extension of one given earlier. The reader is recommended to supply a proof in each case.

4.2.1 Definition. *The function $f : S \subseteq \mathbb{R}^m \to \mathbb{R}^n$ is said to be continuous at $\mathbf{p} \in S$ if, for any sequence $(\mathbf{a}_k)$ in S, $f(\mathbf{a}_k) \to f(\mathbf{p})$ whenever $\mathbf{a}_k \to \mathbf{p}$. The function f is* continuous *if it is continuous at every point of S.*

Notice that if the point $\mathbf{p}$ in S is not a cluster point of S then the condition of Definition 4.2.1 is vacuously satisfied, and so f is continuous at $\mathbf{p}$.

4.2.2 Theorem. *The function $f : S \subseteq \mathbb{R}^m \to \mathbb{R}^n$ is continuous at $\mathbf{p} \in S$ if and only if for each $i = 1, \ldots, n$ the coordinate function $f_i : S \subseteq \mathbb{R}^m \to \mathbb{R}$ is continuous at $\mathbf{p}$.*

4.2.3 Example. The function $f : \mathbb{R}^2 \to \mathbb{R}^3$ defined by

$$f(x, y) = \begin{cases} (x + y, x + 1, y^2 + 2) & \text{when } (x, y) \neq (0, 0), \\ (1, 1, 2) & \text{when } (x, y) = (0, 0) \end{cases}$$

is discontinuous at $(0, 0)$ and continuous elsewhere. The coordinate functions f_2 and f_3 are continuous everywhere, but f_1 is not continuous at $(0, 0)$.

4.2.4 Corollary.   *Let D be an open subset of* $\mathbb{R}^m$. *A function*
$f : D \subseteq \mathbb{R}^m \to \mathbb{R}$ *is continuously differentiable if and only if*
$\operatorname{grad} f : D \subseteq \mathbb{R}^m \to \mathbb{R}^m$ *is continuous.*

Proof.   Immediate from definition.

4.2.5 Corollary.   *Any linear function* $f : \mathbb{R}^m \to \mathbb{R}^n$ *is continuous.*

Proof.   The coordinate functions of a linear function are themselves
linear, and, by Theorem 3.2.2, they are therefore continuous. The
result follows from Theorem 4.2.2.

4.2.6 Definition.   *Given a cluster point* $\mathbf{p}$ *of* $S \subseteq \mathbb{R}^m$, *a point* $\mathbf{q}$ *of* $\mathbb{R}^n$,
and a function $f : S \subseteq \mathbb{R}^m \to \mathbb{R}^n$, *we write* $\lim_{\mathbf{x} \to \mathbf{p}} f(\mathbf{x}) = \mathbf{q}$ *if* $f(\mathbf{a}_k) \to \mathbf{q}$
whenever $\mathbf{a}_k \to \mathbf{p}$, *where the sequence* $(\mathbf{a}_k)$ *lies in* S *and* $\mathbf{a}_k \neq \mathbf{p}$ *for all*
$k \in \mathbb{N}$. (Compare Definitions 2.4.5 and 3.2.18).

4.2.7 Theorem.   *Given a function* $f : S \subseteq \mathbb{R}^m \to \mathbb{R}^n$ *and* $\mathbf{p}$ *a cluster*
point of S, *then* $\lim_{\mathbf{x} \to \mathbf{p}} f(\mathbf{x}) = \mathbf{q}$ *if and only if to each* $\varepsilon > 0$ *there*
corresponds $\delta > 0$ *such that* $\| f(\mathbf{x}) - \mathbf{q} \| < \varepsilon$ *whenever* $\mathbf{x} \in S$ *and*
$0 < \| \mathbf{x} - \mathbf{p} \| < \delta$.

4.2.8 Theorem.   *The function* $f : S \subseteq \mathbb{R}^m \to \mathbb{R}^n$ *is continuous at* $\mathbf{p} \in S$
if and only if $\lim_{\mathbf{x} \to \mathbf{p}} f(\mathbf{x}) = f(\mathbf{p})$.

4.2.9 Theorem.   *The function* $f : S \subseteq \mathbb{R}^m \to \mathbb{R}^n$ *is continuous at* $\mathbf{p} \in S$
if and only if to each $\varepsilon > 0$ *there corresponds* $\delta > 0$ *such that*
$f(\mathbf{x}) \in N(f(\mathbf{p}), \varepsilon)$ *whenever* $\mathbf{x} \in N(p, \delta)$.

We can use Theorem 4.2.9 to relate the above definition of
continuity with a special case of that given in accounts of General
Topology.

4.2.10 Theorem.   *A function* $f : D \subseteq \mathbb{R}^m \to \mathbb{R}^n$, *where D is an open*
subset of $\mathbb{R}^m$, *is continuous if and only if for each open subset V of* $\mathbb{R}^n$
the inverse image

$$f^{-1}(V) = \{ \mathbf{x} \in \mathbb{R}^m \,|\, \mathbf{x} \in D, f(\mathbf{x}) \in V \}$$

is open in $\mathbb{R}^m$.

Proof.   [*i*] Let f be continuous and let V be an open subset of $\mathbb{R}^n$.

We must show that every point $\mathbf{p} \in f^{-1}(V)$ is an interior point of $f^{-1}(V)$. If $f(\mathbf{p}) \in V$ then there exists $\varepsilon > 0$ such that $N(f(\mathbf{p}), \varepsilon) \subseteq V$. By Theorem 4.2.9 there exists $\delta > 0$ such that $f(\mathbf{x}) \in N(f(\mathbf{p}), \varepsilon)$ whenever $\mathbf{x} \in N(\mathbf{p}, \delta)$. Therefore $f(\mathbf{x}) \in V$ whenever $\mathbf{x} \in N(\mathbf{p}, \delta)$ and so $N(\mathbf{p}, \delta) \subseteq f^{-1}(V)$. It follows that $\mathbf{p}$ is an interior point of $f^{-1}(V)$ and, since this is true of every point of $f^{-1}(V)$, that $f^{-1}(V)$ is open in $\mathbb{R}^m$.

[*ii*] Assume that $f^{-1}(V)$ is open in $\mathbb{R}^m$ for each open subset V of $\mathbb{R}^n$. We must show that f is continuous at every point $\mathbf{p} \in D$. Given any $\varepsilon > 0$, $N(f(\mathbf{p}), \varepsilon)$ is open in $\mathbb{R}^n$. Hence $f^{-1}(N(f(\mathbf{p}), \varepsilon))$ is open in $\mathbb{R}^m$. But $\mathbf{p} \in f^{-1}(N(f(p), \varepsilon))$ and so there exists $\delta > 0$ such that $N(\mathbf{p}, \delta) \subseteq f^{-1}(N(f(p), \varepsilon)) \subseteq D$. Therefore $f(\mathbf{x}) \in N(f(\mathbf{p}), \varepsilon))$ whenever $\mathbf{x} \in N(\mathbf{p}, \delta)$. It follows from Theorem 4.2.9 that f is continuous at $\mathbf{p}$ and hence that f is continuous.

Let $f : S \subseteq \mathbb{R}^m \to \mathbb{R}^n$ be a continuous function. The continuity of f at a point $\mathbf{p}$ in its domain implies that to each $\varepsilon > 0$ there corresponds a number $\delta > 0$ such that

$$\|f(\mathbf{x}) - f(\mathbf{p})\| < \varepsilon \qquad \text{whenever } \mathbf{x} \in S \text{ and } \|\mathbf{x} - \mathbf{p}\| < \delta.$$

The choice of the number δ may depend on which point $\mathbf{p}$ in S is being considered. If, however, we can find a δ which applies for *all* possible choices of $\mathbf{p} \in S$ then we say that f is *uniformly continuous*.

4.2.11 Definition.   *A function* $f : S \subseteq \mathbb{R}^m \to \mathbb{R}^n$ *is* uniformly continuous *on S if to each $\varepsilon > 0$ there corresponds $\delta > 0$ such that, for all* $\mathbf{x}, \mathbf{y} \in S$,

$$\|f(\mathbf{x}) - f(\mathbf{y})\| < \varepsilon \qquad \text{whenever } \|\mathbf{x} - \mathbf{y}\| < \delta.$$

The following two theorems concern continuous functions defined on compact domains.

4.2.12 Theorem.   *Let* $f : K \subseteq \mathbb{R}^m \to \mathbb{R}^n$ *be a continuous function on a compact domain K. Then*
    [i] *f is uniformly continuous on K;*
    [ii] *the image set $f(K)$ is compact in $\mathbb{R}^n$.*

Proof.   [*i*] See for example Apostol, *Mathematical Analysis*, Theorem 4-24. [*ii*] Let $(f(\mathbf{a}_k))$, $\mathbf{a}_k \in K$, be a sequence in $f(K)$. Since K is compact, by Theorem 3.2.32 the sequence $(\mathbf{a}_k)$ has a subsequence $(\mathbf{b}_l)$ which converges to a point $\mathbf{p}$ in K. Since f is continuous, $f(\mathbf{b}_l) \rightarrow f(\mathbf{p})$. Therefore the sequence $(f(\mathbf{a}_k))$ has a convergent subsequence, and so, by Theorem 3.2.32, $f(K)$ is compact.

4.2.13 Theorem.   *Let $f:K \subseteq \mathbb{R}^m \rightarrow \mathbb{R}^n$ be a 1–1 continous function defined on a compact domain K. Then the inverse function $f^{-1}:f(K) \subseteq \mathbb{R}^n \rightarrow \mathbb{R}^m$ is continuous.*

Proof.   Let $(\mathbf{q}_k)$ be a sequence in $f(K)$ converging to $\mathbf{q}$ in $f(K)$. Put $f^{-1}(\mathbf{q}_k) = \mathbf{p}_k$ and $f^{-1}(\mathbf{q}) = \mathbf{p}$. We must prove that $\mathbf{p}_k \rightarrow \mathbf{p}$. Suppose not, then there is a subsequence of $(\mathbf{p}_k)$ which does not converge to $\mathbf{p}$. Therefore, since K is compact, there is a subsequence $(\mathbf{p}_{k_i})$ of $(\mathbf{p}_k)$ converging to $\mathbf{r} \in K$, where $\mathbf{r} \neq \mathbf{p}$. Since f is continuous, $f(\mathbf{p}_{k_i}) \rightarrow f(\mathbf{r})$, that is $\mathbf{q}_{k_i} \rightarrow f(\mathbf{r})$. Therefore $f(\mathbf{r}) = \mathbf{q} = f(\mathbf{p})$. Since f is 1–1 it follows that $\mathbf{r} = \mathbf{p}$. This contradiction establishes the result.

The compactness condition cannot be omitted from the statement of Theorem 4.2.13. See Exercise 4.2.6.

Exercises 4.2

1.   Given that the functions $f:S \subseteq \mathbb{R}^m \rightarrow \mathbb{R}^n$, $g:S \subseteq \mathbb{R}^m \rightarrow \mathbb{R}^n$ and $\phi:S \subseteq \mathbb{R}^m \rightarrow \mathbb{R}$ are continuous at $\mathbf{p} \in S$, prove that the functions $f + g$, ϕf, $f \cdot 1g$ and (for the case $n = 3$) $f \times g$ are continuous at $\mathbf{p}$. (Consider the coordinate functions.)

2.   Given $g:E \subseteq \mathbb{R}^l \rightarrow \mathbb{R}^m$ and $f:D \subseteq \mathbb{R}^m \rightarrow \mathbb{R}^n$ such that $g(E) \subseteq D$, prove that if g is continous at $\mathbf{p} \in E$ and f is continous at $g(\mathbf{p}) \in D$, then the composite function $f \circ g$ is continuous at $\mathbf{p}$.

3.   A continuous function defined on a compact set is uniformly continuous. For the continuous function $f(x) = \sqrt{x}$ defined on the compact interval $[0, 1]$, find a suitable δ (depending only on ε) such that the condition of Definition 4.2.11 is satisfied.

4.   Show that the continuous function $f(x) = 1/x$ defined on the open interval $]0, 1[$ is not uniformly continuous.

5.   Prove that if a function is continuous on a set S then it is continuous on every subset T of S.

6.   Let $f:]0, 2\pi] \subseteq \mathbb{R} \rightarrow \mathbb{R}^2$ be the 1–1 continuous function defined by

$$f(t) = (\cos t, \sin t) \qquad 0 < t \leqslant 2\pi.$$

Show that the inverse function f^{-1} is not continuous at $(1, 0)$ in $\mathbb{R}^2$. Compare with Theorem 4.2.13.

Hint: consider the sequence $\mathbf{q}_k = (\cos 1/k, \sin 1/k)$.

4.3   Differentiability

The differentiability of a function $f : D \subseteq \mathbb{R}^m \to \mathbb{R}^n$ at $\mathbf{p} \in D$ has already been defined for the case $m = 1$ in Section 2.10 and the case $n = 1$ in Section 3.3. The definitions were concerned with the possibility of finding a close linear approximation (the differential) to the difference, $f(\mathbf{p} + \mathbf{h}) - f(\mathbf{p})$, near $\mathbf{h} = \mathbf{0}$. There is no difficulty in extending the earlier work to the general situation.

4.3.1 Definition. *Given a function $f : D \subseteq \mathbb{R}^m \to \mathbb{R}^n$, where D is an open subset of $\mathbb{R}^m$, and a point $\mathbf{p} \in D$, the* difference function $\delta_{f,\mathbf{p}} : D_{\mathbf{p}} \subseteq \mathbb{R}^m \to \mathbb{R}^n$ *is defined by*

$$\delta_{f,\mathbf{p}}(\mathbf{h}) = f(\mathbf{p} + \mathbf{h}) - f(\mathbf{p}), \qquad \mathbf{h} \in D_{\mathbf{p}}.$$

Compare with Definitions 2.10.1 and 3.3.3.

4.3.2 Definition. *Let $g : N \subseteq \mathbb{R}^m \to \mathbb{R}^n$ and $g^* : M \subseteq \mathbb{R}^m \to \mathbb{R}^n$ be two functions whose open domains both contain $\mathbf{0}$. We will say that g and g^** closely approximate each other near $\mathbf{0}$ *if there is a function $\eta : N \cap M \subseteq \mathbb{R}^m \to \mathbb{R}^n$ such that*

  [i] $g(\mathbf{h}) - g^*(\mathbf{h}) = \|\mathbf{h}\| \eta(\mathbf{h}), \qquad \mathbf{h} \in N \cap M,$

and

  [ii] $\lim\limits_{\mathbf{h} \to \mathbf{0}} \eta(\mathbf{h}) = \mathbf{0}.$

4.3.3 Definition. *A function $f : D \subseteq \mathbb{R}^m \to \mathbb{R}^n$, defined on an open subset D of $\mathbb{R}^m$, is* differentiable at $\mathbf{p} \in D$ *if the difference function $\delta_{f,\mathbf{p}} : D_{\mathbf{p}} \subseteq \mathbb{R}^m \to \mathbb{R}^n$ can be closely approximated by a linear function $L : \mathbb{R}^m \to \mathbb{R}^n$ near $\mathbf{0}$. The function is said to be* differentiable *if it is differentiable everywhere in D.*

This definition leads to those given in Sections 2.10 and 3.3 in the particular cases considered there. Notice that f is differentiable at $\mathbf{p}$ if and only if there is a linear function $L : \mathbb{R}^m \to \mathbb{R}^n$ and a function $\eta : D_{\mathbf{p}} \subseteq \mathbb{R}^m \to \mathbb{R}^n$ such that

4.3.4 $f(\mathbf{p} + \mathbf{h}) - f(\mathbf{p}) = L(\mathbf{h}) + \|\mathbf{h}\| \eta(\mathbf{h}), \qquad \mathbf{h} \in D_{\mathbf{p}}$

and

4.3.5 $\lim\limits_{\mathbf{h} \to \mathbf{0}} \eta(\mathbf{h}) = \mathbf{0}.$

The following theorem enables us to use the work done in Chapter 3 to study the differentiability of vector-valued functions.

4.3.6 Theorem.   *The function* $f : D \subseteq \mathbb{R}^m \to \mathbb{R}^n$ *is differentiable at* $\mathbf{p} \in D$ *if and only if each of its coordinate functions* $f_i : D \subseteq \mathbb{R}^m \to \mathbb{R}$, $i = 1, \ldots, n$, *is differentiable at* $\mathbf{p} \in D$.

Proof.   Remember that a function $L : \mathbb{R}^m \to \mathbb{R}^n$ is linear if and only if each of its coordinate functions is linear.

Assume firstly that f is differentiable at $\mathbf{p} \in D$. Choose a linear function L and a function η satisfying 4.3.4 and 4.3.5. By considering the ith coordinates of the vectors involved in 4.3.4 we obtain the following expression involving coordinate functions for each $i = 1, \ldots, n$,

4.3.7 $f_i(\mathbf{p} + \mathbf{h}) - f_i(\mathbf{p}) = L_i(\mathbf{h}) + \|\mathbf{h}\| \eta_i(\mathbf{h}), \qquad \mathbf{h} \in D_{\mathbf{p}}.$

Furthermore, from 4.3.5,

4.3.8 $$\lim_{\mathbf{h} \to \mathbf{0}} \eta_i(\mathbf{h}) = 0,$$

since $|\eta_i(\mathbf{h})| \leqslant \|\eta(\mathbf{h})\|$ for each i.

These two expressions, 4.3.7 and 4.3.8, imply that f_i is differentiable at $\mathbf{p}$ for each i.

Conversely, assume that each coordinate function of f is differentiable at $\mathbf{p} \in D$. There therefore exists for each $i = 1, \ldots, n$ a linear function $L_i : \mathbb{R}^m \to \mathbb{R}$ and a function $\eta_i : D_{\mathbf{p}} \subseteq \mathbb{R}^m \to \mathbb{R}$ satisfying 4.3.7 and 4.3.8. Define a linear function $L : \mathbb{R}^m \to \mathbb{R}^n$ by

4.3.9 $L(\mathbf{h}) = (L_1(\mathbf{h}), \ldots, L_n(\mathbf{h})), \qquad \mathbf{h} \in \mathbb{R}^m,$

and a function $\eta : D_{\mathbf{p}} \subseteq \mathbb{R}^m \to \mathbb{R}^n$ by

4.3.10 $\eta(\mathbf{h}) = (\eta_1(\mathbf{h}), \ldots, \eta_n(\mathbf{h})), \qquad \mathbf{h} \in D_{\mathbf{p}}.$

Now 4.3.7 and 4.3.9 imply 4.3.4. Finally, since

$$\|\eta(\mathbf{h})\|^2 = |\eta_1(\mathbf{h})|^2 + \cdots + |\eta_n(\mathbf{h})|^2,$$

4.3.8 and 4.3.10 imply 4.3.5. Therefore f is differentiable at $\mathbf{p} \in D$.

It follows immediately from the definition of differentiability that linear functions are differentiable (they are their own close linear approximation). Theorem 4.3.6 implies that if $f : D \subseteq \mathbb{R}^m \to \mathbb{R}^n$ and $g : D \subseteq \mathbb{R}^m \to \mathbb{R}^n$ and $\phi : D \subseteq \mathbb{R}^m \to \mathbb{R}$ are differentiable at $\mathbf{p} \in D$, then so are the functions $f + g$, $f \cdot g$, ϕf and (provided $n = 3$) so is $f \times g$. It is enough to consider the coordinates of the given combinations and use the result of Exercise 3.3.13.

4.3.11 Corollary. *If $f : D \subseteq \mathbb{R}^m \to \mathbb{R}^n$ is differentiable at $\mathbf{p} \in D$, then f is continuous at $\mathbf{p}$.*

Proof. Since f is differentiable at $\mathbf{p}$ its coordinate functions are differentiable at $\mathbf{p}$. By Theorem 3.3.11 the coordinate functions are continuous at $\mathbf{p} \in D$. Hence by Theorem 4.2.2 f is continuous at $\mathbf{p}$.

4.3.12 Corollary. *Let $f : D \subseteq \mathbb{R}^m \to \mathbb{R}^n$ be differentiable at $\mathbf{p} \in D$. Then there is only one close linear approximation to $\delta_{f, \mathbf{p}}$ near $\mathbf{0}$. It is the linear function $L : \mathbb{R}^m \to \mathbb{R}^n$ given by*

4.3.13 $$L(\mathbf{h}) = (L_{f_1, \mathbf{p}}(\mathbf{h}), \ldots, L_{f_n, \mathbf{p}}(\mathbf{h})), \qquad \mathbf{h} \in \mathbb{R}^m.$$

Proof. If L is any close linear approximation to $\delta_{f, \mathbf{p}}$ near $\mathbf{0}$, then by the first part of the proof of Theorem 4.3.6 and in particular by 4.3.7 and 4.3.8, the coordinate functions of L are the differentials $L_{f_i, \mathbf{p}}$, $i = 1, \ldots, n$. The result follows.

4.3.14 Definition. *Let $f : D \subseteq \mathbb{R}^m \to \mathbb{R}^n$ be differentiable at $\mathbf{p} \in D$. Then the (unique) close linear approximation to $\delta_{f, \mathbf{p}}$ at $\mathbf{0}$ is denoted by $L_{f, \mathbf{p}}$ and is called the* differential *of f at $\mathbf{p}$.*

4.3.15 Theorem. *If the function $f : D \subseteq \mathbb{R}^m \to \mathbb{R}^n$ is differentiable at $\mathbf{p} \in D$, then all the partial derivatives $(\partial f_i / \partial x_j)(\mathbf{p})$ exist and the matrix representing the differential $L_{f, \mathbf{p}}$ with respect to the standard bases of $\mathbb{R}^m$ and $\mathbb{R}^n$ is the $n \times m$ matrix*

4.3.16 $$J_{f, \mathbf{p}} = \left[\frac{\partial f_i}{\partial x_j}(\mathbf{p}) \right] = \begin{bmatrix} \dfrac{\partial f}{\partial x_1}(\mathbf{p}) & \cdots & \dfrac{\partial f_1}{\partial x_m}(\mathbf{p}) \\ \vdots & & \vdots \\ \dfrac{\partial f_n}{\partial x_1}(\mathbf{p}) & \cdots & \dfrac{\partial f_n}{\partial x_m}(\mathbf{p}) \end{bmatrix}.$$

Proof. If f is differentiable at $\mathbf{p}$, then so is each of its coordinate function f_i for $1 \leqslant i \leqslant n$. Therefore by Theorem 3.3.20(i) the partial derivatives $(\partial f_i / \partial x_j)(\mathbf{p})$ exist for each $1 \leqslant i \leqslant n$ and $1 \leqslant j \leqslant m$.

For each $1 \leqslant j \leqslant m$, by 4.3.13 and Theorem 3.3.20(ii),

$$L_{f,\mathbf{p}}(\mathbf{e}_j) = (L_{f_1,\mathbf{p}}(\mathbf{e}_j), \ldots, L_{f_n,\mathbf{p}}(\mathbf{e}_j)) = \left(\frac{\partial f_1}{\partial x_j}(\mathbf{p}), \ldots, \frac{\partial f_n}{\partial x_j}(\mathbf{p})\right).$$

But the coordinates of $L_{f,\mathbf{p}}(\mathbf{e}_j)$ with respect to the standard basis of $\mathbb{R}^n$ form the jth column of the matrix representing $L_{f,\mathbf{p}}$. The result follows.

4.3.17 Definition. *Let* $f : D \subseteq \mathbb{R}^m \to \mathbb{R}^n$ *be differentiable at* $\mathbf{p} \in D$. *The* $n \times m$ *matrix* $J_{f,\mathbf{p}}$ *defined in 4.3.16 is called the* Jacobian *of* f *at* $\mathbf{p}$. *When* $n = m$ *the determinant of the square matrix* $J_{f,\mathbf{p}}$ *is denoted by*

$$\det J_{f,\mathbf{p}} = \frac{\partial(f_1, \ldots, f_m)}{\partial(x_1, \ldots, x_m)}(\mathbf{p}).$$

Notice that the Jacobian matrix of a differentiable function from $\mathbb{R}^m$ to $\mathbb{R}^n$ is an $n \times m$ matrix, as is any matrix representing a linear function from $\mathbb{R}^m$ to $\mathbb{R}^n$.

The vectors $L_{f,\mathbf{p}}(\mathbf{e}_j) \in \mathbb{R}^n, j = 1, \ldots, m$, whose coordinates form the columns of $J_{f,\mathbf{p}}$ are very significant in the study of a function $f : D \subseteq \mathbb{R}^m \to \mathbb{R}^n$ which is differentiable at $\mathbf{p} \in D$. They are denoted by

$$\frac{\partial f}{\partial x_j}(\mathbf{p}) = \left(\frac{\partial f_1}{\partial x_j}(\mathbf{p}), \ldots, \frac{\partial f_n}{\partial x_j}(\mathbf{p})\right) = L_{f,\mathbf{p}}(\mathbf{e}_j).$$

Since $L_{f,\mathbf{p}}$ is linear, it follows that for all $\mathbf{h} \in \mathbb{R}^m$,

$$L_{f,\mathbf{p}}(\mathbf{h}) = h_1 \frac{\partial f}{\partial x_1}(\mathbf{p}) + \cdots + h_m \frac{\partial f}{\partial x_m}(\mathbf{p}),$$

and that there exists a function $\eta : D_{\mathbf{p}} \subseteq \mathbb{R}^m \to \mathbb{R}^n$ such that $\lim_{\mathbf{h} \to \mathbf{0}} \eta(\mathbf{h}) = \mathbf{0}$ and

$$f(\mathbf{p} + \mathbf{h}) - f(\mathbf{p}) = h_1 \frac{\partial f}{\partial x_1}(\mathbf{p}) + \cdots + h_m \frac{\partial f}{\partial x_m}(\mathbf{p}) + \|\mathbf{h}\| \eta(\mathbf{h}), \qquad \mathbf{h} \in D_{\mathbf{p}}.$$

This last expression is an identity between vectors in $\mathbb{R}^n$. It should be compared with the special case considered in Section 3.3.

We have seen (Example 3.3.27) that a function $f : D \subseteq \mathbb{R}^m \to \mathbb{R}^n$ may not be differentiable at $\mathbf{p} \in D$ even though $(\partial f_i / \partial x_j)(\mathbf{p})$ exists for all i and j. The linear transformation represented by the matrix $J_{f,\mathbf{p}}$ of 4.3.16 may not be a close linear approximation to $\delta_{f,\mathbf{p}}$ near $\mathbf{0}$. However, we have the following extension of Theorem 3.3.29.

4.3.18 Theorem. *Consider a function $f : D \subseteq \mathbb{R}^m \to \mathbb{R}^n$ which is defined on an open subset D of $\mathbb{R}^m$. If all the partial derivatives $\partial f_i / \partial x_j$ exist throughout D, $1 \leq i \leq n$, $1 \leq j \leq m$, and if they are all continuous at $\mathbf{p} \in D$, then f is differentiable at $\mathbf{p}$.*

Proof. Exercise.

4.3.19 Definition. *Let D be an open subset of $\mathbb{R}^m$. A function $f : D \subseteq \mathbb{R}^m \to \mathbb{R}^n$ all of whose partial derivatives are continuous in D is said to be* continuously differentiable *on D. Any such function is also said to be a C^1 function.*

It follows from Theorem 4.3.18 that any continuously differentiable function on D is also differentiable.

4.3.20 Example. The function $f : \mathbb{R}^2 \to \mathbb{R}^2$ given by

$$f(x, y) = (|x - y|, x + y), \qquad (x, y) \in \mathbb{R}^2,$$

is continuously differentiable on any open subset of $\mathbb{R}^2$ excluding the line $x = y$.

4.3.21 Example. Consider the linear function $f : \mathbb{R}^3 \to \mathbb{R}^2$ defined by

$$f(\mathbf{x}) = (2x_1 + 2x_2 + x_3, x_1 - x_2) \qquad \mathbf{x} = (x_1, x_2, x_3) \in \mathbb{R}^2.$$

The coordinate functions $f_i : \mathbb{R}^3 \to \mathbb{R}$ are defined by $f_1(\mathbf{x}) = 2x_1 + 2x_2 + x_3$ and $f_2(\mathbf{x}) = x_1 - x_2$. Since the function f is linear it is continuously differentiable at every point $\mathbf{p} \in \mathbb{R}^3$. By taking $\eta(\mathbf{h}) = \mathbf{0}$ for all $\mathbf{h}$, 4.3.5 is certainly satisfied and 4.3.4 becomes

$$f(\mathbf{p} + \mathbf{h}) - f(\mathbf{p}) = f(\mathbf{h}) + \|\mathbf{h}\| \mathbf{0}, \qquad \mathbf{h} \in \mathbb{R}^3.$$

Therefore, for all $\mathbf{p} \in \mathbb{R}^3$, $L_{f, \mathbf{p}} = f$. The Jacobian matrix is

$$J_{f, \mathbf{p}} = \left[\frac{\partial f_i}{\partial x_j} \right](\mathbf{p}) = \begin{bmatrix} 2 & 2 & 1 \\ 1 & -1 & 0 \end{bmatrix}, \qquad \text{for all } \mathbf{p} \in \mathbb{R}^3.$$

As expected, the Jacobian matrix is just the matrix representing f with respect to the standard bases.

4.3.22 Example. The function $f : \mathbb{R}^3 \to \mathbb{R}^2$ given by

$$f(x_1, x_2, x_3) = (x_1^2 + x_2^2 + x_3^2, x_1 + x_2 + x_3)$$

is continuously differentiable in $\mathbb{R}^3$. The Jacobian matrix of f at $\mathbf{p} \in \mathbb{R}^3$ is given by

$$J_{f,\mathbf{p}} = \begin{bmatrix} 2p_1 & 2p_2 & 2p_3 \\ 1 & 1 & 1 \end{bmatrix}.$$

It is sometimes necessary to consider vector-valued functions whose domains are not open. It may not be possible to define the partial derivatives of the coordinate functions of such a function $g : S \subseteq \mathbb{R}^m \to \mathbb{R}^n$ at some points of S. In this case we have the following extension of Definition 4.3.19.

4.3.23 Definition. *A function $g : S \subseteq \mathbb{R}^m \to \mathbb{R}^n$ whose domain S is not an open subset of R^m is said to be a C^1 function if g can be extended to a C^1 function $f : D \subseteq \mathbb{R}^m \to \mathbb{R}^n$ defined on an open set D containing S, such that f and g agree at all points in S.*

We shall need the following generalization of the idea of smoothness introduced in Definitions 2.6.18 and 3.8.14.

4.3.24 Definition. *A function $f : D \subseteq \mathbb{R}^m \to \mathbb{R}^n$ is smooth if f is a C^1 function and if for all $\mathbf{p} \in D$, the Jacobian $J_{f,\mathbf{p}}$ is of maximum possible rank $\min(m, n)$. A function $g : S \subseteq \mathbb{R}^m \to \mathbb{R}^n$ whose domain is not an open subset of $\mathbb{R}^m$ is said to be smooth if g can be extended to a smooth function on an open domain containing S.*

4.3.25 Example. (a) A C^1 function $f : D \subseteq \mathbb{R} \to \mathbb{R}^n$ is smooth if and only if

$$J_{f,\mathbf{p}} = [f'_1(p) \ldots f'_n(p)]^t \neq [0 \ldots 0]^t \qquad \text{for all } p \in D.$$

(b) A C^1 function $f : D \subseteq \mathbb{R}^m \to \mathbb{R}$ is smooth if and only if

$$J_{f,\mathbf{p}} = \left[\frac{\partial f}{\partial x_1}(\mathbf{p}) \ldots \frac{\partial f}{\partial x_m}(\mathbf{p}) \right] \neq [0 \ldots 0] \qquad \text{for all } \mathbf{p} \in D.$$

4.3.26 Example. The linear function $f : \mathbb{R}^3 \to \mathbb{R}^2$ defined in Example 4.3.21 is smooth since f is a C^1 function and rank $J_{f,\mathbf{p}} = 2$ for all $\mathbf{p} \in \mathbb{R}^3$.

4.3.27 Example. The C^1 function $f : \mathbb{R}^3 \to \mathbb{R}^2$ defined in Example 4.3.22 is not smooth since rank $J_{f,\mathbf{p}} = 1$ at any point $\mathbf{p}$ on the straight line $x_1 = x_2 = x_3$ in $\mathbb{R}^3$.

The true significance of smooth functions as defined in 4.3.24 will become clear when we study smooth simple surfaces (Section 8.1).

Let $f : D \subseteq \mathbb{R}^m \to \mathbb{R}^n$ and $g : D \subseteq \mathbb{R}^m \to \mathbb{R}^n$ be two differentiable functions such that $f(\mathbf{x}) = g(\mathbf{x}) + \mathbf{k}$ for some fixed $\mathbf{k} \in \mathbb{R}^n$ and all $\mathbf{x} \in D$. Since $\delta_{f,\mathbf{p}} = \delta_{g,\mathbf{p}}$ for all $\mathbf{p} \in D$, the differentials $L_{f,\mathbf{p}}$ and $L_{g,\mathbf{p}}$ are equal (see Corollary 4.3.12). The converse result is a corollary of the following theorem.

4.3.28 Theorem. *Let $f : D \subseteq \mathbb{R}^m \to \mathbb{R}^n$ be a differentiable function defined on an open path connected subset D of $\mathbb{R}^m$. If the differential $L_{f,\mathbf{p}}$ is the zero function for all $\mathbf{p} \in D$ then f is a constant function.*

Proof. Consider the Jacobian matrix $J_{f,\mathbf{p}}$. Let $1 \leqslant i \leqslant n$. By definition, the elements of the ith row of $J_{f,\mathbf{p}}$ are the coordinates of $(\text{grad } f_i)(\mathbf{p})$ where $f_i : D \subseteq \mathbb{R}^m \to \mathbb{R}$ is the ith coordinate function.

If $L_{f,\mathbf{p}}$ is the zero function for all $\mathbf{p} \in D$, then $J_{f,\mathbf{p}}$ is the zero $n \times m$ matrix. In particular the ith row of $J_{f,\mathbf{p}}$ consists of m zeros for each $\mathbf{p}$. Hence $\text{grad } f_i : D \subseteq \mathbb{R}^m \to \mathbb{R}^m$ is the zero vector field. Therefore, by Theorem 3.8.22, f_i is a constant function for each i. Hence f is a constant function.

4.3.29 Corollary. *Let $f : D \subseteq \mathbb{R}^m \to \mathbb{R}^n$ and $g : D \subseteq \mathbb{R}^m \to \mathbb{R}^n$ be differentiable functions on an open path-connected subset D of $\mathbb{R}^m$. If the differentials $L_{f,\mathbf{p}}$ and $L_{g,\mathbf{p}}$ are equal for all $\mathbf{p} \in D$, then there exists $\mathbf{k} \in \mathbb{R}^n$ such that $f(\mathbf{x}) = g(\mathbf{x}) + \mathbf{k}$ for all $\mathbf{x} \in D$.*

Proof. Exercise.

4.3.30 Example. The identity function $\mathbf{l} : \mathbb{R}^n \to \mathbb{R}^n$ is defined by $\mathbf{l}(\mathbf{x}) = \mathbf{x}$ for all $\mathbf{x} \in \mathbb{R}^n$. Being linear, $\mathbf{l}$ is differentiable and $L_{\mathbf{l},\mathbf{p}} = \mathbf{l}$. Furthermore $J_{\mathbf{l},\mathbf{p}}$ is the $n \times n$ identity matrix I for each $\mathbf{p}$.

Consider a differentiable function $f : D \subseteq \mathbb{R}^n \to \mathbb{R}^n$, where D is an open path-connected subset of $\mathbb{R}^n$. If $L_{f,\mathbf{p}} = \mathbf{l}$ for all $\mathbf{p} \in D$, then $L_{f,\mathbf{p}} = L_{\mathbf{l},\mathbf{p}}$ for all $\mathbf{p}$. Hence, by Corollary 4.3.29, there exists a vector $\mathbf{k} \in \mathbb{R}^n$ such that $f(\mathbf{x}) = \mathbf{x} + \mathbf{k}$ for all $\mathbf{x} \in D$.

Exercises 4.3

1. Let $f : \mathbb{R}^3 \to \mathbb{R}^2$ be defined as in Example 4.3.22 by

$$f(x_1, x_2, x_3) = (x_1^2 + x_2^2 + x_3^2, x_1 + x_2 + x_3).$$

Write down an expression defining $L_{f,\mathbf{p}} : \mathbb{R}^3 \to \mathbb{R}^2$, and determine the remainder term $\|\mathbf{h}\| \eta(\mathbf{h})$ in the expression $f(\mathbf{p} + \mathbf{h}) - f(\mathbf{p}) = L_{f,\mathbf{p}}(\mathbf{h}) + \|\mathbf{h}\| \eta(\mathbf{h})$. Prove that $\eta(\mathbf{h}) \to \mathbf{0}$ as $\mathbf{h} \to \mathbf{0}$.

Answer: $L_{f,\mathbf{p}}(\mathbf{h}) = (2p_1 h_1 + 2p_2 h_2 + 2p_3 h_3, h_1 + h_2 + h_3)$ and
$\|\mathbf{h}\|\eta(\mathbf{h}) = (h_1^2 + h_2^2 + h_3^2, 0) = \|\mathbf{h}\| (\|\mathbf{h}\|, 0).$

2. Repeat Exercise 1 with the function $f : \mathbb{R}^2 \to \mathbb{R}^3$ defined by
   $f(x_1, x_2) = (x_1^3 + x_2^3, e^{x_1 x_2}, x_1 - x_2).$

3. Find the Jacobian matrix $J_{f,\mathbf{p}}$ for each of the following differentiable
   functions $f : \mathbb{R}^3 \to \mathbb{R}^2$:
   (a) $f(x_1, x_2, x_3) = (x_2^2 + 2x_2, 2\sin^2 x_1 x_2 x_3)$
   (b) $f(x_1, x_2, x_3) = ((x_2 + 1)^2, -\cos 2x_1 x_2 x_3)$

4. Find (a) the Jacobian matrix $J_{f,(r,\theta)}$ where the function $f : \mathbb{R}^2 \to \mathbb{R}^2$ is
   defined by
   $$f(r, \theta) = (r\cos\theta, r\sin\theta);$$

   (b) the Jacobian matrix $J_{f,(r,\phi,\theta)}$ where $f : \mathbb{R}^3 \to \mathbb{R}^3$ is defined by

   $$f(r, \phi, \theta) = (r\sin\phi\cos\theta, r\sin\phi\sin\theta, r\cos\phi).$$

   Show also that the determinants of these matrices take the value
   (a) r, (b) $r^2\sin\phi$.

5. Prove that if the function $f : \mathbb{R}^m \to \mathbb{R}^n$ is linear, then $L_{f,\mathbf{p}} = f$. Discuss the
   nature of the Jacobian matrix of f at $\mathbf{p}$.

6. Prove Theorem 4.3.18. (Apply Theorem 3.3.29.)

7. Prove that if $f : D \subseteq \mathbb{R}^m \to \mathbb{R}^n$, $g : D \subseteq \mathbb{R}^m \to \mathbb{R}^n$ and $\phi : D \subseteq \mathbb{R}^m \to \mathbb{R}$ are
   differentiable at $\mathbf{p} \in D$, then so are the functions $f + g$, $f \cdot g$, ϕf and
   (provided $n = 3$) $f \wedge g$. Prove also the following relations on the
   differentials at $\mathbf{p}$.
   $$L_{f+g,\mathbf{p}} = L_{f,\mathbf{p}} + L_{g,\mathbf{p}}$$
   $$L_{f\cdot g,\mathbf{p}}(\mathbf{h}) = f(\mathbf{p}) \cdot L_{g,\mathbf{p}}(\mathbf{h}) + L_{f,\mathbf{p}}(\mathbf{h}) \cdot g(\mathbf{p})$$
   $$L_{\phi f,\mathbf{p}}(\mathbf{h}) = \phi(\mathbf{p})L_{f,\mathbf{p}}(\mathbf{h}) + L_{\phi,\mathbf{p}}(\mathbf{h})f(\mathbf{p})$$
   $$L_{f\wedge g,\mathbf{p}}(\mathbf{h}) = f(\mathbf{p}) \wedge L_{g,\mathbf{p}}(\mathbf{h}) + L_{f,\mathbf{p}}(\mathbf{h}) \wedge g(\mathbf{p}).$$
   Verify these formulae for the case $m = 2$, $n = 3$ and
   $$f(x_1, x_2) = (x_1^2, x_2^2, x_1 x_2)$$
   $$g(x_1, x_2) = (x_2, x_1, 0)$$
   $$\phi(x_1, x_2) = x_1 x_2.$$

8. There are difficulties in using Definition 4.3.3 to define the differentiabi-
   lity of a function $f : S \subset \mathbb{R}^m \to \mathbb{R}^n$ when S is not an open subset of $\mathbb{R}^m$.
   For example, let S be the x, y plane in $\mathbb{R}^3$ and let $f : S \subseteq \mathbb{R}^3 \to \mathbb{R}^3$ be
   defined by $f(\mathbf{x}) = \mathbf{x}$ for all $\mathbf{x} \in S$. Prove that for each $\mathbf{p} \in S$ and each
   $(a, b, c) \in \mathbb{R}^3$ the linear function $L : \mathbb{R}^3 \to \mathbb{R}^3$ defined by $L(x, y, z) = (x + az, y + bz, cz)$, $(x, y, z) \in \mathbb{R}^3$ satisfies
   $$f(\mathbf{p} + \mathbf{h}) - f(\mathbf{p}) = L(\mathbf{h}), \qquad \text{all } \mathbf{h} \in S_\mathbf{p}.$$

Deduce that the 'differential' L is not unique. Where does the proof of Corollary 4.3.12 break down?

4.4 The Chain Rule

The Chain Rule of elementary calculus and Theorems 2.6.11 and 3.6.4 are special cases of the following important result.

4.4.1 Theorem.   *The Chain Rule. Let $g : E \subseteq \mathbb{R}^l \to \mathbb{R}^m$ be defined on an open subset E of $\mathbb{R}^l$ and let $f : D \subseteq \mathbb{R}^m \to \mathbb{R}^n$ be defined on an open subset D of $\mathbb{R}^m$ such that $g(E) \subseteq D$. Define $F : E \subseteq \mathbb{R}^l \to \mathbb{R}^n$ to be the composite function given by*

$$F(\mathbf{t}) = (f \circ g)(\mathbf{t}) = f(g(\mathbf{t})), \qquad \mathbf{t} \in E.$$

Suppose that g is differentiable at $\mathbf{a} \in E$ and that f is differentiable at $g(\mathbf{a}) \in D$. Then F is differentiable at $\mathbf{a}$ and its differential is given by

4.4.2 $$L_{F, \mathbf{a}} = L_{f, g(\mathbf{a})} \circ L_{g, \mathbf{a}}.$$

Furthermore, the Jacobian matrix of F at $\mathbf{a}$ is the matrix product

4.4.3 $$J_{F, \mathbf{a}} = J_{f, g(\mathbf{a})} J_{g, \mathbf{a}}.$$

We prove the theorem by appealing to the following lemma.

4.4.4 Lemma.   *Suppose that $G : \mathbb{R}^l \to \mathbb{R}^m$ is closely approximated near $\mathbf{0} \in \mathbb{R}^l$ by a linear function $M : \mathbb{R}^l \to \mathbb{R}^m$ and that $F : \mathbb{R}^m \to \mathbb{R}^n$ is closely approximated near $\mathbf{0} \in \mathbb{R}^m$ by a linear function $L : \mathbb{R}^m \to \mathbb{R}^n$. Then $F \circ G : \mathbb{R}^l \to \mathbb{R}^n$ is closely approximated near $\mathbf{0} \in \mathbb{R}^l$ by the linear function $L \circ M : \mathbb{R}^l \to \mathbb{R}^n$.*

Proof of Lemma.   There exist functions $\mu : \mathbb{R}^l \to \mathbb{R}^m$ and $\eta : \mathbb{R}^m \to \mathbb{R}^n$ such that

[i] $G(\mathbf{k}) = M(\mathbf{k}) + \|\mathbf{k}\| \mu(\mathbf{k})$ for all $\mathbf{k} \in \mathbb{R}^l$ and $\lim\limits_{\mathbf{k} \to \mathbf{0}} \mu(\mathbf{k}) = \mathbf{0}$

[ii] $F(\mathbf{h}) = L(\mathbf{h}) + \|\mathbf{h}\| \eta(\mathbf{h})$ for all $\mathbf{h} \in \mathbb{R}^m$ and $\lim\limits_{\mathbf{h} \to \mathbf{0}} \eta(\mathbf{h}) = \mathbf{0}$.

It follows from (i) that

4.4.5 $$\frac{\|G(\mathbf{k})\|}{\|\mathbf{k}\|} \leq \frac{1}{\|\mathbf{k}\|} \|M(\mathbf{k})\| + \|\mu(\mathbf{k})\| \text{ for all } \mathbf{k} \neq \mathbf{0}, \mathbf{k} \in \mathbb{R}^l.$$

Therefore (see Exercise 4.4.6)

[iii] there exists an open neighbourhood U of $\mathbf{0} \in \mathbb{R}^l$ such that $\|G(\mathbf{k})\|/\|\mathbf{k}\|$ is bounded for all $\mathbf{k} \in U$, $\mathbf{k} \neq \mathbf{0}$.

To test how close $F \circ G$ and $L \circ M$ are near $\mathbf{0} \in \mathbb{R}^l$, we consider, for $\mathbf{k} \in \mathbb{R}^l$,

$$
\begin{aligned}
F(G(\mathbf{k})) - L(M(\mathbf{k})) &= F(G(\mathbf{k})) - L(G(\mathbf{k})) + L(G(\mathbf{k})) - L(M(\mathbf{k})) \\
&= \|G(\mathbf{k})\| \eta(G(\mathbf{k})) + L(G(\mathbf{k}) - M(\mathbf{k})) \\
&\quad \text{(by (ii) and the linearity of } L) \\
&= \|G(\mathbf{k})\| \eta(G(\mathbf{k})) + \|\mathbf{k}\| L(\mu(\mathbf{k})) \\
&\quad \text{(by (i) and the linearity of } L)
\end{aligned}
$$

Hence, for all $\mathbf{k} \in U$, $\mathbf{k} \neq \mathbf{0}$,

4.4.6 $$\frac{F(G(\mathbf{k})) - L(M(\mathbf{k}))}{\|\mathbf{k}\|} = \frac{\|G(\mathbf{k})\|}{\|\mathbf{k}\|} \eta(G(\mathbf{k})) + L(\mu(\mathbf{k})).$$

Now by [i] $\lim_{\mathbf{k} \to \mathbf{0}} G(\mathbf{k}) = \mathbf{0}$ and therefore by [ii], $\lim_{\mathbf{k} \to \mathbf{0}} \eta(G(\mathbf{k})) = \mathbf{0}$. Similarly, since L is linear, $\lim_{\mathbf{k} \to \mathbf{0}} L(\mu(\mathbf{k})) = \mathbf{0}$. Therefore by [iii], the right-hand side of 4.4.6 tends to $\mathbf{0} \in \mathbb{R}^n$ as $\mathbf{k}$ tends to $\mathbf{0}$. Hence the left-hand side tends to $\mathbf{0}$ and this completes the proof of the lemma.

Proof of Theorem 4.4.1.   The function $\delta_{g, \mathbf{a}} : E_{\mathbf{a}} \subseteq \mathbb{R}^l \to \mathbb{R}^m$ is closely approximated by $L_{g, \mathbf{a}} : \mathbb{R}^l \to \mathbb{R}^m$ near $\mathbf{0} \in \mathbb{R}^l$, and the function $\delta_{f, g(\mathbf{a})} : D_{g(\mathbf{a})} \subseteq \mathbb{R}^m \to \mathbb{R}^n$ is closely approximated by $L_{f, g(\mathbf{a})} : \mathbb{R}^m \to \mathbb{R}^n$ near $\mathbf{0} \in \mathbb{R}^m$. It follows from Lemma 4.4.4 that $\delta_{f, g(\mathbf{a})} \circ \delta_{g, a}$ is closely approximated by $L_{f, g(\mathbf{a})} \circ L_{g, \mathbf{a}}$ near $\mathbf{0} \in \mathbb{R}^l$. A simple calculation (see 3.6.6) shows that

4.4.7 $$\delta_{f \circ g, \mathbf{a}} = \delta_{f, g(\mathbf{a})} \circ \delta_{g, \mathbf{a}}.$$

Therefore $\delta_{f \circ g, \mathbf{a}}$ is closely approximated by the linear function $L_{f, g(\mathbf{a})} \circ L_{g, \mathbf{a}}$ near $\mathbf{0} \in \mathbb{R}^l$. It follows that $F = f \circ g$ is differentiable at $\mathbf{a}$ and that the differential $L_{f \circ g, \mathbf{a}}$ is $L_{f, g(\mathbf{a})} \circ L_{g, \mathbf{a}}$. Expression 4.4.3 follows immediately, thus completing the proof of the theorem.

From 4.4.3 the (i, j) entry of $J_{F, \mathbf{a}}$ is the dot product of the ith row of $J_{f, g(\mathbf{a})}$ and the jth column of $J_{g, \mathbf{a}}$. Denoting variables in E by $\mathbf{t}$ and variables in D by $\mathbf{x}$ we therefore have, for each $\mathbf{a} \in E$,

4.4.8 $$\frac{\partial F_i}{\partial t_j}(\mathbf{a}) = \frac{\partial f_i}{\partial x_1}(g(\mathbf{a}))\frac{\partial g_1}{\partial t_j}(\mathbf{a}) + \cdots + \frac{\partial f_i}{\partial x_m}(g(\mathbf{a}))\frac{\partial g_m}{\partial t_j}(\mathbf{a}).$$

One often meets 4.4.8 in a less precise but perhaps more convenient form. In considering a vector-valued function f of m variables $x_1, \ldots, x_m$, each of which is a function of l variables $t_1, \ldots, t_l$, the expression

4.4.9 $$f(x_1(t_1, \ldots, t_l), \ldots, x_m(t_1, \ldots, t_l))$$

can be thought of as defining f as a function of $t_1, \ldots, t_l$. In the terms set out in Theorem 4.4.1, if we define a function $g : E \subseteq \mathbb{R}^l \to \mathbb{R}^m$ by

$$g(t_1, \ldots, t_l) = (x_1(t_1, \ldots, t_l), \ldots, x_m(t_1, \ldots, t_l)), (t_1, \ldots, t_l) \in E,$$

then 4.4.9 gives the values of $F(t_1, \ldots, t_l)$.

The Chain Rule is then simplified to the identity (compare 4.4.8)

4.4.10 $$\frac{\partial f_i}{\partial t_j} = \frac{\partial f_i}{\partial x_1} \frac{\partial x_1}{\partial t_j} + \cdots + \frac{\partial f_i}{\partial x_m} \frac{\partial x_m}{\partial t_j}.$$

Of course $\partial f_i/\partial t_j$ is really $\partial F_i/\partial t_j$ and $\partial x_k/\partial t_j$ is really $\partial g_k/\partial t_j$. If $\partial f_i/\partial t_j$ is to be evaluated at $\mathbf{t} \in \mathbb{R}^l$ using 4.4.10, then $\partial x_k/\partial t_j$ must be evaluated at $\mathbf{t}$ and $\partial f_i/\partial x_k$ must be evaluated at $(x_1(\mathbf{t}), \ldots, x_m(\mathbf{t}))$.

We now have a complete generalization of the Chain Rule of elementary calculus in its conventional form $dy/dt = (dy/dx)(dx/dt)$. This special case is considered further in our first example.

4.4.11 Example. Let $g : \mathbb{R} \to \mathbb{R}$ be differentiable at $a \in \mathbb{R}$. The differential $L_{g,a}$ is defined by $L_{g,a}(h) = (g'(a))h$ and the Jacobian of g at a is the 1×1 matrix $J_{g,a} = [g'(a)]$. Suppose now that $f : \mathbb{R} \to \mathbb{R}$ is differentiable at $g(a)$; then the Chain Rule implies that $f \circ g$ is differentiable at a. Furthermore, since $J_{f,g(a)} = [f'(g(a))]$ and $J_{f \circ g,a} = [(f \circ g)'(a)]$, expression 4.4.3 is a statement about 1×1 matrices:

$$[(f \circ g)'(a)] = [f'(g(a))][g'(a)],$$

which is effectively the elementary form of the Chain Rule.

4.4.12 Example. Define functions $g : \mathbb{R}^3 \to \mathbb{R}^2$ and $f : \mathbb{R}^2 \to \mathbb{R}^2$ as follows:

$$g(t_1, t_2, t_3) = (t_1 t_2, t_2 t_3), \qquad (t_1, t_2, t_3) \in \mathbb{R}^3$$
$$f(x_1, x_2) = (2x_1 + x_2^2, 3x_1^2 - x_2), \quad (x_1, x_2) \in \mathbb{R}^2.$$

The composite function $F = f \circ g : \mathbb{R}^3 \to \mathbb{R}^2$ is given by

4.4.13 $$\begin{aligned} F(t_1, t_2, t_3) &= f(g(t_1, t_2, t_3)) = f(t_1 t_2, t_2 t_3) \\ &= (2t_1 t_2 + t_2^2 t_3^2, 3t_1^2 t_2^2 - t_2 t_3). \end{aligned}$$

We could of course work out the partial derivatives of F directly from this. Alternatively, by informally putting $x_1 = t_1 t_2$ and $x_2 = t_2 t_3$, we could use 4.4.10 to obtain (with the conventions noted there) for example

$$\frac{\partial f_1}{\partial t_2} = \frac{\partial f_1}{\partial x_1}\frac{\partial x_1}{\partial t_2} + \frac{\partial f_1}{\partial x_2}\frac{\partial x_2}{\partial t_2} = 2t_1 + 2x_2 t_3 = 2t_1 + 2t_2 t_3^2.$$

A little more formally, applying 4.4.3 at $\mathbf{t} = (t_1, t_2, t_3) \in \mathbb{R}^3$,

$$
\begin{aligned}
J_{F,\mathbf{t}} &= \begin{bmatrix} 2 & 2x_2(\mathbf{t}) \\ 6x_1(\mathbf{t}) & -1 \end{bmatrix} \begin{bmatrix} t_2 & t_1 & 0 \\ 0 & t_3 & t_2 \end{bmatrix} \\
&= \begin{bmatrix} 2 & 2t_2 t_3 \\ 6t_1 t_2 & -1 \end{bmatrix} \begin{bmatrix} t_2 & t_1 & 0 \\ 0 & t_3 & t_2 \end{bmatrix} \\
&= \begin{bmatrix} 2t_2 & 2t_1 + 2t_2 t_3^2 & 2t_2^2 t_3 \\ 6t_1 t_2^2 & 6t_1^2 t_2 - t_3 & -t_2 \end{bmatrix}.
\end{aligned}
$$

These entries may be checked directly from 4.4.13.

4.4.14 Example.   Let $g : E \subseteq \mathbb{R}^2 \to \mathbb{R}^2$ and $f : \mathbb{R}^2 \to \mathbb{R}^2$, where $E = \{(t_1, t_2) | t_1 > t_2\}$, be defined as follows:

$$
\begin{aligned}
g(t_1, t_2) &= (\sqrt{(t_1 - t_2)}, t_1 + t_2), & (t_1, t_2) \in E, \\
f(x_1, x_2) &= (\tfrac{1}{2}(x_2 + x_1^2), \tfrac{1}{2}(x_2 - x_1^2)), & (x_1, x_2) \in \mathbb{R}^2.
\end{aligned}
$$

The composite function $F = f \circ g : E \subseteq \mathbb{R}^2 \to \mathbb{R}^2$ is differentiable since both g and f are differentiable throughout their domains. We can calculate the Jacobian matrix $J_{F,\mathbf{t}}$ by putting $x_1 = \sqrt{(t_1 - t_2)}$ and $x_2 = t_1 + t_2$ in the usual informal way. We find that

$$\frac{\partial F_1}{\partial t_1} = \frac{\partial f_1}{\partial x_1}\frac{\partial x_1}{\partial t_1} + \frac{\partial f_1}{\partial x_2}\frac{\partial x_2}{\partial t_1} = x_1 \frac{1}{2\sqrt{(t_1 - t_2)}} + \tfrac{1}{2} \cdot 1 = 1$$

$$\frac{\partial F_1}{\partial t_2} = \frac{\partial f_1}{\partial x_1}\frac{\partial x_1}{\partial t_2} + \frac{\partial f_1}{\partial x_2}\frac{\partial x_2}{\partial t_2} = x_1 \frac{-1}{2\sqrt{(t_1 - t_2)}} + \tfrac{1}{2} \cdot 1 = 0.$$

Similarly $\partial F_2/\partial t_1 = 0$ and $\partial F_2/\partial t_2 = 1$ (check this). Therefore the Jacobian matrix $J_{F,\mathbf{t}}$ is the 2×2 identity matrix I. Thus the differential $L_{F,\mathbf{t}} : \mathbb{R}^2 \to \mathbb{R}^2$ is the identity function, and, by Example 4.3.30, there is a vector $\mathbf{k} \in \mathbb{R}^2$ such that $F(\mathbf{t}) = \mathbf{t} + \mathbf{k}$ for all $\mathbf{t} \in E$. From a direct computation

$$
\begin{aligned}
F(t_1, t_2) &= f(g(t_1, t_2)) = f(\sqrt{(t_1 - t_2)}, t_1 + t_2) \\
&= (\tfrac{1}{2}(t_1 + t_2 + t_1 - t_2), \tfrac{1}{2}(t_1 + t_2 - t_1 + t_2)) = (t_1, t_2),
\end{aligned}
$$

we see that in fact F is the identity function on E. So we have

4.4.15 $F = f \circ g = \mathbf{I}.$

In a sense the functions f and g are inverse to each other. This is a very important idea which we consider in some detail in the next two sections.

In Example 4.4.12 we observed that with the particular functions there defined we could find the partial derivatives of $f \circ g$ without using the Chain Rule, by calculating an expression 4.4.13 for $(f \circ g)(\mathbf{t})$ in terms of $t_1, \ldots, t_n$. This is always possible where f and g are *explicitly* defined. The Chain Rule is of great importance in theoretical situations where such a direct approach is not possible. We use it for example in the following generalization of the elementary Mean-Value Theorem which we shall need in our study of inverse functions.

4.4.16 *Mean-Value Theorem.* Let $f : D \subseteq \mathbb{R}^m \to \mathbb{R}^n$ be a differentiable function whose open domain D contains the points $\mathbf{q}$ and $\mathbf{q} + \mathbf{h}$ and the segment joining them. Then corresponding to any vector $\mathbf{u} \in \mathbb{R}^n$ there is $0 < \theta < 1$ such that

4.4.17 $$(f(\mathbf{q} + \mathbf{h}) - f(\mathbf{q})) \cdot \mathbf{u} = L_{f, \mathbf{q} + \theta \mathbf{h}}(\mathbf{h}) \cdot \mathbf{u}$$

Proof. The function $f^* : D \subseteq \mathbb{R}^m \to \mathbb{R}$ defined by

$$f^*(\mathbf{x}) = (f(\mathbf{x}) - f(\mathbf{q})) \cdot \mathbf{u}, \qquad \mathbf{x} \in D$$

is differentiable (see Exercise 4.3.7). Let $\mathbf{e}_1, \ldots, \mathbf{e}_m$ be the standard basis of $\mathbb{R}^m$. It is easy to check from the definition of partial derivatives that for each $k = 1, \ldots, m$,

4.4.18 $$\frac{\partial f^*}{\partial x_k}(\mathbf{x}) = \frac{\partial f}{\partial x_k}(\mathbf{x}) \cdot \mathbf{u} = L_{f, \mathbf{x}}(\mathbf{e}_k) \cdot \mathbf{u}, \qquad \mathbf{x} \in D.$$

Now consider a function $F : E \subseteq \mathbb{R} \to \mathbb{R}$, where $[0, 1] \subseteq E$ and

$$F(t) = (f(\mathbf{q} + t\mathbf{h}) - f(\mathbf{q})) \cdot \mathbf{u}, \qquad t \in E.$$

Then, for all $t \in E$, $F(t) = f^*(\mathbf{q} + t\mathbf{h})$ and so (by the Chain Rule), F is differentiable. Furthermore, by 4.4.8, for all $t \in E$

$$\begin{aligned}
F'(t) &= h_1 \frac{\partial f^*}{\partial x_1}(\mathbf{q} + t\mathbf{h}) + \cdots + h_m \frac{\partial f^*}{\partial x_m}(\mathbf{q} + t\mathbf{h}) \\
&= (h_1 L_{f, \mathbf{q} + t\mathbf{h}}(\mathbf{e}_1) + \cdots + h_m L_{f, \mathbf{q} + t\mathbf{h}}(\mathbf{e}_m)) \cdot \mathbf{u} \qquad \text{(by 4.4.18)} \\
&= L_{f, \mathbf{q} + t\mathbf{h}}(\mathbf{h}) \cdot \mathbf{u}.
\end{aligned}$$

The elementary Mean-Value Theorem applied to F implies that there exists $0 < \theta < 1$ such that

$$(f(\mathbf{q} + \mathbf{h}) - f(\mathbf{q})) \cdot \mathbf{u} = F(1) - F(0) = F'(\theta) = L_{f,\,\mathbf{q} + \theta\mathbf{h}}(\mathbf{h}) \cdot \mathbf{u}.$$

This completes the proof of the theorem.

4.4.19 *Example*. Taking $n = 1$ and $u = 1$ in Theorem 4.4.16, we find that for any two points $\mathbf{q}$ and $\mathbf{q} + \mathbf{h}$ in the domain of a differentiable function $f : D \subseteq \mathbb{R}^m \to \mathbb{R}$ there exists $0 < \theta < 1$ such that

4.4.20 $f(\mathbf{q} + \mathbf{h}) - f(\mathbf{q}) = L_{f,\,\mathbf{q} + \theta\mathbf{h}}(\mathbf{h})$

$$= h_1 \frac{\partial f}{\partial x_1}(\mathbf{q} + \theta\mathbf{h}) + \cdots + h_m \frac{\partial f}{\partial x_m}(\mathbf{q} + \theta\mathbf{h}).$$

This is merely a restatement of Theorem 3.9.1. Clearly the elementary Mean-Value Theorem corresponds to the case $m = 1$.

4.4.21 *Remark*. There is no hope of identity 4.4.20 being generally true for functions $f : D \subseteq \mathbb{R}^m \to \mathbb{R}^n$. For example, it is easy to construct a differentiable path $f : [a, b] \subseteq \mathbb{R} \to \mathbb{R}^3$ such that at no point $t \in [a, b]$ is the tangent vector $f'(t)$ parallel to $f(b) - f(a)$. (Consider the circular helix of Example 2.1.7.)

Exercises 4.4

1. Define the functions $g : \mathbb{R}^2 \to \mathbb{R}^3$ and $f : \mathbb{R}^3 \to \mathbb{R}^3$ by

$$g(u, v) = (u^2 + v^2, u^2 - v^2, 2uv), \qquad (u, v) \in \mathbb{R}^2$$
$$f(x, y, z) = (x + y, y + z, z + x), \qquad (x, y, z) \in \mathbb{R}^3$$

By direct substitution obtain a formula defining the function $F = f \circ g : \mathbb{R}^2 \to \mathbb{R}^3$. Calculate the entries $(\partial F_i / \partial u)(\mathbf{p})$ and $(\partial F_i / \partial v)(\mathbf{p})$ of the Jacobian matrix $J_{F,\,\mathbf{p}}$, where $\mathbf{p} = (u, v)$, (a) by the Chain Rule 4.4.8, (b) from your formula for $F(u, v)$.

2. Let $g : \mathbb{R}^n \to \mathbb{R}^m$ and $f : \mathbb{R}^m \to \mathbb{R}^n$ be differentiable functions such that $f \circ g$ is the identity function $\mathbf{1} : \mathbb{R}^n \to \mathbb{R}^n$. Regard f as acting on vectors $(x_1, \ldots, x_m) \in \mathbb{R}^m$ and put

$$g(\mathbf{t}) = (x_1(\mathbf{t}), \ldots, x_m(\mathbf{t}))$$

where

$$\mathbf{t} = (t_1, \ldots, t_n) \in \mathbb{R}^n.$$

Prove that

$$\frac{\partial f_i}{\partial x_1}\frac{\partial x_1}{\partial t_j} + \cdots + \frac{\partial f_i}{\partial x_m}\frac{\partial x_m}{\partial t_j} = \begin{cases} 1 & \text{if } i = j \\ 0 & \text{if } i \neq j. \end{cases}$$

For an illustration, see Example 4.4.14.

3. The expressions

$$z = x^2 + y^2 + xy, \qquad x = r\cos\theta, \qquad y = r\sin\theta,$$

effectively define z as a function of r and θ. Calculate $\partial z/\partial r$ from the Chain Rule in the informal notation

$$\frac{\partial z}{\partial r} = \frac{\partial z}{\partial x}\frac{\partial x}{\partial r} + \frac{\partial z}{\partial y}\frac{\partial y}{\partial r}.$$

Similarly calculate $\partial z/\partial\theta$. Check the results by direct substitution.

Reformulate the exercise in precise notation. (Start by defining $f:\mathbb{R}^2 \to \mathbb{R}$ by $f(x, y) = x^2 + y^2 + xy$.)

4. Let f and g be real-valued functions of x and y. The expressions

$$z = (f(x,y), g(x,y)), \qquad x = u(t), \qquad y = v(t), \qquad t \in \mathbb{R},$$

effectively define z as a function of t. Justify the Chain Rule

$$\frac{dz}{dt} = \frac{\partial z}{\partial x}\frac{dx}{dt} + \frac{\partial z}{\partial y}\frac{dy}{dt}$$

and use it in the case $z = (x^2 + y^2, x^2 - y^2)$, $x = \sin t$, $y = \cos t$.

Reformulate the exercise in precise notation.

5. Criticize the following argument. If $z = f(x, y)$ and $y = g(x, t)$, where f and g are real-valued functions, then z is effectively defined as a function of x and t. By the Chain Rule

$$\frac{\partial z}{\partial x} = \frac{\partial z}{\partial x}\frac{\partial x}{\partial x} + \frac{\partial z}{\partial y}\frac{\partial y}{\partial x} = \frac{\partial z}{\partial x} + \frac{\partial z}{\partial y}\frac{\partial y}{\partial x}.$$

Hence

$$\frac{\partial z}{\partial y}\frac{\partial y}{\partial x} = 0.$$

In particular, if $z = x^2 + y$ and $y = x + t$, then $\partial z/\partial y = 1$, $\partial y/\partial x = 1$. Therefore $1 = 0$.

6. In relation to the proofs of Theorem 4.4.1 and Lemma 4.4.4,
   (a) establish identity 4.4.7;
   (b) prove statement (iii) in Lemma 4.4.4.

Hint: in 4.4.5, $\lim_{\mathbf{k} \to \mathbf{0}} \mu(\mathbf{k}) = \mathbf{0}$ and

$$\|M(\mathbf{k})\| \leqslant |k_1|\,\|M(\mathbf{e}_1)\| + \cdots + |k_l|\,\|M(\mathbf{e}_l)\|.$$

7. The following example illustrates that in Lemma 4.4.4 the linearity condition on the functions L and M cannot be dropped. Let G, M, F, L

be functions from $\mathbb{R}$ to $\mathbb{R}$ defined by

$$G(x) = x(1 + x^{1/3}), \qquad M(x) = x$$
$$F(x) = x^{1/3}, \qquad L(x) = x^{1/3}.$$

Prove that G is closely approximated near 0 by M and that F is closely approximated near 0 by L, but that $F \circ G$ is not closely approximated near 0 by $L \circ M$. In fact,

$$\frac{(F \circ G)(h) - (L \circ M)(h)}{|h|} \longrightarrow \infty \qquad \text{as } h \to 0.$$

8. Prove the following special case of the Chain Rule. let $g : E \subseteq \mathbb{R} \to \mathbb{R}^m$ and $f : D \subseteq \mathbb{R}^m \to \mathbb{R}^n$ be differentiable functions, and suppose that $g(E) \subseteq D$. Then $f \circ g : E \subseteq \mathbb{R} \to \mathbb{R}^n$ is differentiable and, for any $p \in E$,

$$(f \circ g)'(p) = \sum_{i=1}^{m} g_i'(p) \frac{\partial f}{\partial x_i}(g(p)).$$

For the case $m = 1$ this result is conveniently remembered as

$$(f \circ g)'(p) = f'(g(p))g'(p),$$

where the real number $g'(p)$ is written after the vector $f'(g(p))$.

9. Let $g : E \subseteq \mathbb{R}^2 \to \mathbb{R}^2$ and $f : D \subseteq \mathbb{R}^2 \to \mathbb{R}^2$ be differentiable functions defined on open sets E and D such that $g(E) \subseteq D$. Let $F : E \subseteq \mathbb{R}^2 \to \mathbb{R}^2$ be the composite function $F = f \circ g$. Regarding g as a function of t_1, t_2 and f as a function of x_1, x_2, establish the determinant condition (in the notation 4.3.17)

$$\frac{\partial(F_1, F_2)}{\partial(t_1, t_2)}(\mathbf{p}) = \frac{\partial(f_1, f_2)}{\partial(x_1, x_2)}(g(\mathbf{p})) \frac{\partial(g_1, g_2)}{\partial(t_1, t_2)}(\mathbf{p}), \qquad \mathbf{p} \in E.$$

4.5   One-to-one functions and their inverses

4.5.1 Definition.   *Let U and V be sets. A function $f : U \to V$ is said to be* one-to-one *(written 1–1) when no two points of U have the same image under f.*

When U and V are subsets of $\mathbb{R}$ a sketch of the graph of $f : U \subseteq \mathbb{R} \to \mathbb{R}$ will reveal the property. The function is 1–1 if and only if every line of the form $y = k$ intersects the graph in at most one point.

4.5.2 Example.   Let f and h be continuous real-valued functions of $\mathbb{R}$ defined by $f(x) = (x - 2)^2 + 1$ and $h(x) = x^3(x - 5)^2/50$. The graphs of f and

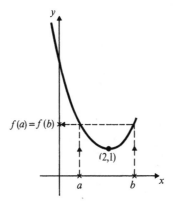

Fig. 4.2 $y = (x - 2)^2 + 1$

h are shown in Figs. 4.2 and 4.3 respectively. Neither f nor h is 1–1, in particular $f(1) = f(3) = 2$ and $h(0) = h(5) = 0$.

4.5.3 Example. Any real number x can be expressed in the form $n + \varepsilon$, where $n \in \mathbb{Z}$ and $0 \leqslant \varepsilon < 1$. The integer n is called the *integer part* of x and is denoted by $[x]$. The discontinuous function $f : \mathbb{R} \to \mathbb{R}$ defined by $f(x) = x - 2[x]$, whose graph is sketched in Fig. 4.4, is 1–1. No two x-values lead to the same y-value.

A function which is both continuous and 1–1 has to be more straightforward than those of the previous examples as the following theorem shows.

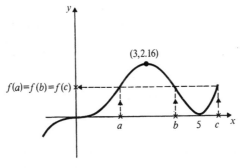

Fig. 4.3 $50y = x^3(x - 5)^2$

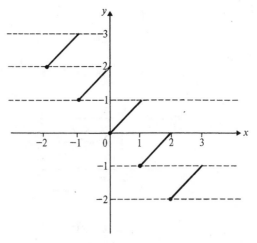

Fig. 4.4 $y = x - 2[x]$

4.5.4 Theorem.   *Let D be an open interval in* $\mathbb{R}$ *and consider a continuous function* $f : D \subseteq \mathbb{R} \to \mathbb{R}$. *The following three conditions are equivalent:*

[i]   *f is* 1–1,
[ii] *f(N) is an open interval wherever N is an open interval in D,*
[iii] *f is strictly monotonic.*

Proof.   We shall prove the three implications [i] ⇒ [ii], [ii] ⇒ [iii] and [iii] ⇒ [i], from which the theorem follows.

[i] ⇒ [ii]. Assuming that f is 1–1, consider an open interval $N \subseteq D$, and put $f(N) = M$. Since f is continuous, the Intermediate Value Theorem implies that M is an interval of some sort. We will show that M is an open interval. Suppose that M contains its greatest lower bound p. Then there exists $c \in N$ such that $f(c) = p$. Find an interval $[a, b] \subseteq N$ such that $c \in \,]a, b[$. Then $f(a) > f(c)$ and $f(b) > f(c)$. (Equality is ruled out since f is 1–1.) Let q be a number such that

$$f(a) > q > f(c) \quad \text{and} \quad f(c) < q < f(b)$$

(see Fig. 4.5(i)).

The Intermediate-Value Theorem implies that there are two points, one between a and c and the other between c and b, at which the function takes the value q, and so f is not 1–1. This contradiction establishes that M does not contain its greatest lower bound.

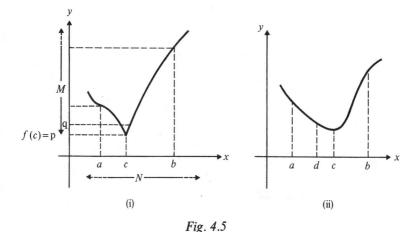

Fig. 4.5

Similarly M does not contain its least upper bound either, and so M is an open interval.

[**ii**] ⇒ [**iii**]. Suppose that the images under f of open intervals are open. If $f : D \subseteq \mathbb{R} \to \mathbb{R}$ is not strictly monotonic, then there exist three points $a < d < b$ in D such that either $f(a) \geq f(d) \leq f(b)$ see Fig. 4.5(ii)) or $f(a) \leq f(d) \geq f(b)$. In the former case the function $f : [a, b] \subseteq \mathbb{R} \to \mathbb{R}$ attains its infimum at some point $c \in]a, b[$. Hence the image under f of the open interval $]a, b[$ contains its greatest lower bound and it is therefore not an open interval. A similar contradiction follows from the second inequality. Therefore $f : D \subseteq \mathbb{R} \to \mathbb{R}$ is strictly monotonic.

(**iii**) ⇒ (**i**).   This is immediate.

4.5.5 *Example*.   Figure 4.6 shows the graph of the exponential function $\exp : \mathbb{R} \to \mathbb{R}$ given by $\exp(x) = e^x$. Clearly exp is strictly increasing and 1–1. To illustrate property [**ii**] of the theorem, $\exp(\mathbb{R}) = \mathbb{R}^+ = \{y \mid y > 0\}$ and $\exp(]0, 1[) =]1, e[$.

4.5.6 *Example*.   The function $f : \mathbb{R} \to \mathbb{R}$ given by $f(x) = (x - 2)^2 + 1$ considered in Example 4.5.2 is not 1–1. In this case we find that $f(\mathbb{R}) = [1, \infty[$ and $f(]1, 3[) = [1, 2[$, for example.

A 1–1 function $f : U \to V$ establishes a 1–1 correspondence between U and $f(U)$. We can therefore define, unambiguously, a function $g : f(U) \subseteq V \to U$ which simply reverses the effect of f. (The reader should consider this process in relation to Fig. 4.5 and notice

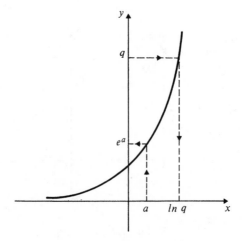

Fig. 4.6 $y = \exp x$; $x = \ln y$

the difficulty of defining a function that reverses the effect of the function suggested by the figure.) More formally we have the following definition.

4.5.7 *Definition.* *Let* $f : U \to V$ *be a* 1–1 *function. The* inverse *of* f *is the function* $g : f(U) \subseteq V \to U$ *defined by*

4.5.8 $g(f(u)) = u,$ $f(u) \in f(U).$

In agreement with arithmetic and matrix notation, the inverse of a function f is sometimes denoted by f^{-1}. Clearly f^{-1} is also a 1–1 function. The functions $f^{-1} \circ f : U \to U$ and $f \circ f^{-1} : f(U) \to f(U)$ are both identity functions, and $(f^{-1})^{-1} = f : U \to V.$

4.5.9 *Example.* The image of the exponential function considered in Example 4.5.5 is $\mathbb{R}^+ = \{y | y > 0\} \subseteq \mathbb{R}$. Its inverse is the natural logarithm function $\ln : \mathbb{R}^+ \to \mathbb{R}$ (see Fig. 4.6).

4.5.10 *Example.* The function $f : \mathbb{R} \to \mathbb{R}$ given by $f(x) = x - 2[x]$ considered in Example 4.5.3 is its own inverse. Using the notation introduced there, for any $x = n + \varepsilon \in \mathbb{R}$,

$$f(f(x)) = f(f(n + \varepsilon)) = f(n + \varepsilon - 2n) = f(-n + \varepsilon)$$
$$= -n + \varepsilon - 2(-n) = n + \varepsilon.$$

Hence $f^{-1} = f : \mathbb{R} \to \mathbb{R}$. This property is by no means as exceptional as it might appear. It can be shown (Exercise 4.5.1) that the graph of the inverse function of any 1–1 function $f : \mathbb{R} \to \mathbb{R}$ is just the reflection of the graph of f in the line $y = x$. Any 1–1 function whose graph is symmetric about this line is therefore its own inverse.

Further consideration of the functions defined in Example 4.5.2 and of their graphs suggests that a function may become 1–1 if its domain is restricted in some way. Theorem 4.5.4 implies that a continuous function $f : D \subseteq \mathbb{R} \to \mathbb{R}$ is 1–1 on an open interval $N \subseteq D$ if and only if f is strictly monotonic on N.

4.5.11 *Example.* We have seen that the function $f : \mathbb{R} \to \mathbb{R}$ defined by the rule $f(x) = (x - 2)^2 + 1$ is not $1 - 1$. The graph of f (Fig. 4.2) indicates that the function is strictly decreasing (and therefore 1–1) on the open interval $]-\infty, 2[$. If we use the same rule to define a function $f_1 :]-\infty, 2[\to \mathbb{R}$ by $f_1(x) = f(x)$ for all $x \in]-\infty, 2[$, then f_1 is 1–1. Furthermore $f_1(]-\infty, 2[) =]1, \infty[$ and its inverse $f_1^{-1} :]1, \infty[\to]-\infty, 2[$ is given by $f_1^{-1}(y) = 2 - \sqrt{(y - 1)}$.

Similarly we can define a function $f_2 :]2, \infty[\subseteq \mathbb{R} \to \mathbb{R}$ by $f_2(x) = f(x)$, $x \in]2, \infty[$. The function f_2 is strictly increasing and is therefore 1–1. The inverse of f_2 is the function $f_2^{-1} :]1, \infty[\to]2, \infty[$ given by $f_2^{-1}(y) = 2 + \sqrt{(y - 1)}$.

We are led to the following definition.

4.5.12 *Definition.* *The function $f : D \subseteq \mathbb{R}^n \to \mathbb{R}^n$ is said to be locally $1-1$ at $\mathbf{p} \in D$ if there is an open set N in $\mathbb{R}^n$ with $\mathbf{p} \in N \subseteq D$ such that f is $1 - 1$ on N. We also say under these circumstances that f is locally invertible at $\mathbf{p}$ and, where there is no possible ambiguity, denote the local inverse by $f^{-1} : f(N) \subseteq \mathbb{R}^n \to \mathbb{R}^n$.*

4.5.13 *Theorem.* *A differentiable function $f : D \subseteq \mathbb{R} \to \mathbb{R}$ is not locally 1–1 at any point where it has a local extremum.*

Proof. Exercise.

4.5.14 *Example.* The function $f : \mathbb{R} \to \mathbb{R}$ considered in Examples 4.5.2 and 4.5.11 is locally 1–1 at all points in $]-\infty, 2[$ and $]2, \infty[$. We have denoted the local inverses on these two intervals by f_1^{-1} and f_2^{-1} respectively. The function is not locally 1–1 at the point 2, where it has a local minimum, since it is not monotonic on any open interval containing 2 (see Fig. 4.2).

4.5.15 *Example.* The function $f : \mathbb{R} \to \mathbb{R}$ defined by $f(x) = \sin x$ is not locally 1–1 at its extremal points $(2n + 1)\pi/2$, $n \in \mathbb{Z}$. See Fig. 4.7. The function is not monotonic in any open interval containing $\pi/2$, for example. It is, however, locally 1–1 between these points. In the notation of Definition 4.5.12 the open interval $]-\pi/2, \pi/2[$ is an appropriate choice for N corresponding to any point which it contains. The local inverse function there is arc sin $:]-1, 1[\rightarrow]-\pi/2, \pi/2[$. Similarly $]\pi/2, 3\pi/2[$ is an appropriate choice for N corresponding to any point in $]\pi/2, 3\pi/2[$, but the local inverse in this case is $g:]-1, 1[\rightarrow]\pi/2, 3\pi/2[$ defined by $g(y) = \pi - \arcsin y$.

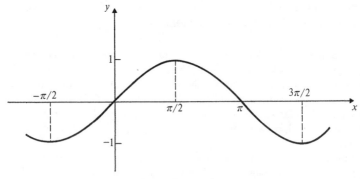

Fig. 4.7 $y = \sin x$

We shall see in Exercise 4.5.5 that a differentiable function $f : \mathbb{R} \rightarrow \mathbb{R}$ may fail to be locally 1–1 at $p \in \mathbb{R}$ even if $f'(p) \neq 0$ since it can happen that *every* open interval containing p contains a point where f has a local extremum. The added condition that f is continuously differentiable is sufficient, however, to ensure that f is locally 1–1 at non-extremal points.

4.5.16 Theorem.   *Let $f : D \subseteq \mathbb{R} \rightarrow \mathbb{R}$ be continuously differentiable and let $p \in D$ be such that $f'(p) \neq 0$. Then there is an open interval N containing p such that f is 1–1 on N, $f(N)$ is an open interval and the function $g : f(N) \subseteq \mathbb{R} \rightarrow \mathbb{R}$ locally inverse to f is also continuously differentiable.*

Furthermore,

4.5.17 $g'(f(x))f'(x) = 1, \qquad x \in N.$

Proof.   Suppose that $f'(p) > 0$. Since $f' : D \subseteq \mathbb{R} \rightarrow \mathbb{R}$ is continuous at p, there exists an open interval N containing p such that

4.5.18 $f'(x) > \tfrac{1}{2}f'(p) > 0$ for all $x \in N$.

Therefore f is strictly increasing on N and so, by Theorem 4.5.4, f is 1–1 on N and $f(N)$ is an open interval. Let the local inverse of f be $g : M \rightarrow N$, where $M = f(N)$.

We prove that g is differentiable. Given y and $y + k$ in M, there are unique points x and $x + h$ in N such that

4.5.19 $\quad y = f(x), \quad x = g(y), \quad y + k = f(x + h), \quad x + h = g(y + k).$

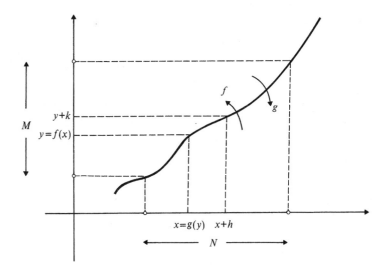

Fig. 4.8

See Fig. 4.8. By the Mean-Value Theorem,

$$k = f(x + h) - f(x) = hf'(x + \theta h) \text{ for some } \theta, 0 < \theta < 1.$$

Hence, by 4.5.18, $|k| \geqslant |h| f'(p)/2$. Therefore

$$\lim_{k \to 0} \frac{g(y + k) - g(y)}{k} = \lim_{k \to 0} \frac{h}{f(x + h) - f(x)}$$

$$= \lim_{h \to 0} \frac{h}{f(x + h) - f(x)} = \frac{1}{f'(x)}.$$

This proves that g is differentiable at $f(x)$ and also establishes the identity 4.5.17. Using that identity and the notation of 4.5.19 again,

$$\lim_{k \to 0} g'(y + k) = \lim_{k \to 0} g'(f(x + h)) = \lim_{k \to 0} \frac{1}{f'(x + h)}$$

$$= \lim_{h \to 0} \frac{1}{f'(x + h)} = g'(y).$$

Therefore g' is continuous and the proof is complete.

Notice that 4.5.17 is the consequence of applying the Chain Rule to the identity function $g \circ f = I : N \to N$. The rule can only be applied, however, if it is known that both functions are differentiable; it would not therefore be of any help to us in the proof of Theorem 4.5.16.

With reference to the notation $y = f(x)$, $x = g(y)$, identity 4.5.17 is often given in the less precise form

$$\frac{dy}{dx}\frac{dx}{dy} = 1.$$

Theorem 4.5.16 guarantees the *existence* of an inverse function under the stated conditions. However, an *explicit* formula for the inverse function is not necessarily available. Sometimes functions are invented expressly to describe the inverse – the 'arc sin' function is an example.

4.5.20 *Example.* The continuously differentiable function $h : \mathbb{R} \to \mathbb{R}$ given by $50h(x) = x^3(x - 5)^2$ was considered in Example 4.5.2 and its graph sketched in Fig. 4.3. Since

$$10h'(x) = x^2(x - 3)(x - 5),$$

the only extremal points of h are 3 and 5 (the point 0 is a point of inflection). By appealing to the above theorem, or by just looking at the graph, we can see that h is locally invertible at 2, for example. An appropriate open interval containing this point is $]-\infty, 3[$, and $h(]-\infty, 3[) =]-\infty, 2.16[$. Let the corresponding local inverse of h be $g :]-\infty, 2.16[\to]-\infty, 3[$. We cannot find a simple formula for g since the quintic equation $x^3(x - 5)^2 = 50y$ has no elementary solution for x in terms of y. However, 4.5.17 gives us information about the derivative of g, for example

$$g'(h(2)) = g'(72) = \frac{1}{h'(2)} = \frac{1}{60}.$$

Notice that h is strictly increasing on $]-\infty, 3[$ even through $h'(0) = 0$. The function h is locally invertible at 0 but the inverse function g is not differentiable at $h(0) = 0$ since if it were we would have, by applying the Chain Rule to $g \circ h = I :]-\infty, 3[\to]-\infty, 3[$, that

4.5.21 $g'(h(0))h'(0) = g'(0).0 = 1.$

The graph of g is the reflection of the graph of $h :]-\infty, 3[\to]-\infty, 2.16[$ in the line $y = x$. The point of inflection of h at 0 becomes a point of infinite slope of g. This supports 4.5.21.

Exercises 4.5

1. (a) Sketch the graphs of the exponential function $y = e^x$, $x \in \mathbb{R}$ and the logarithmic function $y = \ln x$, $x \in \mathbb{R}^+$ on the same paper. Show that the second graph is a reflection of the first in the line $y = x$.
   (b) Prove generally that the graph of a 1–1 function $f : D \subseteq \mathbb{R} \to \mathbb{R}$ and the graph of its inverse function $f^{-1} : f(D) \to D$ are reflections of each other in the line $y = x$.

2. Sketch the graphs of the following functions $f : D \subseteq \mathbb{R} \to \mathbb{R}$. Find the inverse function $f^{-1} : f(D) \to D$, if it exists, and sketch its graph.

   (a) $f(x) = x^3$, $x \in \mathbb{R}$,   (b) $f(x) = x^2$, $x \in \mathbb{R}$,
   (c) $f(x) = x^2$ $x \in \mathbb{R}^+$ (d) $f(x) = x^2$, $x \in \mathbb{R}^-$,
   (e) $f(x) = \cos x$, $x \in \mathbb{R}$,   (f) $f(x) = \cos x$, $x \in]0, \pi[$.

Answer: inverse functions exist in cases (a), (c), (d), (f).

3. Sketch the graphs of the following functions $f : D \subseteq \mathbb{R} \to \mathbb{R}$. Prove that the inverse functions $f^{-1} : f(D) \to D$ exist. In these examples there is no simple formula for the inverse functions. Roughly sketch the graphs of the inverse functions.

   (a) $f(x) = x \ln x$, $x \in]0, 1/e[$,   (b) $f(x) = 3x^5 - 5x^3 + 2$, $x \in]-1, 1[$,
              (c) $f(x) = x - \sin x$, $x \in \mathbb{R}$.

   Prove that the inverse function of Example (b) is not differentiable at the point $f(0) = 2$. At which points is the inverse function of Example (c) differentiable?

4. Prove Theorem 4.5.13. (Consider as domain of f an interval containing an extremal point of f.)

5. Define $f : \mathbb{R} \to \mathbb{R}$ by $f(x) = x^2 \sin(1/x) + x$, $x \neq 0$, and $f(0) = 0$. Sketch its graph. Prove that
   (a) f is differentiable on $\mathbb{R}$, and in particular, $f'(0) = 1$,
   (b) f is not locally 1–1 at 0 (show that every open interval containing 0 contains an extremal point of f),
   (c) f' is discontinuous at 0.
   This example shows that the assumption that f is differentiable and that $f'(p) \neq 0$ does not imply that f is locally invertible at p.

6. Prove that the inverse of a 1–1 continuous function $f : D \subseteq \mathbb{R} \to \mathbb{R}$ is continuous on its domain $f(D)$.

Hint: consider $(f^{-1})^{-1}(N)$, where N is an open subset of D, and apply Theorems 4.5.4(ii) and 4.2.10.

4.6   The Inverse Function Theorem

We now turn to the problem of generalizing Theorem 4.5.16 and
consider what properties of a function $f : D \subseteq \mathbb{R}^n \to \mathbb{R}^n$ will guarantee
that f has a 'well behaved' local inverse at some point $\mathbf{p}$ in the
open domain D. Since the effect of f near $\mathbf{p}$ is closely described by
the linear function $L_{f,\mathbf{p}}$, one would expect a close relationship
between the local invertibility of f near $\mathbf{p}$ and the invertibility of $L_{f,\mathbf{p}}$.
Now $L_{f,\mathbf{p}}$ is invertible if and only if the Jacobian matrix $J_{f,\mathbf{p}}$ is non-
singular or equivalently if and only if $\det J_{f,\mathbf{p}} \neq 0$. Notice that when
$n = 1$ this last condition reduces to $f'(p) \neq 0$, a property which was part
of the hypothesis of Theorem 4.5.16. The generalization we are
looking for involves the assumptions that $\det J_{f,\mathbf{p}} \neq 0$ and that f is
continuously differentiable (in order to avoid the problem illustrated
in Exercise 4.5.5). The result is the celebrated Inverse Function
Theorem – one of the most important theorems of vector calculus.
We prove it by first showing that under the above assumptions there
is a neighbourhood of $\mathbf{p}$ on which f is 1–1 and smooth (that is,
$\det J_{f,\mathbf{q}} \neq 0$ for all $\mathbf{q}$ in the neighbourhood). We need two lemmas.
The first one is a fundamental result about linear functions.

4.6.1 *Lemma.* *If the linear function* $T : \mathbb{R}^n \to \mathbb{R}^n$ *is an isomorphism
(and so is invertible) then there exists* $\mu > 0$ *such that, for all* $\mathbf{x} \in \mathbb{R}^n$,

4.6.2 $\qquad \|T(\mathbf{x})\| \leqslant \mu \|\mathbf{x}\|$ *and* $\|T^{-1}(\mathbf{x})\| \geqslant \dfrac{1}{\mu}\|\mathbf{x}\|.$

Proof.   Let $\mathbf{e}_1, \ldots, \mathbf{e}_n$ be the standard basis of $\mathbb{R}^n$. For any $\mathbf{x} \in \mathbb{R}^n$

$$T(\mathbf{x}) = T(x_1 \mathbf{e}_1 + \cdots + x_n \mathbf{e}_n) = x_1 T(\mathbf{e}_1) + \cdots + x_n T(\mathbf{e}_n).$$

The triangle inequality implies that

$$\|T(\mathbf{x})\| \leqslant |x_1| \|T(\mathbf{e}_1)\| + \cdots + |x_n| \|T(\mathbf{e}_n)\|$$
$$= (|x_1|, \ldots, |x_n|) \cdot (\|T(\mathbf{e}_1)\|, \ldots, \|T(\mathbf{e}_n)\|),$$

and therefore it follows from the Cauchy–Schwarz inequality that

$$\|T(\mathbf{x})\|^2 \leqslant \sum_1^n |x_i|^2 \sum_1^n \|T(\mathbf{e}_i)\|^2.$$

Setting $\mu^2 = \sum_1^n \|T(e_i)\|^2$ gives the first part of 4.6.2.

The second inequality follows by replacing $\mathbf{x}$ by $T^{-1}(\mathbf{x})$ in the first
one.

The second lemma shows that the continuous differentiability of f leads to a continuous variation of the differentials $L_{f,\mathbf{x}}$.

4.6.3 Lemma. *Let $f : D \subseteq \mathbb{R}^n \to \mathbb{R}^n$ be a C^1 function and let $\mathbf{p}$ be a point in the open set D. Then to each $\varepsilon > 0$ there corresponds a neighbourhood $N(\mathbf{p}, \delta) \subseteq D$ such that for all $\mathbf{q} \in N(\mathbf{p}, \delta)$,*

[i] $\|(L_{f,\mathbf{q}} - L_{f,\mathbf{p}})(\mathbf{h})\| \leqslant \varepsilon \|\mathbf{h}\|$ *for all $\mathbf{h} \in \mathbb{R}^n$,*

[ii] $\|f(\mathbf{q} + \mathbf{h}) - f(\mathbf{q}) - L_{f,\mathbf{p}}(\mathbf{h})\| \leqslant \varepsilon \|\mathbf{h}\|$

for all $\mathbf{h} \in \mathbb{R}^n$ such that $\mathbf{q} + \mathbf{h} \in N(\mathbf{p}, \delta)$.

Proof. For all $\mathbf{q} \in D$ and $\mathbf{h} \in \mathbb{R}^n$,

$$L_{f,\mathbf{q}}(\mathbf{h}) - L_{f,\mathbf{p}}(\mathbf{h}) = h_1\left(\frac{\partial f}{\partial x_1}(\mathbf{q}) - \frac{\partial f}{\partial x_1}(\mathbf{p})\right) + \cdots + h_n\left(\frac{\partial f}{\partial x_n}(\mathbf{q}) - \frac{\partial f}{\partial x_n}(\mathbf{p})\right).$$

Therefore for all such $\mathbf{q}$ and $\mathbf{h}$,

4.6.4 $$\|L_{f,\mathbf{q}}(\mathbf{h}) - L_{f,\mathbf{p}}(\mathbf{h})\| \leqslant |h_1| \left\|\frac{\partial f}{\partial x_1}(\mathbf{q}) - \frac{\partial f}{\partial x_1}(\mathbf{p})\right\| + \cdots$$
$$+ |h_n| \left\|\frac{\partial f}{\partial x_n}(\mathbf{q}) - \frac{\partial f}{\partial x_n}(\mathbf{p})\right\|.$$

The partial derivatives of f are continuous in D. Therefore, corresponding to any $\varepsilon > 0$ there is a neighbourhood $N(\mathbf{p}, \delta) \subseteq D$ such that for all $i = 1, \ldots, n$

$$\left\|\frac{\partial f}{\partial x_i}(\mathbf{q}) - \frac{\partial f}{\partial x_i}(\mathbf{p})\right\| \leqslant \frac{\varepsilon}{n}$$ whenever $\mathbf{q} \in N(\mathbf{p}, \delta)$.

[*i*] By the choice of δ and by 4.6.4

$$\|L_{f,\mathbf{q}}(\mathbf{h}) - L_{f,\mathbf{p}}(\mathbf{h})\| \leqslant \frac{\varepsilon}{n}(|h_1| + \cdots + |h_n|) \leqslant \varepsilon \|\mathbf{h}\|$$

whenever $\mathbf{q} \in N(\mathbf{p}, \delta)$.

[*ii*] For any $\mathbf{q}$ and $\mathbf{q} + \mathbf{h}$ in $N(\mathbf{p}, \delta)$ let

$$\mathbf{u} = f(\mathbf{q} + \mathbf{h}) - f(\mathbf{q}) - L_{f,\mathbf{p}}(\mathbf{h}).$$

The Mean-Value Theorem 4.4.16 implies that there exists $0 < \theta < 1$

such that

$$(f(\mathbf{q} + \mathbf{h}) - f(\mathbf{q})) \cdot \mathbf{u} = L_{f, \mathbf{q} + \theta\mathbf{h}}(\mathbf{h}) \cdot \mathbf{u}.$$

Hence

$$(f(\mathbf{q} + \mathbf{h}) - f(\mathbf{q}) - L_{f, \mathbf{p}}(\mathbf{h})) \cdot \mathbf{u} = (L_{f, \mathbf{q} + \theta\mathbf{h}}(\mathbf{h}) - L_{f, \mathbf{p}}(\mathbf{h})) \cdot \mathbf{u}.$$

Therefore, substituting for $\mathbf{u}$ and applying the Cauchy–Schwarz inequality,

$$\|f(\mathbf{q} + \mathbf{h}) - f(\mathbf{q}) - L_{f, \mathbf{p}}(\mathbf{h})\| \leqslant \|(L_{f, \mathbf{q} + \theta\mathbf{h}} - L_{f, \mathbf{p}})(\mathbf{h})\|.$$

The result follows from (i).

We can now prove our first main result, that if a C^1 function is smooth at a point, then it is smooth and 1–1 in a neighbourhood of the point.

4.6.5 Theorem.   *Let $f : D \subseteq \mathbb{R}^n \to \mathbb{R}^n$ be a C^1 function on an open set D in $\mathbb{R}^n$ and let $\mathbf{p}$ be a point in D such that $\det J_{f, \mathbf{p}} \neq 0$. Then there exists neighbourhood $N(\mathbf{p}, \delta) \subseteq D$ such that*
  [i] $\det J_{f, \mathbf{q}} \neq 0$ *for all $\mathbf{q} \in N(\mathbf{p}, \delta)$,*
  [ii] $f : N(\mathbf{p}, \delta) \subseteq \mathbb{R}^n \to \mathbb{R}^n$ *is 1–1.*

Proof.   Since $L_{f, \mathbf{p}}$ is invertible, by Lemma 4.6.1 there exists $\mu > 0$ such that for all $\mathbf{x} \in \mathbb{R}^n$

4.6.6 $$\|L_{f, \mathbf{p}}^{-1}(\mathbf{x})\| \leqslant \mu \|\mathbf{x}\| \qquad \text{and} \qquad \|L_{f, \mathbf{p}}(\mathbf{x})\| \geqslant \frac{1}{\mu} \|\mathbf{x}\|$$

We shall prove that the neighbourhood $N(\mathbf{p}, \delta)$ which corresponds in Lemma 4.6.3 to $\varepsilon = 1/2\mu$ has the properties stated in the theorem.

  [*i*] Suppose that for some $\mathbf{q} \in N(p, \delta)$, $\det J_{f, \mathbf{q}} = 0$. Then there exists $\mathbf{h} \neq \mathbf{0}$ in $\mathbb{R}^n$ such that $L_{f, \mathbf{q}}(\mathbf{h}) = 0$. But, by Lemma 4.6.3(i), since $\varepsilon = 1/2\mu$

$$\| - L_{f, \mathbf{p}}(\mathbf{h})\| = \|(L_{f, \mathbf{q}} - L_{f, \mathbf{p}})(\mathbf{h})\| \leqslant \frac{1}{2\mu} \|\mathbf{h}\|.$$

This contradicts 4.6.6.

  [*ii*] Suppose similarly that there are two distinct points $\mathbf{q}$ and $\mathbf{q} + \mathbf{h}$ in $N(\mathbf{p}, \delta)$ such that $f(\mathbf{q} + \mathbf{h}) = f(\mathbf{q})$. Then by Lemma 4.6.3[ii],

$$\| - L_{f, \mathbf{p}}(\mathbf{h})\| = \|f(\mathbf{q} + \mathbf{h}) - f(\mathbf{q}) - L_{f, \mathbf{p}}(\mathbf{h})\| \leqslant \frac{1}{2\mu} \|\mathbf{h}\|,$$

and again we have a contradiction.

Theorem 4.6.5 tells us that if the Jacobian of a continuously differentiable function f is non-singular at a point $\mathbf{p}$ in the domain of f, then f is locally invertible at $\mathbf{p}$. It tells us nothing about the function g inverse to f. The remarkable result which we now prove is that the set $f(N(\mathbf{p},\delta))$ contains a neighbourhood V of $f(\mathbf{p})$ on which the inverse function g is itself continuously differentiable. Lemma 4.6.3 again gives the appropriate information about the function f to facilitate the proof.

4.6.7 Theorem.   The Inverse Function Theorem.   *Let* $f : D \subseteq \mathbb{R}^n \to \mathbb{R}^n$ *be continuously differentiable on an open set D in $\mathbb{R}^n$ and let* $\det J_{f,\,\mathbf{p}} \neq 0$ *at a point* $\mathbf{p} \in D$. *Then*
[i] *there exists an open set U in $\mathbb{R}^n$ containing $\mathbf{p}$ and an open set V in $\mathbb{R}^n$ containing $f(\mathbf{p})$ such that $U \subseteq D$, $f(U) = V$ and f is 1–1 on U;*
[ii] *the function $g : V \subseteq \mathbb{R}^n \to \mathbb{R}^n$ locally inverse to f on U is continuously differentiable;*
[iii] $J_{g,\,f(\mathbf{p})} J_{f,\,\mathbf{p}} = I.$

Proof.   Preliminaries.   Denote the linear function $L_{f,\,\mathbf{p}}$ by L. Since $\det J_{f,\,\mathbf{p}} \neq 0$ the function L has an inverse $L^{-1} : \mathbb{R}^n \to \mathbb{R}^n$. Let $\mu > 0$ be such that (see 4.6.2)

4.6.8 $$\| L^{-1}(\mathbf{z}) \| \leqslant \mu \| \mathbf{z} \|, \qquad \text{for all } \mathbf{z} \in \mathbb{R}^n.$$

Let $N(\mathbf{p}, \delta) \subseteq D$ be the neighbourhood (see Lemma 4.6.3 with $\varepsilon = 1/2\mu$) such that whenever $\mathbf{q}$ and $\mathbf{q} + \mathbf{h}$ lie in $N(\mathbf{p}, \delta)$,

4.6.9 $$\| f(\mathbf{q} + \mathbf{h}) - f(\mathbf{q}) - L(\mathbf{h}) \| \leqslant \frac{1}{2\mu} \| \mathbf{h} \|.$$

It was shown in Theorem 4.6.5 that f is 1–1 on $N(\mathbf{p}, \delta)$ and that $L_{f,\,\mathbf{q}}$ is an isomorphism on $\mathbb{R}^n$ for all $\mathbf{q} \in N(\mathbf{p},\delta)$.

We shall show that there exists a neighbourhood $V = N(f(\mathbf{p}), \sigma)$ of $f(\mathbf{p})$ in $\mathbb{R}^n$ such that for each $\mathbf{y}$ in V there is a (unique) $\mathbf{x} \in N(\mathbf{p}, \delta)$ such that $\mathbf{y} = f(\mathbf{x})$. This is equivalent to showing that

4.6.10 $$V \subseteq f(N(\mathbf{p}, \delta)).$$

Once this is established part [i] of the theorem will follow since the set $U = f^{-1}(V) \cap N(\mathbf{p}, \delta)$ is open in D (and therefore in $\mathbb{R}^n$) and 4.6.10 implies that $f(U) = V$. See Fig. 4.9.

For each $\mathbf{y} \in N(f(\mathbf{p}), \sigma)$ we shall find $\mathbf{x} \in N(\mathbf{p}, \delta)$ such that $\mathbf{y} = f(\mathbf{x})$ by a method of successive approximation. Starting with $\mathbf{x}_1 = \mathbf{p}$ we find a sequence $(\mathbf{x}_k)$ in $N(\mathbf{p}, \delta)$ such that $\mathbf{y} - f(\mathbf{x}_k) \to \mathbf{0}$ in $f(N(\mathbf{p}, \delta))$. If

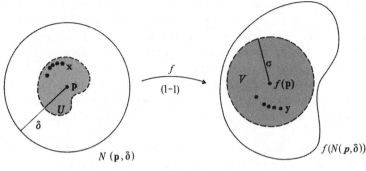

Fig. 4.9

$(\mathbf{x}_k)$ converges to $\mathbf{x}$ in $N(\mathbf{p}, \delta)$, then the continuity of f implies that $\mathbf{y} = f(\mathbf{x})$. The number σ has to be small enough to guarantee that for each $\mathbf{y} \in N(f(\mathbf{p}), \sigma)$ the corresponding sequence $(\mathbf{x}_k)$ converges in $N(\mathbf{p}, \delta)$.

Before deciding what σ is to be, it is useful to consider how the sequence $(\mathbf{x}_k)$ might be chosen. Suppose we already have $\mathbf{x}_k$ in $N(\mathbf{p}, \delta)$. How should we choose $\mathbf{x}_{k+1} = \mathbf{x}_k + \mathbf{h}$ in $N(\mathbf{p}, \delta)$ so that $f(\mathbf{x}_{k+1})$ is a better approximation to $\mathbf{y}$ than is $f(\mathbf{x}_k)$? Since f is differentiable, we have

$$f(\mathbf{x}_k + \mathbf{h}) \approx f(\mathbf{x}_k) + L_{f, \mathbf{x}_k}(\mathbf{h}).$$

But $L_{f, \mathbf{x}_k}(\mathbf{h})$ is not far from $L(\mathbf{h})$ (see Lemma 4.6.3) and it therefore seems plausible to choose $\mathbf{h}$ such that

$$\mathbf{y} = f(\mathbf{x}_k) + L(\mathbf{h}).$$

Taking $\mathbf{h} = L^{-1}(\mathbf{y} - f(\mathbf{x}_k))$ we are led to define

4.6.11 $\mathbf{x}_{k+1} = \mathbf{x}_k + L^{-1}(\mathbf{y} - f(\mathbf{x}_k)).$

We need to check of course that $\mathbf{x}_{k+1}$ is also in $N(\mathbf{p}, \delta)$ before proceeding to the next stage.

Notice that if the terms of the sequence $(\mathbf{x}_k)$ satisfy 4.6.11 for each $k \in \mathbb{N}$ and if $(\mathbf{x}_k)$ converges to $\mathbf{x}$ in $N(\mathbf{p}, \delta)$, then, since L^{-1} and f are continuous,

$$\mathbf{x} = \mathbf{x} + L^{-1}(\mathbf{y} - f(\mathbf{x})).$$

But L^{-1} is an isomorphism. Therefore $\mathbf{y} = f(\mathbf{x})$.

In studying the convergence of $(\mathbf{x}_k)$ we shall consider vectors $\mathbf{x}_{k+1} - \mathbf{x}_k$, for each $\mathbf{k} \in \mathbb{N}$. There exists $\mathbf{z}_k \in \mathbb{R}^n$ for each $k \in \mathbb{N}$ such that

4.6.12 $f(\mathbf{x}_{k+1}) - f(\mathbf{x}_k) = L(\mathbf{x}_{k+1} - \mathbf{x}_k) + \|\mathbf{x}_{k+1} - \mathbf{x}_k\|\mathbf{z}_k,$

where by 4.6.9,

4.6.13 $$\|\mathbf{z}_k\| \leqslant \frac{1}{2\mu}.$$

It follows from 4.6.12 that for each $k \in \mathbb{N}$,

4.6.14 $\mathbf{x}_{k+1} - \mathbf{x}_k = L^{-1}(f(\mathbf{x}_{k+1}) - f(\mathbf{x}_k)) - \|\mathbf{x}_{k+1} - \mathbf{x}_k\|L^{-1}(\mathbf{z}_k).$

Proof of [i]   Let $\sigma = \delta/4\mu$ and choose $\mathbf{y} \in N(f(\mathbf{p}), \sigma)$. With initial value $\mathbf{x}_1 = \mathbf{p} \in N(\mathbf{p}, \delta)$ we consider the next two terms of a sequence $(\mathbf{x}_k)$ defined by the recurrence relation 4.6.11. Putting $k = 1$ and $\mathbf{x}_1 = \mathbf{p}$ in 4.6.11 we obtain

$$\|\mathbf{x}_2 - \mathbf{p}\| = \|L^{-1}(\mathbf{y} - f(\mathbf{p}))\|,$$

and from 4.6.8

$$\|L^{-1}(\mathbf{y} - f(\mathbf{p}))\| \leqslant \mu\|\mathbf{y} - f(\mathbf{p})\| < \mu\sigma = \frac{\delta}{4}.$$

Hence

4.6.15 $$\|\mathbf{x}_2 - \mathbf{x}_1\| = \|\mathbf{x}_2 - \mathbf{p}\| < \frac{\delta}{4}$$

and therefore $\mathbf{x}_2 \in N(\mathbf{p}, \delta)$.

Since $\mathbf{x}_2 \in D$, the recurrence relation 4.6.11 defines $\mathbf{x}_3$, and

$$\begin{aligned}
\mathbf{x}_3 - \mathbf{x}_2 &= (\mathbf{x}_2 + L^{-1}(\mathbf{y} - f(\mathbf{x}_2))) - (\mathbf{x}_1 + L^{-1}(\mathbf{y} - f(\mathbf{x}_1)))\\
&= (\mathbf{x}_2 - \mathbf{x}_1) - L^{-1}(f(\mathbf{x}_2) - f(\mathbf{x}_1))\\
&= -\|\mathbf{x}_2 - \mathbf{x}_1\|L^{-1}(\mathbf{z}_1) \qquad \text{(by 4.6.14)}.
\end{aligned}$$

Therefore by 4.6.8 and 4.6.13

4.6.16 $$\|\mathbf{x}_3 - \mathbf{x}_2\| \leqslant \tfrac{1}{2}\|\mathbf{x}_2 - \mathbf{x}_1\| < \frac{\delta}{8}.$$

Now by the triangle inequality, and by 4.6.16 and 4.6.15.

$$\|\mathbf{x}_3 - \mathbf{p}\| \leqslant \|\mathbf{x}_3 - \mathbf{x}_2\| + \|\mathbf{x}_2 - \mathbf{x}_1\| < (\tfrac{1}{8} + \tfrac{1}{4})\delta < \tfrac{1}{2}\delta.$$

The above argument can be used step by step to show that for all $k \geqslant 2$

4.6.17 $$\|\mathbf{x}_k - \mathbf{x}_{k-1}\| < \frac{\delta}{2^k}$$

and that

4.6.18 $\qquad \|\mathbf{x}_k - \mathbf{p}\| < \left(\dfrac{1}{2^k} + \cdots + \dfrac{1}{2^2}\right)\delta < \tfrac{1}{2}\delta.$

Furthermore from 4.6.17 and 4.6.18, for all $k > l \geqslant 1$,

4.6.19 $\qquad \|\mathbf{x}_k - \mathbf{x}_l\| \leqslant \|\mathbf{x}_k - \mathbf{x}_{k-1}\| + \cdots + \|\mathbf{x}_{l+1} - \mathbf{x}_l\|$

$$\leqslant \left(\frac{1}{2^k} + \cdots + \frac{1}{2^{l+1}}\right)\delta < \frac{\delta}{2^l}.$$

Hence $(\mathbf{x}_k)$ is a Cauchy sequence in $N(\mathbf{p}, \delta)$ which therefore converges to say $\mathbf{x} \in \mathbb{R}^n$. By 4.6.18, $\|\mathbf{x} - \mathbf{p}\| \leqslant \tfrac{1}{2}\delta$ and therefore $\mathbf{x} \in N(\mathbf{p}, \delta)$. As observed earlier, 4.6.11 now implies that $f(\mathbf{x}) = \mathbf{y}$.

The conclusions of part [i] are therefore satisfied by $V = N(f(\mathbf{p}), \sigma)$ and $U = f^{-1}(V) \cap N(\mathbf{p}, \delta)$ (see the remarks following 4.6.9).

Remark. Our proof can be expressed in terms of a function $\phi : B \subseteq \mathbb{R}^n \to \mathbb{R}^n$ where B is a closed ball, centre $\mathbf{p}$, and

$$\phi(\mathbf{x}) = \mathbf{x} + L^{-1}(\mathbf{y} - f(\mathbf{x})), \qquad \mathbf{x} \in B$$

(compare 4.6.11). One can prove that, subject to certain restrictions,

$$\|\phi(\mathbf{x}') - \phi(\mathbf{x})\| \leqslant \tfrac{1}{2}\|\mathbf{x}' - \mathbf{x}\|, \qquad \mathbf{x}, \mathbf{x}' \in B,$$

and $\phi(B) \subset B$. The function ϕ is a *contraction mapping* on B and as such has a unique fixed point $\mathbf{x}$ in B (such that $\phi(\mathbf{x}) = \mathbf{x}$). Therefore $\mathbf{y} = f(\mathbf{x})$. For details of this approach see for example Rudin, *Principles of Analysis.*

Proof of [ii] *and* [iii].   Let U and V be the open subsets of $\mathbb{R}^n$ considered in part [i]. We know that $f : U \to V$ is 1–1 and maps U onto V. Let $g : V \to U$ be the local inverse of f on U. Suppose that $\mathbf{y}$ and $\mathbf{y} + \mathbf{k}$ are both in V. Then there exist unique $\mathbf{x}$ and $\mathbf{x} + \mathbf{h}$ in U such that

4.6.20 $\qquad \begin{aligned} f(\mathbf{x}) &= \mathbf{y}, & \mathbf{x} &= g(\mathbf{y}) \\ f(\mathbf{x} + \mathbf{h}) &= \mathbf{y} + \mathbf{k}, & \mathbf{x} + \mathbf{h} &= g(\mathbf{y} + \mathbf{k}). \end{aligned}$

Since f is differentiable at $\mathbf{x}$ there is a function $\eta : D_{\mathbf{x}} \subseteq \mathbb{R}^n \to \mathbb{R}^n$ such that (by 4.6.20).

4.6.21 $\qquad \mathbf{k} = f(\mathbf{x} + \mathbf{h}) - f(\mathbf{x}) = L_{f,\mathbf{x}}(\mathbf{h}) + \|\mathbf{h}\|\eta(\mathbf{h})$

and
$$\lim_{\mathbf{h} \to \mathbf{0}} \eta(\mathbf{h}) = \mathbf{0}.$$

By Theorem 4.6.5[i], $L_{f,\mathbf{x}}$ is an isomorphism on $\mathbb{R}^n$.
Let $S = L_{f,\mathbf{x}}^{-1} : \mathbb{R}^n \to \mathbb{R}^n$. Then applying S to 4.6.21

$$S(\mathbf{k}) = \mathbf{h} + S(\|\mathbf{h}\| \eta(\mathbf{h})).$$

Therefore

4.6.22 $\qquad g(\mathbf{y} + \mathbf{k}) - g(\mathbf{y}) = \mathbf{h} = S(\mathbf{k}) + \|\mathbf{k}\| S\left(-\dfrac{\|\mathbf{h}\|}{\|\mathbf{k}\|} \eta(\mathbf{h}) \right).$

But by 4.6.9

$$\|L(\mathbf{h})\| - \|f(\mathbf{x} + \mathbf{h}) - f(\mathbf{x})\| \leqslant \frac{1}{2\mu} \|\mathbf{h}\|.$$

So by 4.6.8 and 4.6.21,

$$\frac{1}{\mu} \|\mathbf{h}\| - \|\mathbf{k}\| \leqslant \frac{1}{2\mu} \|\mathbf{h}\|,$$

that is

$$\|\mathbf{k}\| \geqslant \frac{1}{2\mu} \|\mathbf{h}\|.$$

Hence, since S is continuous at $\mathbf{0}$,

4.6.23 $\qquad \lim_{\mathbf{k} \to \mathbf{0}} S\left(-\dfrac{\|\mathbf{h}\|}{\|\mathbf{k}\|} \eta(\mathbf{h}) \right) = \mathbf{0}.$

Expressions 4.6.22 and 4.6.23 imply that g is differentiable at $\mathbf{y}$
with differential $L_{g,\mathbf{y}} = S = L_{f,\mathbf{x}}^{-1}$. Therefore $J_{g,\mathbf{y}}$ is non-singular and

4.6.24 $\qquad J_{g,\mathbf{y}} = (J_{f,\mathbf{x}})^{-1}, \qquad\qquad \mathbf{x} \in U, \ y = f(\mathbf{x}) \in V.$

Part [iii] of the theorem follows taking $\mathbf{x} = \mathbf{p}$ and $\mathbf{y} = f(\mathbf{p})$.

The (i, j) entry of $(J_{f,\mathbf{x}})^{-1}$ is a rational combination of $(\partial f_k/\partial x_l)(\mathbf{x})$,
$k, l = 1, \ldots, n$ (Cramer's Rule).

From 4.6.24, $(\partial g_i/\partial y_i)(\mathbf{y}) = (\partial g_i/\partial y_j)(f(\mathbf{x}))$ is therefore a continuous
function of $\mathbf{x}$. But $\mathbf{x} = g(\mathbf{y})$ is a continuous function of $\mathbf{y}$ and therefore

$$\frac{\partial g_i}{\partial y_j}(f(g(\mathbf{y}))) = \frac{\partial g_i}{\partial y_j}(\mathbf{y})$$

is a continuous function of **y**. Hence g is continuously differentiable on V.

4.6.25 Remark. Part [iii] of the theorem is a consequence of applying the Chain Rule (Theorem 4.4.1) to the identity function $\mathbf{l} = g \circ f : U \to U$.

4.6.26 Example. The function $f : \mathbb{R}^2 \to \mathbb{R}^2$ given by

$$f(\mathbf{r}) = f(x, y) = (e^x \cos y, e^x \sin y), \qquad (x, y) \in \mathbb{R}^2$$

is continuously differentiable in $\mathbb{R}^2$, and

$$J_{f,\mathbf{r}} = \begin{bmatrix} e^x \cos y & -e^x \sin y \\ e^x \sin y & e^x \cos y \end{bmatrix}.$$

Hence

$$\det J_{f,\mathbf{r}} = e^x \neq 0,$$

and so f is locally invertible at every point in $\mathbb{R}^2$. Indeed, given any $\mathbf{p} \in \mathbb{R}^2$ there is a formula for the inverse of f which applies when the domain of f is restricted to a sufficiently small open set containing $\mathbf{p}$.

For example, f is invertible on the open rectangle

$$R_1 = \{(x, y) \in \mathbb{R}^2 \mid -1 < x < 1, -\tfrac{1}{2}\pi < y < \tfrac{1}{2}\pi\},$$

and $f(R_1) = A$, as illustrated in Fig. 4.10.

The inverse of $f : R_1 \to A$ is $g : A \to R_1$ given by the formula $g(u, v) = (x, y) \in R_1$, where

$$x = \ln \sqrt{(u^2 + v^2)}, \qquad y = \arctan \frac{v}{u}, \qquad (u, v) \in A.$$

Again, f is invertible on

$$R_2 = \{(x, y) \in \mathbb{R}^2 \mid -1 < x < 1, \tfrac{3}{2}\pi < y < \tfrac{5}{2}\pi\},$$

and $f(R_2) = A$. The inverse of $f : R_2 \to A$ is $g^* : A \to R_2$ given by the formula $g^*(u, v) = (x, y) \in R_2$, where

$$x = \ln \sqrt{(u^2 + v^2)}, \qquad y = \left(\arctan \frac{v}{u}\right) + 2\pi, \qquad (u, v) \in A.$$

This example illustrates the *local* nature of the Inverse Function Theorem. Although $\det J_{f,\mathbf{r}} \neq 0$ for all $\mathbf{r} \in \mathbb{R}^2$ and f is locally invertible at every point in $\mathbb{R}^2$, the function f is not *globally* invertible. By this we mean that f is not invertible on the whole of its domain. This is clearly so because f is not 1–1 on $\mathbb{R}^2$.

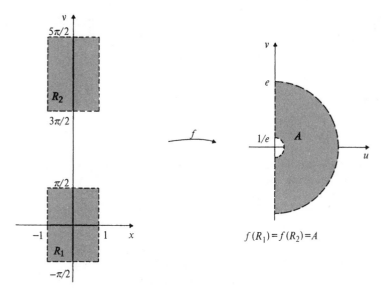

Fig. 4.10 $f(x, y) = (e^x \cos y, e^x \sin y)$

4.6.27 Example. The function $f : \mathbb{R}^2 \to \mathbb{R}^2$ given by

$$f(\mathbf{r}) = f(x, y) = (y \sin x, x + y + 1), \qquad (x, y) \in \mathbb{R}^2$$

is continuously differentiable in $\mathbb{R}^2$, and

$$J_{f,\mathbf{r}} = \begin{bmatrix} y \cos x & \sin x \\ 1 & 1 \end{bmatrix}.$$

Therefore $\det J_{f,\mathbf{r}} = y \cos x - \sin x$. In particular, if $\mathbf{a} = (0, 1)$, then $\det J_{f,\mathbf{a}} = 1$ and so f is locally invertible at $\mathbf{a}$. Put $\mathbf{b} = f(\mathbf{a}) = (0, 2)$. By the Inverse Function Theorem there exist open sets U and V with $\mathbf{a} \in U$ and $\mathbf{b} \in V$ such that f is invertible on U and the inverse of $f : U \to V$ is a continuously differentiable function $g : V \to U$. Although the existence of g is guaranteed, there is no simple formula that describes it. However, the local behaviour of g near the point $\mathbf{b}$ is known through the Jacobian matrix

$$J_{g,\mathbf{b}} = (J_{f,\mathbf{a}})^{-1} = \begin{bmatrix} 1 & 0 \\ 1 & 1 \end{bmatrix}^{-1} = \begin{bmatrix} 1 & 0 \\ -1 & 1 \end{bmatrix}.$$

Note the following interpretation of Example 4.6.27: the equations

$$u = y \sin x, \qquad v = x + y + 1$$

can be solved uniquely for x and y with (x, y) sufficiently close to

(0, 1) in terms of u and v for values of (u, v) sufficiently close to (0, 2), and the solutions $x = g_1(u, v)$, $y = g_2(u, v)$ are continuously differentiable.

Exercises 4.6

1. Let $f : \mathbb{R}^2 \to \mathbb{R}^2$ be defined by

$$f(\mathbf{r}) = f(x, y) = (x^2 - y^2, 2xy), \qquad (x, y) \in \mathbb{R}^2.$$

(a) Calculate the Jacobian matrix $J_{f,\mathbf{r}}$ and prove that f is locally invertible except possibly at the origin.

(b) Let $U \subseteq \mathbb{R}^2$ be a neighbourhood of the point $\mathbf{p} = (1, 1) \in \mathbb{R}^2$ as in Theorem 4.6.7 (so that f is 1–1 on U), and denote the inverse of $f : U \to \mathbb{R}^2$ by $g : f(U) \to U$. Find the Jacobian matrix $J_{g, f(\mathbf{p})}$.

Hint: apply Theorem 4.6.7(iii).

(c) Let $U^* \subseteq \mathbb{R}^2$ be a neighbourhood of the point $\mathbf{q} = (-1, -1) \in \mathbb{R}^2$ as in Theorem 4.6.7 and denote the inverse of $f : U^* \to \mathbb{R}^2$ by $g^* : f(U^*) \to U^*$. Find the Jacobian matrix $J_{g^*, f(\mathbf{q})}$.

(d) Observe that $f(\mathbf{p}) = f(\mathbf{q}) = (0, 2)$, but that $J_{g, (0, 2)} \neq J_{g^*, (0, 2)}$. Conclude that although f is locally invertible at every $\mathbf{r} \in \mathbb{R}^2 \backslash \{0\}$, it is not globally invertible on $\mathbb{R}^2 \backslash \{0\}$.

Direct proof: f is not 1–1 on $\mathbb{R}^2 \backslash \{0\}$.

(e) Put $f(x, y) = (u, v)$ and find an expression for $g(u, v)$, $(u, v) \in f(U)$ and an expression for $g^*(u, v)$, $(u, v) \in f(U)$.

Answer: the equations $u = x^2 - y^2$, $v = 2xy$ are solved by

$$x = \pm \left[\tfrac{1}{2}(u + \sqrt{(u^2 + v^2)}\,)\right]^{1/2}, \qquad y = \pm \left[\tfrac{1}{2}(-u + \sqrt{(u^2 + v^2)}\,)\right]^{1/2}.$$

The plus or minus sign must be chosen in the definitions of g and g^* such that $g(0, 2) = (1, 1)$ and $g^*(0, 2) = (-1, -1)$.

(f) Is f locally invertible at (0, 0)?

Hint: is there a neighbourhood of (0, 0) on which f is 1–1?

2. Without solving the equations

$$u = x^2 - y^2, \qquad v = 2xy,$$

prove that they effectively define x and y (a) as continuously differentiable functions of u and v in a neighbourhood of $(u, v) = (0, 2)$ such that $x = 1$, $y = 1$ when $(u, v) = (0, 2)$, and also (b) as continuously differentiable functions of u and v in a neighbourhood of $(u, v) = (0, 2)$ such that $x = -1$, $y = -1$ when $(u, v) = (0, 2)$. By considering appropriate Jacobian matrices show that in case (a),

$$\frac{\partial x}{\partial v}(0, 2) = \tfrac{1}{4}$$

and in case (b),

$$\frac{\partial x}{\partial v}(0, 2) = -\tfrac{1}{4}.$$

(This is a rephrasing of part of Exercise 1, and illustrates a possible ambiguity encountered in solving functional equations.)

3. Prove that the equations

$$u = y\sin x, \qquad v = x + y + 1$$

effectively define x and y as continuously differentiable functions of u and v in a neighbourhood of $(u, v) = (0, 2)$ with (x, y) taking values in a neighbourhood of $(1, 0)$. If the functional relationship is described by $(x, y) = g^*(u, v)$, calculate the Jacobian matrix $J_{g^*, (0, 2)}$. Compare with Example 4.6.27.

4. Let $f : \mathbb{R}^2 \to \mathbb{R}^2$ be defined by

$$f(x, y) = (x^3 + y^3, x^3 - y^3), \qquad (x, y) \in \mathbb{R}^2.$$

Prove that the Jacobian matrix $J_{f, (0, 0)}$ is the zero matrix. Show that nevertheless f is globally invertible on $\mathbb{R}^2$.

Hint: prove that f is 1–1 on $\mathbb{R}^2$.
Show also that the inverse function is not differentiable at $f(0, 0) = (0, 0)$.

5. Compute the Jacobian matrix $J_{f, (x, y)}$ and discuss the local behaviour of the functions $f : \mathbb{R}^2 \to \mathbb{R}^2$, where
   (a) $f(x, y) = (x\cos y, x\sin y)$,
   (b) $f(x, y) = (x^2 + 2xy + y^2, 3x + 3y)$,
   (c) $f(x, y) = \dfrac{1}{\sqrt{2}}(x + y, x - y)$.

4.7 Implicitly defined functions

A rule such as

4.7.1 $\qquad f(x) = \sqrt{(1 - x^2)}, \qquad x \in [-1, 1]$

defines a function f explicitly. By this we mean that at each point in its domain the value of f is calculated by direct substitution in the formula. The graph of f is also explicitly determined. In the present example it is the set of points

$$S = \{(x, \surd(1 - x^2)) \in \mathbb{R}^2 \mid x \in [-1, 1]\}$$

(see Fig. 4.11(i)).

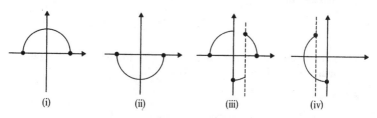

Fig. 4.11 (i) $y = f(x)$; (ii) $y = f^*(x)$; (iii) $y = f^\dagger(x)$

The graph S is a subset of the circle

4.7.2 $$C = \{(x, y) \in \mathbb{R}^2 \mid x^2 + y^2 - 1 = 0\},$$

for if we substitute $y = f(x) = \surd(1 - x^2)$, then the equation

$$x^2 + y^2 - 1 = 0$$

is satisfied whenever $x \in [-1, 1]$. This equation is said to define the function f *implicitly*. If we put $F(x, y) = x^2 + y^2 - 1$, $(x, y) \in \mathbb{R}^2$, then we have

$$F(x, f(x)) = 0, \qquad x \in [-1, 1].$$

This means that the graph of f is a subset of the level set of F corresponding to the value 0. Generalizing this example, we are led to the following definition.

4.7.3 Definition.   *Given a real-valued function $F : D \subseteq \mathbb{R}^m \to \mathbb{R}$, where $m \geqslant 2$, if there exists a real-valued function $f : D^* \subseteq \mathbb{R}^{m-1} \to \mathbb{R}$ such that for all $(x_1, \ldots x_{m-1}) \in D^*$,*

4.7.4 $$F(x_1, \ldots, x_{m-1}, f(x_1, \ldots, x_{m-1})) = 0,$$

then the function f is said to be defined implicitly *on D^* by the equation*

4.7.5 $$F(x_1, \ldots, x_m) = 0.$$

The condition 4.7.4 can be expressed in the form

4.7.6 $$F(\mathbf{x}, f(\mathbf{x})) = 0 \qquad \mathbf{x} \in D^* \subseteq \mathbb{R}^{m-1}.$$

We observe that equation 4.7.5 defines a real-valued function f implicitly if the level set of F corresponding to the value 0 contains the graph of f as a subset.

4.7.7 *Example*.   An equation of the form 4.7.5 does not in general lead to the implicit definition of just one function. For example, if

$$F(x, y) = x^2 + y^2 - 1, \qquad (x, y) \in \mathbb{R}^2,$$

then the level set of F corresponding to 0 is the circle C of 4.7.2 which contains an infinity of different graphs. We have illustrated just three in Fig. 4.11(i)–(iii). They are the graphs of the following functions:

(i) $f(x) = \sqrt{(1 - x^2)}$, $x \in [-1, 1]$;

(ii) $f^*(x) = -\sqrt{(1 - x^2)}$, $x \in [-1, 1]$;

(iii) $f^\dagger(x) = \begin{cases} \sqrt{(1 - x^2)}, & x \in [-1, 0[; \\ -\sqrt{(1 - x^2)}, & x \in [0, \frac{1}{2}[; \\ \sqrt{(1 - x^2)}, & x \in [\frac{1}{2}, 1]. \end{cases}$

All functions defined implicitly by the equation $x^2 + y^2 - 1 = 0$ have as their domain a subset of the interval $[-1, 1]$.

The curve of Fig. 4.11(iv) illustrates a subset of the level set C which is not the graph of a function, because for any $x \in]-1, -\frac{1}{2}]$ there are two values of y such that $(x, y) \in C$. We observe, however, that this curve is the graph of a function when the roles of the x and y coordinates are interchanged, for it consists of the points $(g(y), y)$ where

4.7.8 $g(y) = -\sqrt{(1 - y^2)}$, $y \in [-1, \frac{1}{2}\sqrt{3}]$.

In this way the equation $F(x, y) = 0$ may also be regarded as implicitly defining the function $g : [-1, \frac{1}{2}\sqrt{3}] \subseteq \mathbb{R} \to \mathbb{R}$ given in 4.7.8.

4.7.9 *Example*.   The equation

4.7.10 $x_1^2 + \frac{1}{2}x_2^2 + x_3^2 - 1 = 0$

implicitly defines (among others) the function $f : D \subseteq \mathbb{R}^2 \to \mathbb{R}$, where

$$D = \{(x_1, x_2) \in \mathbb{R}^2 \,|\, x_1^2 + \tfrac{1}{2}x_2^2 \leqslant 1\}$$

and

$$f(x_1, x_2) = -\sqrt{(1 - x_1^2 - \tfrac{1}{2}x_2^2)}, \qquad (x_1, x_2) \in D.$$

The graph of f is part of the surface of the ellipsoid which is defined by Equation 4.7.10.

It is in the nature of the equation $x^2 + y^2 - 1 = 0$ considered above that the functions it defines implicitly can also be described explicitly. It is simply a case of 'solving' for y (or for x) and making a selection of the possibilities. The following example illustrates an implicitly defined function that has no simple explicit presentation.

4.7.11 *Example.* Consider the function $F : \mathbb{R}^2 \to \mathbb{R}$ defined by

4.7.12 $$F(x, y) = x^5 + y^5 - 5x^3 y.$$

The level set of F corresponding to the value 0,

$$S = \{(x, y) \in \mathbb{R}^2 \,|\, F(x, y) = 0\},$$

is illustrated in Fig. 4.12.

Here we cannot 'solve' for y as a function of x (nor for x as function of y) and yet the equation

4.7.13 $$x^5 + y^5 - 5x^3 y = 0$$

implicitly defines a number of real-valued functions on various subsets of $\mathbb{R}$. Our sketch of the solution set of the equation suggests that there are three different continuously differentiable functions defined implicitly by the equation 4.7.13 in a neighbourhood of $x = 1$. By putting $x = 1$ in 4.7.13 we find that the corresponding values of y are approximately 1.44, 0.20 and -1.54.

Our examples illustrate the *Implicit Function Theorem* 4.7.14, which asserts that if an equation $F(x, y) = 0$ has a solution $x = a$, $y = b$, then under not too severe restrictions the equation implicitly defines in a neighbourhood of $x = a$ a unique continuously differentiable function that takes the value b at $x = a$. Its generalization has many applications, as we shall see in the next section.

4.7.14 *Theorem.* *Let* $F : D \subseteq \mathbb{R}^2 \to \mathbb{R}$ *be a real-valued continuously differentiable function defined in a neighbourhood D of a point* $(a, b) \in \mathbb{R}^2$. *Suppose that*

    [i] $F(a, b) = 0$,

    [ii] $\dfrac{\partial F}{\partial y}(a, b) \neq 0$.

Then there exists a neighbourhood N of $a \in \mathbb{R}$, *a neighbourhood M of* $b \in \mathbb{R}$, *and a continuously differentiable function* $f : N \subseteq \mathbb{R} \to \mathbb{R}$, *such that*

    [1] $f(a) = b$, *and* $f(N) \subseteq M$;

    [2] *for each $x \in N$ the equation $F(x, y) = 0$ is uniquely solved by*

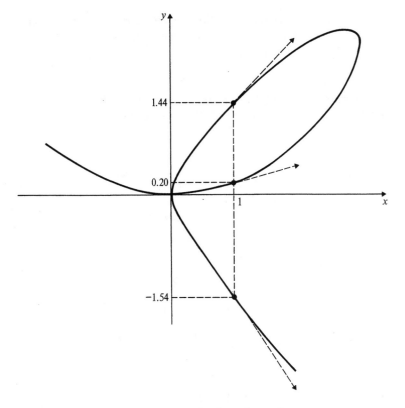

Fig. 4.12 $\{(x, y)\in\mathbb{R}^2\,|\,x^5 + y^5 - 5x^3y = 0\}$

$y = f(x)\in M$, *provided that the possible values of y are restricted to lie in M. Moreover, the derivative of f is given by*

4.7.15
$$f'(t) = -\frac{\partial F}{\partial x}(t, f(t))\bigg/ \frac{\partial F}{\partial y}(t, f(t)), \qquad t\in N.$$

Theorem 4.7.14 is the two-dimensional version of the Implicit Function Theorem 4.8.1, which we shall state and prove presently.

4.7.16 *Remarks.* **[i]** Theorem 4.7.14 guarantees the existence of certain solutions of the equation $F(x, y) = 0$ subject to certain conditions, but it does not provide an explicit formula for the solutions. In view of Example 4.7.11, a general formula cannot be expected. (A theorem such as 4.7.14 is called an *existence theorem*.)

[ii] Theorem 4.7.14 has the following geometrical interpretation. Subject to the stated conditions, there exists an open rectangle $R = N \times M$ centred at (a, b),

$$R = \{(x, y) \in \mathbb{R}^2 \mid x \in N \subseteq \mathbb{R}, \; y \in M \subseteq \mathbb{R}\},$$

and a continuously differentiable function $f : N \to M$ such that subject to the restriction $(x, y) \in R$, the level set $F(x, y) = 0$ is the

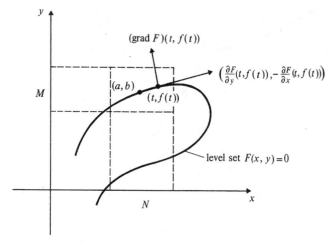

Fig. 4.13 Level set $F(x, y) = 0$ is graph of f in rectangle $N \times M$

graph of f. See Fig. 4.13. Since

$$(\text{grad } F)(t, f(t)) = \left(\frac{\partial F}{\partial x}(t, f(t)), \; \frac{\partial F}{\partial y}(t, f(t)) \right),$$

the tangent line to the level set (graph) at $(t, f(t))$ is in the direction

4.7.17 $\qquad \left(\dfrac{\partial F}{\partial y}(t, f(t)), \; -\dfrac{\partial F}{\partial x}(t, f(t)) \right), \qquad t \in N.$

The slope of this line is indeed the slope of the graph of f at $(t, f(t))$ as given by 4.7.15, for if 4.7.15 is satisfied, then the vector 4.7.17 is a non-zero scalar multiple of $(1, f'(t))$.

[iii] Once the existence of a differentiable function $f : N \subseteq \mathbb{R} \to \mathbb{R}$ satisfying $F(t, f(t)) = 0$, $t \in N$ has been established, expression 4.7.15 for its derivative is obtained by the Chain Rule as follows.

Put $g(t) = (t, f(t))$, $t \in N$. Then $(F \circ g)(t) = F(g(t)) = 0$. Hence on differentiating we obtain

$$0 = \frac{\partial F}{\partial x}(t, f(t)).1 + \frac{\partial F}{\partial y}(t, f(t)).f'(t), \qquad t \in N$$

and Formula 4.7.15 follows. The requirement $(\partial F/\partial y)(t, f(t)) \neq 0, t \in N$ is a consequence of the continuity of $\partial F/\partial y$ and the condition $(\partial F/\partial y)(a, b) \neq 0$, as stated in the theorem.

The following example illustrates the application of 4.7.15.

4.7.18 Example. [i] Let f be the function defined implicitly near $x = 1$ by the equation $F(x, y) = 0$, where

$$F(x, y) = x^5 + y^5 - 5x^3 y \qquad (x, y) \in \mathbb{R}^2$$

and by the condition $f(1) = 1.44$ (see Example 4.7.11). Find the slope of the tangent vector to the graph of f at $x = 1$.

[ii] Repeat with the functions defined implicitly near $x = 1$ by $F(x, y) = 0$ and taking respectively the values 0.20 and -1.54 at $x = 1$.

Solution. [i] we have for $(x, y) \in \mathbb{R}^2$

$$\frac{\partial F}{\partial x}(x, y) = 5x^4 - 15x^2 y, \qquad \frac{\partial F}{\partial y}(x, y) = 5y^4 - 5x^3.$$

Hence, by Formula 4.7.15, the slope of the tangent vector to the graph of f at $x = 1$ is

$$f'(1) = -\frac{\partial F}{\partial x}(1, f(1)) \Big/ \frac{\partial F}{\partial y}(1, f(1))$$

$$= -\frac{5 - 15(1.44)}{5(1.44)^2 - 5} = 1.01$$

correct to 2 decimal places.

Alternatively, the Chain Rule of elementary calculus can be applied directly as follows. The function f satisfies the equation

$$x^5 + (f(x))^5 - 5x^3 f(x) = 0$$

for all x in a neighbourhood of 1. Differentiating with respect to x we obtain

$$5x^4 + 5(f(x))^4 f'(x) - 15x^2 f(x) - 5x^3 f'(x) = 0,$$

and hence $f'(1)$ is calculated by substituting $x = 1$ and $f(x) = 1.44$.

[ii] Similar calculations lead to tangent vectors with slopes 0.40 and -1.22.

The tangent vectors for the three cases are indicated in Fig. 4.12.

Exercises 4.7

1. Consider the folium of Descartes defined by the equation $F(x, y) = 0$, where

    $$F(x, y) = x^3 + y^3 - 3xy, \qquad (x, y) \in \mathbb{R}^2.$$

    (See Example 3.1.7 and Fig. 3.2.)
    Find the slope of the tangent vector to the graph of the function f at $x = \frac{2}{3}$ defined implicitly by the equation $F(x, y) = 0$ and the condition
    (a) $f(x) = \frac{4}{3}$; (b) $f(x) = (-2 + \sqrt{6})/3$; (c) $f(x) = (-2 - \sqrt{6})/3$.
    Sketch the curve and the tangent vectors.

2. (a) Let $F(x, y) = x + y^2 + \sin(xy)$. Prove that in a sufficiently small neighbourhood of $(0, 0)$ the equation $F(x, y) = 0$ defines implicitly a continuously differentiable function g such that $g(0) = 0$ and $F(x, y) = 0$ if and only if $x = g(y)$. Show also that $g'(0) = 0$. (Apply Theorem 4.7.14 with the roles of x and y reversed.)
    (b) According to part (a) of this exercise, the condition $x + y^2$ $+ \sin(xy) = 0$, with (x, y) sufficiently close to $(0, 0)$, implies that $x = g(y)$. Does it also imply that $y = f(x)$ for some continuously differentiable function f such that $f(0) = 0$?

    Answer: no. Observe that near $(0, 0)$ the equation $F(x, y) = 0$ behaves like $x + y^2 + xy = 0$.

3. As in Exercise 1, let $F(x, y) = x^3 + y^3 - 3xy$. Observe that (a) $F(0, 0) = 0$, but (b) $(\partial F/\partial y)(0, 0) = (\partial F/\partial x)(0, 0) = 0$. Therefore condition (ii) of Theorem 4.7.14 fails. Note from your sketch of the curve $F(x, y) = 0$ (or from Fig. 3.2) that the conclusion (2) of Theorem 4.7.14 also fails. The curve crosses itself at $(0, 0)$ and therefore the equation $F(x, y) = 0$ is clearly not solved *uniquely* in the form $y = f(x)$ or $x = g(y)$ near the origin. The sketch suggests that there are two continuously differentiable function, f and g such that $f(0) = g(0) = 0$, and near $(0, 0)$ if $F(x, y) = 0$ then either $y = f(x)$ or $x = g(y)$. Approximations to the functions f and g are given by $f(x) = \frac{1}{3}x^2$, $g(y) = \frac{1}{3}y^2$.

4. Describe explicitly all continuously differentiable functions f defined in a neighbourhood N of 1 and such that $F(x, f(x)) = 0$ for all $x \in N$, where
    (a) $F(x, y) = y^2 - x^2 - 1$;
    (b) $F(x, y) = x - 2\cos y$.
    In each case calculate $f'(1)$. Illustrate your answers with a sketch.

    Answer: (b) $f(x) = 2k\pi \pm \arccos \frac{1}{2}x$, $k \in \mathbb{N}$, $f'(1) = \pm 1/\sqrt{3}$.

4.8 Implicit Function Theorems

In this section we first discuss the Implicit Function Theorem in a form which directly generalizes Theorem 4.7.14. We then briefly consider further possible generalizations.

4.8.1 *The Implicit Function Theorem.* *Let $F : D \subseteq \mathbb{R}^m \to \mathbb{R}$ be a real-valued continuously differentiable function defined in a neighbourhood D of a point $(p_1, \ldots, p_m) \in \mathbb{R}^m$, $m \geq 2$. Suppose that*

[i] $F(p_1, \ldots, p_m) = 0$, *and*

[ii] $\dfrac{\partial F}{\partial x_m}(p_1, \ldots, p_m) \neq 0$.

Then there exists a neighbourhood N of $(p_1, \ldots, p_{m-1}) \in \mathbb{R}^{m-1}$, a neighbourhood M of $p_m \in \mathbb{R}$, and a continuously differentiable function $f : N \subseteq \mathbb{R}^{m-1} \to \mathbb{R}$ such that

(1) $f(p_1, \ldots, p_{m-1}) = p_m$, *and* $f(N) \subseteq M$,

(2) *for each $(x_1, \ldots, x_{m-1}) \in N$ the equation $F(x_1, \ldots, x_{m-1}, x_m) = 0$ is uniquely solved by $x_m = f(x_1, \ldots, x_{m-1}) \in M$, provided that the possible values of x_m are restricted to lie in M.*

Proof. For simplicity we give the proof for the case $m = 3$. The general case is proved along similar lines.

The idea of the proof is to construct from the function $F : D \subseteq \mathbb{R}^3 \to \mathbb{R}$ a related function to which the Inverse Function Theorem 4.6.7 can be applied. Towards this end, define $F^* : D \subseteq \mathbb{R}^3 \to \mathbb{R}^3$ by

4.8.2 $\quad F^*(x_1, x_2, x_3) = (x_1, x_2, F(x_1, x_2, x_3)), \qquad (x_1, x_2, x_3) \in D.$

Notice that $F^*(x_1, x_2, x_3)$ lies in the x_1, x_2 plane if and only if $F(x_1, x_2, x_3) = 0$. In particular, by condition (i),

4.8.3 $\qquad\qquad\qquad F^*(p_1, p_2, p_3) = (p_1, p_2, 0).$

The function F^* is continuously differentiable on D and its Jacobian matrix at $\mathbf{p} \in D$ is

$$J_{F^*, \mathbf{p}} = \begin{bmatrix} 1 & 0 & 0 \\ 0 & 1 & 0 \\ \dfrac{\partial F}{\partial x_1}(\mathbf{p}) & \dfrac{\partial F}{\partial x_2}(\mathbf{p}) & \dfrac{\partial F}{\partial x_3}(\mathbf{p}) \end{bmatrix}.$$

By condition (ii),

$$\det J_{F^*, \mathbf{p}} = \frac{\partial F}{\partial x_3}(\mathbf{p}) \neq 0,$$

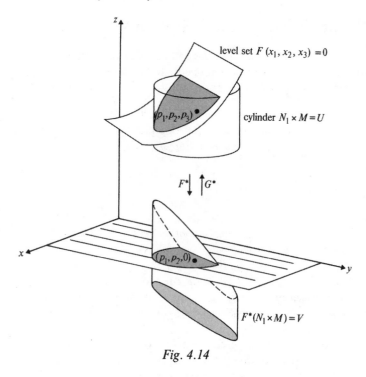

Fig. 4.14

and the Inverse Function Theorem may be applied to F^* in a neighbourhood of **p**. The theorem implies that **p** lies in an open set $U \subseteq D$ in $\mathbb{R}^3$ such that F^* is invertible with C^1 inverse on U, and such that $F^*(U) = V$ is an open subset of $\mathbb{R}^3$ containing $F^*(p) = (p_1, p_2, 0)$. We may assume, by taking a further subset if necessary, that $U = N_1 \times M$ where N_1 is a neighbourhood of (p_1, p_2) in $\mathbb{R}^2$ and M is a neighbourhood of p_3 in $\mathbb{R}$. See Fig. 4.14.

Denote the C^1 inverse of $F^* : U \to V$ by $G^* : V \to U$. It is clear from 4.8.2 that G^* has the form

4.8.4

$$G^*(x_1, x_2, x_3) = (x_1, x_2, G(x_1, x_2, x_3)) \in N_1 \times M, (x_1, x_2, x_3) \in V,$$

where $G : V \subseteq \mathbb{R}^3 \to \mathbb{R}$ is a C^1 function. By 4.8.4,

4.8.5 $G(V) \subseteq M.$

Since V is an open subset of $\mathbb{R}^3$ containing $(p_1, p_2, 0)$, there is a neighbourhood N of (p_1, p_2) in $\mathbb{R}^2$ such that $N \times \{0\} \subseteq V$.

We can now define a function $f : N \subseteq \mathbb{R}^2 \to \mathbb{R}$ by

4.8.6 $\qquad f(x_1, x_2) = G(x_1, x_2, 0) \qquad (x_1, x_2) \in N.$

The function f is well defined since $(x_1, x_2, 0) \in N \times \{0\} \subseteq V$ whenever $(x_1, x_2) \in N$. Furthermore, since G is a C^1 function, f is a C^1 function. We shall show that f has the required properties (1) and (2).

(1) Firstly, $f(p_1, p_2) = p_3$ since by 4.8.6,

$$(p_1, p_2, f(p_1, p_2)) = (p_1, p_2, G(p_1, p_2, 0))$$
$$= G^*(p_1, p_2, 0) = (p_1, p_2, p_3).$$

Secondly, since $f(N) = G(N \times \{0\})$ and $N \times \{0\} \subseteq V$, expression 4.8.5 implies that

4.8.7 $\qquad\qquad\qquad f(N) \subseteq M.$

(2) It is clear from the definition of F^* that $V \subseteq N_1 \times \mathbb{R}$ and therefore, since $N \times \{0\} \subseteq V$ we have

4.8.8 $\qquad\qquad N \subseteq N_1 \quad \text{and} \quad N \times M \subseteq N_1 \times M.$

Therefore, for each $(x_1, x_2) \in N$, by 4.8.7, the vector $(x_1, x_2, f(x_1, x_2))$ lies in the domain of F. Now for each $(x_1, x_2) \in N$,

4.8.9 $\quad (x_1, x_2, F(x_1, x_2, f(x_1, x_2))) = (x_1, x_2, F(x_1, x_2, G(x_1, x_2, 0))).$
$$= (F^* \circ G^*)(x_1, x_2, 0) = (x_1, x_2, 0).$$

Hence $x_3 = f(x_1, x_2) \in M$ solves $F(x_1, x_2, x_3) = 0$ for all $(x_1, x_2) \in N$.

Finally, to prove uniqueness, suppose that for some $(a, b) \in N$ there exists $c \in M$ and $d \in M$ such that

$$F(a, b, c) = F(a, b, d).$$

Then

4.8.10 $\qquad\qquad\qquad F^*(a, b, c) = F^*(a, b, d).$

Since F^* is 1–1 on $N \times M \subseteq N_1 \times M$, we have $c = d$. This completes the proof of the theorem.

4.8.11 *Example.* Let $F : \mathbb{R}^3 \to \mathbb{R}$ be given by $F(x, y, z) = x^2 + y^2 + z^2 - 1$. Then
(a) $F(\frac{1}{2}, \frac{1}{2}, -\frac{1}{2}\sqrt{3}) = 0$ and
(b) $\dfrac{\partial F}{\partial z}(\frac{1}{2}, \frac{1}{2}, -\frac{1}{2}\sqrt{3}) = -\sqrt{3}.$

By the Implicit Function Theorem 4.8.1 the equation $F(x, y, z) = 0$

uniquely defines a continuously differentiable function $f : N \subseteq \mathbb{R}^2 \to \mathbb{R}$ in a neighbourhood N of $(\frac{1}{2}, \frac{1}{2})$ in $\mathbb{R}^2$ such that $f(\frac{1}{2}, \frac{1}{2}) = -\frac{1}{2}\sqrt{3}$ and $F(x, y, f(x, y)) = 0$ for all $(x, y) \in N$.

There is an explicit formula for the function f which we obtain by solving for z in the equation $x^2 + y^2 + z^2 - 1 = 0$. The appropriate solution is

$$f(x, y) = -\sqrt{(1 - x^2 - y^2)},$$

which satisfies the required condition $f(\frac{1}{2}, \frac{1}{2}) = -\frac{1}{2}\sqrt{3}$. The explicit formula for f shows that the neighbourhood N of $(\frac{1}{2}, \frac{1}{2})$ on which f is defined must be restricted at least by the condition $x^2 + y^2 < 1$.

A second solution for z of the equation $x^2 + y^2 + z^2 - 1 = 0$ leads to the function given by

$$g(x, y) = +\sqrt{(1 - x^2 - y^2)},$$

where we have $g(\frac{1}{2}, \frac{1}{2}) = \frac{1}{2}\sqrt{3}$.

The graphs of the functions f and g lie on the surface of the sphere $x^2 + y^2 + z^2 = 1$, centre the origin, radius 1 (Fig. 4.15).

As a final illustration, note that

$$\text{(i) } F(1, 0, 0) = 0 \qquad \text{but} \qquad \text{(ii) } \frac{\partial F}{\partial z}(1, 0, 0) = 0.$$

Thus condition (ii) of the Implicit Function Theorem 4.8.1 fails at the point $(1, 0, 0)$. The conclusions of the theorem also fail, for there is no neighbourhood N of the point $(1, 0)$ on which a differentiable function $h : N \subseteq \mathbb{R}^2 \to \mathbb{R}$ can be defined such that $F(x, y, h(x, y)) = 0$ for all $(x, y) \in N$. A glance at Fig. 4.15 should suggest why this is so. Note, however, that $(\partial F/\partial x)(1, 0, 0) = 2 \neq 0$. By switching the roles of the first and third coordinates we can infer the existence of a continuously differentiable function $h : N \subseteq \mathbb{R}^2 \to \mathbb{R}$ defined in a neighbourhood N of $(0, 0)$ such that $h(0, 0) = 1$ and $F(h(y, z), y, z) = 0$ for all $(y, z) \in N$. The function h can be defined explicitly (Exercise 4.8.2).

4.8.12 *Example.* The equation

4.8.13 $\qquad\qquad x + y + z - \sin(xyz) = 0, \qquad\qquad (x, y, z) \in \mathbb{R}^3,$

is satisfied by $(x, y, z) = (0, 0, 0)$. We can apply the Implicit Function Theorem 4.8.1 to justify the process of solving for z as a function of x and y close to the origin. Define $F : \mathbb{R}^3 \to \mathbb{R}$ by $F(x, y, z) = x + y + z - \sin(xyz)$, $(x, y, z) \in \mathbb{R}^3$. Then F is continuously differentiable on $\mathbb{R}^3$ and

$$\text{(i) } F(0, 0, 0) = 0 \quad \text{and} \quad \text{(ii) } \frac{\partial F}{\partial z}(0, 0, 0) = 1 \neq 0.$$

Hence there exists a neighbourhood $N \subseteq \mathbb{R}^2$ of $(0, 0)$, a neighbourhood $M \subseteq \mathbb{R}$ of 0, and a continuously differentiable function $f : N \subseteq \mathbb{R}^2 \to \mathbb{R}$ such

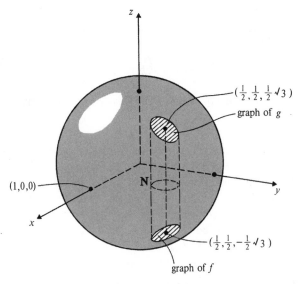

$$(\tfrac{1}{2}, \tfrac{1}{2}, \tfrac{1}{2}\sqrt{3})$$

graph of g

$(1,0,0)$

$$(\tfrac{1}{2}, \tfrac{1}{2}, -\tfrac{1}{2}\sqrt{3})$$

graph of f

Fig. 4.15

that Equation 4.8.13 is uniquely solved by $z = f(x, y)$ in the open cylinder $N \times M$.

The function f is defined implicitly and there is no explicit formula for it. However, given $(a, b) \in N$ it will of course be possible to calculate $f(a, b)$ by solving Equation 4.8.13 with $x = a$, $y = b$ by a numerical method.

We can obtain information about the partial derivatives of f as follows. Define $g : N \subseteq \mathbb{R}^2 \to \mathbb{R}^3$ by

$$g(u, v) = (u, v, f(u, v)), \qquad (u, v) \in N.$$

Then

4.8.14 $\qquad (F \circ g)(u, v) = F(u, v, f(u, v)) = 0, \qquad (u, v) \in N.$

Thus the function $H = F \circ g : N \subseteq \mathbb{R}^2 \to \mathbb{R}$ is the zero function. Therefore, using the Chain Rule in the form 4.4.3, (with suitably adjusted notation) we obtain,

4.8.15 $\qquad 0 = J_{H,(a,b)} = J_{F,g(a,b)} J_{g,(a,b)} \qquad (a, b) \in N.$

The matrix identity 4.8.15 is equivalent to the system

$$0 = \frac{\partial H}{\partial u}(a, b) = \frac{\partial F}{\partial x}(a, b, f(a, b)) \cdot 1 + \frac{\partial F}{\partial y}(a, b, f(a, b)) \cdot 0$$

$$+ \frac{\partial F}{\partial z}(a, b, f(a, b)) \frac{\partial f}{\partial u}(a, b)$$

$$0 = \frac{\partial H}{\partial v}(a, b) = \frac{\partial F}{\partial x}(a, b, f(a, b)) \cdot 0 + \frac{\partial F}{\partial y}(a, b, f(a, b)) \cdot 1$$

$$+ \frac{\partial F}{\partial z}(a, b, f(a, b)) \frac{\partial f}{\partial v}(a, b).$$

These identities give us expressions for $(\partial f/\partial u)(a, b)$ and $(\partial f/\partial v)(a, b)$. See Exercise 4.8.3.

The Implicit Function Theorem 4.8.1 concerns the solution of one equation $F(x_1, \ldots, x_m) = 0$ for the "unknown" x_m. There is a generalization that deals with the solution of k equations in $x_1, \ldots, x_m$ for k unknowns, where $2 \leqslant k \leqslant m - 1$. The following theorem provides a typical illustration. It concerns conditions under which two equations in $x_1, \ldots, x_5$ can be solved for x_4 and x_5 in terms of x_1, x_2 and x_3.

4.8.16 Theorem. *Let* $H : D \subseteq \mathbb{R}^5 \to \mathbb{R}$ *and* $K : D \subseteq \mathbb{R}^5 \to \mathbb{R}$ *be real-valued continuously differentiable functions defined in a neighbourhood* D *of a point* $\mathbf{p} = (p_1, \ldots, p_5) \in \mathbb{R}^5$. *Suppose that*
  [i] $H(p_1, \ldots, p_5) = 0$ *and* $K(p_1, \ldots, p_5) = 0$, *and*
  [ii] *the determinant*

$$\frac{\partial(H, K)}{\partial(x_4, x_5)}(\mathbf{p}) = \det \begin{bmatrix} \dfrac{\partial H}{\partial x_4}(\mathbf{p}) & \dfrac{\partial H}{\partial x_5}(\mathbf{p}) \\ \dfrac{\partial K}{\partial x_4}(\mathbf{p}) & \dfrac{\partial K}{\partial x_5}(\mathbf{p}) \end{bmatrix}$$

is non-zero.
  *Then there exists a neighbour*hood N *of* $(p_1, p_2, p_3) \in \mathbb{R}^3$, *a neighbourhood* M *of* $(p_4, p_5) \in \mathbb{R}^2$, *and real-valued continuously differentiable functions* $f : N \subseteq \mathbb{R}^3 \to \mathbb{R}$, *and* $g : N \subseteq \mathbb{R}^3 \to \mathbb{R}$ *such that*
  (1) $f(p_1, p_2, p_3) = p_4$, $\quad g(p_1, p_2, p_3) = p_5$
and

$$(f(x_1, x_2, x_3), g(x_1, x_2, x_3)) \in M \text{ for all } (x_1, x_2, x_3) \in N,$$
  (2) *for each* $(x_1, x_2, x_3) \in N$ *the equations*

$$H(x_1, \ldots, x_5) = 0, \qquad K(x_1, \ldots, x_5) = 0$$

are simultaneously solved uniquely by

$$x_4 = f(x_1, x_2, x_3), \qquad x_5 = g(x_1, x_2, x_3),$$

provided that (x_4, x_5) *is restricted to lie in* M.

Sketch of Proof.   Apply the Inverse Function Theorem 4.6.7 to the continuously differentiable function $F^* : D \subseteq \mathbb{R}^5 \to \mathbb{R}^5$ defined by

4.8.17
$$\begin{aligned} F^*(\mathbf{x}) &= F^*(x_1, x_2, x_3, x_4, x_5) \\ &= (x_1, x_2, x_3, H(\mathbf{x}), K(\mathbf{x})), \qquad \mathbf{x} \in D, \end{aligned}$$

so that, in particular, by condition (i),

4.8.18
$$F^*(\mathbf{p}) = (p_1, p_2, p_3, 0, 0).$$

It will be found that

$$\det J_{F^*, \mathbf{p}} = \frac{\partial(H, K)}{\partial(x_4, x_5)}(\mathbf{p}) \neq 0,$$

by condition (ii). Therefore by the Inverse Function Theorem 4.6.7 there exist open sets U and V in $\mathbb{R}^5$ containing $\mathbf{p}$ and $F^*(\mathbf{p})$ respectively such that $U \subseteq D$, $F^*(U) = V$, F^* is 1–1 on U, and the function $G^* : V \to U$ locally inverse to F^* on U is continuously differentiable.

From the form of 4.8.17 it is clear that

4.8.19 $G^*(y_1, y_2, y_3, y_4, y_5) = (y_1, y_2, y_3, G_4^*(\mathbf{y}), G_5^*(\mathbf{y})), \qquad \mathbf{y} \in V,$

and, in particular, that

4.8.20 $G^*(p_1, p_2, p_3, 0, 0) = (p_1, p_2, p_3, p_4, p_5).$

The proof is completed by a method similar to that of Theorem 4.8.1.

The following example illustrates the solution of two equations in three unknowns:

4.8.21 *Example.*   The point $\mathbf{p} = (\tfrac{1}{2}, \tfrac{1}{4}(1 + \sqrt{5}), \tfrac{1}{4}(1 - \sqrt{5}))$ satisfies the equations

4.8.22
$$\begin{cases} x + y + z = 1, \\ x^2 + y^2 + z^2 = 1. \end{cases}$$

We show that these equations implicitly define y and z as continuously differentiable functions of x in a neighbourhood of $x = \tfrac{1}{2}$ that take the values $\tfrac{1}{4}(1 + \sqrt{5})$ and $\tfrac{1}{4}(1 - \sqrt{5})$ respectively at $x = \tfrac{1}{2}$. Define $H : \mathbb{R}^3 \to \mathbb{R}$ and $K : \mathbb{R}^3 \to \mathbb{R}$ by

$$H(x, y, z) = x + y + z - 1, \quad K(x, y, z) = x^2 + y^2 + z^2 - 1,$$
$$(x, y, z) \in \mathbb{R}^3.$$

Then H, K are continuously differentiable on $\mathbb{R}^3$ and

$$\frac{\partial(H, K)}{\partial(y, z)}(x, y, z) = \det \begin{bmatrix} 1 & 1 \\ 2y & 2z \end{bmatrix} = 2(z - y).$$

This is non-zero at **p**. Therefore, by The General Implicit Function Theorem, there exists a neighbourhood N of $\frac{1}{2} \in \mathbb{R}$, a neighbourhood M of $(\frac{1}{4}(1 + \sqrt{5}), \frac{1}{4}(1 - \sqrt{5}) \in \mathbb{R}^2$ and continuously differentiable functions $f : N \subseteq \mathbb{R} \to \mathbb{R}$ and $g : N \subseteq \mathbb{R} \to \mathbb{R}$ such that $f(\frac{1}{2}) = \frac{1}{4}(1 + \sqrt{5}), g(\frac{1}{2}) = \frac{1}{4}(1 - \sqrt{5})$ and equations 4.8.22 are uniquely solved by

$$y = f(x), \qquad z = g(x), \qquad x \in N$$

in the open cylinder $N \times M$.

It turns out that equations 4.8.22 can be solved explicitly for y and z as functions of x. The formulae

$$y = -\tfrac{1}{2}(x - 1) + \tfrac{1}{2}\sqrt{(1 + 2x - 3x^2)}, \qquad z = 1 - x - y$$

provide solutions of 4.8.22 in a neighbourhood of $x = \frac{1}{2}$.

The point $(1, 0, 0)$ also satisfies equations 48.22, but, as

$$\frac{\partial(H, K)}{\partial(y, z)}(1, 0, 0) = 0,$$

for this case the Implicit Function Theorem is inconclusive. In fact equations 4.8.22 do not implicitly define y and z as real-valued functions of x in a neighbourhood of $(1, 0, 0)$, for there are no solutions of 4.8.22 with $x > 1$.

Exercises 4.8

1.   (a) Prove that the smooth surface S in $\mathbb{R}^3$ whose equation is $xyz + e^{xz} = 1$ is the graph G of a continuously differentiable function in a neighbourhood of the point $(1, 0, 0) \in S$.

Hint: apply the Implicit Function Theorem 4.8.1 to the function $F(x, y, z) = xyz + e^{xz} - 1$ in a neighbourhood of $(1, 0, 0)$. There is an explicit formula for the graph G: it is the plane $z = 0$.

Does there exist a continuously differentiable function g with $g(0, 0) = 1$ such that $x = g(y, z)$ for all (x, y, z) in S close to $(1, 0, 0)$?

Hint: since $(\partial F / \partial x)(1, 0, 0) = 0$, the Implicit Function Theorem is inconclusive. The following alternative method establishes that there does not even exist a continuous function g with the stated properties. Suppose the contrary. Then $\lim_{z \to 0} g(0, z) = g(0, 0) = 1$. But also for $y = 0$ and all z close to 0

$$g(0, z).0.z + \exp(g(0, z).z) = 1$$

Deduce the contradiction $\lim_{z \to 0} g(0, z) = 0$.

(b) Is the surface S of part (a) the graph of a continuously differentiable function in a neighbourhood of the point (i) $(0, 0, 1) \in S$, (ii) $(0, 1, 0) \in S$?

Hint: The Implicit Function Theorem is inconclusive. By inspection, the plane $z = 0$ lies in S and contains $(0, 1, 0)$. But does this plane *describe* S in a neighbourhood of $(0, 1, 0)$? Try another plane.

2. Let $F(x, y, z) = x^2 + y^2 + z^2 - 1$. As suggested in Example 4.8.11, find a continuously differentiable function $h : N \subseteq \mathbb{R}^2 \to \mathbb{R}$ defined in a neighbourhood N of $(0, 0)$ such that $h(0, 0) = 1$ and $F(h(y, z), y, z) = 0$ for all $(y, z) \in N$.

3. Consider the continuously differentiable function $F : \mathbb{R}^3 \to \mathbb{R}$ defined by $F(x, y, z) = x + y + z - \sin(xyz)$. In Example 4.8.12 we proved the existence of a neighbourhood $N \subseteq \mathbb{R}^2$ of $(0, 0)$, a neighbourhood $M \subseteq \mathbb{R}$ of 0 and a continuously differentiable function $f : M \to N$ such that

$$F(u, v, f(u, v)) = 0, \qquad (u, v) \in N.$$

Prove that

$$\frac{\partial f}{\partial u}(0, 0) = \frac{\partial f}{\partial v}(0, 0) = -1.$$

4. Exercise 3 can be stated as follows. By the Implicit Function Theorem the equation $x + y + z - \sin(xyz) = 0$ is uniquely solved in a neighbourhood of $(0, 0, 0)$ by $z = f(x, y)$, where f is a continuously differentiable function. Find $(\partial f / \partial x)(0, 0)$ and $(\partial f / \partial y)(0, 0)$ from the identity

$$x + y + f(x, y) - \sin(xyf(x, y)) = 0.$$

5. For each of the following functions $F : \mathbb{R}^3 \to \mathbb{R}$, show that the equation $F(x, y, z) = 0$ defines implicitly a continuously differentiable function $z = f(x, y)$ in a neighbourhood of the given point (a, b, c). Find $(\partial f / \partial x)(a, b)$ and $(\partial f / \partial y)(a, b)$.
   (a) $F(x, y, z) = x^3 + y^3 + z^3 - xyz - 2$ at $(1, 1, -1)$;
   (b) $F(x, y, z) = x^2 + z^3 - z - xy \sin z$ at $(1, 1, 0)$.

Answer: (a) $-2, -2$;   (b) $1, 0$.

6. Show that the equation $F(x, y, z) = 0$ in Exercise 5(a) also defines implicitly y as a function of x, z, and x as a function of y, z in a neighbourhood of the given point (a, b, c).
   Is the same true in Exercise 5(b)?

7. Let $F : \mathbb{R}^2 \to \mathbb{R}$ be defined by $F(x_1, x_2) = x_1^9 - x_2^3$. Then (a) $F(0, 0) = 0$ and (b) $(\partial F / \partial x_1)(0, 0) = (\partial F / \partial x_2)(0, 0) = 0$. Hence condition (ii) of the Implicit Function Theorem 4.8.1 is not satisfied. Show that nevertheless the conclusions of the theorem hold, since the equation $F(x_1, x_2) = 0$ is uniquely solved by $x_2 = f(x_1) = x_1^3$.

8. Define the functions $H : \mathbb{R}^4 \to \mathbb{R}$ and $K : \mathbb{R}^4 \to \mathbb{R}$ by

$$H(x, y, z, t) = x^2 - yz, \qquad K(x, y, z, t) = xy - zt.$$

Prove by the method of Theorem 4.8.16 that in a neighbourhood of $(-2, 2, 2, -2)$ the equations $H(x, y, z, t) = 0$, $K(x, y, z, t) = 0$ implicitly define z and t as continuously differentiable functions of x and y taking the values $z = 2$ and $t = -2$ at $(x, y) = (-2, 2)$. (Show that

$$\frac{\partial(H, K)}{\partial(z, t)}(-2, 2, 2, -2) \neq 0.$$

There is an explicit formula for z and t as functions of x and y.)
Are z and t defined as functions of x and y near $(0, 1, 0, 2)$?
Prove that

(a) $\dfrac{\partial(H, K)}{\delta(x, z)}(0, 1, 0, 2) \neq 0$ (b) $\dfrac{\partial(H, K)}{\delta(x, z)}(\tfrac{1}{2}, 1, \tfrac{1}{4}, 2) \neq 0.$

Obtain explicit expressions for x and z as continuously differentiable functions of y and t in a neighbourhood of $(y, t) = (1, 2)$ such that

(i) $x = z = 0$ at $(y, t) = (1, 2)$,
(ii) $x = \tfrac{1}{2}$, $z = \tfrac{1}{4}$ at $(y, t) = (1, 2)$.

Answer: (ii) $x = y^2/t$, $z = y^3/t^2$. The answer to (i) is different!

9. Prove that if $\mathbf{q}$ is a point of a smooth surface S in $\mathbb{R}^m$, then the normal line to S at $\mathbf{q}$ (see Definition 3.8.18) is independent of which smooth function $f : D \subseteq \mathbb{R}^m \to \mathbb{R}$ is used to define S.

Hint: consider local graph like behaviour and follow the argument of Section 3.8.

10. Let $\mathbf{p}$ be a point in a surface S in $\mathbb{R}^m$. Prove that there is a $1-1$ C^1 function $g : N(\mathbf{0}, \delta) \subseteq \mathbb{R}^{m-1} \to \mathbb{R}^m$ such that $g(\mathbf{0}) = \mathbf{p}$ and $g(N(\mathbf{0}, \delta)) \subseteq S$. It follows that S is locally a copy of $\mathbb{R}^{m-1}$.

4.9   Local extrema subject to constraints. The method of Lagrange multipliers

In Section 3.12 we considered a method of determining the local extrema (local maxima and minima) of a differentiable function $f : D \subseteq \mathbb{R}^m \to \mathbb{R}$, whose domain D is an open subset of $\mathbb{R}^m$. An important class of problems concerns the determination of the extreme values of f when its domain is restricted to a proper subset of D by the imposition of 'constraints'. Our first example is a simple illustration. It can be solved by the method of Section 3.12. However, in the general case an appeal to the Implicit Function Theorem is necessary.

4.9.1 *Example.* Find the dimensions of a rectangular box such that the surface area of its base and vertical sides is a minimum subject to the condition that the box has prescribed volume V.

Solution. The surface area of the base and sides of a rectangular box of base measurements $x > 0$, $y > 0$ and height $z > 0$ is given by the formula

4.9.2 $$g(x, y, z) = xy + 2xz + 2yz, \qquad (x, y, z) \in D,$$

where the domain of the function $g : D \subseteq \mathbb{R}^3 \to \mathbb{R}$ is the open set

$$D = \{(x, y, z) \in \mathbb{R}^3 \mid x > 0, y > 0, z > 0\}.$$

The volume xyz of the box is subject to the constraint

4.9.3 $$xyz = V \qquad (x, y, z) \in D.$$

We wish to find the minimum value of $g : D \subseteq \mathbb{R}^3 \to \mathbb{R}$ subject to the constraint 4.9.3. This is easily done by solving 4.9.3 for z and substituting in 4.9.2. Hence the required surface area of a box of volume V and base measurements $x > 0$, $y > 0$ is given by

4.9.4 $$g\left(x, y, \frac{V}{xy}\right) = s(x, y) = xy + \frac{2V}{y} + \frac{2V}{x}, \qquad x > 0, \quad y > 0.$$

The identity 4.9.4 defines the function $s : E \subseteq \mathbb{R}^2 \to \mathbb{R}$, where

$$E = \{(x, y) \in \mathbb{R}^2 \mid x > 0, y > 0\}.$$

We require the minimum value of s on E, and this can be found by an application of Corollary 3.12.4. The critical points of s are solutions of the equations

4.9.5 $$0 = \frac{\partial s}{\partial x}(x, y) = y - \frac{2V}{x^2} \quad \text{and} \quad 0 = \frac{\partial s}{\partial y}(x, y) = x - \frac{2V}{y^2}, \ (x, y) \in E.$$

The unique solution of 4.9.5, $x = y = (2V)^{1/3}$, determines a minimum of s. Therefore the dimensions of the rectangular box of volume V and minimum surface area of base and vertical sides are $(2V)^{1/3}$, $(2V)^{1/3}$, $\frac{1}{2}(2V)^{1/3}$.

4.9.6 *Remarks.* [i] It is instructive to interpret the solution of Example 4.9.1 geometrically as follows. By application of the results of Example 3.12.29 we see that the function $g : D \subseteq \mathbb{R}^3 \to \mathbb{R}$ given by 4.9.2 has no local maxima or minima. The constraint 4.9.3 restricts the domain of g to a surface S in $\mathbb{R}^3$, where $S \subseteq D$. On this surface the function g attains a minimum at the point

$$((2V)^{1/3}, (2V)^{1/3}, \tfrac{1}{2}(2V)) \in S.$$

In view of the above comment we speak of the *extremal points of g on S* when the domain of g is restricted to S.

[ii] The constraint 4.9.3 allows us to solve for z and to redefine the surface area of the box in terms of the function $s : E \subseteq \mathbb{R}^2 \to \mathbb{R}$ on which the methods of Section 3.12 are applicable. The method would fail with a constraint such as, for example,

$$x + y + z - \sin(xyz) = 0,$$

for then it is not possible to solve for z or for x or for y in terms of the other two variables.

A different type of difficulty arises with a constraint such as

$$x^2 + \tfrac{1}{2}y^2 + z^2 - 1 = 0.$$

Here it is possible to solve for z in terms of x and y, but there is more than one case to be considered. (See Example 4.7.9.)

These difficulties are avoided by the so-called method of Lagrange multiplier. The theorem on which the method is based concerns the nature of the local extrema of a real-valued function $g : D \subseteq \mathbb{R}^m \to \mathbb{R}$ when its domain is restricted to a subset of D upon which some function $F : D \subseteq \mathbb{R}^m \to \mathbb{R}$ takes only the value 0. For example, the constraint 4.9.3 on the domain of the function g of Example 4.9.1 is expressible in this form by defining $F(x, y, z) = xyz - V$. Define the subset S of $D \subseteq \mathbb{R}^m$ by

$$S = \{\mathbf{x} \in D \,|\, F(\mathbf{x}) = 0\}.$$

We wish to consider the local extrema of the function g when its domain is restricted to S. To avoid exceptional cases, assume that the functions F and g are both smooth at points in S. Let $\mathbf{p}$ be a point in S. The set

$$T = \{\mathbf{x} \in D \,|\, g(\mathbf{x}) = g(\mathbf{p})\}$$

is the boundary between subsets T^+ and T^- of $\mathbb{R}^m$, where

$$T^+ = \{\mathbf{x} \in D \,|\, g(\mathbf{x}) > g(\mathbf{p})\} \qquad \text{and} \qquad T^- = \{\mathbf{x} \in D \,|\, g(\mathbf{x}) < g(\mathbf{p})\}.$$

Notice that $\mathbf{p}$ lies in both S and T. If $\mathbf{p}$ is an extremal point of

$g : S \subseteq \mathbb{R}^m \to \mathbb{R}$, then S cannot cross over T at $\mathbf{p}$ from T^+ to T^-. This means that the level set S of F and the level set T of g must be tangential at $\mathbf{p}$ and therefore that $(\operatorname{grad} F)(\mathbf{p})$ and $(\operatorname{grad} g)(\mathbf{p})$ are in the same direction (normal to S and T). The situation is illustrated for $m = 2$ in Fig. 4.16(i) and (ii). We therefore expect, under suitable conditions, to find all the extremal points of g on S among those points $\mathbf{p}$ at which $(\operatorname{grad} g)(\mathbf{p})$ and $(\operatorname{grad} F)(\mathbf{p})$ are linearly dependent. Figure 4.16(iii) shows however that some such points may not be extremal points. Here S and T are tangential at $\mathbf{p}$ but S crosses over T from T^+ to T^-.

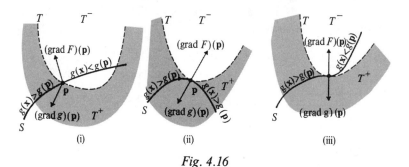

Fig. 4.16

4.9.7 Theorem. *Let $g : D \subseteq \mathbb{R}^m \to \mathbb{R}$ be a differentiable function and let $F : D \subseteq \mathbb{R}^m \to \mathbb{R}$ be a C^1 function defined on an open set $D \subseteq \mathbb{R}^m$, where $m \geq 2$. Let S be the level set of F corresponding to the value 0, that is,*

$$S = \{(x_1, \ldots, x_m) \in D \,|\, F(x_1, \ldots, x_m) = 0\}.$$

Suppose that $\mathbf{p} = (p_1, \ldots, p_m) \in S$ satisfies the condition

$$(\operatorname{grad} F)(\mathbf{p}) \neq \mathbf{0}.$$

If $\mathbf{p}$ is an extremal point of $g : S \subseteq \mathbb{R}^m \to \mathbb{R}$ (where the domain of g is restricted to S), then there exists $\lambda \in \mathbb{R}$ such that

4.9.8 $(\operatorname{grad} g)(\mathbf{p}) = \lambda (\operatorname{grad} F)(\mathbf{p}).$

Proof. Let $\mathbf{p}$ be an extremal point of $g : S \subseteq \mathbb{R}^m \to \mathbb{R}$. Expression 4.9.8 is trivially satisfied by $\lambda = 0$ if $(\operatorname{grad} g)(\mathbf{p}) = \mathbf{0}$ (any such point is a critical point of g on D). Assume then that $(\operatorname{grad} g)(\mathbf{p}) \neq \mathbf{0}$. We establish 4.9.8 by contradiction. Assume that $(\operatorname{grad} g)(\mathbf{p})$ and $(\operatorname{grad} F)(\mathbf{p})$ are linearly independent. Let $\mathbf{u} = (u_1, \ldots, u_m) \in \mathbb{R}^m$ be a vector such that $\{(\operatorname{grad} F)(\mathbf{p}), \mathbf{u}\}$ is an orthogonal basis of the plane spanned by $(\mathrm{grad}\,F)(\mathbf{p})$ and $(\mathrm{grad}\,g)(\mathbf{p})$. Then

4.9.9 $(\mathrm{grad}\,F)(\mathbf{p})\cdot\mathbf{u}=0,$

4.9.10 $(\mathrm{grad}\,g)(\mathbf{p})\cdot\mathbf{u}\neq 0.$

We now construct a curve in S whose tangent at $\mathbf{p}$ is in the direction of $\mathbf{u}$. The condition $(\mathrm{grad}\,F)(\mathbf{p})\neq\mathbf{0}$ means that we can assume $(\partial F/\partial x_m)(\mathbf{p})\neq 0$. (Otherwise $(\partial F/\partial x_i)(\mathbf{p})\neq 0$ for some i, $1\leqslant i\leqslant m-1$, and we interchange the roles of the x_i and x_m coordinates.) The function F satisfies the conditions of the Implicit Function Theorem 4.8.1. Therefore there exists a neighbourhood N of $(p_1,\ldots,p_{m-1})\in\mathbb{R}^{m-1}$, and a continuously differentiable function $f:N\subseteq\mathbb{R}^{m-1}\to\mathbb{R}$ such that

4.9.11 $F(x_1,\ldots,x_{m-1},f(x_1,\ldots,x_{m-1}))=0,$

$$\text{for all } (x_1,\ldots,x_{m-1})\in N.$$

Consider the differentiable function $\alpha:E\subseteq\mathbb{R}\to\mathbb{R}^m$ defined by

4.9.12

$$\alpha(t)=(p_1+tu_1,\ldots,p_{m-1}+tu_{m-1},f(p_1+tu_1,\ldots,p_{m-1}+tu_{m-1})),t\in E,$$

whose open domain E contains 0 and is such that $(p_1+tu_1,\ldots,p_{m-1}+tu_{m-1})$ lies in N for all $t\in E$.

The image of α lies in S (by 4.9.11) and $\alpha(0)=\mathbf{p}$. In fact $\alpha(E)$ is the projection parallel to the x_m-axis of the line $\{\mathbf{p}+t\mathbf{u}\in\mathbb{R}^m|t\in\mathbb{R}\}$ onto S. We will show that $\alpha'(0)=\mathbf{u}$. From 4.9.12, for some $w\in\mathbb{R}$,

4.9.13 $\alpha'(0)=(u_1,\ldots,u_{m-1},w).$

Now $F(\alpha(t))=0$ for all $t\in E$. Hence $(F\circ\alpha)'(0)=0$ and so, by the Chain Rule and 4.9.13

4.9.14 $(\mathrm{grad}\,F)(\mathbf{p})\cdot(u_1,\ldots,u_{m-1},w)=0.$

Subtracting 4.9.14 from 4.9.9 gives

4.9.15 $(\mathrm{grad}\,F)(\mathbf{p})\cdot(0,\ldots,0,u_m-w)=0.$

Therefore, since $(\partial F/\partial x_m)(\mathbf{p})\neq 0$, $u_m=w$. It follows from 4.9.13 that $\alpha'(0)=\mathbf{u}$.

The required contradiction now follows. Since $\mathbf{p}$ is an assumed extremal point of g on S and since $\alpha(E)\subseteq S$ and $\alpha(0)=\mathbf{p}$, the function $(g\circ\alpha):E\subseteq\mathbb{R}\to\mathbb{R}$ has an extremal point at 0. But by the

Chain Rule and 4.9.10,

$$(g \circ \alpha)'(0) = (\text{grad } g)(\mathbf{p}) \cdot \alpha'(0) = (\text{grad } g)(\mathbf{p}) \cdot \mathbf{u} \neq 0.$$

This contradiction completes the proof of the theorem.

4.9.16 Remark. *The method of the Lagrange multiplier.* According to Theorem 4.9.7 there are two possibilities for an extremal point $\mathbf{p}$ of the function $g : S \subseteq \mathbb{R}^m \to \mathbb{R}$. *Either*

4.9.17 $\qquad F(\mathbf{p}) = 0 \qquad$ and $\qquad (\text{grad } F)(\mathbf{p}) = \mathbf{0},$

or

4.9.18 $\qquad \begin{cases} F(\mathbf{p}) = 0, \quad (\text{grad } F)(\mathbf{p}) \neq \mathbf{0} \text{ and} \\ (\text{grad } g)(\mathbf{p}) = \lambda(\text{grad } F)(\mathbf{p}), \end{cases}$

for some $\lambda \in \mathbb{R}$. The constant λ is called the *Lagrange multiplier* or *undetermined multiplier*. In the course of solution its value is usually left undetermined.

A solution $\mathbf{p}$ of either equations 4.9.17 or equations 4.9.18 is called a *critical point of g on S*. The set of critical points of g on S contains the extremal points of g on S. However, a critical point is not necessarily an extremal point. See Example 4.9.30 and 4.9.33.

4.9.19 Example. Let us apply the method of Lagrange multiplier to find the extreme values of

4.9.20 $\qquad g(x, y, z) = xy + 2xz + 2yz, \qquad x > 0, y > 0, z > 0$

subject to the constraint $F(x, y, z) = 0$, where

4.9.21 $\qquad\qquad\qquad F(x, y, z) = xyz - V,$

and V is a non-zero constant. (Compare Example 4.9.1.)

Here the restricted domain of g is

$$S = \{(x, y, z) \in \mathbb{R}^3 | F(x, y, z) = 0, x > 0, y > 0, z > 0\}.$$

We first calculate the gradient functions of g and F:

4.9.22 $\qquad\qquad (\text{grad } g)(x, y, z) = (y + 2z, x + 2z, 2x + 2y),$

and

4.9.23 $\qquad\qquad\qquad (\text{grad } F)(x, y, z) = (yz, xz, xy).$

By Remark 4.9.16, the critical points of g on S fall into two classes. *Class* 4.9.17: the set of solutions of the equations

4.9.24 $\qquad xyz - V = 0, \qquad yz = xz = xy = 0, \qquad (x, y, z) \in S.$

There are no solutions.

Class 4.9.18: the set of solutions for some $\lambda \in \mathbb{R}$ of the system

4.9.25 $xyz - V = 0,$

4.9.26 $y + 2z = \lambda yz$

4.9.27 $x + 2z = \lambda xz$

4.9.28 $2x + 2y = \lambda xy$

where $(x, y, z) \in S$ and $(yz, xz, xy) \neq (0, 0, 0)$.

In solving the system 4.9.25–28, note first that $\lambda \neq 0$, for otherwise $x = y = -2z$ and $x = -y$, and so $x = y = z = V = 0$, a contradiction. Next, multiply equations 4.9.26, 27, 28 by x, y, z respectively. Then

4.9.29 $\lambda xyz = \lambda V = xy + 2xz = xy + 2yz = 2xz + 2yz,$ $(x, y, z) \in S.$

It follows easily that the equations 4.9.29 have the unique solution

$$x = y = 2z = (2V)^{1/3}.$$

The function $g : S \subseteq \mathbb{R}^3 \to \mathbb{R}$ takes the value $3(4V^2)^{1/3}$ at the critical point $((2V)^{1/3}, (2V)^{1/3}, \frac{1}{2}(2V)^{1/3})$. It is a local minimum.

The nature of a critical point of g on S can be determined by reducing the extremum problem with constraint to one without constraint, and then applying the method of Section 3.12. For example, in Example 4.9.19 the constraint $xyz - V = 0$ enables us to express the values that g takes on S as a function of x and y only. (See 4.9.4.) Here we can solve for z explicitly as a function of x and y. In more complicated examples we need to appeal to the Implicit Function Theorem to give us an implicit expression for one of the variables in terms of the others. Frequently the physical nature of a problem makes a mathematical analysis unnecessary.

The following two examples illustrate the possibility that a critical point may be neither a maximum nor a minimum.

4.9.30 *Example.*   Consider the critical points of the function g given by

4.9.31 $g(x, y, z) = x + y + z,$ $(x, y, z) \in \mathbb{R}^3$

on the restricted domain $S = \{(x, y, z) \in \mathbb{R}^3 \,|\, F(x, y, z) = 0\}$, where

$$F(x, y, z) = x^2 - y^2 - x - y - z.$$

It will be found that there is a unique critical point $(0, 0, 0)$. We show that this is neither a minimum nor a maximum as follows. Note that

$$(a, b, a^2 - b^2 - a - b) \in S, \qquad a, b \in \mathbb{R},$$

and that, by 4.9.31,

4.9.32 $g(a, b, a^2 - b^2 - a - b) = a^2 - b^2.$

Now $g(0, 0, 0) = 0$, and, by 4.9.32,

$$g(a, 0, a^2 - a) = a^2, \qquad g(0, b, -b^2 - b) = -b^2.$$

It follows that g takes positive and negative values at points in S arbitrarily close to its critical point $(0, 0, 0)$. Therefore $(0, 0, 0)$ is not an extremal point of g on S.

4.9.33 Example. Find the minimum of the function g given by

$$g(x, y) = (x - 1)^2 + (y + 1)^2, \qquad (x, y) \in \mathbb{R}^2$$

on the restricted domain $S = \{(x, y) \in \mathbb{R}^2 \,|\, F(x, y) = 0\}$, where

$$F(x, y) = x^2 + y^2 - 2xy.$$

Applying the method of Lagrange multiplier, we calculate

$$(\operatorname{grad} g)(x, y) = (2x - 2, 2y + 2)$$

and

$$(\operatorname{grad} F)(x, y) = (2x - 2y, 2y - 2x).$$

From 4.9.17 we obtain critical points of g on S as solutions of the equations

$$x^2 + y^2 - 2xy = 0 \qquad \text{and} \qquad 2x - 2y = 2y - 2x = 0.$$

This system is solved by $x = y = a$, for any constant a. Now

$$g(a, a) = 2a^2 + 2,$$

and it follows that the critical point $(0, 0)$ is a minimum of g on S whereas the critical points (a, a), $a \neq 0$, are neither maxima nor minima.

The conditions 4.9.18 yield no further critical points.

Example 4.9.33 is easily solved without recourse to Theorem 4.9.7 (Exercise 4.9.4). Our purpose was to use it as an illustration that the conditions 4.9.17 of the method of the Lagrange multiplier must not be overlooked in the search for local maxima and minima of a function on a restricted domain.

Theorem 4.9.7 can be extended to functions whose domain is subject to more than one constraint. The generalization concerns a real-valued function $g : D \subseteq \mathbb{R}^m \to \mathbb{R}$ defined on an open set $D \subseteq \mathbb{R}^m$, whose domain is restricted by $r (< m)$ constraints to the set

$$S = \{\mathbf{x} \in \mathbb{R}^m \,|\, F_1(\mathbf{x}) = 0, \ldots, F_r(\mathbf{x}) = 0\}$$

where the functions F_i, $i = 1, \ldots, r$, are continuously differentiable on D. Suppose that $\mathbf{p}$ is a point in S such that the vectors $(\operatorname{grad} F_1)(\mathbf{p}), \ldots, (\operatorname{grad} F_r)(\mathbf{p})$ are linearly independent. Then if $\mathbf{p}$ is an extremal point of g on S, there exist numbers (Lagrange multipliers)

$\lambda_1, \ldots, \lambda_r$ such that

$$(\text{grad } g)(\mathbf{p}) = \lambda_1(\text{grad } F_1)(\mathbf{p}) + \cdots + \lambda_r(\text{grad } F_r)(\mathbf{p}).$$

We omit the proof, but here, to end the section, is an illustration.

4.9.34 Example. The intersection of the sphere

$$\{(x, y, z) \in \mathbb{R}^3 \mid x^2 + y^2 + z^2 = 1\}$$

$$\{(x, y, z) \in \mathbb{R}^3 \mid x + y + z = 1]$$

is a circle C. Find the shortest distance from the point $(0, 3, 3)$ to the circle C.

Solution. The square of the distance from $(0, 3, 3)$ to a point (x, y, z) on C is

$$g(x, y, z) = x^2 + (y - 3)^2 + (z - 3)^2.$$

We require the minimum value of g when its domain is restricted to the set

$$S = \{(x, y, z) \in \mathbb{R}^3 \mid F_1(x, y, z) = 0, F_2(x, y, z) = 0\}$$

where

$$F_1(x, y, z) = x^2 + y^2 + z^2 - 1$$
$$F_2(x, y, z) = x + y + z - 1.$$

We calculate the gradients:

$$(\text{grad } g)(x, y, z) = (2x, 2y - 6, 2z - 6),$$
$$(\text{grad } F_1)(x, y, z) = (2x, 2y, 2z),$$
$$(\text{grad } F_2)(x, y, z) = (1, 1, 1).$$

Hence $(\text{grad } F_1)(x, y, z)$ and $(\text{grad } F_2)(x, y, z)$ are linearly independent, provided that (x, y, z) is not a multiple of $(1, 1, 1)$. In our problem the possibility $x = y = z = a$ does not arise, for the equations $F_1(a, a, a) = 0$ and $F_2(a, a, a) = 0$ cannot be simultaneously satisfied. Therefore the generalization of Theorem 4.9.7 applies, and the minimum of g on S occurs among the points $(x, y, z) \in \mathbb{R}^3$ which simultaneously solve the equations $F_i(x, y, z) = 0$, $i = 1, 2$ and

$$(\text{grad } g)(x, y, z) = \lambda(\text{grad } F_1)(x, y, z) + \mu(\text{grad } F_2)(x, y, z)$$

for some constants λ, μ.

The equations are as follows:

4.9.35 $x^2 + y^2 + z^2 - 1 = 0$ and $x + y + z - 1 = 0$

4.9.36 $2x = \lambda(2x) + \mu$

4.9.37 $2y - 6 = \lambda(2y) + \mu$

4.9.38 $2z - 6 = \lambda(2z) + \mu.$

Eliminating μ from 4.9.36, 37 and from 4.9.37, 38 we obtain

$$(\lambda - 1)(x - y) = 3 \quad \text{and} \quad (\lambda - 1)(y - z) = 0.$$

Hence $\lambda - 1 \neq 0$ and $y = z$. The equations 4.9.35 reduce to

$$x^2 + 2y^2 = 1 \quad \text{and} \quad x + 2y = 1.$$

This leads to the solutions

$$(x, y, z) = (1, 0, 0) \quad \text{and} \quad (x, y, z) = (-\tfrac{1}{3}, \tfrac{2}{3}, \tfrac{2}{3}).$$

It will be found that $g(1, 0, 0) = 19$ and $g(-\tfrac{1}{3}, \tfrac{2}{3}, \tfrac{2}{3}) = 11$. Therefore the shortest distance from the point $(0, 3, 3)$ to the circle C is $\sqrt{11}$, and the point on C closest to $(0, 3, 3)$ is $(-\tfrac{1}{3}, \tfrac{2}{3}, \tfrac{2}{3})$. The second solution $(x, y, z) = (1, 0, 0)$ gives the point on C furthest from $(0, 3, 3)$.

Exercises 4.9

1. Find by the method of the Lagrange multiplier the dimensions of the rectangular parallelopiped of largest volume that can be inscribed in the ellipsoid $x^2/a^2 + y^2/b^2 + z^2/c^2 = 1$.

    Hint: find the maximum of $g(x, y, z) = 8xyz$ subject to the constraint $F(x, y, z) = 0$, where $F(x, y, z) = x^2/a^2 + y^2/b^2 + z^2/c^2 - 1$.

    Answer: $x = a/\sqrt{3},\ y = b/\sqrt{3},\ z = c/\sqrt{3}$.

2. (a) Find the maximum and minimum distance from the origin $(0, 0)$ to the ellipsoid $5x^2 - 6xy + 5y^2 = 8$.

    Hint: consider the critical points of $x^2 + y^2$ subject to the ellipsoidal constraint.

    Answer: max 2, min 1.

    (b) Find by the method of Lagrange multiplier the shortest distance from the point $(1, 0)$ to the parabola $y^2 = 4x$. Check your answer by a method of substitution.

    Answer: 1.

3. Complete the calculations of Example 4.9.30.

4. Solve Example 4.9.33 by a direct method.

    Hint: $x^2 + y^2 - 2xy = 0$ if and only if $x = y$.

5. Find by the method of Example 4.9.34 the point P nearest the origin on the line of intersection of the two planes $2x + 3y + z = 6$ and $x + 2y + 2z = 4$. Verify that the direction of OP is orthogonal to the line.

    Answer: $(\tfrac{12}{13}, \tfrac{17}{13}, \tfrac{3}{13})$.

6. Verify that the point $\mathbf{p} = (0, \frac{1}{2}, 0)$ is a critical point of

$$g(x, y, z) = xz + (2y - 1)^3$$

subject to the constraint

$$x^2 + 4y^2 + z^2 - 4xy + 2x - 4y + 1 = 0.$$

Show also that it is neither a local maximum nor a local minimum.

Hint: show that in any neighbourhood of $\mathbf{p}$ there are points subject to the constraint at which g takes positive and negative values.

7. Verify that the critical point of g on S calculated in Example 4.9.19 is a local minimum.

8. The points P and P^* lie on non-intersecting smooth surfaces S and S^* in $\mathbb{R}^3$. Prove that, when the distance PP^* has a local minimum or maximum, the line PP^* is normal to both surfaces.

Hint: define the surfaces as level sets of smooth functions F and F^*. Use (x, y, z) to denote points in S and (u, v, w) to denote points in S^*. Determine the critical points of

$$g(x, y, z, u, v, w) = (x - u)^2 + (y - v)^2 + (z - w)^2$$

on $S \times S^*$, given that $S \cap S^*$ is empty, that is, $(x, y, z) \neq (u, v, w)$. By the extension of Theorem 4.9.7, at a critical point there exist numbers λ and μ such that

$$(\mathrm{grad}\, g)(x, y, z, u, v, w) = \lambda(\mathrm{grad}\, F)(x, y, z) + \mu(\mathrm{grad}\, F^*)(u, v, w)$$

Deduce that $(x - u, y - v, z - w)$ is a scalar multiple of $(\mathrm{grad}\, F)(x, y, z)$ and of $(\mathrm{grad}\, F^*)(u, v, w)$.

9. Find the shortest distance from the hyperbola $x^2 - y^2 = 3$ to the line $y = 2x$.

Hint: treat the square of the distance between points (x, y) on the hyperbola and (u, v) on the line as a function of x, y, u, v.

Answer: $3/\sqrt{5}$.

10. Find by the method of the Lagrange multiplier the critical points of

$$g(x, y) = x + y^4$$

subject to the constraint $F(x, y) = 0$, where

$$F(x, y) = y^2 - x^3.$$

Answer: condition 4.9.17 gives the critical point $(0, 0)$ where g takes its minimum value 0 subject to the constraint $y^2 = x^3$. Condition 4.9.18 gives no further critical points. A sketch will reveal why this is so.

11. Solve Exercise 10 by a direct method.

4.10   The curl of a vector field in $\mathbb{R}^3$

In this section we consider vector fields on an open subset D of $\mathbb{R}^3$. We introduce the vector field curl F which, as we shall see, plays an important role in the Integral Vector Calculus. It will be convenient to use the standard notation **i, j, k** for the vectors $(1, 0, 0)$, $(0, 1, 0)$, $(0, 0, 1)$ respectively. For any vector field $G : D \subseteq \mathbb{R}^3 \to \mathbb{R}^3$ whose coordinate functions are G_1, G_2, G_3, and for any $\mathbf{r} \in D$,

$$G(\mathbf{r}) = (G_1(\mathbf{r}), G_2(\mathbf{r}), G_3(\mathbf{r})) = G_1(\mathbf{r})\mathbf{i} + G_2(\mathbf{r})\mathbf{j} + G_3(\mathbf{r})\mathbf{k}.$$

The corresponding vector field identity is

$$G = (G_1, G_2, G_3) = G_1\mathbf{i} + G_2\mathbf{j} + G_3\mathbf{k}.$$

4.10.1 *Definition.*   Let $F : D \subseteq \mathbb{R}^3 \to \mathbb{R}^3$ be a differentiable vector field. Define the vector field curl $F : D \subseteq \mathbb{R}^3 \to \mathbb{R}^3$ in terms of its coordinate functions by

4.10.2 $$\operatorname{curl} F = \left(\frac{\partial F_3}{\partial y} - \frac{\partial F_2}{\partial z}, \frac{\partial F_1}{\partial z} - \frac{\partial F_3}{\partial x}, \frac{\partial F_2}{\partial x} - \frac{\partial F_1}{\partial y} \right).$$

The vector field curl F can also be defined in terms of the formal determinant

4.10.3 $$\operatorname{curl} F = \det \begin{bmatrix} \mathbf{i} & \mathbf{j} & \mathbf{k} \\ \dfrac{\partial}{\partial x} & \dfrac{\partial}{\partial y} & \dfrac{\partial}{\partial z} \\ F_1 & F_2 & F_3 \end{bmatrix},$$

since expanding by the first row gives

$$\operatorname{curl} F = \left(\frac{\partial F_3}{\partial y} - \frac{\partial F_2}{\partial z} \right)\mathbf{i} + \left(\frac{\partial F_1}{\partial z} - \frac{\partial F_3}{\partial x} \right)\mathbf{j} + \left(\frac{\partial F_2}{\partial x} - \frac{\partial F_1}{\partial y} \right)\mathbf{k},$$

which agrees with 4.10.2.

Comparison of 4.10.3 with the vector product formula 1.2.21 suggests the formal expression

$$\operatorname{curl} F = \left(\frac{\partial}{\partial x}, \frac{\partial}{\partial y}, \frac{\partial}{\partial z} \right) \times F.$$

In terms of the formal operator 'del' defined by

4.10.4 $$\nabla = \left(\frac{\partial}{\partial x}, \frac{\partial}{\partial y}, \frac{\partial}{\partial z} \right)$$

we have

$$\operatorname{curl} F = \nabla \times F.$$

Notice also that for a differentiable function $f : D \subseteq \mathbb{R}^3 \to \mathbb{R}$

$$\nabla f = \left(\frac{\partial}{\partial x}, \frac{\partial}{\partial y}, \frac{\partial}{\partial z} \right) f = \operatorname{grad} f.$$

4.10.5 *Example.* Let $F(\mathbf{r}) = F(x, y, z) = (yz, xz, xy)$. Then

$$(\operatorname{curl} F)(\mathbf{r}) = \det \begin{bmatrix} \mathbf{i} & \mathbf{j} & \mathbf{k} \\ \dfrac{\partial}{\partial x} & \dfrac{\partial}{\partial y} & \dfrac{\partial}{\partial z} \\ yz & xz & xy \end{bmatrix} = (x - x)\mathbf{i} + (y - y)\mathbf{j} + (z - z)\mathbf{k} = \mathbf{0}.$$

Let $F : D \subseteq \mathbb{R}^3 \to \mathbb{R}^3$ be a C^1 vector field. Then the vector $(\operatorname{curl} F)(\mathbf{r})$ provides a measure of the 'rotational' effect of the field F in a neighbourhood of $\mathbf{r} \in D$. In order to give this statement a precise meaning we require integration theory and in particular Stokes' Theorem. For the moment let us give it plausibility with the following examples.

4.10.6 *Example.* Consider a general Euclidean motion in space consisting of the combination of a translation and a rotation. Referred to a fixed right-handed coordinate system centred at a point O, let P be the point with position vector $\mathbf{r} = (x, y, z)$. Then the velocity of P—call it $F(\mathbf{r})$—is the vector sum of the velocity of rotation $\boldsymbol{\omega} \times \mathbf{r}$ relative to O and the velocity $\mathbf{v}_0$ of the point at O at the instant of measurement (Example 1.2.24)

$$\mathbf{v}_P = \boldsymbol{\omega} \times \mathbf{r} + \mathbf{v}_0 = F(\mathbf{r}).$$

Since $\mathbf{v}_0$ is independent of x, y, z, we have

$$\begin{aligned} (\operatorname{curl} F)(\mathbf{r}) &= \det \begin{bmatrix} \mathbf{i} & \mathbf{j} & \mathbf{k} \\ \dfrac{\partial}{\partial x} & \dfrac{\partial}{\partial y} & \dfrac{\partial}{\partial z} \\ \omega_2 z - \omega_3 y & \omega_3 x - \omega_1 z & \omega_1 y - \omega_2 x \end{bmatrix} \\ &= (\omega_1 + \omega_1)\mathbf{i} + (\omega_2 + \omega_2)\mathbf{j} + (\omega_3 + \omega_3)\mathbf{k} \\ &= 2\boldsymbol{\omega}. \end{aligned}$$

The point P was chosen arbitrarily, and so we have shown that the curl of the velocity field of a Euclidean motion takes the constant value equal to twice the angular velocity vector **ω**. Therefore the curl provides a direct measure of rotational effect.

4.10.7 *Example.* We consider the velocity fields given by
[i] $F(x, y, z) = (-y, x, 0)$ and

[ii] $F^*(x, y, z) = \left(\dfrac{-y}{x^2 + y^2}, \dfrac{x}{x^2 + y^2}, 0\right),$

where $(x, y) \neq (0, 0)$, and compare their curl.
  We have

$$(\text{curl } F)(\mathbf{r}) = \det \begin{bmatrix} \mathbf{i} & \mathbf{j} & \mathbf{k} \\ \dfrac{\partial}{\partial x} & \dfrac{\partial}{\partial y} & \dfrac{\partial}{\partial z} \\ -y & x & 0 \end{bmatrix} = 2\mathbf{k}.$$

The field F is expressible in the form $F(\mathbf{r}) = \mathbf{k} \times \mathbf{r}$, and this corresponds to a rotation about the z-axis with unit angular speed. (Example 4.10.6). The field is illustrated in Fig. 4.17(i).
  Next consider curl F^*. Putting $\lambda = x^2 + y^2$, we obtain, for $(x, y) \neq (0, 0)$,

$$(\text{curl } F^*)(\mathbf{r}) = \left(\frac{\partial}{\partial x}\left(\frac{x}{\lambda}\right) + \frac{\partial}{\partial y}\left(\frac{y}{\lambda}\right)\right)\mathbf{k} = \left(\frac{2}{\lambda} - \frac{1}{\lambda^2}(2x^2 + 2y^2)\right)\mathbf{k} = \mathbf{0}.$$

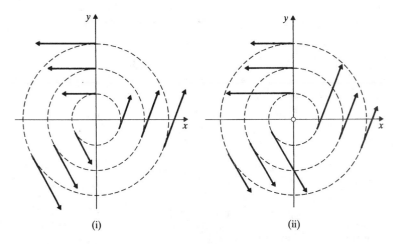

(i)   (ii)

Fig. 4.17 (i) $F(x, y, z) = (-y, x, 0)$

(ii) $F^*(x, y, z) = \left(\dfrac{-y}{x^2 + y^2}, \dfrac{x}{x^2 + y^2}, 0\right)$

(z axis perpendicular to paper)

The essential difference between the vector fields F and F^* lies in the norm of the vectors $F(\mathbf{r})$ and $F^*(\mathbf{r})$ associated with the point $\mathbf{r}$. Put $\rho = \sqrt{(x^2 + y^2)} = \sqrt{\lambda}$. Then

$$\|F(\mathbf{r})\| = \rho, \quad \text{and} \quad \|F^*(\mathbf{r})\| = \frac{\rho}{\lambda} = \frac{1}{\rho}.$$

For the vector field F^* the apparent counterclockwise rotation about the z-axis (Fig. 4.17(ii)) is balanced by an apparent clockwise effect produced by the diminution of speed with increasing distance ρ from the z-axis. The fact that curl F^* is the zero function indicates that the *local* rotational effect at any point is zero.

4.10.8 Definition.   *A vector field $F : D \subseteq \mathbb{R}^3 \to \mathbb{R}^3$ such that* $(\operatorname{curl} F)(\mathbf{r}) = \mathbf{0}$ *for all $\mathbf{r} \in D$ is said to be* irrotational.

The following theorem provides an important class of irrotational fields.

4.10.9 Theorem.   *Let $f : D \subseteq \mathbb{R}^3 \to \mathbb{R}$ have continuous second-order partial derivatives. Then the field* grad f *is irrotational, that is,* curl (grad f) *is the zero function on D.*

Proof.   We find on performing the usual determinant expansion that

$$\operatorname{curl}((\operatorname{grad} f)(\mathbf{r})) = \det \begin{bmatrix} \mathbf{i} & \mathbf{j} & \mathbf{k} \\ \dfrac{\partial}{\partial x} & \dfrac{\partial}{\partial y} & \dfrac{\partial}{\partial z} \\ \dfrac{\partial f}{\partial x}(\mathbf{r}) & \dfrac{\partial f}{\partial y}(\mathbf{r}) & \dfrac{\partial f}{\partial z}(\mathbf{r}) \end{bmatrix} = \mathbf{0},$$

since the mixed second-order partial derivatives $\dfrac{\partial^2 f}{\partial x_i \, \partial x_j}$ and $\dfrac{\partial^2 f}{\partial x_j \, \partial x_i}$ are equal, where x_i and x_j stand for any two of x, y and z. Therefore grad f is an irrotational field.

There is a partial converse to Theorem 4.10.9: if $(\operatorname{curl} F)(\mathbf{r}) = \mathbf{0}$ for all $\mathbf{r} \in D$ and *if D is an open ball* then F is the gradient of a scalar field f on D (see Exercise 9.4.8). The full converse of Theorem 4.10.9 is not true. For example, the vector field defined on the domain $D = \{(x, y, z) \in \mathbb{R}^3 \mid (x, y) \neq (0, 0)\}$ by

$$F^*(x, y, z) = \left(\frac{-y}{x^2 + y^2}, \frac{x}{x^2 + y^2}, 0\right), \qquad (x, y) \neq (0, 0)$$

is irrotational (Example 4.10.7), but it is not a gradient field, as we shall later see (Exercise 7.6.5).

Exercises 4.10

1. For each of the following differentiable vector fields $F : D \subseteq \mathbb{R}^3 \to \mathbb{R}^3$ calculate $(\text{curl } F)(x, y, z)$.
   (a) $F(x, y, z) = (xy, yz, zx)$, $(x, y, z) \in \mathbb{R}^3$;
   (b) $F(\mathbf{r}) = \mathbf{r}$,   where $\mathbf{r} = (x, y, z) \in \mathbb{R}^3$;
   (c) $F(\mathbf{r}) = \mathbf{r}/r$, $\mathbf{r} \in \mathbb{R}^3 \backslash \{\mathbf{0}\}$,   where $r = \|\mathbf{r}\|$;
   (d) $F(x, y, z) = (1, \frac{1}{2}x^2, 0)$, $(x, y, z) \in \mathbb{R}^3$;
   (e) $F(x, y, z) = (\frac{1}{4}x, \frac{1}{4}, 0)$, $(x, y, z) \in \mathbb{R}^3$.
   Which of the fields are irrotational?

Answers: (a) $(\text{curl } F)(x, y, z) = (-y, -z, -x)$. The fields (b), (c) and (e) are irrotational.

2. Let $F : D \subseteq \mathbb{R}^3 \to \mathbb{R}^3$ be a differentiable vector field and let $f : D \subseteq \mathbb{R}^3 \to \mathbb{R}$ be a differentiable scalar field. Prove that

$$\text{curl } (fF) = f \text{ curl } F + (\text{grad} f) \times F.$$

Hint: apply 4.10.3 to the product fF and perform routine differentiation.

3. A vector field $F : \mathbb{R}^3 \backslash \{\mathbf{0}\} \to \mathbb{R}^3$ is said to have spherical symmetry if there is a real-valued function $f : \mathbb{R} \backslash \{0\} \to \mathbb{R}$ such that

$$F(\mathbf{r}) = f(r)\mathbf{r}, \qquad \mathbf{r} \neq \mathbf{0}.$$

For example, the field of Exercise 1(c) has spherical symmetry. Prove that every spherically symmetric differentiable vector field is irrotational.

4. Prove that the irrotational field $F : \mathbb{R}^3 \to \mathbb{R}^3$ given by $F(x, y, z) = (yz, xz, xy)$ is the gradient of scalar field.

Hint: if $F = \text{grad} f$ then

$$\frac{\partial f}{\partial x}(x, y, z) = F_1(x, y, z) = yz.$$

Therefore f must be such that $f(x, y, z) = xyz + g(y, z)$ for some function g of y and z. By considering similarly the functions $\partial f/\partial y = F_2$ and $\partial f/\partial z = F_3$ deduce that $F = \text{grad} f$, where $f(x, y, z) = xyz$.

5. Which of the vector fields F of Exercise 1 are gradient fields?

Answer: the fields (a) and (d) cannot be gradient fields, since they are not
irrotational. In the other cases $F = \text{grad} f$, where $f(x, y, z)$ is given by
(b) $\frac{1}{2}(x^2 + y^2 + z^2) = \frac{1}{2}r^2$; (c) r; (e) $\frac{1}{8}x^2 + \frac{1}{4}y$.

4.11   The divergence of a vector field

4.11.1 Definition.   *Let $F : D \subseteq \mathbb{R}^m \to \mathbb{R}^m$ be a differentiable vector
field. We define the* divergence *of F to be the scalar field
$\text{div} F : D \subseteq \mathbb{R}^m \to \mathbb{R}$ given by*

$$(\text{div} F)(\mathbf{p}) = \frac{\partial F_1}{\partial x_1}(\mathbf{p}) + \cdots + \frac{\partial F_m}{\partial x_m}(\mathbf{p}), \qquad\qquad \mathbf{p} \in D,$$

where $F_1, \ldots, F_m$ are the coordinate functions of F.

Expressed in terms of scalar fields,

4.11.2 $$\text{div} F = \frac{\partial F_1}{\partial x_1} + \cdots + \frac{\partial F_m}{\partial x_m}.$$

For the case $m = 1$ div $F = F'$, and the operator div reduces to
elementary differentiation. The operator ∇ is used to express 4.11.2
as a formal dot product

$$\text{div} F = \left(\frac{\partial}{\partial x_1}, \ldots, \frac{\partial}{\partial x_m} \right) \cdot (F_1, \ldots, F_m) = \nabla \cdot F.$$

4.11.3 Example.   Define $F : \mathbb{R}^3 \to \mathbb{R}^3$ by $F(\mathbf{r}) = \mathbf{r}$, $\mathbf{r} \in \mathbb{R}^3$. Then $F_1(\mathbf{r}) = x$,
$F_2(\mathbf{r}) = y$, $F_3(\mathbf{r}) = z$ and

$$(\text{div} F)(\mathbf{r}) = \frac{\partial F_1}{\partial x}(\mathbf{r}) + \frac{\partial F_2}{\partial y}(\mathbf{r}) + \frac{\partial F_3}{\partial z}(\mathbf{r}) = 1 + 1 + 1 = 3.$$

We can extend the result of Example 4.11.3 as follows.

4.11.4 Example.   Define $\mathbb{R}^m \to \mathbb{R}^m$ by $F(\mathbf{r}) = r^k \mathbf{r}$, $\mathbf{r} \in \mathbb{R}^m$, where k is a
fixed non-negative real number and $r = \|\mathbf{r}\|$. For all $i = 1, \ldots, m$,
$F_i(\mathbf{r}) = r^k x_i$. Hence

$$\frac{\partial F_i}{\partial x_i}(\mathbf{r}) = x_i k r^{k-1} \frac{\partial r}{\partial x_i} + r^k = k r^{k-2} x_i^2 + r^k.$$

Therefore

$$(\text{div} F)(\mathbf{r}) = k r^{k-2}(x_1^2 + \cdots + x_m^2) + m r^k = (m + k)r^k.$$

The same formula holds when k is negative, provided $\mathbf{r} \neq \mathbf{0}$.

If a vector field in $\mathbb{R}^3$ corresponds to the velocity field of a gas then $(\text{div } F)(\mathbf{r})$ measures the rate of expansion per unit volume of the gas at the point $\mathbf{r}$. Thus, roughly speaking, $(\text{div } F)(\mathbf{r})$ measures the rate at which the gas 'diverges' from the point $\mathbf{r}$. The following example provides a quantitative illustration in a simple analogous two-dimensional case. The three-dimensional case is given later in terms of Gauss' Divergence Theorem (see Exercise 10.5.6).

4.11.5 *Example.* Imagine a plane fluid flow, where the velocity field is described relative to suitable coordinates by $F(x, y) = (\frac{1}{4}x, \frac{1}{4})$, $(x, y) \in \mathbb{R}^2$. (See Fig. 4.18.) At the point (x, y) the horizontal and vertical components of velocity have magnitude $F_1(x, y) = \frac{1}{4}x$ and $F_2(x, y) = \frac{1}{4}$ respectively. The speed of the fluid at (x, y) is $\frac{1}{4}\sqrt{(x^2 + 1)}$ and this increases with increasing x. In fact, each point of the fluid experiences an acceleration which is measured by the vector $(\frac{1}{4}, 0)$.

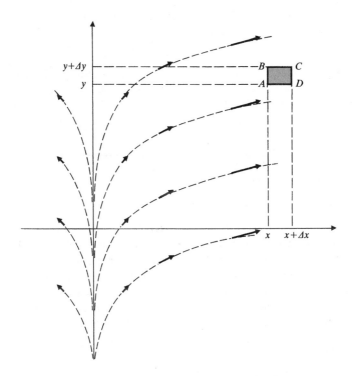

Fig. 4.18 Vector field $F(x, y) = (\frac{1}{4}x, \frac{1}{4})$

Consider now the passage of the fluid through a small rectangle ABCD as shown in Fig. 4.18. The velocity field is such that there is an inward flow across the edges AB and AD and an outward flow across CD and BC. The rate at which the fluid flows across the vertical edges AB and CD is governed by the horizontal component of velocity, and the flow across the horizontal edges AD and BC is governed by the vertical component.

The quatitative measurements are as follows:

Rate of horizontal flow	Inward across AB	$F_1(x, y) \, \Delta y$
	Outward across CD	$F_1(x + \Delta x, y) \, \Delta y$
Rate of vertical flow	Inward across AD	$F_2(x, y) \Delta x$
	Outward across BC	$F_2(x, y + \Delta y) \, \Delta x$

Hence:

Net rate of outward flow from rectangle ABCD

$$= (F_1(x + \Delta x, y) - F_1(x, y)) \, \Delta y + (F_2(x, y + \Delta y) - F_2(x, y)) \, \Delta x$$

$$= \left(\frac{F_1(x + \Delta x, y) - F_1(x, y)}{\Delta x} + \frac{F_2(x, y + \Delta y) - F_2(x, y)}{\Delta y} \right) \Delta x \, \Delta y.$$

Dividing this expression by $\Delta x \, \Delta y$, we obtain an expression for the rate of outward flow (expansion) per unit area from the rectangle. Now let Δx and Δy shrink to zero. We obtain

$$\text{rate of expansion per unit area 'at } (x, y)\text{'} = \frac{\partial F_1}{\partial x}(x, y) + \frac{\partial F_2}{\partial y}(x, y)$$

$$= (\text{div } F)(x, y).$$

In our present example $\dfrac{\partial F_1}{\partial x}(x, y) = \tfrac{1}{4}$, $\dfrac{\partial F_2}{\partial y}(x, y) = 0$, and so div F is the constant function that takes the value $\tfrac{1}{4}$.

As a fluid expands, so its density decreases. Thus the number $(\text{div } F)(\mathbf{r})$ measures the rate of change of the density of a fluid at the point $\mathbf{r}$, where the vector function F describes the velocity field of the fluid. If $(\text{div } F)(\mathbf{r})$ is positive, the fluid is expanding at $\mathbf{r}$ and its density is decreasing, whereas if $(\text{div } F)(\mathbf{r})$ is negative, the fluid is being compressed at $\mathbf{r}$ and its density is increasing. A (velocity) field F such that $(\text{div } F)(\mathbf{r}) = 0$ for all $\mathbf{r}$ is said to be *incompressible*.

Exercises 4.11

1. Calculate $(\text{div } F)(x, y, z)$ for the following differentiable vector fields $F : \mathbb{R}^3 \to \mathbb{R}^3$:

(a) $F(x, y, z) = (2x^2yz, xy^2z, xyz^2)$;
(b) $F(x, y, z) = (2x^2yz, -xy^2z, -xyz^2)$.

Answers: (a) $8xyz$; (b) 0.

2. A differentiable vector field $F : D \subseteq \mathbb{R}^m \to \mathbb{R}^m$ is called *divergence-free* or *solenoidal* on D if $(\text{div } F)(\mathbf{p}) = 0$ for all $\mathbf{p} \in D$. For example, the field F of Exercise 1(b) is solenoidal on $\mathbb{R}^3$. Prove that if $F : D \subseteq \mathbb{R}^3 \to \mathbb{R}^3$ is a C^2 vector field, then curl F is solenoidal on D. (Prove div curl $F = 0$ by routine calculations from definition.)

3. Let $f : D \subseteq \mathbb{R}^3 \to \mathbb{R}$ be a C^2 scalar field. Prove that

$$(\text{div grad} f)(\mathbf{r}) = \left(\frac{\partial^2 f}{\partial x^2} + \frac{\partial^2 f}{\partial y^2} + \frac{\partial^2 f}{\partial z^2} \right)(\mathbf{r}), \qquad \mathbf{r} = (x, y, z) \in D.$$

A scalar field f is said to be *harmonic* if div grad $f = 0$. Prove that the following fields are harmonic:
(a) $f(x, y, z) = x^2 + y^2 - 2z^2$;
(b) $f(\mathbf{r}) = 1/r$, $\mathbf{r} \neq \mathbf{0}$;
(c) $f(x, y, z) = z/r^3$, $r \neq 0$.

Show also that the scalar fields defined by $f(\mathbf{r}) = 1/r^k$, $\mathbf{r} \neq \mathbf{0}$, $k \neq 1$ are not harmonic.

4. Given that $f : D \subseteq \mathbb{R}^3 \to \mathbb{R}^3$ and $g : D \subseteq \mathbb{R}^3 \to \mathbb{R}^3$ are C^2 scalar fields, prove that div (grad $f \times$ grad g) = 0.

4.12 Some formulae involving grad, div, and curl

In this section we present some of the more important formulae of vector analysis that involve the operators grad, div, and curl. In the formulae F and G denote differentiable vector fields and f and g denote differentiable scalar fields, all of which are defined on an open set $D \subseteq \mathbb{R}^m$. The vector fields fF and $F \times G$ and the scalar fields fg and $F \cdot G$ have their usual meaning (Definition 1.5.4). Where curl is involved we assume that $m = 3$.

A number of formulae are given in an alternative form using the operator ∇. Recall that

$$\text{grad} f = \nabla f, \qquad \text{curl } F = \nabla \times F \qquad \text{and} \qquad \text{div } F = \nabla \cdot F.$$

In formulae 4.12.10–4.12.12 it is assumed that F and f are C^2 functions. This ensures the equality of the mixed second-order partial derivatives such as, for example, $\dfrac{\partial^2 F_1}{\partial x\,\partial y}$ and $\dfrac{\partial^2 F_1}{\partial y\,\partial x}$. The symbol 0 stands for the zero function.

In formula 4.12.12, the symbol ∇^2 stands for the important *Laplace operator,* which is defined by

$$\nabla^2 f = \frac{\partial^2 f}{\partial x_1^2} + \cdots + \frac{\partial^2 f}{\partial x_m^2} : D \subseteq \mathbb{R}^m \to \mathbb{R}$$

when ∇^2 acts on a scalar field f, and by

$$\nabla^2 F = (\nabla^2 F_1, \ldots, \nabla^2 F_m) : D \subseteq \mathbb{R}^m \to \mathbb{R}^m$$

when ∇^2 acts on a vector field F.

Note that

$$\nabla^2 f = \left(\frac{\partial}{\partial x_1}, \ldots, \frac{\partial}{\partial x_m}\right) \cdot \left(\frac{\partial f}{\partial x_1}, \ldots, \frac{\partial f}{\partial x_m}\right) = \nabla \cdot \nabla f = \operatorname{div}\operatorname{grad} f.$$

	Formula	Expression in terms of ∇
4.12.1	$\operatorname{grad}(f+g) = \operatorname{grad} f + \operatorname{grad} g$	$\nabla(f+g) = \nabla f + \nabla g$
4.12.2	$\operatorname{grad}(cf) = c\operatorname{grad} f,\ c \in \mathbb{R}$	$\nabla(cf) = c\nabla f$
4.12.3	$\operatorname{curl}(F+G) = \operatorname{curl} F + \operatorname{curl} G$	$\nabla \times (F+G) = \nabla \times F + \nabla \times G$
4.12.4	$\operatorname{curl}(cF) = c\operatorname{curl} F,\ c \in \mathbb{R}$	$\nabla \times (cF) = c\nabla \times F$
4.12.5	$\operatorname{div}(F+G) = \operatorname{div} F + \operatorname{div} G$	$\nabla \cdot (F+G) = \nabla \cdot F + \nabla \cdot G$
4.12.6	$\operatorname{div} cF = c\operatorname{div} F$	$\nabla \cdot (cF) = c(\nabla \cdot F)$
4.12.7	$\operatorname{grad}(fg) = f\operatorname{grad} g + g\operatorname{grad} f$	$\nabla(fg) = f\nabla g + g\nabla f$
4.12.8	$\operatorname{curl}(fF) = f\operatorname{curl} F + (\operatorname{grad} f) \times F$	$\nabla \times (fF) = f\nabla \times F + \nabla f \times F$
4.12.9	$\operatorname{div}(F \times G) = (\operatorname{curl} F)\cdot G - (\operatorname{curl} G)\cdot F$	$\nabla \cdot (F \times G) = (\nabla \times F)\cdot G - (\nabla \times G)\cdot F$
4.12.10	$\operatorname{curl}\operatorname{grad} f = 0$	$\nabla \times (\nabla f) = 0$
4.12.11	$\operatorname{div}\operatorname{curl} F = 0$	$\nabla \cdot (\nabla \times F) = 0$
4.12.12	$\operatorname{curl}\operatorname{curl} F = \operatorname{grad}\operatorname{div} F - \nabla^2 F$	$\nabla \times (\nabla \times F) = \nabla(\nabla \cdot F) - \nabla^2 F$

4.13   Towards integral vector calculus

It is inevitable that the integral calculus must draw on material from the differential calculus. In the elementary theory derivatives and integrals are related through the Fundamental Theorem, which states that if f is a continuously differentiable function defined on an interval $[a, b]$ then the integral over $[a, b]$ of the *derivative* of f is given by the formula

$$\int_a^b f'(x)\,\mathrm{d}x = f(b) - f(a).$$

The key results of integral vector calculus again concern the integration of functions which in a sense are derivatives of C^1 functions.

The three main theorems of this book concern

   [1]   the integration of grad f along a path (the Fundamental Theorem);

   [2]   the integration of curl F over an oriented surface (Stokes' Theorem);

   [3]   the integration of div F through a volume (Divergence Theorem).

The remainder of this book is devoted to the development and applications of these theorems.

Path integrals in $\mathbb{R}^n$

5.1 Introduction

The first type of integral we shall study is the integral of a real-valued function $f : S \subseteq \mathbb{R}^n \to \mathbb{R}$ (scalar field) along a path $\alpha : [a, b] \subseteq \mathbb{R} \to \mathbb{R}^n$. We use lower-case Greek letters to describe paths of integration in $\mathbb{R}^n$, thus distinguishing them clearly from the function f to be integrated.

We begin by reviewing the definition of the (Riemann) integral of a bounded function $f : [a, b] \subseteq \mathbb{R} \to \mathbb{R}$ over the compact interval $[a, b]$. The definition involves partitions of the interval $[a, b]$ into subintervals. Let one such partition $\mathcal{P}$ be given by $a = x_0 < x_1 < \cdots < x_k = b$. We associate with $\mathcal{P}$ a *Riemann sum* of the form

5.1.1
$$R(f, \mathcal{P}) = \sum_{i=1}^{k} f(p_i)(x_i - x_{i-1})$$

where p_i is any point chosen in the ith subinterval $[x_{i-1}, x_i]$ of $\mathcal{P}$.

Denoting the infimum and the supremum of f on the subinterval $[x_{i-1}, x_i]$ by m_i and M_i respectively and defining the *lower* and *upper Riemann sums* of f corresponding to the partititon $\mathcal{P}$ by

5.1.2
$$L(f, \mathcal{P}) = \sum_{i=1}^{k} m_i(x_i - x_{i-1}) \quad \text{and} \quad U(f, \mathcal{P}) = \sum_{i=1}^{k} M_i(x_i - x_{i-1}),$$

we have the inequality

5.1.3
$$L(f, \mathcal{P}) \leqslant \sum_{i=1}^{k} f(p_i)(x_i - x_{i-1}) \leqslant U(f, \mathcal{P}).$$

It follows (see Exercise 5.1.4) that for any partitions $\mathcal{P}$ and $\mathcal{Q}$ of $[a, b]$

$$L(f, \mathcal{P}) \leqslant U(f, \mathcal{Q}).$$

Let $L(f)$ and $U(f)$ denote respectively the supremum of the set of lower sums $L(f, \mathcal{P})$ and the infimum of the set of upper sums

$U(f, \mathcal{P})$. Then for any partitions $\mathcal{P}$ and $\mathcal{Q}$ of $[a, b]$,

5.1.4 $$L(f, \mathcal{P}) \leqslant L(f) \leqslant U(f) \leqslant U(f, \mathcal{Q}).$$

5.1.5 Definition. *The function $f : [a, b] \subseteq \mathbb{R} \to \mathbb{R}$ is said to be* (Riemann) integrable over $[a, b]$ *if $L(f) = U(f)$. The common value is then called the* (definite) integral *of f over $[a, b]$ and is denoted by* $\int_a^b f(x) \, dx$.

Notice that f is integrable over $[a, b]$ is and only if to each $\varepsilon > 0$ there correspond partitions $\mathcal{P}$ and $\mathcal{Q}$ of $[a, b]$ such that

$$U(f, \mathcal{Q}) - L(f, \mathcal{Q}) < \varepsilon.$$

5.1.6 Example. Let $f : [0, b] \subseteq \mathbb{R} \to \mathbb{R}$ be given by $f(x) = x^3$ for all $x \in [0, b]$. Consider the partition $\mathcal{P}$ of $[0, b]$ given, for some $k \in \mathbb{N}$, by

$$0 < \frac{b}{k} < \frac{2b}{k} < \cdots < \frac{(k-1)b}{k} < \frac{kb}{k} = b.$$

Then the ith subinterval in $\mathcal{P}$ is $[(i-1)b/k, ib/k]$ which has length b/k. Since f is monotonic increasing we have, with the above notation, for each $i = 1, \ldots, k$,

$$m_i = \left(\frac{(i-1)b}{k} \right)^3 \quad \text{and} \quad M_i = \left(\frac{ib}{k} \right)^3.$$

Therefore

$$L(f, \mathcal{P}) = \sum_{i=1}^{k} \frac{(i-1)^3 b^3}{k^3} \cdot \frac{b}{k} = \frac{b^4}{k^4} \sum_{i=1}^{k} (i-1)^3$$

and

$$U(f, \mathcal{P}) = \sum_{i=1}^{k} \frac{i^3 b^3}{k^3} \cdot \frac{b}{k} = \frac{b^4}{k^4} \sum_{i=1}^{k} i^3$$

It follows that

$$U(f, \mathcal{P}) - L(f, \mathcal{P}) = \frac{b^4}{k}.$$

Since this can be made arbitrarily small by taking k large enough, the function f is integrable over $[0, b]$.

Applying the well known formula

$$1^3 + 2^3 + \cdots + n^3 = \tfrac{1}{2} n^2 (n+1)^2$$

we find that

$$L(f, \mathscr{P}) = \tfrac{1}{4}b^4\left(\frac{k-1}{k}\right)^2 \quad \text{and} \quad U(f, \mathscr{P}) = \tfrac{1}{4}b^4\left(\frac{k+1}{k}\right)^2.$$

Hence $L(f) \geqslant \tfrac{1}{4}b^4$ and $U(f) \leqslant \tfrac{1}{4}b^4$. Therefore the value of the integral $\int_0^b x^3 \, dx$ is $\tfrac{1}{4}b^4$.

There are bounded functions $f : [a, b] \subseteq \mathbb{R} \to \mathbb{R}$ which are not integrable (see Exercise 5.1.7). However, all *continuous* functions are integrable and, in particular (the Fundamental Theorem), if $f : [a, b] \subseteq \mathbb{R} \to \mathbb{R}$ is a differentiable function whose derivative f' is continuous then

5.1.7 $$\int_a^b f'(x) \, dx = f(b) - f(a).$$

Taking $f(x) = \tfrac{1}{4}x^4$ and $a = 0$ in 5.1.7 leads to the integral evaluated in Example 5.1.4.

There is an alternative definition of the integrability of the function $f : [a, b] \subseteq \mathbb{R} \to \mathbb{R}$ in terms of the mesh of partitions of $[a, b]$.

5.1.8 Definition.   *Let $\mathscr{P}$ be a partition of $[a, b]$ given by $a = x_0 < x_1 < \cdots < x_k = b$. The* mesh *of $\mathscr{P}$ is the length of the longest subinterval in $\mathscr{P}$. It is defined and denoted by*

$$\mu(\mathscr{P}) = \max \{(x_i - x_{i-1}) \mid i = 1, \ldots, k\}.$$

The following theorem tells us in what sense the integral of a function is the limit of the approximating Riemann sums $R(f, \mathscr{P})$ as the mesh of $\mathscr{P}$ tends to 0. Remember that the value of $R(f, \mathscr{P})$ depends upon both the partition $\mathscr{P}$ and the choice of the points p_i in the subintervals of $\mathscr{P}$.

5.1.9 Theorem.   *A bounded function $f : [a, b] \subseteq \mathbb{R} \leftarrow \mathbb{R}$ is integrable with integral $I = \int_a^b f(x) \, dx$ if and only if, whenever $\mathscr{P}_1, \mathscr{P}_2, \mathscr{P}_3, \ldots$ is a sequence of partitions of $[a, b]$ whose mesh tends to 0, $R(f, \mathscr{P}_n)$ tends to I, no matter what points are chosen in the subintervals of the partitions $\mathscr{P}_n$ in evaluating the Riemann sums $R(f, \mathscr{P}_n)$.*

5.1.10 Example.   Consider a straight rod AB of uniform cross sectional area 1 cm² and of length b cm. Given that the density of the material of the rod at a point x cm from A is x^3 gm/cm³, find the mass of the rod.

Consider a partition $\mathscr{P}$ of the rod into a large number of short sections with partition points at $0 = x_0 < x_1 < \cdots < x_k = b$ centimetres from A.

For each $i = 1, \ldots, k$ choose a point p_i cm from A where $p_i \in [x_{i-1}, x_i]$. The length of the ith subsection is $(x_i - x_{i-1})$ cm and the density of it is approximately p_i^3 gm/cm^3. The mass of this section is therefore approximately $p_i^3(x_i - x_{i-1})$ gm. The total mass of the rod is correspondingly approximated by

5.1.11 $$\sum_{i=1}^{k} p_i^3(x_i - x_{i-1}) \text{ gm.}$$

This approximation tends to the true mass of the rod as the mesh of the partition $\mathscr{P}$ of the rod tends to 0. But expression 5.1.11 is a Riemann sum, $R(f, \mathscr{P})$, of the function $f(x) = x^3$. Hence by Example 5.1.4, the mass of the rod is $\frac{1}{4}b^4$ gm.

We have spelt out Example 5.1.10 in some detail in order to emphasise that the Riemann sum 5.1.1 is found by first finding the length of each subinterval, second 'weighting' each length by the value of the function at some point on it, and finally adding up these weighted lengths. The integral $\int_a^b f(x)\,dx$ to which the sum approximates is, as the notation suggests, a continuous form of this process. Of course in general, since the function will take both positive and negative values, the weights may be positive or negative.

The evaluation of the mass of the rod in Example 5.1.10 does not depend upon the rod being straight. In later sections we shall extend the definition of the Riemann integral to the integral of real valued functions over curves in $\mathbb{R}^n$ and of vector valued functions along oriented (directed) simple arcs in $\mathbb{R}^n$. But first we define the integral of a real-valued function along a path as a natural extension of the integral considered in this section.

Exercises 5.1

1. Let $f:[0, 1] \subseteq \mathbb{R} \to \mathbb{R}$ be the continuous function defined by $f(x) = x^2$, $x \in [0, 1]$. Let $\mathscr{P}$ be the partition $0 = x_0 < x_1 < \cdots < x_{k-1} < x_k = 1$ of $[0, 1]$ into k equal pieces; so $x_i = i/k$, $i = 0, \ldots, k$. Show that

    $$L(f, \mathscr{P}) = \sum_{i=1}^{k} \frac{(i-1)^2}{k^3} \quad \text{and} \quad U(f, \mathscr{P}) = \sum_{i=1}^{k} \frac{i^2}{k^3}.$$

    Deduce that f is integrable over $[0, 1]$, and that $\int_0^1 f(x)\,dx = \frac{1}{3}$.

Note: $\sum_{i=1}^{k} i^2 = \frac{1}{6}k(k+1)(2k+1)$.

2. Prove similarly from first principles that $\int_0^b x \, dx = \frac{1}{2}b^2$, and that $\int_a^b x^2 \, dx = \frac{1}{3}(b^3 - a^3)$.

3. Let $\mathscr{P}$ and $\mathscr{P}^*$ be partitions of $[a, b]$ given respectively by $a = x_0 < x_1 < \cdots < x_{k-1} < x_k = b$ and $a = x_0^* < x_1^* < \cdots < x_{m-1}^* < x_m^* = b$. The partition $\mathscr{P}^*$ is said to be a *refinement* of $\mathscr{P}$ if each point x_i, $i = 1, \ldots, k - 1$, occurs among the points $x_1^*, \ldots, x_{m-1}^*$. Prove that if $\mathscr{P}^*$ is a refinement of the partition $\mathscr{P}$ of $[a, b]$ then for any bounded function $f : [a, b] \subseteq \mathbb{R} \to \mathbb{R}$,

$$L(f, \mathscr{P}) \le L(f, \mathscr{P}^*) \le U(f, \mathscr{P}^*) \le U(f, \mathscr{P}).$$

4. Let $f : [a, b] \subseteq \mathbb{R} \to \mathbb{R}$ be a bounded function. Prove that for any partitions $\mathscr{P}$ and $\mathscr{Q}$ of $[a, b]$,

$$L(f, \mathscr{P}) \le U(f, \mathscr{Q}).$$

Hint: consider a refinement $\mathscr{P}^*$ of both $\mathscr{P}$ and $\mathscr{Q}$ which contains all the points of $\mathscr{P}$ and of $\mathscr{Q}$.

5. Prove that a bounded function $f : [a, b] \subseteq \mathbb{R} \to \mathbb{R}$ is Riemann integrable over $[a, b]$ if and only if to any $\varepsilon > 0$ there corresponds a partition $\mathscr{P}$ of $[a, b]$ such that $U(f, \mathscr{P}) - L(f, \mathscr{P}) < \varepsilon$.

Hint: consider a refinement of the partitions $\mathscr{P}$ and $\mathscr{Q}$ referred to in the text.

6. Prove from first principles that a bounded monotonic function $f : [a, b] \subseteq \mathbb{R} \to \mathbb{R}$ is integrable over $[a, b]$.

Hint: let $\mathscr{P}$ be the partition $a = x_0 < x_1 < \cdots < x_k = b$ of $[a, b]$ into k equal pieces. Prove that

$$U(f, \mathscr{P}) - L(f, \mathscr{P}) = \frac{|f(b) - f(a)|}{k}.$$

7. Let $f : [0, 1] \subseteq \mathbb{R} \to \mathbb{R}$ be the bounded function defined by

$$f(x) = \begin{cases} 0 & \text{when } x \text{ is rational,} \\ 1 & \text{when } x \text{ is irrational.} \end{cases}$$

(a) Prove that f is not integrable over $[0, 1]$. Show in particular that $L(f) = 0$ and $U(f) = 1$.

Note: an interval $[p, q]$ where $p < q$ has both rational and irrational points.

(b) Show that for any sequence $\mathscr{P}_1, \mathscr{P}_2, \mathscr{P}_3, \ldots$ of partitions of $[0, 1]$ such that $\mu(\mathscr{P}_n) \to 0$ there are choices of points in the subintervals such that $R(f, \mathscr{P}_n) \to 0$. Why does this not contradict Theorem 5.1.9?

5.2 Integral of a scalar field along a path

The integral of a real-valued function (scalar field) $f : S \subseteq \mathbb{R}^n \to \mathbb{R}$ along a path $\alpha : [a, b] \subseteq \mathbb{R} \to \mathbb{R}^n$ in S is defined in much the same way as the integral of a real-valued function over an interval, which was reviewed in Section 5.1. In that case the starting point is the Riemann sum 5.1.1 where the length $(x_i - x_{i-1})$ of each subinterval is weighted by multiplying it by the value $f(p_i)$ of the function at some point p_i within it. Similarly, integrating a real-valued function $f : S \subseteq \mathbb{R}^n \to \mathbb{R}$ along a path $\alpha : [a, b] \subseteq \mathbb{R} \to \mathbb{R}^n$ in S is a process of accumulating weighted lengths of path as $\alpha(t)$ traces, and perhaps retraces, its image.

5.2.1 *Example.* Imagine a fish swimming, open mouthed, through the sea gathering plankton as it goes. The total mass of plankton that the fish gathers in a given time interval depends upon the path α which describes its motion and the density f of plankton at each point it passes through. In a short time interval the mass gathered is approximately the distance covered multiplied by the linear density of plankton at any one point passed through in that short time. The mass gathered along the whole path α can be approximated by finding a partition of the time interval and adding up all the local approximations. See Fig. 5.1. By considering partitions of arbitrarily small mesh we obtain in the limit the true mass of plankton gathered. This is the integral of the density function f along the path α.

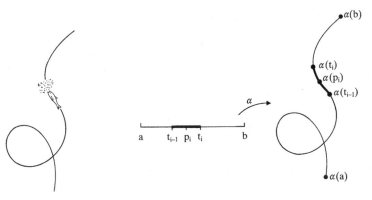

Fig. 5.1

We now establish a formal definition of the path integral of a real-valued function which was illustrated in the above example.

Let $\alpha : [a, b] \subseteq \mathbb{R} \to \mathbb{R}^n$ be a C^1 path in $\mathbb{R}^n$. Any partition $a = t_0 < t_1 < \cdots < t_k = b$ of $[a, b]$, call it $\mathcal{P}$, leads to a sequence of points $\alpha(t_0), \alpha(t_1), \ldots, \alpha(t_k)$ which α passes through. (See Fig. 5.1.)

Let $\lambda(t)$ be the length of the path $\alpha : [a, t] \subseteq \mathbb{R} \to \mathbb{R}^n$. Then the distance covered in following the path from $\alpha(t_{i-1})$ to $\alpha(t_i)$ is $\lambda(t_i) - \lambda(t_{i-1})$. Choose $p_i \in [t_{i-1}, t_i]$ for each $i = 1, \ldots, k$.

Suppose now that $f : S \subseteq \mathbb{R}^n \to \mathbb{R}$ is a function which is bounded on the image of the path α. We consider the (weighted) sum

5.2.2 $$R_\alpha(f, \mathcal{P}) = \sum_{i=1}^{k} f(\alpha(p_i))(\lambda(t_i) - \lambda(t_{i-1})).$$

Notice the similarlity between 5.2.2 and the Riemann sum 5.1.1. The expression 5.2.2 is a typical *Riemann sum of the function f along the path* α.

By our hypothesis, on the function f the function $f \circ \alpha : [a, b] \subseteq \mathbb{R} \to \mathbb{R}$ is bounded. Denote the infimum and the supremum of $f \circ \alpha$ on the subinterval $[t_{i-1}, t_i]$ by m_i and M_i respectively. So

$$m_i = \inf \{ f(\alpha(t)) \in \mathbb{R} \mid t_{i-1} \leqslant t \leqslant t_i \}$$

and

$$M_i = \sup \{ f(\alpha(t)) \in \mathbb{R} \mid t_{i-1} \leqslant t \leqslant t_i \}.$$

We define the *lower* and *upper Riemann sums of f along* α *corresponding to the partition* $\mathcal{P}$ *of* $[a, b]$ by

5.2.3 $$L_\alpha(f, \mathcal{P}) = \sum_{i=1}^{k} m_i(\lambda(t_i) - \lambda(t_{i-1}))$$

and

5.2.4 $$U_\alpha(f, \mathcal{P}) = \sum_{i=1}^{k} M_i(\lambda(t_i) - \lambda(t_{i-1})).$$

We then have the inequality

$$L_\alpha(f, \mathcal{P}) \leqslant \sum_{i=1}^{k} f(\alpha(p_i))(\lambda(t_i) - \lambda(t_{i-1})) \leqslant U_\alpha(f, \mathcal{P}).$$

More generally, for any two partitions $\mathcal{P}$ and $\mathcal{Q}$ of $[a, b]$,

$$L_\alpha(f, \mathcal{P}) \leqslant U_\alpha(f, \mathcal{Q}).$$

This is proved, as in the elementary theory, by considering

refinements of partitions. Define

5.2.5 $L_\alpha(f) = \sup\{L(f, \mathcal{P}) \mid \mathcal{P} \text{ a partition of } [a, b]\}$

and

5.2.6 $U_\alpha(f) = \inf\{U(f, \mathcal{P}) \mid \mathcal{P} \text{ a partition of } [a, b]\}.$

Then, for any partitions $\mathcal{P}$ and $\mathcal{Q}$ of $[a, b]$,

5.2.7 $L_\alpha(f, \mathcal{P}) \leqslant L_\alpha(f) \leqslant U_\alpha(f) \leqslant U_\alpha(f, \mathcal{Q}).$

The function f is said to be (*Riemann*) *integrable along the path* α if $L_\alpha(f) = U_\alpha(f)$. The common value is then called the *integral of* f *along* α and is denoted by

5.2.8 $\displaystyle\int_\alpha f \, ds.$

It follows from 5.2.7 that f is integrable along α if and only if to each $\varepsilon > 0$ there correspond partitions $\mathcal{P}$ and $\mathcal{Q}$ of $[a, b]$ such that

5.2.9 $U_\alpha(f, \mathcal{P}) - L_\alpha(f, \mathcal{Q}) < \varepsilon.$

In the notation 5.2.8, the quantity ds suggests an element of path-length if we describe the path-length function by $s = \lambda(t)$, as in Section 2.7.

5.2.10 *Example.* In Example 5.2.1, if the fish traverses a path $\alpha : [a, b] \subseteq \mathbb{R} \to \mathbb{R}^3$ and if the linear density of plankton is given by a function $f : \mathbb{R}^3 \to \mathbb{R}$ then the total mass of plankton gathered is the path integral $\int_\alpha f \, ds$.

5.2.11 *Example.* Let $\alpha : [-1, 1] \subseteq \mathbb{R} \to \mathbb{R}^2$ be a C^1 path given by $\alpha(t) = (\cos 2\pi t, \sin 2\pi t)$, a path which passes twice around the unit circle in $\mathbb{R}^2$ in a counterclockwise direction. Then, with the above notation

$$\lambda(t) = \int_{-1}^{t} \|\alpha'(t)\| \, dt = \int_{-1}^{t} 2\pi \, dt = 2\pi(1 + t).$$

Let $f : \mathbb{R}^2 \to \mathbb{R}$ be given by $f(x, y) = x^2$. Then f is bounded on the image of α (unit circle). Consider a partition $\mathcal{P}$ of $[-1, 1]$, say $-1 = t_0 < t_1 < \cdots < t_k = 1$. Taking $p_i \in [t_{i-1}, t_i]$ for each $i = 1, \ldots, k$, form the Riemann sum

$$R_\alpha(f, \mathcal{P}) = \sum_{i=1}^{k} f(\alpha(p_i))(\lambda(t_i) - \lambda(t_{i-1}))$$

$$= \sum_{i=1}^{k} (\cos^2 2\pi p_i) 2\pi(t_i - t_{i-1}).$$

But this gives the Riemann sums of the function $2\pi \cos^2 2\pi t$ over the interval $[-1, 1]$. Since

$$\int_{-1}^{1} 2\pi (\cos^2 2\pi t)\, dt = 2\pi,$$

therefore

$$\int_{\alpha} f\, ds = 2\pi.$$

The path integral along a path $\alpha : [a, b] \subseteq \mathbb{R} \to \mathbb{R}^n$ has the familiar linear properties

5.2.12 $\quad \displaystyle\int_{\alpha} (f + g)\, ds = \int_{\alpha} f\, ds + \int_{\alpha} g\, ds \quad$ *and* $\quad \displaystyle\int_{\alpha} kf\, ds = k \int_{\alpha} f\, ds$

for all functions $f : D \subseteq \mathbb{R}^n \to \mathbb{R}$ and $g : D \subseteq \mathbb{R}^n \to \mathbb{R}$ that are integrable along α, and for all $k \in \mathbb{R}$.

The following theorem provides a useful formula for the evaluation of path integrals.

5.2.13 Theorem. *Let $\alpha : [a, b] \subseteq \mathbb{R} \to \mathbb{R}^n$ be a C^1 path and let $f : S \subseteq \mathbb{R}^n \to \mathbb{R}$ be a function whose domain S contains the image of α. If the composite function $f \circ \alpha : [a, b] \subseteq \mathbb{R} \to \mathbb{R}$ is integrable over $[a, b]$ then f is integrable along α, and its integral is given by*

5.2.14 $\qquad \displaystyle\int_{\alpha} f\, ds = \int_{a}^{b} f(\alpha(t)) \|\alpha'(t)\|\, dt.$

As a special case of Theorem 5.2.13 we have the following important result.

5.2.15 Corollary. *Let $f : S \subseteq \mathbb{R}^n \to \mathbb{R}$ be a continuous function and let $\alpha : [a, b] \subseteq \mathbb{R} \to \mathbb{R}^n$ be a C^1 path in S. Then f is integrable along α, and its integral is given by* 5.2.14.

Proof of Theorem 5.2.13. Let $a = t_0 < t_1 < \cdots < t_k = b$ be a partition $\mathscr{P}$ of $[a, b]$. Consider the Riemann sum of f along α

5.2.16 $\qquad R_{\alpha}(f, \mathscr{P}) = \displaystyle\sum_{i=1}^{k} f(\alpha(p_i))(\lambda(t_i) - \lambda(t_{i-1}))$

where for each $i = 1, \ldots, k$, the point p_i is chosen arbitrarily in the interval $[t_{i-1}, t_i]$.

Remember that $\lambda(t)$ is the length of the path $\alpha : [a, t] \subseteq \mathbb{R} \to \mathbb{R}^n$, so

5.2.17 $$\lambda(t) = \int_a^t \|\alpha'(t)\| \, dt.$$

Since $\|\alpha'\|$ is continuous, $\lambda'(t) = \|\alpha'(t)\|$. By the Mean-Value Theorem, there exists $q_i \in [t_{i-1}, t_i]$ for each $i = 1, \ldots, k$ such that

$$\lambda(t_i) - \lambda(t_{i-1}) = \|\alpha'(q_i)\|(t_i - t_{i-1}).$$

Substituting in 5.2.16, we obtain

$$R_\alpha(f, \mathscr{P}) = \sum_{i=1}^k f(\alpha(p_i))\|\alpha'(q_i)\|(t_i - t_{i-1}).$$

Hence

5.2.18 $$R_\alpha(f, \mathscr{P}) = \sum_{i=1}^k f(\alpha(p_i))\|\alpha'(p_i)\|(t_i - t_{i-1}) + r$$

where

5.2.19 $$r = \sum_{i=1}^k f(\alpha(p_i))(\|\alpha'(q_i)\| - \|\alpha'(p_i)\|)(t_i - t_{i-1}).$$

Now the first term in the expression 5.2.18 for $R_\alpha(f, \mathscr{P})$ is a typical Riemann sum over $[a, b]$ of the function $g : [a, b] \subseteq \mathbb{R} \to \mathbb{R}$ given by

5.2.20 $$g(t) = f(\alpha(t))\|\alpha'(t)\|, \qquad t \in [a, b].$$

Since by hypothesis the function $f \circ \alpha$ is integrable over $[a, b]$ and since $\|\alpha'\|$ is continuous (because α' is continuous), the function g is integrable over $[a, b]$. Therefore (in the notation of Definition 5.1.5)

$$L(g) = U(g) = \int_a^b g(t) \, dt = \int_a^b f(\alpha(t))\|\alpha'(t)\| \, dt.$$

We shall complete the proof by deducing from 5.2.18 and 5.2.19 that $L_\alpha(f) = U_\alpha(f) = L(g)$.

Consider first the remainder term r in the Riemann sum 5.2.18. Since $\|\alpha'\|$ is continuous on $[a, b]$ it is also uniformly continuous. Therefore to any given $\varepsilon > 0$ there corresponds $\delta > 0$ such that

$$|\|\alpha'(q)\| - \|\alpha'(p)\|| < \varepsilon \qquad \text{whenever } |q - p| < \delta.$$

Suppose now that the partition $\mathscr{P}$ of $[a, b]$ has mesh less than δ. Then for each $i = 1, \ldots, k$

$$-\varepsilon < \|\alpha'(q_i)\| - \|\alpha'(p_i)\| < \varepsilon$$

for all choices of $p_i, q_i \in [t_{i-1}, t_i]$.

Let M be an upper bound of the function $|f \circ \alpha|$ on $[a, b]$. Then the remainder term r given in 5.2.19 is bounded as follows for all partitions $\mathscr{P}$ of mesh less than δ:

5.2.21 $$-M\varepsilon(b - a) < r < M\varepsilon(b - a).$$

Consider next the Riemann sums $R_\alpha(f, \mathscr{P})$ given by 5.2.18. Taking lower Riemann sums, it follows from 5.2.20 and 5.2.21 that

$$L_\alpha(f, \mathscr{P}) > L(g, \mathscr{P}) - M\varepsilon(b - a)$$

for all partitions $\mathscr{P}$ with $\mu(\mathscr{P}) < \delta$. Hence the least upper bound $L_\alpha(f)$ of the lower sums $L_\alpha(f, \mathscr{P})$ satisfies

$$L_\alpha(f) > L(g, \mathscr{P}) - M\varepsilon(b - a)$$

provided $\mu(\mathscr{P}) < \delta$. From this it follows that the least upper bound $L(g)$ of the lower sums $L(g, \mathscr{P})$ satisfies

$$L_\alpha(f) + M\varepsilon(b - a) \geqslant L(g).$$

Similarly

$$U_\alpha(f) - M\varepsilon(b - a) \leqslant U(g).$$

Now $L(g) = U(g) = \int_a^b g(t)\, dt$, and so, for all $\varepsilon > 0$,

$$L_\alpha(f) + M\varepsilon(b - a) \geqslant \int_a^b g(t)\, dt \geqslant U_\alpha(f) - M\varepsilon(b - a).$$

Therefore

$$L_\alpha(f) \geqslant \int_a^b g(t)\, dt \geqslant U_\alpha(f).$$

But $L_\alpha(f) \leqslant U_\alpha(f)$, by 5.2.7. Hence $L_\alpha(f) = U_\alpha(f)$, and so f is integrable along α and the integral is

$$\int_\alpha f\, ds = \int_a^b g(t)\, dt = \int_a^b f(\alpha(t)) \|\alpha'(t)\|\, dt.$$

5.2.22 *Remark.* Again, put $s = \lambda(t)$. Since, by 5.2.17,

$ds/dt = \|\alpha'(t)\|$, Theorem 5.2.13 establishes that

$$\int_\alpha f \, ds = \int_a^b f(\alpha(t)) \frac{ds}{dt} \, dt,$$

which further justifies our notation for the path integral.

5.2.23 Example. In Example 5.2.11 we found from first principles the integral of the function $f(x, y) = x^2$ along the path $\alpha(t) = (\cos 2\pi t, \sin 2\pi t)$, where $t \in [-1, 1]$. We can now find the path integral more directly. Since $\|\alpha'(t)\| = 2\pi$ for all t,

$$\int_\alpha f \, ds = \int_{-1}^1 f(\alpha(t))\|\alpha'(t)\| \, dt = \int_{-1}^1 (\cos 2\pi t)^2 2\pi \, dt = 2\pi.$$

The reader should notice in comparing Theorem 5.2.13 and Definition 2.7.2 that the length of a C^1 path $\alpha : [a, b] \subseteq \mathbb{R} \to \mathbb{R}^n$ is the path integral of the constant function $f(x) = 1$ along α. In short

5.2.24
$$l(\alpha) = \int_\alpha ds.$$

This is to be expected since, for any partition $\mathscr{P}$ of $[a, b]$, from 5.2.16

$$R_\alpha(1, \mathscr{P}) = \sum_{i=1}^k (\lambda(t_i) - \lambda(t_{i-1})),$$

a Riemann sum of the path integral, is also an expression for the length of α.

The following result generalizes Corollary 5.2.15 to *piecewise C^1* paths.

5.2.25 Theorem. *Let $\alpha : [a, b] \subseteq \mathbb{R} \to \mathbb{R}^n$ be a piecewise C^1 path and let $f : D \subseteq \mathbb{R}^n \to \mathbb{R}$ be a continuous function whose domain D contains the image of α. Then the path integral of f along α exists and is given by*

$$\int_\alpha f \, ds = \int_a^b f(\alpha(t))\|\alpha'(t)\| \, dt,$$

where this integral is interpreted, in the spirit of Definition 2.6.35, as a sum of integrals.

Proof. Exercise.

It follows that if $\beta:[a, b] \subseteq \mathbb{R} \to \mathbb{R}^n$ and $\gamma:[b, c] \subseteq \mathbb{R} \to \mathbb{R}^n$ are C^1 paths such that $\beta(b) = \gamma(b)$, and if $\alpha:[a, c] \subseteq \mathbb{R} \to \mathbb{R}^n$ is defined by $\alpha \mid [a, b] = \beta$ and $\alpha \mid [b, c] = \gamma$ then α is a piecewise C^1 path, and

5.2.26
$$\int_\alpha f \, \mathrm{d}s = \int_\beta f \, \mathrm{d}s + \int_\gamma f \, \mathrm{d}s$$

for any continuous function f whose domain contains the image of α.

5.2.27 *Example.* Let $\alpha:[-1, 1] \subseteq \mathbb{R} \to \mathbb{R}^2$ be the piecewise C^1 path defined by

$$\alpha(t) = (t, |t|), \qquad t \in [-1, 1].$$

and let $f(x, y) = x^2 y$. Let $\alpha \mid [-1, 0] = \beta$ and $\alpha \mid [0, 1] = \gamma$. Then, since $\|\alpha'(t)\| = \sqrt{2}$, $t \neq 0$,

$$\int_\alpha f \, \mathrm{d}s = \int_\beta f \, \mathrm{d}s + \int_\gamma f \, \mathrm{d}s$$

$$= \int_{-1}^0 -t^3 \sqrt{2} \, \mathrm{d}t + \int_0^1 t^3 \sqrt{2} \, \mathrm{d}t = \tfrac{1}{2}\sqrt{2}.$$

The way that length increases as a path traces its image is matched by the way in which the path integral accumulates values. In particular, when a path traces its image and then retraces it in the opposite direction there is a doubling up in the path integral rather than a cancelling out. For example, the path $\alpha(t) = (t^2, t^4)$, $t \in [-1, 1]$ traces the parabolic arc $y = x^2$, $0 \leqslant x \leqslant 1$ twice, and $l(\alpha)$ is twice the length of the arc.

Again, if the feeding fish turns tail and retraces its route in the opposite direction it will gather as much plankton as it did on the outward journey. Here is a further illustration.

5.2.28 *Example.* Let S be the unit semi-circle $\{(x, y) \mid x^2 + y^2 = 1, y \geqslant 0\}$ in the upper half-plane in $\mathbb{R}^2$ and let $f:S \subseteq \mathbb{R}^2 \to \mathbb{R}$ be the function given by $f(x, y) = y^2$.

[i] The path $\alpha:[0, \pi] \subseteq \mathbb{R} \to \mathbb{R}^2$ given by $\alpha(t) = (\cos t, \sin t)$ for all t traces S once from $(1, 0)$ to $(-1, 0)$. Since $\|\alpha'(t)\| = 1$ for all $t \in [0, \pi]$,

$$\int_\alpha f \, \mathrm{d}s = \int_0^\pi \sin^2 t \, \mathrm{d}t = \tfrac{1}{2}\pi.$$

[ii] The path $\beta:[-\sqrt{\pi}, 0] \subseteq \mathbb{R} \to \mathbb{R}^2$ given by $\beta(t) = (\cos t^2, \sin t^2)$ for all t traces S once from $(-1, 0)$ to $(1, 0)$. Since $\|\beta'(t)\| = 2 |t|$ for all

$t \in [-\sqrt{\pi}, 0]$,

$$\int_\beta f \, ds = \int_{-\sqrt{\pi}}^0 2|t| \sin^2 (t^2) \, dt = -\int_{-\sqrt{\pi}}^0 2t \sin^2 (t^2) \, dt = \tfrac{1}{2}\pi.$$

[iii] The path $\gamma : [-\sqrt{\pi}, \sqrt{\pi}] \subseteq \mathbb{R} \to \mathbb{R}^2$ given by $\gamma(t) = (\cos t^2, \sin t^2)$ for all t traces S from $(-1, 0)$ to $(1, 0)$ and back again. This time the path integral is

$$\int_\gamma f \, ds = \int_{-\sqrt{\pi}}^{\sqrt{\pi}} 2|t| \sin^2 (t^2) \, dt = \int_0^{\sqrt{\pi}} 4t^2 \sin^2 (t^2) \, dt = \pi.$$

[iv] The path $\delta : [-2\sqrt{\pi}, 2\sqrt{\pi}] \subseteq \mathbb{R} \to \mathbb{R}^2$ given by $\delta(t) = (\cos \tfrac{1}{4}t^2, \sin \tfrac{1}{4}t^2)$ for all t also traces S from $(-1, 0)$ to $(1, 0)$ and back again. We obtain

$$\int_\delta f \, ds = \int_{-2\sqrt{\pi}}^{2\sqrt{\pi}} \tfrac{1}{2}|t| \sin^2 (\tfrac{1}{4}t^2) \, dt = \int_0^{2\sqrt{\pi}} t \sin^2 (\tfrac{1}{4}t^2) \, dt = \pi.$$

In Example 5.2.28 the equality of the path integrals of f along α and β and along γ and δ illustrates the following two theorems. Notice that the paths α and β are both 1–1. The paths γ and δ, although not 1–1, are equivalent, since $\delta = \gamma \circ \phi$, where ϕ is the smooth function from $[-2\sqrt{\pi}, 2\sqrt{\pi}]$ on to $[-\sqrt{\pi}, \sqrt{\pi}]$ given by $\phi(u) = \tfrac{1}{2}u$.

5.2.29 Theorem.   *Let $\alpha : [a, b] \subseteq \mathbb{R} \to \mathbb{R}^n$ and $\beta : [c, d] \subseteq \mathbb{R} \to \mathbb{R}^n$ be two 1–1 piecewise C^1 paths having the same image C in $\mathbb{R}^n$. Then*

[i]   *α and β have the same length,*

[ii]   *for any continuous function $f : C \subseteq \mathbb{R}^n \to \mathbb{R}$, the path integrals of f along α and β are the same.*

Proof.   Let $a = t_0 < t_1 < \cdots < t_k = b$ be a partition $\mathscr{P}$ of $[a, b]$. Since β is 1–1, there are unique points $u_0, u_1, \ldots, u_k$ in $[c, d]$ such that

5.2.30 $\alpha(t_i) = \beta(u_i), \qquad i = 0, \ldots, k.$

Furthermore, by Theorem 4.2.13, β^{-1} is continuous and therefore $\beta^{-1} \circ \alpha : [a, b] \to [c, d]$ is continuous. Since $\beta^{-1} \circ \alpha$ is also 1–1 it is monotonic, and so the points $u_0, u_1, \ldots, u_k$ form either an increasing or a decreasing partition $\mathscr{Q}$ of $[c, d]$, depending on whether β traces C from $\alpha(a)$ to $\alpha(b)$ or from $\alpha(b)$ to $\alpha(a)$. By a similar process, any partition $\mathscr{Q}$ of $[c, d]$ gives rise to a unique partition $\mathscr{P}$ of $[a, b]$.

In this way we establish a 1–1 correspondence between partitions $\mathscr{P}$ of $[a, b]$ and partitions $\mathscr{Q}$ of $[c, d]$ such that 5.2.30 is satisfied.

[i] For any pair of partitions $\mathscr{P}$ and $\mathscr{Q}$ that match in the above sense,

$$\sum_{i=1}^{k} \|\alpha(t_i) - \alpha(t_{i-1})\| = \sum_{i=1}^{k} \|\beta(u_i) - \beta(u_{i-1})\|.$$

Therefore, by the Mean-Value Theorem

$$\sum_{i=1}^{k} \|\alpha'(p_i)\|(t_i - t_{i-1}) = \sum_{i=1}^{k} \|\beta'(q_i)\| (u_i - u_{i-1})$$

for suitable $p_i \in \,]t_{i-1}, t_i[$ and $q_i \in \,]u_{i-1}, u_i[$. Hence, by the definition of the length of a path, $l(\alpha) = l(\beta)$.

[ii] Again let $\mathscr{P}$ and $\mathscr{Q}$ be matching partitions in the sense of 5.2.30. Suppose that $c = u_0 < u_1 < \cdots < u_k = d$. Then, by [i], for each $i = 1, \ldots, k$ the lengths of the paths

$$\alpha : [t_{i-1}, t_i] \subseteq \mathbb{R} \to \mathbb{R}^n \quad \text{and} \quad \beta : [u_{i-1}, u_i] \subseteq \mathbb{R} \to \mathbb{R}^n$$

are the same. Let this length be l_i. Then

5.2.31 $\quad R_\alpha(f, \mathscr{P}) = \displaystyle\sum_{i=1}^{k} f(\alpha(p_i))l_i = \sum_{i=1}^{k} f(\beta(q_i))l_i = R_\beta(f, \mathscr{Q}),$

where $p_i \in [t_{i-1}, t_i]$ and $q_i \in [u_{i-1}, u_i]$ are chosen so that $\alpha(p_i) = \beta(q_i)$. It follows from 5.2.31 and the definition of the path integral that

$$\int_\alpha f \, \mathrm{d}s = \int_\beta f \, \mathrm{d}s.$$

A simple adjustment in the proof leads to the same conclusion when the partition $\mathscr{Q}$ that matches $\mathscr{P}$ is given by the decreasing sequence $d = u_0 > u_1 > \cdots > u_k = c$.

The following important theorem provides another case where the integrals of f along two related paths are equal.

5.2.32 Theorem. *Let* $\alpha : [a, b] \subseteq \mathbb{R} \to \mathbb{R}^n$ *and* $\beta : [c, d] \subseteq \mathbb{R} \to \mathbb{R}^n$ *be equivalent* C^1 *paths in* $\mathbb{R}^n$ *with image the curve* C. *Then for any continuous function* $f : C \subseteq \mathbb{R}^n \to \mathbb{R}$ *the path integrals of* f *along* α *and* β *are the same.*

Proof.   Exercise 5.2.2. Recall (Definition 2.6.9) that if β is equivalent to α there exists a differentiable function ϕ such that $\beta = \alpha \circ \phi$ and ϕ' takes either positive values only or negative values only.

We close this section by illustrating Theorem 5.2.29 for the case $n = 1$.

5.2.33 *Example.* Let $f:[a, b] \subseteq \mathbb{R} \to \mathbb{R}$ be a continuous function. The *identity path* $\alpha:[a, b] \subseteq \mathbb{R} \to \mathbb{R}$ given by $\alpha(t) = t$ is a 1–1 C^1 path in $\mathbb{R}$. Since $|\alpha'(t)| = 1$, we have

$$\int_\alpha f \, \mathrm{d}s = \int_a^b f(t) \, \mathrm{d}t.$$

Suppose that $\beta:[c, d] \subseteq \mathbb{R} \to \mathbb{R}$ is a 1–1 C^1 path in $\mathbb{R}$ whose image is $[a, b]$. Then

$$\int_\beta f \, \mathrm{d}s = \int_c^d f(\beta(u)) \, |\beta'(u)| \, \mathrm{d}u.$$

Hence by Theorem 5.2.29

5.2.34 $$\int_a^b f(t) \, \mathrm{d}t = \int_c^d f(\beta(u)) \, |\beta'(u)| \, \mathrm{d}u.$$

This is a particular case of the change of variable rule for elementary integrals. Note that since a 1–1 path in $\mathbb{R}$ is strictly monotonic, the path β is either strictly increasing or strictly decreasing. Suppose the latter. Then $\beta(c) = b$, $\beta(d) = a$ and $\beta'(u) \leq 0$ for all $u \in [c, d]$. The change of variable $t = \beta(u)$ then gives

$$\int_a^b f(t) \, \mathrm{d}t = \int_d^c f(\beta(u)) \frac{\mathrm{d}t}{\mathrm{d}u} \, \mathrm{d}u = \int_c^d f(\beta(u))(-\beta'(u)) \, \mathrm{d}u,$$

which agrees with 5.2.34.

Exercises 5.2

1. Let $f:\mathbb{R}^2 \to \mathbb{R}$ be defined by $f(x, y) = x^2 y^2$. Evaluate the path integrals $\int_\alpha f \, \mathrm{d}s$, $\int_\beta f \, \mathrm{d}s$, and $\int_\gamma f \, \mathrm{d}s$, where α, β, and γ are the C^1 paths defined by

$$\begin{aligned}
\alpha(t) &= (\cos t, \sin t), & t &\in [0, 4\pi], \\
\beta(t) &= (\cos 2t, \sin 2t), & t &\in [-2\pi, 0], \\
\gamma(t) &= (\cos t, -\sin t), & t &\in [0, 2\pi].
\end{aligned}$$

Note that α, β, and γ all have the same image. Describe how these paths trace their image and relate this to the values of the path integrals of f along α, β, and γ.

Answer: 2π, 2π, π. The image C of α, β, and γ is the unit circle

$x^2 + y^2 = 1$. The paths α and β trace C twice counterclockwise, and γ traces C once clockwise.

2.   Let $\alpha : [a, b] \subseteq \mathbb{R} \to \mathbb{R}^n$ and $\beta : [c, d] \subseteq \mathbb{R} \to \mathbb{R}^n$ be C^1 paths such that $\beta = \alpha \circ \phi$ where ϕ is a C^1 function from $[c, d]$ *onto* $[a, b]$.
(a) Prove that α and β have the same image.
(b) Given that ϕ is monotonic, prove that for any continuous function $f : S \subseteq \mathbb{R}^n \to \mathbb{R}$ such that im $\alpha \subseteq S$,

$$\int_\beta f \, ds = \int_\alpha f \, ds.$$

Hint: in the integral of f along β, put $\beta(u) = \alpha(\phi(u))$, $u \in [c, d]$, and use the Chain Rule $\beta'(u) = \alpha'(\phi(u))\phi'(u)$.
(c) Illustrate part (b) with the paths α and β in Exercise 1. Does the result remain valid when ϕ is not necessarily monotonic?

Answer: in Exercise 1, $\beta = \alpha \circ \phi$, where $\phi : [-2\pi, 0] \to [0, 4\pi]$ is defined by $\phi(t) = 2t + 4\pi$. The result does not remain valid for non-monotonic ϕ. For example, if $\phi : [c, d] \to [a, b]$ traces the interval $[a, b]$ twice, then $\int_\beta f \, ds = 2 \int_\alpha f \, ds$. Verify this for the case $\phi(u) = u^2$, $u \in [-1, 1]$, $\alpha(t) = (t, t)$, $t \in [0, 1]$, and $f(x, y) = 1$, all x, y.

3.   Show that the C^1 paths δ and μ defined by

$$\delta(t) = (\cos \pi(t^2), \sin \pi(t^2)) \qquad\qquad t \in [-1, 2],$$
$$\mu(t) = (-\cos \pi(2t - t^2), \sin \pi(2t - t^2)), \qquad t \in [-1, 2],$$

have the same image, and describe how it is traced. Let $f : \mathbb{R}^2 \to \mathbb{R}$ be defined by $f(x, y) = x^2 + y^2$. Predict the values of the path integrals $\int_\delta f \, ds$ and $\int_\mu f \, ds$, and check by integrating.
    Verify also that $\mu = \delta \circ \phi$, where ϕ is the monotonic decreasing function from $[-1, 2]$ to $[-1, 2]$ defined by $\phi(u) = 1 - u$.

4.   Let $f : S \subseteq \mathbb{R}^n \to \mathbb{R}$ be continuous, and let $\alpha : [a, b] \subseteq \mathbb{R} \to \mathbb{R}^n$ be a piecewise C^1 path in S. Define the *mean value of f along the path* α by

$$\bar{f}_\alpha = \frac{1}{l(\alpha)} \int_\alpha f \, ds,$$

where $l(\alpha)$ is the length of α.
(a) Calculate $\bar{f}_\alpha$ in the following cases.

(i)	$\alpha(t) = (\cos t, \sin t)$,	$t \in [0, \pi]$,	$f(x, y) = x$
(ii)	$\alpha(t) = (\cos t, \sin t)$,	$t \in [0, \pi]$,	$f(x, y) = y$,
(iii)	$\alpha(t) = (\cos t^2, \sin t^2)$,	$t \in [-\sqrt{\pi}, \sqrt{\pi}]$,	$f(x, y) = y$,
(iv)	$\alpha(t) = (t, |t|)$,	$t \in [-1, 1]$,	$f(x, y) = x^2 y$.

Answers: (i) 0, (ii) $2/\pi$, (iii) $2/\pi$, (iv) 1/4.

(b) The equality of the values $\bar{f}$ in Exercises 4(a) (ii) and (iii) is no coincidence. Consider the path $\alpha(t) = (\cos t^2, \sin t^2)$, $t \in [-\sqrt{\pi}, \sqrt{\pi}]$ as composed of two 1–1 C^1 pieces: $\beta = \alpha \mid [-\sqrt{\pi}, 0]$ and $\gamma = \alpha \mid [0, \sqrt{\pi}]$. Compare the integral of f along these paths with the integral of f along the path 4(a) (ii) in the manner of Exercise 2.

5. Let α, β, and γ be the 1–1 C^1 paths from $(1, 1, 1)$ to $(0, 0, 0)$ defined by

$$\alpha(t) = (-t, -t, -t), \qquad t \in [-1, 0],$$
$$\beta(t) = (t^2, t^2, t^2), \qquad t \in [-1, 0],$$
$$\gamma(t) = \begin{cases} (t^2, t^2, 1), & t \in [-1, 0] \\ (0, 0, 1 - t^2), & t \in [0, 1]. \end{cases}$$

Evaluate the path integrals $\int_\alpha f \, ds$, $\int_\beta f \, ds$, and $\int_\gamma f \, ds$, where $f(x, y, z) = xy + z - 1$.

Answers: $-\sqrt{3}/6$, $-\sqrt{3}/6$, $(2\sqrt{2} - 3)/6$. The equality of the first two integrals illustrates Theorem 5.2 29. See also Exercise 2.

6. Let $f : S \subseteq \mathbb{R}^n \to \mathbb{R}$ be continuous on S, and let $\alpha : [a, b] \subseteq \mathbb{R} \to \mathbb{R}^n$ and $\beta : [c, d] \subseteq \mathbb{R} \to \mathbb{R}^n$ be piecewise C^1 paths in S such that $\alpha(b) = \beta(c)$. Prove the following integral formulae concerning the inverse path α^- and the product path $\alpha\beta$

$$\int_{\alpha^-} f \, ds = \int_\alpha f \, ds,$$
$$\int_{\alpha\beta} f \, ds = \int_\alpha f \, ds + \int_\beta f \, ds.$$

5.3   Integral of a vector field along a path

In this section we define the integral of a continuous vector field $F : S \subseteq \mathbb{R}^m \to \mathbb{R}^m$ along a path in S. Recall that a necessary and sufficient condition for the continuity of a vector field is that its coordinate functions are continuous.

5.3.1 *Example.* The function $F : \mathbb{R}^2 \to \mathbb{R}^2$ given by $F(x, y) = \frac{1}{4}(-y, x)$ defines a continuous vector field on $\mathbb{R}^2$. The coordinate functions of F are $F_i : \mathbb{R}^2 \to \mathbb{R}$, where $F_1(x, y) = -\frac{1}{4}y$ and $F_2(x, y) = \frac{1}{4}x$.

When the domain of a vector field F is a subset of $\mathbb{R}^2$ then we can sketch the field by drawing for various values of x, y the vector

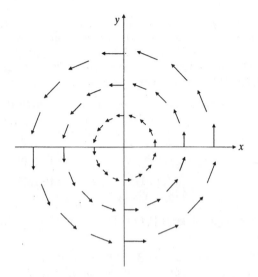

Fig. 5.2 Field $F(x, y) = \frac{1}{4}(-y, x)$

$F(x, y)$ based at $(x, y) \in \mathbb{R}^2$. Figure 5.2 is such a sketch of the field defined in Example 5.3.1. In this case, for each (x, y), the vector $F(x, y)$ is at right angles to the straight line through $(0, 0)$ and (x, y), since the dot product

5.3.2 $(-y, x) \cdot (x, y) = -yx + xy$

is zero for all x, y. The overall effect of the vector field is a counterclockwise flow.

   In many applications, when m is 2 or 3, vector fields are used to model the effect of a force field on a particle. Examples include gravitational (or magnetic or electrostatic) fields acting on a point mass (pole, charge). In such a model, the path integral (to be defined below) corresponds to the work done by the force field on a particle when the particle moves, or is moved, from one position to another. We discuss this application first as a way of introducing the definition of the integral. In order to explain it we need to introduce some terms from mechanics.

   Let **F** be a force acting on a particle in space. Then **F** is a vector quantity—it has both magnitude and direction. In Standard International (SI) units the magnitude of **F**, which we denote by $\|\mathbf{F}\|$, is measured in newtons.

   Let **u** be any non-zero vector and let the angle between **F** and **u**

be θ. Then **F** is uniquely expressible as the sum of two mutually
perpendicular vectors

$$\mathbf{F} = \mathbf{F_u} + \mathbf{F_u^{\perp}},$$

where $\mathbf{F_u}$ is a scalar multiple of the vector **u**. See Fig. 5.3. The
vector $\mathbf{F_u}$ is called the *(perpendicular) projection of* **F** *in direction* **u**.
From Fig. 5.3 it is clear that the magnitude of $\mathbf{F_u}$ is $\|\mathbf{F}\| \cos \theta$ when
$0 \leqslant \theta \leqslant \frac{1}{2}\pi$, and $-\|\mathbf{F}\| \cos \theta$ when $\frac{1}{2}\pi \leqslant \theta \leqslant \pi$. The quantity
$\|\mathbf{F}\| \cos \theta$ is called the *component of* **F** *in direction* **u**. The
component of **F** in a particular direction may therefore be positive,
negative, or (when **F** is perpendicular to the given direction) zero.

Consider now a particle moving in a constant force field, that is
wherever the particle is in space it experiences the same (constant)
force **F**. In this case, when the particle is moved in a given direction
along a straight line, the *work* done on it by the force field is
defined to be the product of the component of **F** in the direction of
motion and the distance the particle moves. In SI units work is
measured in newton metres—one newton metre being called a
joule. Notice that the work done, in joules, may be positive,
negative, or zero.

Now consider the work done by a variable force field on a particle
that moves through the field along a curve. We can construct a
mathematical model of this situation by choosing rectangular axes
for space and an axis for time, and using the scales prescribed by SI
units. The force field leads to a vector field $F : S \subseteq \mathbb{R}^3 \to \mathbb{R}^3$ and the
moving particle to a path $\alpha : [a, b] \subseteq \mathbb{R} \to \mathbb{R}^3$ whose image lies in S.
At time t seconds the particle has position vector $\alpha(t)$ and
experiences a force of $\|F(\alpha(t))\|$ newtons in the direction of
$F(\alpha(t))$. We assume that F is continuous and that α is C^1. Denote

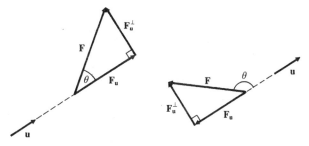

Fig. 5.3

the length of the path $\alpha : [a, t] \subseteq \mathbb{R} \to \mathbb{R}^3$ by $\lambda(t)$. Then
$\lambda'(t) = \|\alpha'(t)\|$.

Let $a = t_0 < t_1 < \cdots < t_k = b$ be a partition of $[a, b]$. We
approximate the work done by the field on the particle by adding up
approximations to the work done in each of the time intervals
$[t_{i-1}, t_i]$. Let the length of the path $\alpha : [t_{i-1}, t_i] \subseteq \mathbb{R} \to \mathbb{R}^3$ be l_i for
each $i = 1, \ldots, k$. So $l_i = \lambda(t_i) - \lambda(t_{i-1})$. By the Mean-Value
Theorem there exists $p_i \in [t_{i-1}, t_i]$ such that

5.3.3 $$l_i = \lambda'(p_i)(t_i - t_{i-1}) = \|\alpha'(p_i)\|(t_i - t_{i-1}).$$

If $l_i = 0$, then the work done in $[t_{i-1}, t_i]$ is 0 joules. When $l_i \neq 0$ the
vector $\alpha'(p_i)$ is non-zero and gives the direction of motion of the
particle at time p_i seconds. In this case, let the angle between
$F(\alpha(p_i))$ and $\alpha'(p_i)$ be $\theta(p_i)$. See Fig. 5.4.

At time p_i seconds, the component of the force on the particle in
the direction of its motion is $\|F(\alpha(p_i))\| \cos \theta(p_i)$ newtons. The
work done by the field on the particle in the time interval $[t_{i-1}, t_i]$ is
therefore approximately

$$\|F(\alpha(p_i))\| \cos \theta(p_i).l_i \text{ joules.}$$

The total work done by the field on the particle in the time interval
$[a, b]$ is approximated (in joules) by

5.3.4 $$\sum_{i=1}^{k} \|F(\alpha(p_i))\| \cos \theta(p_i).l_i.$$

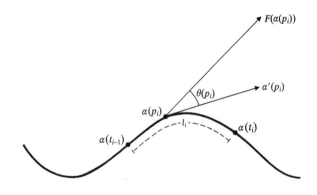

Fig. 5.4

Using 5.3.3 and 1.2.17, this approximation becomes

5.3.5 $$\sum_{i=1}^{k} F(\alpha(p_i)) \cdot \alpha'(p_i)(t_i - t_{i-1}).$$

The work done by the force field on the particle in the time interval $[a, b]$ is defined to be (in joules) the limit of this expression as the mesh of the partition tends to 0. But 5.3.5 is a Riemann sum of the function $F(\alpha(t)) \cdot \alpha'(t)$. Hence the work done by the field on the particle in the time interval $[a, b]$ is given (in joules) by

$$\int_a^b F(\alpha(t)) \cdot \alpha'(t) \, dt.$$

This integral is very important in both the theory and applications of vector analysis.

5.3.6 Definition. *Let $F : S \subseteq \mathbb{R}^n \to \mathbb{R}^n$ be a continuous vector field, and let $\alpha : [a, b] \subseteq \mathbb{R} \leftarrow \mathbb{R}^n$ be a C^1 path in S. The path integral of F along α is defined and denoted by*

5.3.7 $$\int_\alpha F \cdot d\alpha = \int_a^b F(\alpha(t)) \cdot \alpha'(t) \, dt.$$

5.3.8 Remark. In the following examples, and whenever the path integral of F along α occurs, it will be helpful to remember that the Riemann sums for the integral $\int_\alpha F \cdot d\alpha$ are formed from the expression 5.3.4. They are the sums of lengths along the path, where each length is weighted by the component of the vector field *in the direction in which the length is traversed.*

5.3.9 Example. Consider the vector field $F : \mathbb{R}^2 \to \mathbb{R}^2$ given by $F(x, y) = \frac{1}{4}(-y, x)$, and sketched in Fig. 5.2.

    [i] Let $\alpha : [-1, 1] \subseteq \mathbb{R} \to \mathbb{R}^2$ be the path $\alpha(t) = (t, t^2)$ which traces a segment of the parabola $y = x^2$ from $(-1, 1)$ to $(1, 1)$. See Fig. 5.5. Then $F(\alpha(t)) = \frac{1}{4}(-t^2, t)$ and $\alpha'(t) = (1, 2t)$. The path integral of F along α is

$$\int_\alpha F \cdot d\alpha = \frac{1}{4} \int_{-1}^1 (-t^2, t) \cdot (1, 2t) \, dt = \frac{1}{4} \int_{-1}^1 t^2 \, dt = 1/6.$$

    [ii] Let $\beta : [0, 3\pi/2] \subseteq \mathbb{R} \to \mathbb{R}^2$ be the path $\beta(t) = (\cos t, -\sin t)$ which traces the unit circle from $(1, 0)$ to $(0, 1)$ in a clockwise direction. The component of $F(\beta(t))$ in direction $\beta'(t)$ is negative for all t (see Fig. 5.6),

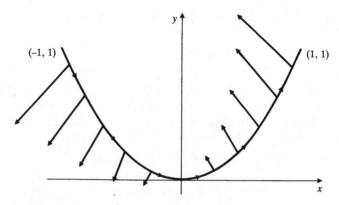

Fig. 5.5 Path α and vector field $F(x, y) = \frac{1}{4}(-y, x)$

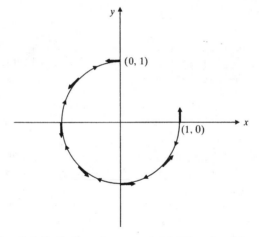

Fig. 5.6 Path β and vector field $F(x, y) = \frac{1}{4}(-y, x)$

and we expect the path integral to be negative. Indeed

$$\int_\beta F \cdot \mathrm{d}\beta = \frac{1}{4}\int_0^{3\pi/2} (\sin t, \cos t) \cdot (-\sin t, -\cos t) \, \mathrm{d}t = -3\pi/8.$$

[iii] Let $\gamma : [-1, 1] \subseteq \mathbb{R} \to \mathbb{R}^2$ be the path $\gamma(t) = (t, t)$ which runs along the straight line $y = x$ from $(-1, -1)$ to $(1, 1)$. The vectors $F(\gamma(t))$ and $\gamma'(t)$ are orthogonal for all t, and so the component of $F(\gamma(t))$ in direction $\gamma'(t)$ is always 0 (see Fig. 5.7). The path integral is

$$\int_\gamma F \cdot \mathrm{d}\gamma = \frac{1}{4}\int_{-1}^1 (-t, t) \cdot (1, 1) \, \mathrm{d}t = 0.$$

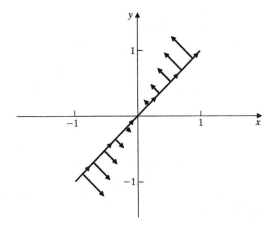

Fig. 5.7 Path γ and vector field $F(x, y) = \frac{1}{4}(-y, x)$

We saw in Section 5.2 first that the path integral of a *scalar field* is independent of the direction in which a path traces its image and second that it is accumulative, in the sense that if part of the image is traced twice in opposite directions then there is a doubling up in the path integral. In contrast, the physical interpretation of the path integral of a *vector field* and the content of Remark 5.3.8 suggest that reversing the direction of the path will change the sign of the integral, and that if part of the image is traced twice in opposite directions then there will be a cancelling out in the path integral. This is because when the direction of tracing is reversed, the component of the vector field in the direction of tracing changes sign.

5.3.10 *Example* As in Example 5.3.9 let $F : \mathbb{R}^2 \to \mathbb{R}^2$ be the vector field $F(x, y) = \frac{1}{4}(-y, x)$.
[i] The path $\mu : [\frac{1}{2}\pi, 2\pi] \subseteq \mathbb{R} \to \mathbb{R}^2$ given by $\mu(t) = (\cos t, \sin t)$ traces the segment of the unit circle sketched in Fig. 5.6 from $(0, 1)$ to $(1, 0)$ in a counterclockwise direction. The component of the field in the direction of tracing is now always positive. The path integral is

$$\int_\mu F \cdot d\mu = \frac{1}{4} \int_{\frac{1}{2}\pi}^{2\pi} (-\sin t, \cos t) \cdot (-\sin t, \cos t) \, dt = 3\pi/8.$$

Compare this answer with the path integral of F along β as defined in Example 5.3.9 [ii].

[ii] The path $\rho : [-1, 2] \subseteq \mathbb{R} \to \mathbb{R}^2$ defined by $\rho(t) = (\cos \frac{1}{2}\pi t^2, \sin \frac{1}{2}\pi t^2)$ traces a quadrant of the unit circle clockwise from $(0, 1)$ to $(1, 0)$ and then back to $(0, 1)$ before continuing to $(1, 0)$ in a counterclockwise direction. We expect the path integrals along μ and ρ to be the same, and in fact

$$\int_\rho F \cdot d\rho = \frac{1}{4} \int_{-1}^{2} (-\sin \tfrac{1}{2}\pi t^2, \cos \tfrac{1}{2}\pi t^2) \cdot (-\pi t \sin \tfrac{1}{2}\pi t^2, \pi t \cos \tfrac{1}{2}\pi t^2)\, dt$$

$$= \frac{1}{4} \int_{-1}^{2} \pi t\, dt = 3\pi/8.$$

The following theorem sums up the 'direction dependence' of the path integral of a vector field. It should be seen in contrast to Exercise 5.2.2. Note that the condition given there that ϕ be monotonic is now dropped.

5.3.11 Theorem.   *Let $F : S \subseteq \mathbb{R}^n \to \mathbb{R}^n$ be a continuous vector field. Let $\alpha : [a, b] \subseteq \mathbb{R} \to \mathbb{R}^n$ and $\beta : [c, d] \subseteq \mathbb{R} \to \mathbb{R}^n$ be two C^1 paths in S, and suppose that there is a C^1 function $\phi : [c, d] \to [a, b]$ such that $\beta = \alpha \circ \phi$.*
   [i] *If $\phi(c) = a$ and $\phi(d) = b$ then $\int_\alpha F \cdot d\alpha = \int_\beta F \cdot d\beta$.*
   [ii] *If $\phi(c) = b$ and $\phi(d) = a$ then $\int_\alpha F \cdot d\alpha = -\int_\beta F \cdot d\beta$.*

Proof.   By the Chain Rule,

$$\int_\beta F \cdot d\beta = \int_c^d F(\beta(u)) \cdot \beta'(u)\, du$$

$$= \int_c^d [F(\alpha(\phi(u)) \cdot \alpha'(\phi(u))]\phi'(u)\, du$$

$$= \int_{\phi(c)}^{\phi(d)} F(\alpha(t)) \cdot \alpha'(t)\, dt.$$

The result follows.

The following example illustrates the two parts of Theorem 5.3.11.

5.3.12 Example.   In Examples 5.3.9 and 5.3.10 the vector field $F(x, y) = \frac{1}{4}(-y, x)$ was integrated along paths β, μ, and ρ.
   [i] If $\phi : [-1, 2] \to [\frac{1}{2}\pi, 2\pi]$ is defined by $\phi(u) = \frac{1}{2}\pi u^2$ then $\rho = \mu \circ \phi$, and $\phi(-1) = \frac{1}{2}\pi$, and $\phi(2) = 2\pi$. The path integrals of F along ρ and μ were found to be equal.
   [ii] If $\phi : [0, 3\pi/2] \to [\frac{1}{2}\pi, 2\pi]$ is defined by $\phi(u) = 2\pi - u$ then $\beta = \mu \circ \phi$, and $\phi(0) = 2\pi$, and $\phi(3\pi/2) = \frac{1}{2}\pi$. The path integrals of F along β and μ were found to be $-3\pi/8$ and $3\pi/8$ respectively.

The path integrals along a path $\alpha : [a, b] \subseteq \mathbb{R} \to \mathbb{R}^n$ of vector fields F and G have the familiar linear properties

5.3.13 $$\int_\alpha (F + G) \cdot \mathrm{d}\alpha = \int_\alpha F \cdot \mathrm{d}\alpha + \int_\alpha G \cdot \mathrm{d}\alpha$$

$$\text{and} \qquad \int_\alpha kF \cdot \mathrm{d}\alpha = k \int_\alpha F \cdot \mathrm{d}\alpha$$

for all $k \in \mathbb{R}$, whenever the image of α lies in the domains of F and G.

Definition 5.3.6 of the path integral of a continuous vector field $F : S \subseteq \mathbb{R}^n \to \mathbb{R}^n$ along a path $\alpha : [a, b] \subseteq \mathbb{R} \to \mathbb{R}^n$ in S is given in terms of a C^1 path α. However, expression 5.3.7 makes sense even if α is piecewise C^1. We can use it to define the path integral in that case also.

5.3.14 Example. Let $F : \mathbb{R}^2 \to \mathbb{R}^2$ be the continuous vector field $F(x, y) = (x^2 + y^2, 1)$ and let $\alpha : [-1, 1] \subseteq \mathbb{R} \to \mathbb{R}^2$ be the piecewise C^1 path $\alpha(t) = (t, |t|)$. Then

$$\alpha'(t) = (1, -1) \qquad \text{for all} \qquad t \in [-1, 0[$$

and

$$\alpha'(t) = (1, 1) \qquad \text{for all} \qquad t \in]0, 1].$$

Hence

$$\int_\alpha F \cdot \mathrm{d}\alpha = \int_{-1}^0 (2t^2, 1) \cdot (1, -1)\, \mathrm{d}t + \int_0^1 (2t^2, 1) \cdot (1, 1)\, \mathrm{d}t = 4/3.$$

Finally in this section we describe two other common notations for the path integral of a vector field that will be very useful in later sections. Let $F : S \subseteq \mathbb{R}^2 \to \mathbb{R}^2$ be a continuous vector field with coordinate functions given by $F = (P, Q)$ and let $\alpha : [a, b] \subseteq \mathbb{R} \to \mathbb{R}^2$ be a (piecewise) C^1 path in S having coordinate functions $\alpha(t) = (x(t), y(t))$, $t \in [a, b]$. Then the path integral of F along α is

$$\int_\alpha F \cdot \mathrm{d}\alpha = \int_a^b P(x(t), y(t)) \frac{\mathrm{d}x}{\mathrm{d}t}\, \mathrm{d}t + \int_a^b Q(x(t), y(t)) \frac{\mathrm{d}y}{\mathrm{d}t}\, \mathrm{d}t.$$

For this reason the path integral is also denoted by

5.3.15 $$\int_\alpha P(x, y)\, \mathrm{d}x + Q(x, y)\, \mathrm{d}y \quad \text{or simply by} \quad \int_\alpha P\, \mathrm{d}x + Q\, \mathrm{d}y.$$

Denoting the vector (x, y) by $\mathbf{r}$ and putting $(dx, dy) = d\mathbf{r}$ leads to the alternative notation

5.3.16
$$\int_\alpha F \cdot d\mathbf{r}$$

for the path integral of F along α.

In general, for a path α in $\mathbb{R}^n$

5.3.17 $\displaystyle\int_\alpha F_1(x_1, \dots, x_n)\, dx_1 + \cdots + F_n(x_1, \dots, x_n)\, dx_n = \int_\alpha F \cdot d\mathbf{r}$

is the path integral $\int_\alpha F \cdot d\alpha$, where F is the vector field in $\mathbb{R}^n$ given by $F = (F_1, \dots, F_n)$.

5.3.18 *Example.* **[i]** For any path α in $\mathbb{R}^2$, $\int_\alpha -y\, dx + x\, dy$ is the path integral along α of the vector field $F(x, y) = (-y, x)$.

[ii] For any path α in $\mathbb{R}^3$, $\int_\alpha (x^2 + y)\, dz$ is the path integral along α of the vector field $F(x, y, z) = (0, 0, x^2 + y)$. The expression $\int_\alpha (x^2 + y)\, dz$ is an abbreviation for $\int_\alpha 0\, dx + 0\, dy + (x^2 + y)\, dz$.

Exercises 5.3

1. Let $\alpha : [0, 1] \subseteq \mathbb{R} \to \mathbb{R}^3$ be the path defined by
    $$\alpha(t) = (1 + t, 1 - t, t^2), \qquad t \in [0, 1].$$
    Evaluate the path integral $\int_\alpha F \cdot d\alpha$, where $F : \mathbb{R}^3 \to \mathbb{R}^3$ is the vector field given by
    (a) $F(x, y, z) = (xy, yz, zx)$,
    (b) $F(x, y, z) = (xyz, 0, 0)$,
    (c) $F(x, y, z) = (0, 0, xyz)$.

Answers: (a) 89/60, (b) 2/15, (c) 1/6.

2. Let L be a straight line through the origin in $\mathbb{R}^2$, and let α be a piecewise C^1 path in L. Let F be the vector field in $\mathbb{R}^2$ defined by
    $$F(x, y) = (y, -x), \qquad (x, y) \in \mathbb{R}^2.$$
    Prove that $\int_\alpha F \cdot d\alpha = 0$. Illustrate with a sketch.

Hint: put $\alpha(t) = (f(t), mf(t))$, m constant.

3. Let $\alpha : [a, b] \subseteq \mathbb{R} \to \mathbb{R}^2$ be a piecewise C^1 path such that $\alpha(a) = \alpha(b)$. Let F be the vector field in $\mathbb{R}^2$ defined by
    $$F(x, y) = (y, x), \qquad (x, y) \in \mathbb{R}^2.$$
    Prove that the path integral of F along α is zero.

Hint: put $\alpha(t) = (x(t), y(t))$, $t \in [a, b]$, and prove that

$$\int_\alpha y \, dx + x \, dy = 0.$$

4. Evaluate the path integral

$$\int_\alpha yz \, dx + xz \, dy + xy \, dz$$

along the path $\alpha(t) = (1 + t, -1 + t, 1 + 2t)$, $t \in [0, 1]$.

Answer: 1.

5. Let $F: \mathbb{R}^3 \to \mathbb{R}^3$ be the field given by

$$F(x, y, z) = (x, yz, xyz), \qquad (x, y, z) \in \mathbb{R}^3.$$

Find the work done by the field on a particle as it moves from the
origin to the point $(1, 1, 1)$:
(a) along the path $\alpha(t) = (t, t^2, t)$, $t \in [0, 1]$; (b) along the path
$\beta(t) = (t^2, t, t)$, $t \in [0, 1]$. (Calculate $\int_\alpha F \cdot d\alpha$ from the formula 5.3.7.)

Answers: (a) 11/10; (b) 31/30. This exercise illustrates that the work done
by a field on a particle in moving from one point to another depends on
te route chosen.

6. Let $F: \mathbb{R}^3 \to \mathbb{R}^3$ be any continuous vector field. Show that the total
work done by the field F on a particle as it moves along the path
$\alpha(t) = (t^2, 0, t^4)$, $t \in [-1, 1]$ is zero.

Hint: the particle moves from $(1, 0, 1)$ to the origin and then moves back
to $(1, 0, 1)$ along the same route. Express the work done in terms of
two integrals whose sum is zero.

7. Let $F: S \subseteq \mathbb{R}^n \to \mathbb{R}^n$ be a continuous vector field, and let
$\alpha: [a, b] \subseteq \mathbb{R} \to \mathbb{R}^n$ be a 1–1 C^1 path in S such that $\alpha'(t)$ is non-zero for
all $t \in [a, b]$. Prove that the path integral of F along α can be expressed
as the path integral of a scalar field along α.

Solution: Put $T(\alpha(t)) = \alpha'(t)/\|\alpha'(t)\|$, $t \in [a, b]$, so $T(\alpha(t))$ is the unit
vector in the direction of $\alpha'(t)$ (unit tangent to α at t). Consider the
real-valued function $f(\alpha(t)) = (F \cdot T)(\alpha(t))$, $t \in [a, b]$. Note that if α is
not 1–1, say $\alpha(t_1) = \alpha(t_2)$ for $t_1 \neq t_2$, there is a problem in the definition
of $f(\alpha(t_1))$ when $\alpha'(t_1) \neq \alpha'(t_2)$.

8. Let $F: D \subseteq \mathbb{R}^n \to \mathbb{R}$ be a continuous vector field, and let
$\alpha: [a, b] \subseteq \mathbb{R} \to \mathbb{R}^n$ and $\beta: [c, d] \subseteq \mathbb{R} \to \mathbb{R}^n$ be piecewise C^1 paths in D
such that $\alpha(b) = \beta(c)$. Prove the following integral formulae

concerning the inverse path α^- and the product path $\alpha\beta$

$$\int_{\alpha^-} F \cdot \mathrm{d}(\alpha^-) = -\int_{\alpha} F \cdot \mathrm{d}\alpha,$$

$$\int_{\alpha\beta} F \cdot \mathrm{d}(\alpha\beta) = \int_{\alpha} F \cdot \mathrm{d}\alpha + \int_{\beta} F \cdot \mathrm{d}\beta.$$

9. Let $\alpha:[0, 1] \subseteq \mathbb{R} \to \mathbb{R}^2$ and $\beta:[-1, 1] \subseteq \mathbb{R} \to \mathbb{R}^2$ be the C^1 paths in $\mathbb{R}^2$ defined by $\alpha(t) = (t, t)$, $t \in [0, 1]$, and $\beta(u) = (u^2, u^2)$, $u \in [-1, 1]$. Verify that $\beta = \alpha \circ \phi$ where the C^1 function $\phi:[-1, 1] \to [0, 1]$ is given by $\phi(u) = u^2$, $u \in [-1, 1]$. Let $F:\mathbb{R}^2 \to \mathbb{R}^2$ be the vector field $F(x, y) = (1, 0)$ for all $(x, y) \in \mathbb{R}^2$. Show that

$$\int_{\alpha} F \cdot \mathrm{d}\alpha = 1, \qquad \text{and} \qquad \int_{\beta} F \cdot \mathrm{d}\beta = 0.$$

Why does this not contradict Theorem 5.3.11?

5.4   The Fundamental Theorem of Calculus

The Fundamental Theorem of elementary calculus establishes a method of returning from a derivative to the original function. Specifically, if $f:[a, b] \subseteq \mathbb{R} \to \mathbb{R}$ is continuously differentiable then

5.4.1
$$\int_a^b f'(t) \, \mathrm{d}t = f(b) - f(a).$$

We have seen how closely the properties of the gradient reflect the properties of the elementary derivative and we would therefore expect an identity corresponding to 5.4.1 in the higher dimensional setting. It concerns path integrals of the vector field $\mathrm{grad}\, f$.

5.4.2   *The Fundamental Theorem of Calculus.* *Let* $f:D \subseteq \mathbb{R}^n \to \mathbb{R}$ *be a continuously differentiable function defined on an open subset D of* $\mathbb{R}^n$, *and let* $\mathbf{p}$ *and* $\mathbf{q}$ *be two points of D. If there exists a piecewise* C^1 *path* α *from* $\mathbf{p}$ *to* $\mathbf{q}$ *in D then*

5.4.3
$$\int_{\alpha} (\mathrm{grad}\, f) \cdot \mathrm{d}\alpha = f(\mathbf{q}) - f(\mathbf{p}).$$

Proof.   Suppose first that $\alpha:[a, b] \subseteq \mathbb{R} \to \mathbb{R}^n$ is a C^1 path in D. Then the composite function $f \circ \alpha:[a, b] \subseteq \mathbb{R} \to \mathbb{R}$ is also C^1. Hence by the Chain Rule 3.8.7

$$(f \circ \alpha)'(t) = (\text{grad}\,f)(\alpha(t)) \cdot \alpha'(t), \qquad t \in [a, b].$$

Therefore

$$\int_a^b (f \circ \alpha)'(t)\,dt = \int_a^b (\text{grad}\,f)(\alpha(t)) \cdot \alpha'(t)\,dt.$$

That is,

5.4.4 $$f(\alpha(b)) - f(\alpha(a)) = \int_\alpha (\text{grad}\,f) \cdot d\alpha.$$

This expression is equivalent to 5.4.3 if α runs from **p** to **q**. The case where α is a piecewise C^1 path is now easily proved by expressing the interval $[a, b]$ as a union of subintervals on which α is C^1.

The proof of this important theorem is remarkably short and perhaps does not explain why the result is true. However, if $a = t_0 < t_1 < \cdots < t_k = b$ is a partition of $[a, b]$ then, since $\text{grad}\,f : D \subseteq \mathbb{R}^n \to \mathbb{R}^n$ is a continuous vector field the Mean-Value Theorem 3.9.1 establishes the following approximation for each $i = 1, \ldots, k$

$$f(\alpha(t_i)) - f(\alpha(t_{i-1})) \approx (\text{grad}\,f)(\alpha(t_{i-1})) \cdot (\alpha(t_i) - \alpha(t_{i-1})).$$

Therefore, if α is a C^1 path,

5.4.5 $f(\alpha(t_i)) - f(\alpha(t_{i-1})) \approx (\text{grad}\,f)(\alpha(t_{i-1})) \cdot \alpha'(t_{i-1})(t_i - t_{i-1}).$

Summing the left-hand side of 5.4.5 from $i = 1$ to k leads to the left-hand side of 5.4.4. Summing the right-hand side of 5.4.5 from $i = 1$ to k gives a Riemann sum approximating the right-hand side of 5.4.4.

Two striking consequences of Theorem 5.4.2 are first that the integrals of $\text{grad}\,f$ along all paths from **p** to **q** are the same and second that the integral of $\text{grad}\,f$ along any path from **p** to **p** is 0. Both these properties are very significant, as we shall show in the next section. Notice that Theorem 5.4.2 applies to any piecewise C^1 path α running from **p** to **q**. In particular, α need not be 1–1.

5.4.6 *Example.* Let $f : \mathbb{R}^2 \to \mathbb{R}$ be defined by

$$f(x, y) = xy + x^2,$$

and let $\alpha : [-1, 2] \subseteq \mathbb{R} \to \mathbb{R}^2$ be the path $\alpha(t) = (t^2, t^4)$. Then

$$(\text{grad}\,f)(x, y) = (y + 2x, x),$$

and

$$\int_\alpha \operatorname{grad} f \cdot d\alpha = \int_{-1}^2 (\operatorname{grad} f)(\alpha(t)) \cdot \alpha'(t)\, dt = \int_{-1}^2 (t^4 + 2t^2,\, t^2) \cdot (2t,\, 4t^3)\, dt$$

$$= \int_{-1}^2 (6t^5 + 4t^3)\, dt = 78.$$

Putting $\alpha(-1) = (1,\, 1) = \mathbf{p}$ and $\alpha(2) = (4,\, 16) = \mathbf{q}$ we obtain

$$f(\mathbf{q}) - f(\mathbf{p}) = 80 - 2 = 78.$$

Notice that the path α is not 1–1. The vector $\alpha(t)$ runs along the parabola $y = x^2$ from $(1, 1)$ to $(0, 0)$ (as t increases from -1 to 0), and then back along the parabola from $(0, 0)$ to $(1, 1)$ (as t increases from 0 to 1), and then along the parabola to $(4, 16)$. The contribution to the integral due to the doubling back $(1, 1) \to (0, 0) \to (1, 1)$ is zero since

$$\int_{-1}^1 (\operatorname{grad} f)(\alpha(t)) \cdot \alpha'(t)\, dt = \int_{-1}^1 (6t^5 + 4t^3)\, dt = 0.$$

5.4.7 Example.   Consider the function $f : \mathbb{R}^2 \backslash \{(0,\, 0)\} \to \mathbb{R}$ given by

$$f(x, y) = \frac{1}{x^2 + y^2} \qquad (x, y) \neq (0, 0),$$

and let $\alpha : [0,\, 2\pi] \subseteq \mathbb{R} \to \mathbb{R}^2$ be the path $\alpha(t) = (2 \cos t,\, \sqrt{2} \sin t)$ whose image traces the ellipse $x^2 + 2y^2 = 4$ in a counterclockwise direction from $(2, 0)$ to $(2, 0)$. See Fig. 5.8. Then

$$(\operatorname{grad} f)(x, y) = \left(\frac{-2x}{(x^2 + y^2)^2},\, \frac{-2y}{(x^2 + y^2)^2} \right) \qquad (x, y) \neq (0, 0)$$

and

$$\int_\alpha \operatorname{grad} f \cdot d\alpha = \int_0^{2\pi} \frac{4 \sin t \cos t}{(2 + 2 \sin^2 t)^2}\, dt = \int_0^{2\pi} \frac{2 \sin 2t}{(3 - \cos 2t)^2}\, dt$$

$$= [-(3 - \cos 2t)^{-1}]_0^{2\pi} = 0.$$

Using the same argument, the integral along α restricted to $[0, \pi/2]$, $[\pi/2, \pi]$, $[\pi, 3\pi/2]$, $[3\pi/2, 2\pi]$ is $1/4$, $-1/4$, $1/4$ and $-1/4$ respectively. The relationship between these values might be expected from the symmetry of Fig. 5.8.

Surprisingly, the Fundamental Theorem in the case $n = 1$ reduces, not to the elementary Fundamental Theorem 5.4.1, but to the elementary Substitution Rule. For, when $n = 1$, we have a C^1 function $f : D \subseteq \mathbb{R} \to \mathbb{R}$ and a piecewise C^1 'path' $\alpha : [a,\, b] \subseteq \mathbb{R} \to \mathbb{R}$

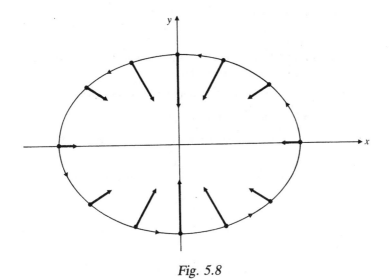

Fig. 5.8

in D. Then Theorem 5.4.2 implies that

5.4.8 $$\int_a^b f'(\alpha(t))\alpha'(t)\,dt = f(\alpha(b)) - f(\alpha(a)).$$

This usually is given in the form (obtained by 'letting $u = \alpha(t)$ and $du = \alpha'(t)\,dt$')

5.4.9 $$\int_a^b f'(\alpha(t))\alpha'(t)\,dt = \int_{\alpha(a)}^{\alpha(b)} f'(u)\,du.$$

The Fundamental Theorem 5.4.1 is obtained by letting α be the identity path $\alpha(t) = t$ in 5.4.8. Notice that the same result is obtained by letting f be the identity function $f(t) = t$ in 5.4.8.

Exercises 5.4

1. Let $f: \mathbb{R}^2 \to \mathbb{R}$ be defined by $f(x, y) = x^2 y + x^3$. Let $\mathbf{p} = (1, 0)$ and $\mathbf{q} = (-1, 0)$. Verify the Fundamental Theorem

$$\int_\alpha (\operatorname{grad} f) \cdot d\alpha = f(\mathbf{q}) - f(\mathbf{p})$$

for the following C^1 paths α from $\mathbf{p}$ to $\mathbf{q}$:
   (a) $\alpha(t) = (\cos t, \sin t)$, $t \in [0, \pi]$;
   (b) $\alpha(t) = (\cos t, \sin t)$, $t \in [0, 3\pi]$;

(c) $\alpha(t) = (t, 0),$ $t \in [1, -1];$
(d) $\alpha(t) = (\sin t, \sin 2t),$ $t \in [\pi/2, 3\pi/2];$
(e) $\alpha(t) = (\sin t, \sin 2t),$ $t \in [\pi/2, 7\pi/2].$
In each case sketch how the path α traces its image, and mark the
vector $(\mathrm{grad}\, f)(\alpha(t))$ at various points $\alpha(t)$.

Answer: in (e) the path α makes $1\frac{1}{2}$ circuits of a figure 8 curve.

2.   Let $F : \mathbb{R}^3 \to \mathbb{R}^3$ be the vector field defined by

$$F(x, y, z) = (2xyz, x^2 z, x^2 y), \qquad (x, y, z) \in \mathbb{R}^3.$$

Let $\alpha : [a, b] \subseteq \mathbb{R} \to \mathbb{R}^3$ be an arbitrary piecewise C^1 path in $\mathbb{R}^3$ such
that $\alpha(a) = (1, 1, 1)$ and $\alpha(b) = (-1, 2, 2)$. Prove that

$$\int_\alpha F \cdot \mathrm{d}\alpha = 3.$$

Hint: note that $F = \mathrm{grad}\, f$, where $f(x, y, z) = x^2 yz$. Apply the Fundamental
Theorem 5.4.2.

3.   Solve Exercise 5.3.3 by the above method.

5.5   Potential functions. Conservative fields.

We shall find throughout the rest of this book that the theory of
vector fields is rich in physical applications, and conversely that
studying the applications helps considerably in the understanding of
the pure theorems. The terms and results of this section, for
example, follow the close relationship in mechanics between work
and energy.

5.5.1 *Example.* Let $F : \mathbb{R}^3 \to \mathbb{R}^3$ be the vector field $F(x, y, z) =$
$(0, 0, -mg)$ which models the gravitational force on a particle of mass m
kilogrammes. For any piecewise C^1 path $\alpha : [a, b] \subseteq \mathbb{R} \to \mathbb{R}^3$, the path
integral of F along $\alpha = (\alpha_1, \alpha_2, \alpha_3)$ is

$$\int_\alpha F \cdot \mathrm{d}\alpha = \int_a^b (0, 0, -mg) \cdot (\alpha_1'(t), \alpha_2'(t), \alpha_3'(t))\, \mathrm{d}t$$

5.5.2

$$= -mg \int_a^b \alpha_3'(t)\, \mathrm{d}t = mg\alpha_3(a) - mg\alpha_3(b).$$

This example shows that, for any two points $\mathbf{p}$ and $\mathbf{q}$ in $\mathbb{R}^3$, the work done
by the gravitational field on the particle when it is moved down any
piecewise C^1 path from $\mathbf{p}$ to $\mathbf{q}$ is just the drop in the particle's potential
energy, $mgp_3 - mgq_3$.

It is significant in Example 5.5.1 that the path integral of the vector field along the path $\alpha : [a, b] \subseteq \mathbb{R} \to \mathbb{R}^2$ depends only upon the end points $\alpha(a)$ and $\alpha(b)$. We have seen this behaviour in the Fundamental Theorem 5.4.2 when integrating a gradient field along a path.

5.5.3 *Example*.   Consider the vector field $F : \mathbb{R}^3 \to \mathbb{R}^3$ given by $F(x, y, z) = (yz, xz, xy)$. Then $F = \operatorname{grad} f$ where $f(x, y, z) = xyz$. For any piecewise C^1 path $\alpha : [a, b] \subseteq \mathbb{R} \to \mathbb{R}^3$, the path integral of F along $\alpha = (\alpha_1, \alpha_2, \alpha_3)$ is, by 5.4.3,

$$\int_\alpha F \cdot d\alpha = \int_\alpha \operatorname{grad} f \cdot d\alpha = f(\alpha(b)) - f(\alpha(a))$$

$$= \alpha_1(b)\alpha_2(b)\alpha_3(b) - \alpha_1(a)\alpha_2(a)\alpha_3(a).$$

Again we find that the path integral of F along α depends only upon the end points of α.

The following theorem explores the nature of vector fields which have the property that their path integral along any piecewise C^1 path α depends only upon the end points of α. The theorem tells us in particular that a continuous vector field has this property if and only if it is a gradient field.

5.5.4 *Definition*.   *A subset S of $\mathbb{R}^n$ is* path connected *if for each pair of points $\mathbf{p} \in S$ and $\mathbf{q} \in S$ there is a path from $\mathbf{p}$ to $\mathbf{q}$ whose image lies in S.*

5.5.5 *Example*.   The annulus $\{(x, y) \in \mathbb{R}^2 \mid 1 \leqslant x^2 + y^2 \leqslant 4\}$ in $\mathbb{R}^2$ is path connected (Fig. 5.9(i)). The set $\{(x, y) \in \mathbb{R}^2 \mid xy > 1\}$ in $\mathbb{R}^2$ is not path connected (Fig. 5.9(ii)).

5.5.6 *Theorem*.   *Let $F : D \subseteq \mathbb{R}^n \to \mathbb{R}^n$ be a continuous vector field on an open subset D of $\mathbb{R}^n$. Then the following four properties are equivalent.*
   [i] *For any two piecewise C^1 paths α and α^* in D from $\mathbf{p}$ to $\mathbf{q}$*

$$\int_\alpha F \cdot d\alpha = \int_{\alpha^*} F \cdot d\alpha^*.$$

That is, the path integral of F along any piecewise C^1 path from $\mathbf{p}$ to $\mathbf{q}$ depends only upon the endpoints $\mathbf{p}$ and $\mathbf{q}$.
   [ii] *There exists a C^1 function $f : D \subseteq \mathbb{R}^n \to \mathbb{R}$ such that $F = \operatorname{grad} f$.*

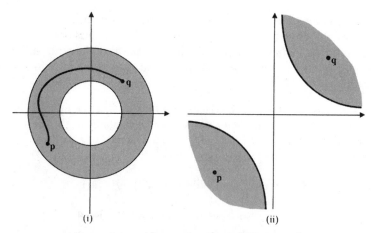

Fig. 5.9 (i)   Path connected subset of $\mathbb{R}^2$
(ii)   Not path connected

[iii] *There exists a C^1 function $h : D \subseteq \mathbb{R}^n \to \mathbb{R}$ such that for any*
$\mathbf{p} \in D$ *and* $\mathbf{q} \in D$ *and any piecewise C^1 path α from $\mathbf{p}$ to $\mathbf{q}$*

$$\int_\alpha F \cdot d\alpha = h(\mathbf{p}) - h(\mathbf{q}).$$

[iv] *For any $\mathbf{p} \in D$ and any piecewise C^1 path γ in D from $\mathbf{p}$ to $\mathbf{p}$,*

$$\int_\gamma F \cdot d\gamma = 0.$$

Proof.   We shall prove the theorem assuming that D is path
connected. For the general case see Exercises 5.5.9 and 5.5.10.

   [i] *implies* **[ii]**. Assume property **[i]**. Choose a 'base point' $\mathbf{k}$ in
D. Since D is assumed path connected we can define a function
$f : D \subseteq \mathbb{R}^n \to \mathbb{R}$ by

5.5.7 $f(\mathbf{r}) = \int_\phi F \cdot d\phi$

where ϕ is any C^1 path from $\mathbf{k}$ to $\mathbf{r}$ in D. We prove that
$F = \operatorname{grad} f$. Choose any point $\mathbf{p}$ in D and let $\mathbf{e}_1 = (1, 0, \dots, 0) \in$
$\mathbb{R}^n$. Since D is an open set, there exists an interval $I \subseteq \mathbb{R}$, centre 0,
such that $\mathbf{p} + u\mathbf{e}_1 \in D$ for all $u \in I$. Choose a particular $u \neq 0$ in I and
consider the path $\alpha : [0, 1] \subseteq \mathbb{R} \to \mathbb{R}^n$ defined by $\alpha(t) = \mathbf{p} + tu\mathbf{e}_1$.
Then α is a C^1 path in D from $\mathbf{p}$ to $\mathbf{p} + u\mathbf{e}_1$. The path integral of

$F = (F_1, \ldots, F_n)$ along α is

$$\int_\alpha F \cdot \mathrm{d}\alpha = \int_0^1 F(\mathbf{p} + t u \mathbf{e}_1) \cdot u \mathbf{e}_1 \, \mathrm{d}t = u \int_0^1 F_1(\mathbf{p} + t u \mathbf{e}_1) \, \mathrm{d}t.$$

By the Integral Mean-Value Theorem, there exists u^* between 0 and u such that

5.5.8 $$\int_\alpha F \cdot \mathrm{d}\alpha = u F_1(\mathbf{p} + u^* \mathbf{e}_1).$$

Let β and γ be piecewise C^1 paths in D from $\mathbf{k}$ to $\mathbf{p}$ and from $\mathbf{k}$ to $\mathbf{p} + u\mathbf{e}_1$ respectively. See Fig. 5.10. Then, by property **[i]** and Exercise 5.3.8,

$$\int_\gamma F \cdot \mathrm{d}\gamma = \int_{\beta\alpha} F \cdot \mathrm{d}(\beta\alpha) = \int_\beta F \cdot \mathrm{d}\beta + \int_\alpha F \cdot \mathrm{d}\alpha.$$

Hence, from 5.5.7,

$$f(\mathbf{p} + u\mathbf{e}_1) = f(\mathbf{p}) + \int_\alpha F \cdot \mathrm{d}\alpha.$$

From 5.5.8,

$$\frac{f(\mathbf{p} + u\mathbf{e}_1) - f(\mathbf{p})}{u} = F_1(\mathbf{p} + u^* \mathbf{e}_1).$$

Since F is continuous at $\mathbf{p}$, the coordinate function $F_1 : D \subseteq \mathbb{R}^n \to \mathbb{R}$

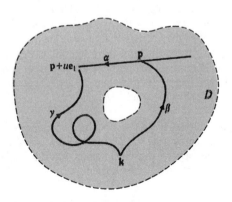

Fig. 5.10

is continuous at **p**. Therefore

$$\frac{\partial f}{\partial x_1}(\mathbf{p}) = \lim_{u \to 0} \frac{f(\mathbf{p} + u\mathbf{e}_1) - f(\mathbf{p})}{u} = \lim_{u \to 0} F_1(\mathbf{p} + u^*\mathbf{e}_1) = F_1(\mathbf{p}).$$

In the same way, taking other unit directions $\mathbf{e}_i$, we can show that all the partial derivatives of f exist, are continuous and satisfy

$$(F_1, \ldots, F_n) = \left(\frac{\partial f}{\partial x_1}, \ldots, \frac{\partial f}{\partial x_n} \right).$$

So $F = \operatorname{grad} f$.

[ii] *implies* **[iii]**. Assume **[ii]**. Put $h = -f$. Let $\alpha : [a, b] \subseteq \mathbb{R} \to \mathbb{R}^n$ be a piecewise C^1 path in D from $\alpha(a) = \mathbf{p}$ to $\alpha(b) = \mathbf{q}$. Then, by the Fundamental Theorem 5.4.2,

$$\int_\alpha F \cdot d\alpha = \int_\alpha \operatorname{grad} f \cdot d\alpha = f(\mathbf{q}) - f(\mathbf{p}) = h(\mathbf{p}) - h(\mathbf{q}).$$

[iii] *implies* **[iv]**. Set $\mathbf{p} = \mathbf{q}$ in (iii).

[iv] *implies* **[i]**. Let $\alpha : [a, b] \subseteq \mathbb{R} \to \mathbb{R}^n$ and $\alpha^* : [a^*, b^*] \subseteq \mathbb{R} \to \mathbb{R}^n$ be two piecewise C^1 paths in D, both running from **p** to **q**. The product path $\gamma = \alpha^* \alpha^-$ runs from **p** to **p**. Therefore, if property **[iv]** holds then

$$0 = \int_\gamma F \cdot d\gamma = \int_{\alpha^*} F \cdot d\alpha^* - \int_\alpha F \cdot d\alpha,$$

by Exercise 5.3.8. Hence condition **[iv]** implies condition **[i]**.

5.5.9 *Example.* In Example 5.5.1 we considered the vector field $F : \mathbb{R}^3 \to \mathbb{R}^3$ given by $F(x, y, z) = (0, 0, -mg)$. We showed in 5.5.2 that F satisfies property **[i]** of the above theorem and that a suitable function $h : \mathbb{R}^3 \to \mathbb{R}$ satisfying property **[iii]** is the potential energy function $h(x, y, z) = mgz$. In terms of the proof of the theorem, this function is obtained by taking a base point anywhere in the x, y plane. The reader can verify (property (ii)) that $F = \operatorname{grad}(-h)$.

5.5.10 *Example.* Similarly, in Example 5.5.3, we showed that the vector field $F : \mathbb{R}^3 \to \mathbb{R}^3$ given by $F(x, y, z) = (yz, xz, xy)$ satisfies property **[i]** of the theorem. In this case we showed in effect that if **p** and **q** lie in $\mathbb{R}^3$ and if α is a piecewise C^1 path from **p** and **q** then

$$\int_\alpha F \cdot d\alpha = h(\mathbf{p}) - h(\mathbf{q})$$

where $h : \mathbb{R}^3 \to \mathbb{R}$ is given by $h(x, y, z) = -xyz$.

Not all continuous vector fields have the properties given in the theorem.

5.5.11 *Example.* Consider the vector field $F : \mathbb{R}^2 \to \mathbb{R}^2$ given by $F(x, y) = \frac{1}{4}(-y, x)$. This is the field sketched in Fig. 5.2.

The C^1 paths $\alpha : [0, \pi/2] \subseteq \mathbb{R} \to \mathbb{R}^2$ and $\beta : [0, 3\pi/2] \subseteq \mathbb{R} \to \mathbb{R}^2$ given by

$$\alpha(t) = (\cos t, \sin t) \qquad \text{and} \qquad \beta(t) = (\cos t, -\sin t)$$

run respectively counterclockwise and clockwise around the unit circle in $\mathbb{R}^2$ from $(1, 0)$ to $(0, 1)$. Now

$$\int_\alpha F \cdot d\alpha = \int_0^{\pi/2} \frac{1}{4}(-\sin t, \cos t) \cdot (-\sin t, \cos t) \, dt = \frac{\pi}{8}$$

and

$$\int_\beta F \cdot d\beta = \int_0^{3\pi/2} \frac{1}{4}(\sin t, \cos t) \cdot (-\sin t, -\cos t) \, dt = -\frac{3\pi}{8}.$$

Since these path integrals are different, property **[i]**, and therefore properties **[ii]**, **[iii]** and **[iv]** of the theorem do not hold. In particular, the field F is not a gradient field.

With the above examples in mind we make the following definition.

5.5.12 *Definition.* *Let* $F : D \subseteq \mathbb{R}^n \to \mathbb{R}^n$ *be a continuous vector field. If there exists a function* $h : D \subseteq \mathbb{R}^n \to \mathbb{R}$ *such that for all* $\mathbf{p} \in D$ *and* $\mathbf{q} \in D$ *and all piecewise* C^1 *paths* α *from* $\mathbf{p}$ *to* $\mathbf{q}$ *in* D

5.5.13 $$\int_\alpha F \cdot d\alpha = h(\mathbf{p}) - h(\mathbf{q})$$

then h *is said to be a* potential function *for F.*

5.5.14 *Example.* **[i]** $h(x, y, z) = mgz$ is a potential function for the vector field $F(x, y, z) = (0, 0, -mg)$. See Example 5.5.9.

[ii] $h(x, y, z) = -xyz$ is a potential function for the vector field $F(x, y, z) = (yz, xz, xy)$. See Example 5.5.10.

We can now restate part of Theorem 5.5.6 in terms of the potential function.

5.5.15 *Theorem.* *Let* $F : D \subseteq \mathbb{R}^n \to \mathbb{R}^n$ *be a continuous vector field in an open set* D. *Then there exists a potential function for F if and only if F is a gradient field* $F = \operatorname{grad} f$ *for some* C^1 *function*

$f : D \subseteq \mathbb{R}^n \to \mathbb{R}$. *In this case the function* $h = -f$ *is a potential function for* F, *and* $F = -\mathrm{grad}\, h$.

Potential functions for a continuous vector field, when they exist, are not unique.

5.5.16 Theorem.   *Let* $h : D \subseteq \mathbb{R}^n \to \mathbb{R}$ *be a potential function for a continuous vector field* $F : D \subseteq \mathbb{R}^n \to \mathbb{R}^n$ *on a path-connected subset* D *of* $\mathbb{R}^n$. *Then*

[i] *the function* $g : D \subseteq \mathbb{R}^n \to \mathbb{R}$ *is also a potential function for* F *if and only if* $g = h + c$ *for some* $c \in \mathbb{R}$,

[ii] *if* $g : D \subseteq \mathbb{R}^n \to \mathbb{R}$ *is a potential function for* F *then every level set of* h *is a level set of* g *and vice versa.*

Proof.   Exercise.

5.5.17 Definition.   *Let* $h : D \subseteq \mathbb{R}^n \to \mathbb{R}$ *be a potential function for a continuous vector field* $F : D \subseteq \mathbb{R}^n \to \mathbb{R}^n$. *For each* $c \in \mathbb{R}$ *the level set* $h^{-1}(c)$ *is called an* equipotential set *of the field* F.

Knowing the equipotential sets of a continuous gradient field F enables us to evaluate path integrals painlessly. For example, if the sets are as given in Fig. 5.11 then the path integral of F along any piecewise C^1 path from **p** to **q** is $1 - (-2) = 3$.

Furthermore, if h is a potential function for F then, according to Theorem 3.8.8, $(\mathrm{grad}\, h)(\mathbf{p})$ is orthogonal at **p** to the equipotential set through **p** and points in the direction of *increasing* h. Since $F = -\mathrm{grad}\, h$, the vector $F(\mathbf{p})$ is therefore also orthogonal at **p** to

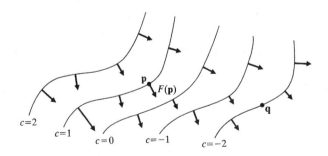

Fig. 5.11 Equipotential sets $h^{-1}(c)$ of field $F = -\mathrm{grad}\, h$

the equipotential set through **p**, but points in the direction in which
h decreases.

The significance of potential functions in applications is
underlined by considering the energy of a particle of mass m
kilogrammes moving through space under the influence of a force
field. With respect to SI units, let the C^2 path $\alpha : [a, b] \subseteq \mathbb{R} \to \mathbb{R}^3$
model the motion of the particle. At time t in $[a, b]$, the derivatives
$\alpha'(t)$ and $\alpha''(t)$ correspond to the particle's velocity and acceleration
respectively. The norm $\|\alpha'(t)\|$ corresponds to the particle's speed
and $\frac{1}{2}m \|\alpha'(t)\|^2$ to its kinetic energy. We shall need to know that,
since $\|\alpha'(t)\|^2 = \alpha'(t) \cdot \alpha'(t)$,

$$\frac{\mathrm{d}}{\mathrm{d}t} \|\alpha'(t)\|^2 = 2\alpha'(t) \cdot \alpha''(t).$$

Suppose that the continuous vector field $F : \mathbb{R}^3 \to \mathbb{R}^3$ models the
force field. By Newton's Second Law of Motion, for all $t \in [a, b]$

$$F(\alpha(t)) = m\alpha''(t).$$

The path integral of F along α is therefore

5.5.18 $$\int_\alpha F \cdot \mathrm{d}\alpha = m \int_a^b \alpha''(t) \cdot \alpha'(t) \, \mathrm{d}t$$

$$= \frac{1}{2}m \|\alpha'(b)\|^2 - \frac{1}{2}m \|\alpha'(a)\|^2.$$

The work done by the field on the particle is therefore the increase
in the kinetic energy of the particle.

Suppose now that there exists a potential function $h : \mathbb{R}^3 \to \mathbb{R}$ for
the vector field F. Then, by Definition 5.5.12,

$$\int_\alpha F \cdot \mathrm{d}\alpha = h(\alpha(a)) - h(\alpha(b)).$$

It follows from 5.5.18 that

5.5.19 $\quad h(\alpha(a)) + \frac{1}{2}m \|\alpha'(a)\|^2 = h(\alpha(b)) + \frac{1}{2}m \|\alpha'(b)\|^2.$

In particular, as in Example 5.5.9, when F models the
gravitational field acting on the particle then 5.5.19 is the *Principle
of Conservation of Energy* which states that the sum of the kinetic
and potential energies of the particle remains constant along its
path.

More generally, we have shown that the Principle of

Conservation of Energy 5.5.19 applies to all force fields which admit a potential function.

5.5.20 Definition.   *A continuous vector field* $F : D \subseteq \mathbb{R}^n \to \mathbb{R}^n$ *is said to be* conservative *if there is a potential function* $h : D \subseteq \mathbb{R}^n \to \mathbb{R}$ *for F.*

5.5.21 Theorem.   *Let* $F : D \subseteq \mathbb{R}^n \to \mathbb{R}^n$ *be a continuous vector field defined on an open set D in* $\mathbb{R}^n$*. Then*

    [i] *the field F is conservative if and only if the work done in moving a particle around any closed contour in D is zero*;

    [ii] *the field F is conservative if and only if* $F = \operatorname{grad} f$ *for some* C^1 *function* $f : D \subseteq \mathbb{R}^n \to \mathbb{R}$.

Proof.   Immediate from Theorem 5.5.6.

Part **[i]** or part **[ii]** of Theorem 5.5.21 is sometimes taken as the definition of a conservative field.

5.5.22 Example.   The vector field

$$F(x, y, z) = (2xyz + z, x^2z + 1, x^2y + x), \qquad (x, y, z) \in \mathbb{R}^3$$

is conservative. To show this we find a function $f : \mathbb{R}^3 \to \mathbb{R}$ such that $F = \operatorname{grad} f$, that is

$$\frac{\partial f}{\partial x}(x, y, z) = 2xyz + z, \qquad \frac{\partial f}{\partial y}(x, y, z) = x^2z + 1, \qquad \frac{\partial f}{\partial z}(x, y, z) = x^2y + x$$

for all $(x, y, z) \in \mathbb{R}^3$. Integrating the first identity, we obtain

$$f(x, y, z) = x^2yz + xz + g(y, z)$$

where g is some function of y and z still to be determined.

    Similarly, integrating the second and third identities, we obtain

$$f(x, y, z) = x^2yz + y + h(x, z)$$
$$f(x, y, z) = x^2yz + xz + k(x, y)$$

Hence by choosing

$$g(y, z) = y, \qquad h(x, z) = xz, \qquad k(x, y) = y,$$

we find

$$f(x, y, z) = x^2yz + xz + y.$$

5.5.23 Example. The vector field

$$F(x, y) = (2y, x + y), \qquad (x, y) \in \mathbb{R}^2$$

is not conservative. Proceeding as in Example 5.5.22, let us attempt to find a function $f : \mathbb{R}^2 \to \mathbb{R}$ sucn that $F = \operatorname{grad} f$. Then, for all $(x, y) \in \mathbb{R}^2$,

$$\frac{\partial f}{\partial x}(x, y) = 2y \quad \text{and} \quad \frac{\partial f}{\partial y}(x, y) = x + y.$$

Hence

$$f(x, y) = 2xy + g(y) \quad \text{and} \quad f(x, y) = xy + \tfrac{1}{2}y^2 + h(x)$$

for suitable functions g and h. Clearly no such functions can be found, and therefore F is not conservative.

5.5.24 Example. The vector field

$$F(x, y) = \tfrac{1}{4}(-y, x), \qquad (x, y) \in \mathbb{R}^2$$

is not conservative. This was shown in Example 5.5.11 by comparing two path integrals from $(1, 0)$ to $(0, 1)$. Alternatively, one could demonstrate that F is not a gradient field by the method of Example 5.5.23.

5.5.25 Example. The gravitational field due to a particle of mass M is modelled by the vector field $F : \mathbb{R}^3 \backslash \{\mathbf{0}\} \to \mathbb{R}^3$ where

5.5.26 $$F(\mathbf{r}) = -\frac{GM}{r^3}\mathbf{r}, \qquad \mathbf{r} \neq \mathbf{0}.$$

Here $\mathbf{r} = (x, y, z)$ is measured relative to axes with origin at the particle and $r = \|\mathbf{r}\| = \sqrt{(x^2 + y^2 + z^2)}$. The vector $F(\mathbf{r})$ indicates the magnitude and direction of the force of attraction exerted by the mass M (supposed at the origin) on a particle of unit mass with position vector $\mathbf{r}$. See Fig. 5.12. The constant G is called the *gravitational constant*. Expression 5.5.26 is Newton's *Inverse Square Law*. It is so called because the magnitude of the attractive force at $\mathbf{r}$ is proportional to the inverse of the square of r:

$$\|F(\mathbf{r})\| = \frac{GM}{r^3}\|\mathbf{r}\| = \frac{GM}{r^2}.$$

The gravitational field F is conservative, because $F = \operatorname{grad} f$, where the function $f : \mathbb{R}^3 \backslash \{\mathbf{0}\} \to \mathbb{R}$ is defined by

5.5.27 $$f(\mathbf{r}) = \frac{GM}{r}, \qquad \text{for all } r \neq 0.$$

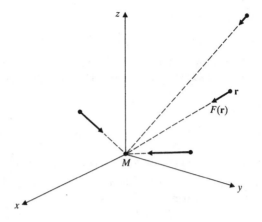

Fig. 5.12

This follows from the calculation

$$\frac{\partial f}{\partial x}(\mathbf{r}) = -\frac{GM}{r^2}\frac{\partial r}{\partial x} = -\frac{GM}{r^3}x,$$

and similar expressions for $(\partial f/\partial y)(\mathbf{r})$ and $(\partial f/\partial z)(\mathbf{r})$.

Finally in this section we consider an example illustrating how the property of a vector field being conservative may depend on the domain on which the field is defined.

5.5.28 Example. In $\mathbb{R}^2$ let A be the half axis $\{(x, 0) \in \mathbb{R}^2 \mid x \leq 0\}$. Then for each $(x, y) \in \mathbb{R}^2 \backslash A$ there is a unique $r > 0$ and a unique θ in the range $-\pi < \theta < \pi$ such that

5.5.29 $x = r \cos \theta$ and $y = r \sin \theta.$

See Fig. 5.13. Let D be the open set $\{(r, \theta) \in \mathbb{R}^2 \mid -\pi < \theta < \pi, r > 0\}$. Define the function $G : D \subseteq \mathbb{R}^2 \to \mathbb{R}^2$ by

$$G(r, \theta) = (r \cos \theta, r \sin \theta), \qquad (r, \theta) \in D.$$

Then G is 1–1 from D onto $\mathbb{R}^2 \backslash A$. Therefore the expressions 5.5.29 establish r and θ as functions of x and y on the domain $\mathbb{R}^2 \backslash A$ and taking values in D. Clearly $r = \sqrt{(x^2 + y^2)}$, but the expression for θ is a bit more troublesome to write down (see Exercise 5.5.8(a)). In any case, since

$$\det J_{G,(r,\theta)} = \det \begin{bmatrix} \cos \theta & -r \sin \theta \\ \sin \theta & r \cos \theta \end{bmatrix} = r,$$

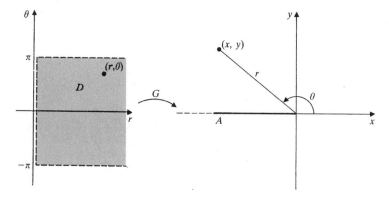

Fig. 5.13

and $r \neq 0$ on D, it follows by the Inverse Function Theorem 4.6.7 that r and θ are differentiable functions of x and y on $\mathbb{R}^2 \backslash A$. Differentiating both r and θ partially with respect to x, we obtain from 5.5.29

$$1 = \frac{\partial r}{\partial x}\cos\theta - r\frac{\partial\theta}{\partial x}\sin\theta,$$

$$0 = \frac{\partial r}{\partial x}\sin\theta + r\frac{\partial\theta}{\partial x}\cos\theta.$$

Therefore $\partial\theta/\partial x = -y/(x^2 + y^2)$ and, by a similar argument, $\partial\theta/\partial y = x/(x^2 + y^2)$.

Let $f : \mathbb{R}^2 \backslash A \to \mathbb{R}$ be the real-valued function defined by $f(x, y) = \theta$, where θ is the polar angle given by 5.5.29, and $-\pi < \theta < \pi$. Then $\text{grad}\, f = (\partial\theta/\partial x, \partial\theta/\partial y)$. Hence $F = \text{grad}\, f : \mathbb{R}^2 \backslash A \to \mathbb{R}^2$ is the conservative vector field such that for each $(x, y) \in \mathbb{R}^2 \backslash A$,

5.5.30
$$F(x, y) = \left(\frac{-y}{x^2 + y^2}, \frac{x}{x^2 + y^2}\right).$$

See Fig. 5.14 for a sketch of this field. The potential function is $h(x, y) = -\theta$.

Notice that if the domain of F is extended to $\mathbb{R}^2 \backslash \{\mathbf{0}\}$ and 5.5.30 is taken as defining F over this new domain then F is not a conservative field. To see this consider the closed path $\alpha : [0, 2\pi] \subseteq \mathbb{R} \to \mathbb{R}^2$ defined by $\alpha(t) = (\cos t, \sin t)$, $t \in [0, 2\pi]$. Then

$$\int_\alpha F \cdot d\alpha = \int_0^{2\pi} F(\alpha(t)) \cdot \alpha'(t)\, dt = \int_0^{2\pi} (-\sin t, \cos t) \cdot (-\sin t, \cos t)\, dt$$

$$= \int_0^{2\pi} dt = 2\pi.$$

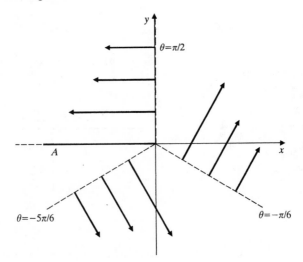

Fig. 5.14 Field $F(x, y) = \left(\dfrac{-y}{x^2 + y^2}, \dfrac{x}{x^2 + y^2} \right)$ and equipotential sets $\theta = $ constant

Since the image of α lies in $\mathbb{R}^2 \backslash \{\mathbf{0}\}$, Theorem 5.5.21(i) tells us that F is not a conservative field over $\mathbb{R}^2 \backslash \{\mathbf{0}\}$. We are prevented from considering such a path α in $\mathbb{R}^2 \backslash A$ by the cut along the negative x-axis. See Fig. 5.15.

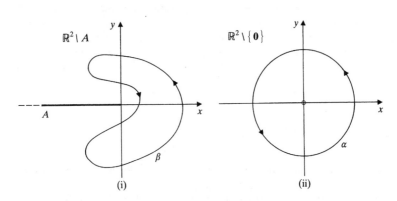

Fig. 5.15 (i) $\displaystyle\int_\beta F \cdot \mathrm{d}\beta = 0$; (ii) $\displaystyle\int_\alpha F \cdot \mathrm{d}\alpha = 2\pi$

Exercises 5.5

1. Let $F : \mathbb{R}^2 \to \mathbb{R}^2$ be the vector field defined by

$$F(x, y) = (2xy, x^2 + 1), \qquad (x, y) \in \mathbb{R}^2.$$

Verify that F is conservative, and find a potential function h for F (such that $F = -\text{grad } h$).

Answer: $h(x, y) = -x^2y - y$

Verify also by direct calculation that $\int_\alpha F \cdot d\alpha = 0$ where α is the closed figure 8 path defined by

$$\alpha(t) = (\sin t, \sin 2t), \qquad t \in [0, 2\pi].$$

2. Sketch the vector field $F : \mathbb{R}^2 \to \mathbb{R}^2$ defined by

$$F(x, y) = (x^2 + y^2, 0), \qquad (x, y) \in \mathbb{R}^2.$$

Note in particular that the field is constant on any circle centred at the origin.

Let α be the closed circular path in $\mathbb{R}^2$ defined by

$$\alpha(t) = (\cos t, \sin t), \qquad t \in [0, 2\pi].$$

Prove that $\int_\alpha F \cdot d\alpha = 0$. Show, however, that F is not a conservative field
(a) by the method of Example 5.5.23,
(b) by constructing a closed piecewise C^1 path β such that $\int_\beta F \cdot d\beta \neq 0$.

Answer: possible closed paths β are (i) the composite 'half-moon' path consisting of the semicircular path $(\cos t, \sin t)$, $t \in [0, \pi]$ followed by the path $(t, 0)$, $t \in [-1, 1]$ along the x-axis, or (ii) a composite path tracing the unit square, vertices $(0, 0)$, $(0, 1)$, $(1, 1)$, $(1, 0)$, for example as follows: $(t, 0)$, $t \in [0, 1]$, then $(1, t)$, $t \in [0, 1]$, then $(1 - t, 1)$, $t \in [0, 1]$, then $(0, 1 - t)$, $t \in [0, 1]$. The integrals $\int_\beta F \cdot d\beta$ take the values (i) $-4/3$, (ii) -1.

3. Let α be any piecewise C^1 path in $\mathbb{R}^3$ running from $(1, -1, 1)$ to $(2, 0, 3)$. Prove that

$$\int_\alpha yz \, dx + xz \, dy + xy \, dz = 1.$$

This generalizes Exercise 5.3.4.

Hint: the field $F(x, y, z) = (yz, xz, xy)$ is conservative.

4. Prove that the vector field $F(x, y, z) = (xy, yz, zx)$ is not conservative.

5. For each of the following conservative fields F find a function f such that $F = \text{grad } f$.
(a) $F(x, y, z) = (yz + 2xz, xz + 1, x^2 + xy + 2)$;
(b) $F(x, y, z) = (x/r, y/r, z/r)$, where $r = \sqrt{(x^2 + y^2 + z^2)} \neq 0$.

Answers: (a) $xyz + x^2z + y + 2z$; (b) r.

6.   Define the vector field $F:\mathbb{R}^3\backslash\{\mathbf{0}\}\rightarrow\mathbb{R}^3$ by

$$F(\mathbf{r}) = kr^m\mathbf{r} \qquad m \in \mathbb{Z},$$

where k is a constant. Prove that F is conservative and find a suitable potential function for F.

Hint: generalize Example 5.5.25 (where $m = -3$). The case $m = -2$ needs special treatment, logarithms!

7.   Let $F:\mathbb{R}^2\rightarrow\mathbb{R}^2$ be the vector field given by

$$F(x, y) = (x + y^2, 2xy), \qquad (x, y) \in \mathbb{R}^2.$$

Let α be a piecewise C^1 path in $\mathbb{R}^2$ running from $(0, 0)$ to (a, b). Prove that the path integral $\int_\alpha F \cdot d\alpha$ is independent of the choice of α, and find its value.

Answer: $\frac{1}{2}a^2 + ab^2$. Note that $F = \operatorname{grad} f$, where $f(x, y) = \frac{1}{2}x^2 + xy^2$.

Show also that the path integral $\int_\alpha G \cdot d\alpha$, where

$$G(x, y) = (x + y, 2xy), \qquad (x, y) \in \mathbb{R}^2,$$

depends on the choice of α.

Answer: compare, for example, the straight line path from $(0, 0)$ to (a, b) and the polygonal path from $(0, 0)$ to $(a, 0)$ to (a, b). The corresponding values of the line integral are $\frac{1}{2}a^2 + \frac{1}{2}ab + \frac{2}{3}ab^2$ and $a^2 + ab^2$ respectively. The field F is conservative, but the field G is not conservative.

8.   Let $F:\mathbb{R}^2\backslash\{\mathbf{0}\}\rightarrow\mathbb{R}^2$ be the vector field defined by

$$F(x, y) = \left(\frac{-y}{x^2 + y^2}, \frac{x}{x^2 + y^2}\right), \qquad (x, y) \neq (0, 0).$$

(a) Prove by direct integration that F is conservative on the restricted domain given by the half plane $\{(x, y) \in \mathbb{R}^2 \mid x > 0\}$.

Answer: $F = \operatorname{grad} f$, where $f(x, y) = \arctan(y/x)$, $x > 0$.

Compare with Example 5.5.28. In fact F is conservative on the extended domain $\mathbb{R}^2\backslash A$, where A is the half axis $\{(x, 0) \in \mathbb{R}^2 \mid x \leqslant 0\}$, and $F = \operatorname{grad} f$ where

$$f(x, y) = \begin{cases} \arctan(y/x) & \text{when } x > 0, \\ \frac{1}{2}\pi & \text{when } x = 0, y > 0, \\ \pi + \arctan(y/x) & \text{when } x < 0, y > 0, \\ -\frac{1}{2}\pi & \text{when } x = 0, y < 0, \\ -\pi + \arctan(y/x) & \text{when } x < 0, y < 0. \end{cases}$$

Verify that this agrees with the definition $f(x, y) = \theta$, $(x, y) \in \mathbb{R}^2\backslash A$, given in Example 5.5.28.

(b) Given $(a, b) \neq (0, 0)$ let H be the half line $\{t(a, b) \in \mathbb{R}^2 \mid t \leqslant 0\}$. Prove that F is conservative on the domain $\mathbb{R}^2\backslash H$.

(c) Let α be a C^1 path running once round the circle centre (a, b) with radius $c < \sqrt{(a^2 + b^2)}$. Prove that $\int_\alpha F \cdot d\alpha = 0$.

9.   Let $D = R \cup L$ be the union of open half-planes in $\mathbb{R}^2$

$$R = \{(x, y) \in \mathbb{R}^2 \mid x > 0\} \quad \text{and} \quad L = \{(x, y) \in \mathbb{R}^2 \mid x < 0\}.$$

Let $F : D \subseteq \mathbb{R}^2 \to \mathbb{R}^2$ be the continuous vector field defined by

$$F(x, y) = \begin{cases} (y, x) & \text{when} & (x, y) \in R, \\ (2x, 0) & \text{when} & (x, y) \in L. \end{cases}$$

Prove that the field F is conservative. Verify that the four properties of Theorem 5.5.6 are satisfied. In particular, express the field F in the form $F = \text{grad } f$.

Answer: F is the gradient of the C^1 function $f : D \subseteq \mathbb{R}^2 \to \mathbb{R}$ defined by

$$f(x, y) = \begin{cases} xy & \text{when} & (x, y) \in R, \\ x^2 & \text{when} & (x, y) \in L. \end{cases}$$

10.   Let D be the disjoint union of two path connected open subsets A and B in $\mathbb{R}^n$. There is then no path in D from a point of A to a point of B. Prove Theorem 5.5.6 for this case.

Hint: in proving that (i) implies (ii) choose a base point in each of the path components A and B.

11.   With the open subset D of $\mathbb{R}^2$ and the conservative field F defined as in Exercise 9, find two potential functions g and h for F that do not differ by a constant. Why does this not contradict Theorem 5.5.16(i)?

Answer: choose for example $g = -f$, where f is as defined in Exercise 9, and define

$$h(x, y) = \begin{cases} -xy & \text{when} & (x, y) \in R \\ -x^2 + 1 & \text{when} & (x, y) \in L. \end{cases}$$

The set D is not path connected.

12.   Let $F : D \subseteq \mathbb{R}^2 \to \mathbb{R}^2$ be a C^1 vector field defined on an open disc D in $\mathbb{R}^2$. Prove that a necessary and sufficient condition for F to be conservative is that $\partial F_1/\partial y = \partial F_2/\partial x$ on D, where F_1, F_2 are the coordinate functions of F.

Proof of sufficiency. assume that $\partial F_1/\partial y = \partial F_2/\partial x$. Fix (a, b) in D and define

$$f_1(x, y) = \int_a^x F_1(t, y)\, dt, \qquad \text{for all } (x, y) \in D.$$

Show that $\partial f_1/\partial x = F_1$ and that $g = F_2 - \partial f_1/\partial y$ is independent of x. Define

$$f_2(y) = \int_b^y g(t)\, dt,$$

and put $f(x, y) = f_1(x, y) + f_2(y)$. Then $F = \operatorname{grad} f$.

13. Let $F: D \subseteq \mathbb{R}^3 \to \mathbb{R}^3$ be a C^1 vector field defined on an open ball D in $\mathbb{R}^3$. Prove that F is conservative if and only if the following functional relations hold on D:

$$\frac{\partial F_1}{\partial y} = \frac{\partial F_2}{\partial x}, \frac{\partial F_1}{\partial z} = \frac{\partial F_3}{\partial x}, \frac{\partial F_2}{\partial z} = \frac{\partial F_3}{\partial y}.$$

14. Show that a field is conservative if and only if the work done by the field in moving a particle is recoverable by returning the particle to its original position.

15. Show that a conservative C^2 vector field is irrotational.

Proof: Theorem 4.10.9.

16. Let D be a path connected subset of $\mathbb{R}^m$. Let $f: D \subseteq \mathbb{R}^m \to \mathbb{R}^n$ be continuous. Show that $f(D)$ is path connected.

6

Line integrals in $\mathbb{R}^n$

6.1 Line integrals of scalar fields and vector fields

In Section 5.2 we considered the integral of a real-valued function (scalar field) $f:S \subseteq \mathbb{R}^n \to \mathbb{R}$ along a path $\alpha:[a, b] \subseteq \mathbb{R} \to \mathbb{R}^n$ in S. The value of the integral depends on the way the path traces and retraces its image. In contrast, elementary Riemann integration concerns the integral of a real-valued function over a compact interval $[a, b] \subseteq \mathbb{R}$ which is regarded as a subset in $\mathbb{R}$. As a direct generalization of this we shall now consider the integral of a scalar field in $\mathbb{R}^n$ over certain subsets of $\mathbb{R}^n$ that generalize the compact intervals in $\mathbb{R}$, in fact simple arcs.

6.1.1 Definition. *Let $f:S \subseteq \mathbb{R}^n \to \mathbb{R}$ be a continuous scalar field and let C be a simple arc in S. The* integral of f *over C is defined and denoted by*

$$\int_C f \, ds = \int_\alpha f \, ds = \int_a^b f(\alpha(t)) \, \|\alpha'(t)\| \, dt$$

where $\alpha:[a, b] \subseteq \mathbb{R} \to \mathbb{R}^n$ is any simple parametrization of C.

The above integral is often called the *line integral of f over C*, although 'curve integral' or 'arc integral' would perhaps be a better description. Definition 6.1.1 is unambiguous, since by Theorem 5.2.29, for any two simple parametrizations α and β of C the path integrals of f along α and β take the same value. It is important to note that only 1–1 parametrizations of C should be used in evaluating the line integral of f over C. The discussion in Section 5.2 shows how this integral may differ in value from the integral of f along a path that traces C more than once.

6.1.2 Example. Let C be the simple arc $y = |x|$, $-1 \leqslant x \leqslant 1$, and let $f:\mathbb{R}^2 \to \mathbb{R}$ be the continuous scalar field defined by $f(x, y) = x^2 y$. The line integral of f over C can be evaluated as a path integral by using any simple parametrization of C. Choose for example the parametrization

$$\alpha(t) = \begin{cases} (t^2, t^2) & \text{when} & -1 \le t \le 0 \\ (-t^2, t^2) & \text{when} & 0 \le t \le 1. \end{cases}$$

Then $\|\alpha'(t)\| = 2\sqrt{2}\,|t|$ and hence

$$\oint_C f \, ds = \int_\alpha f \, ds = \int_{-1}^0 t^6(-2\sqrt{2}t)\,dt + \int_0^1 t^6(2\sqrt{2}t)\,dt = \tfrac{1}{2}\sqrt{2}.$$

More simply, regarding C as a *piecewise simple arc*, we can evaluate the integral of f over C using the piecewise simple parametrization $\beta(t) = (t, |t|)$, $-1 \le t \le 1$. Then $\|\beta'(t)\| = \sqrt{2}$, $t \ne 0$, and

$$\int_C f \, ds = \int_{-1}^0 -t^3 \sqrt{2}\,dt + \int_0^1 t^3 \sqrt{2}\,dt = \tfrac{1}{2}\sqrt{2}.$$

Notice that α traces the arc C from $(1, 1)$ to $(-1, 1)$, whereas β traces it in the opposite direction.

We now turn to the line integral of a vector field. We observed in Section 5.3 that the integral of a vector field along a path depends on the direction in which the path traces its image. Reversing the direction changes the sign of the integral. Accordingly we define the line integral of a vector field along an oriented simple arc. We shall need to know that if α and β are properly equivalent simple parametrizations of a simple arc C (so they trace C in the same direction) then the integrals of a vector field along α and along β are equal.

6.1.3 Theorem. *Let C be a simple arc in $\mathbb{R}^n$ and let $F : S \subseteq \mathbb{R}^n \to \mathbb{R}^n$ be a continuous vector field whose domain S contains C. Let $\alpha : [a, b] \subseteq \mathbb{R} \to \mathbb{R}^n$ and $\beta : [c, d] \subseteq \mathbb{R} \to \mathbb{R}^n$ be properly equivalent simple parametrizations of C. Then*

$$\int_\alpha F \cdot d\alpha = \int_\beta F \cdot d\beta.$$

Proof. We shall prove Theorem 6.1.4 for the important special case where C is a *smooth* simple arc and α and β are properly equivalent smooth simple parametrizations of C. Let $\phi : [c, d] \to [a, b]$ be the differentiable function such that $\beta = \alpha \circ \phi$. By the Chain Rule 2.6.11

$$\beta'(u) = \alpha'(\phi(u))\phi'(u), \qquad u \in [c, d].$$

Taking coordinate functions, we obtain, for $i = 1, \ldots, n$,

6.1.4 $\qquad \beta_i'(u) = \alpha_i'(\phi(u))\phi'(u) \qquad\qquad u \in [c, d]$.

Now fix u in $[c, d]$. Since β is smooth, β' is continuous and non-zero, and so at least one of the coordinate function β_k', $1 \leq k \leq n$ is continuous and non-zero in some neighbourhood of u. Since α' is continuous it follows by 6.1.4 that ϕ' is continuous at u. Hence ϕ is C^1. By Theorem 5.3.11 the integrals of F along α and along β are equal.

The proof of the general result 6.1.3 which we omit, is based on the comparison of Riemann sums.

Before giving the formal definition of the line integral of a vector field along an oriented simple arc C we introduce a useful notation. Recall (2.7.19) that if a path α traces a curve from **a** to **b** then the inverse path α^- traces it from **b** to **a**.

6.1.5. Definition. *Let C be a simple arc in $\mathbb{R}^n$ simply parametrized by α, and let $[\alpha]$ be the set of all simple parametrizations of C which are properly equivalent to α.*

[i] *The oriented simple arc $\{C, [\alpha]\}$ will be denoted either by C^+ or by C^-. If $C^+ = \{C, [\alpha]\}$ then $C^- = \{C, [\alpha^-]\}$ and vice versa.*

[ii] *Let $C^+ = \{C, [\alpha]\}$. Any parametrization in the equivalence class $[\alpha]$ is called a* simple parametrization *of C^+.*

6.1.6. Example. In Example 6.1.2, if $C^+ = \{C, [\alpha]\}$ then $C^- = \{C, [\beta]\}$.

6.1.7 Definition. *Let $F : S \subseteq \mathbb{R}^n \to \mathbb{R}^n$ be a continuous vector field, and let C^+ be an oriented simple arc in S. The* integral of F along C^+ *is defined and denoted by*

$$\int_{C^+} F \cdot d\mathbf{r} = \int_\alpha F \cdot d\alpha$$

where α is any simple parametrization of C^+.

The above integral is often called the *line integral of F along C*. Notice that we speak of integrating a scalar field f *over* a simple arc C, but a vector field F *along* an oriented arc C^+. Theorem 6.1.3 ensures that Definition 6.1.7 is unambiguous. It follows from Exercise 5.3.8 that

6.1.8 $\qquad\qquad\qquad \displaystyle\int_{C^-} F \cdot d\mathbf{r} = -\int_{C^+} F \cdot d\mathbf{r}$.

6.1.9 *Example* Define $F:\mathbb{R}^2 \to \mathbb{R}^2$ by $F(x, y) = (xy, -x)$, and let C^+ be the parabolic arc $y = x^2$, $-1 \leqslant x \leqslant 0$ oriented from $(-1, 1)$ to $(0, 0)$. The line integral of F along C^+ can be evaluated as a path integral by using any simple parametrization of C^+. Choose for example the parametrization $\alpha(t) = (t, t^2)$, $-1 \leqslant t \leqslant 0$. Hence

$$\int_{C^+} F \cdot d\mathbf{r} = \int_\alpha F \cdot d\alpha = \int_{-1}^0 (t^3, -t) \cdot (1, 2t) \, dt = -\tfrac{11}{12}.$$

Also the integral of F along C^- is, by 6.1.8,

$$\int_{C^-} F \cdot d\mathbf{r} = \tfrac{11}{12}.$$

It can be evaluated as the path integral $\int_\beta F \cdot d\beta$ where β is the simple parametrization of C^- given by $\beta(u) = (-u, u^2)$, $0 \leqslant u \leqslant 1$.

There is a useful relationship between the line integral of a vector field along an oriented smooth simple arc C^+ in $\mathbb{R}^n$ and the line integral of a related scalar field over C.

Let $\alpha:[a, b] \subseteq \mathbb{R} \to \mathbb{R}^n$ be a simple parametrization of C^+. Since α is smooth, for each $t \in [a, b]$ the vector $\alpha'(t)$, tangent to α at t, is non-zero. We can use α to define a continuous unit tangent field $T:C \subseteq \mathbb{R}^n \to \mathbb{R}^n$ by putting, for each $\mathbf{r} \in C$,

6.1.10 $\quad T(\mathbf{r}) = \dfrac{\alpha'(t)}{\|\alpha'(t)\|} \qquad$ where $\mathbf{r} = \alpha(t)$, $\qquad t \in [a, b]$.

It follows from Section 2.6 that any other simple parametrization of C^+ leads to the same unit tangent field T. The oriented smooth curve $C^+ = \{C, [\alpha]\}$ could equally be defined by the pair (C, T). Consequently the following definition is unambiguous.

6.1.11 *Definition.* *Let C^+ be an oriented smooth simple arc in $\mathbb{R}^n$, and let $\alpha:[a, b] \subseteq \mathbb{R} \to \mathbb{R}^n$ be any simple parametrization of C^+. The vector field $T:C \subseteq \mathbb{R}^n \to \mathbb{R}^n$ defined in 6.1.10 is called the* unit tangent field *to the oriented curve C^+. For any $\mathbf{r} \in C$, the unit vector $T(\mathbf{r})$ is called the* unit tangent *to C^+ at $\mathbf{r}$.*

Suppose now that $F:C \subseteq \mathbb{R}^n \to \mathbb{R}^n$ is a continuous vector field on C. For each $\mathbf{r} \in C$ the component of $F(\mathbf{r})$ in the direction $T(\mathbf{r})$ is the dot product $F(\mathbf{r}) \cdot T(\mathbf{r}) \in \mathbb{R}$. We therefore have a continuous scalar field $F \cdot T:C \subseteq \mathbb{R}^n \to \mathbb{R}$ which can be integrated over C. We call $F \cdot T$ the *component of F along C^+* (where T is the unit tangent field to C^+).

6.1.12 Theorem. *Let C^+ be an oriented smooth simple arc in $\mathbb{R}^n$ with unit tangent field T. Let $F: C \subseteq \mathbb{R}^n \to \mathbb{R}^n$ be a continuous vector field on C. Then*

$$\int_{C^+} F \cdot d\mathbf{r} = \int_C F \cdot T \, ds.$$

That is, the integral of F along C^+ is equal to the integral of the component of F along C^+ over the arc C.

Proof. Let $\alpha: [a, b] \subseteq \mathbb{R} \to \mathbb{R}^n$ be any simple parametrization of C^+. then

$$\int_{C^+} F \cdot d\mathbf{r} = \int_a^b F(\alpha(t)) \cdot \alpha'(t) \, dt$$

$$= \int_a^b F(\alpha(t)) \cdot T(\alpha(t)) \, \|\alpha'(t)\| \, dt = \int_C F \cdot T \, ds.$$

It is clear that if $T: C \subseteq \mathbb{R}^n \to \mathbb{R}^n$ is the unit tangent field to an oriented smooth simple arc C^+ then $-T: C \subseteq \mathbb{R}^n \to \mathbb{R}^n$ is the unit tangent field to C^-. Applying Theorem 6.1.12 to C^-, we obtain

$$\int_{C^-} F \cdot d\mathbf{r} = \int_C F \cdot (-T) \, ds = -\int_{C^+} F \cdot d\mathbf{r},$$

which confirms 6.1.8. The above relation explains why the integral of a vector field along a smooth simple arc changes sign when its orientation is reversed: the corresponding component scalar fields are $F \cdot T$ and $F \cdot (-T)$.

The work of this section generalizes easily to piecewise C^1 and piecewise smooth simple arcs. We leave the details to the reader.

6.1.13 Example. Let C be the piecewise smooth simple arc $y = |x|$, $-1 \leq x \leq 1$. Let $f: \mathbb{R}^2 \to \mathbb{R}$ be the scalar field defined by $f(x, y) = x^2 y$. The integral of f over C can be evaluated by using any simple parametrization of C. Choose for example the parametrization $\alpha(t) = (t, |t|)$, $-1 \leq t \leq 1$. Since $\|\alpha'(t)\| = \sqrt{2}$, $t \neq 0$, we obtain

$$\int_C f \, ds = \int_{-1}^0 t^2(-t)\sqrt{2} \, dt + \int_0^1 t^2(t)\sqrt{2} \, dt = \tfrac{1}{2}\sqrt{2}.$$

Now let C^+ be the same arc oriented from $(1, 1)$ to $(-1, 1)$ and let $F: \mathbb{R}^2 \to \mathbb{R}^2$ be the vector field defined by $F(x, y) = (xy, -x)$. The path α

defined above is a simple parametrization of C^-. We can use it to evaluate
the integral of F along C^+ as follows. Since

$$\int_\alpha F \cdot d\alpha = \int_{-1}^0 (-t^2, -t) \cdot (1, -1)\, dt + \int_0^1 (t^2, -t) \cdot (1, 1)\, dt = -1,$$

it follows that

$$\int_{C^+} F \cdot d\mathbf{r} = -\int_{C^-} F \cdot d\mathbf{r} = -\int_\alpha F \cdot d\alpha = 1.$$

A sketch of the oriented arc C^+ and the vector field F at points on C will
confirm that the integral of F along C^+ is a positive quantity.

A simple parametrization α of a piecewise smooth simple arc C in
$\mathbb{R}^n$ determines a unit tangent field $T : C \subseteq \mathbb{R}^n \to \mathbb{R}^n$ defined and
continuous at all but a finite number of points of C. The pair (C, T)
could be taken as the definition of the oriented piecewise smooth
simple arc $C^+ = \{C, [\alpha]\}$, and the expression

$$\int_{C^+} F \cdot d\mathbf{r} = \int_C F \cdot T\, ds$$

could be taken as a definition of the line integral of f along (C, T).

It is interesting to note in this connection that if G is any
continuous vector field on C we can define the integral of F over the
pair (C, G) as $\int_C F \cdot G\, ds$. For an application using a unit normal
field to a curve see Theorem 7.6.13.

The alternative notation for the integral of a vector field along a
path, which we described at the end of Section 5.3, is also used for
line integrals. Let $F = (F_1, \ldots, F_n)$ be a vector field on $\mathbb{R}^n$. If C^+ is
an oriented simple arc in $\mathbb{R}^n$ parametrized by $\mathbf{r}(t) =
(x_1(t), \ldots, x_n(t))$, $t \in [a, b]$, then we have

$$\int_{C^+} F \cdot d\mathbf{r} = \int_a^b F(x_1(t), \ldots, x_n(t)) \cdot (x_1'(t), \ldots, x_n'(t))\, dt$$

6.1.14
$$= \int_a^b F_1(\mathbf{r}(t)) x_1'(t)\, dt + \cdots + \int_a^b F_n(\mathbf{r}(t)) x_n'(t)\, dt$$

$$= \int_{C^+} F_1\, dx_1 + \cdots + \int_{C^+} F_n\, dx_n,$$

where

6.1.15
$$\int_{C^+} f\, dx_k = \int_{C^+} (0, \ldots, 0, f, 0, \ldots, 0) \cdot d\mathbf{r},$$

the function f standing in the kth coordinate place. The integral 6.1.14 is also often denoted by

$$\int_{C^+} F \cdot d\mathbf{r} = \int_{C^+} F_1 \, dx_1 + F_2 \, dx_2 + \cdots + F_n \, dx_n.$$

The skill in evaluating the integral 6.1.14 lies in finding a suitable parametrization of C^+. If C^+ is piecewise smooth, it is often best to parametrize each smooth part separately.

6.1.16 *Example.* Let C^+ be the piecewise smooth curve in $\mathbb{R}^2$ formed by straight line segments PQ, QR and oriented from P to R, where P, Q, R are the points $(0, -1)$, $(1, -1)$, $(1, 1)$. See Fig. 6.1. Let $F : \mathbb{R}^2 \to \mathbb{R}^2$ be the vector field defined by $F(x, y) = \frac{1}{2}(-xy, y)$. Using the natural parametrizations

$$(x(t), y(t)) = (t, -1), \qquad\qquad 0 \le t \le 1 \qquad \text{along } PQ$$

and

$$(x(y), y(t)) = (1, -1 + t), \qquad\qquad 0 \le t \le 2 \qquad \text{along } QR$$

we obtain

$$\int_{C^+} F \cdot d\mathbf{r} = \frac{1}{2} \int_0^1 (t, -1) \cdot (1, 0) \, dt + \frac{1}{2} \int_0^2 (1 - t, -1 + t) \cdot (0, 1) \, dt$$

$$= \frac{1}{2} \int_0^1 t \, dt + \frac{1}{2} \int_0^2 (-1 + t) \, dt = \frac{1}{4} + 0 = \frac{1}{4}.$$

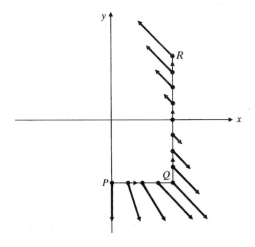

Fig. 6.1 Vector field $F(x, y) = \frac{1}{2}(-xy, y)$ along PQR

The calculations confirm what is clear from Fig. 6.1:

[i] that the line integral of F along PQ is positive (since the component of F along PQ is positive, except at P);

[ii] that the line integral of F along QR is zero.

6.1.17 Example. Let C^+ be the unit semicircle in the lower half-plane $y \leqslant 0$, centred at $(2, 0)$, and oriented from $(1, 0)$ to $(3, 0)$. Let $F(x, y) = (-xy, y)$. Then

$$\int_{C^+} F \cdot dr = \int_{C^+} (-xy)\, dx + y\, dy = \int_{C^+} (-xy)\, dx + \int_{C^+} y\, dy.$$

The equation $(x-2)^2 + y^2 = 1$, $y \leqslant 0$ for C suggests the parametrization of C^+

$$(x(t), y(t)) = (2 - \cos t, -\sin t) \qquad t \in [0, \pi].$$

We can evaluate separately

$$\int_{C^+} (-xy)\, dx = -\int_0^\pi x(t)y(t)x'(t)\, dt$$

$$= \int_0^\pi (2 - \cos t)\sin^2 t\, dt = \pi,$$

$$\int_{C^+} y\, dy = \int_0^\pi y(t)y'(t)\, dt = \int_0^\pi (-\sin t)(-\cos t)\, dt = 0.$$

Therefore

$$\int_{C^+} F \cdot dr = \pi.$$

Exercises 6.1.

1. Sketch the following simple arcs and find their lengths.
    (a) Semicircle $x^2 + y^2 = 4$, $x \geqslant 0$;
    (b) parabolic arc $y = x^2$, $0 \leqslant x \leqslant 1$;
    (c) semicubical parabola $y^2 = x^3$, $-1 \leqslant x \leqslant 1$;

Answers: (a) 2π; (b) $\frac{1}{2}\sqrt{5} + \frac{1}{4}\ln(2 + \sqrt{5})$; (c) $\frac{2}{27}(13\sqrt{13} - 8)$.

2. Let $f : \mathbb{R}^2 \to \mathbb{R}$ be the scalar field defined by $f(x, y) = x^2 y$. Evaluate the line integral $\int_C f\, ds$ of f over the following simple arcs C.
    (a) The unit semicircle $x^2 + y^2 = 1$, $y \geqslant 0$;
    (b) the unit semicircle $x^2 + y^2 = 1$, $y \leqslant 0$;
    (c) the unit semicircle $x^2 + y^2 = 1$, $x \geqslant 0$;
    (d) the two sides of a unit square which connect the point $(1, 1)$ in $\mathbb{R}^2$ with the points $(1, 0)$ and $(0, 1)$.

Hint for (d): consider the $1-1C^1$ paths $(1, t)$, $t \in [0, 1]$ and $(t, 1)$, $t \in [0, 1]$.

Answers: (a) $\frac{2}{3}$; (b) $-\frac{2}{3}$; (c) 0; (d) $\frac{5}{6}$.

3. Sketch the simple arc C in $\mathbb{R}^2$ composed of the parabolic arc $y = x^2$, $-1 \leqslant x \leqslant 0$, and the line segment $y = x$, $0 \leqslant x \leqslant 2$.
    (a) Find a C^1 parametrization of C.
    (b) Calculate $\int_C f \, ds$, where f is the scalar field defined by $f(x, y) = x$.

Answers: (a) The method of Example 2.6.36 gives the C^1 parametrization $\beta : [-1, 2^{1/3}] \subseteq \mathbb{R} \rightarrow \mathbb{R}^2$, where

$$\beta(t) = \begin{cases} (t^3, t^6) & \text{when} \quad -1 \leqslant t \leqslant 0, \\ (t^3, t^3) & \text{when} \quad 0 \leqslant t \leqslant 2^{1/3}. \end{cases}$$

(b) $2\sqrt{2} - ((\sqrt{5})^3 - 1)/12$.

4. Let $F : \mathbb{R}^2 \rightarrow \mathbb{R}^2$ be the vector field defined by $F(x, y) = (-y, x)$. Evaluate $\int_{C^+} F \cdot d\mathbf{r}$ where
    (a) C^+ is the unit semicircle $\{(x, y) \in \mathbb{R}^2 \mid x^2 + y^2 = 1, y \geqslant 0\}$ oriented from $(1, 0)$ to $(-1, 0)$;

Hint: use the parametrization $(x(t), y(t)) = (\cos t, \sin t)$, $t \in [0, \pi]$.

    (b) C^+ is the same semicircle oriented from $(-1, 0)$ to $(1, 0)$;
    (c) C^+ is the piecewise smooth simple arc formed by straight line segments from $P(0, 0)$ to $Q(-1, 1)$ to $R(0, 2)$;
    (d) C^+ is the straight line segment from $P(0, 0)$ to $R(0, 2)$.

Answers: (a) π; (b) $-\pi$; (c) -2; (d) 0.

5. Let C^+ be the oriented semicircle as in Exercise 4(a). Compute the line integral

$$\int_{C^+} -y \, dx + x \, dy$$

by using (a) the C^1 parametrization $(\cos(t^2), \sin(t^2))$, $t \in [0, \sqrt{\pi}]$, (b) the parametrization $(-t, \sqrt{(1 - t^2)})$, $t \in [-1, 1]$. This parametrization is not C^1. It leads to an improper integral which converges to the line integral.

6. Compute the line integral $\int_{C^+} (x^2 - y) \, dx + (y^2 + x) \, dy$ where C^+ is the parabolic arc $y = x^2 + 1$, $0 \leqslant x \leqslant 1$ oriented from $(0, 1)$ to $(1, 2)$.

Answer: 2.

7. A particle moves in a vector field given by

$$F(x, y, z) = (y, z, x), \qquad (x, y, z) \in \mathbb{R}^3.$$

The work done on the particle by F as it moves along an oriented simple arc C^+ from one end to the other is given by the line integral $\int_{C^+} F \cdot d\mathbf{r}$, where the orientation of C^+ agrees with the direction of motion of the particle.

Calculate the work done on the particle by F as the particle moves
   (a) along straight line segments from $P(0, 0, 0)$ to $Q(0, 1, 1)$ to $R(-1, -1, -2)$;
   (b) along the straight line segment from $P(0, 0, 0)$ to $R(-1, -1, -2)$.

Hint: (a) from Q to R a convenient parametrization is $(-t, 1 - 2t, 1 - 3t)$, $t \in [0, 1]$.

Answers: (a) 6; (b) $\frac{5}{2}$.

8.   Let $f: \mathbb{R}^2 \to \mathbb{R}$ be the scalar field defined by $f(x, y) = x^2 y$, and let C^+ be the unit semicircle $x^2 + y^2 = 1$, $y \geqslant 0$, oriented from $(1, 0)$ to $(-1, 0)$. Compare the integral $\int_C f \, ds$ of f over C (Definition 6.1.1) and the integral $\int_{C^+} f \, dx$.

Answer: $\frac{2}{3}$, $-\pi/8$. The second integral is the integral over C of the component of the vector field $F = (f, 0)$ along C^+ (Fig. 6.2).

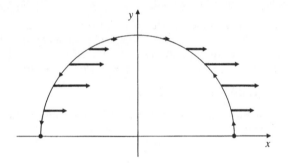

Fig. 6.2 Oriented semicircle C^+ and vector field $(x^2 y, 0)$

9.   Let C^+ be a smooth simple arc, oriented from $\mathbf{v}$ to $\mathbf{w}$ in $\mathbb{R}^n$, let $\sigma: [c, d] \subseteq \mathbb{R} \to \mathbb{R}^n$ be a smooth parametrization of C^+, and let $\alpha: [a, b] \subseteq \mathbb{R} \to \mathbb{R}^n$ be *any* (not necessarily 1-1) C^1 path in C from $\mathbf{v}$ to $\mathbf{w}$.
   (a) Prove that (i) there exists a unique C^1 function ϕ from $[a, b]$ on to $[c, d]$ such that $\alpha = \sigma \circ \phi$, and that (ii) if α is a smooth parametrization of C^+ then $\phi'(p) > 0$ for all $p \in [a, b]$.
   (b) Show that, for any continuous vector field $F: C \subseteq \mathbb{R}^n \to \mathbb{R}^n$,

$$\int_{C^+} F \cdot d\mathbf{r} = \int_\alpha F \cdot d\alpha.$$

Hints: (a) Let $\phi = \sigma^{-1} \circ \alpha$. So $\alpha = \sigma \circ \phi$. Adapt the proof of Theorem 2.6.23 to show that ϕ is differentiable. (b) Using the Chain Rule, Exercise 4.4.8, show that

$$\int_\alpha F \cdot d\alpha = \int_\sigma F \cdot d\sigma = \int_{C^+} F \cdot d\mathbf{r}.$$

This result is illustrated in the following exercise.

10. Let C^+ be the parabolic arc $y = x^2$, $0 \leqslant x \leqslant 1$ oriented from $(1, 1)$ to $(0, 0)$. Show that

(a) the path $\sigma: [-1, 0] \subseteq \mathbb{R} \to \mathbb{R}^2$ defined by

$$\sigma(u) = (-u, u^2), \qquad u \in [-1, 0]$$

is a smooth parametrization of C^+;

(b) the path $\alpha: [0, 3\pi/2] \subseteq \mathbb{R} \to \mathbb{R}^2$ defined by

$$\alpha(t) = (\cos^2 t, \cos^4 t), \qquad t \in [0, 3\pi/2]$$

is a C^1 path that traces C three times, with initial point $(1, 1)$ and terminal point $(0, 0)$.

Verify that

$$\int_{C^+} F \cdot d\mathbf{r} = \int_\sigma F \cdot d\sigma = \int_\alpha F \cdot d\alpha$$

where $F: \mathbb{R}^2 \to \mathbb{R}^2$ is the vector field defined by $F(x, y) = (xy, x)$.

11. Let $f: [a, b] \subseteq \mathbb{R} \to \mathbb{R}$ be a continuous real-valued function. Interpret the Riemann integral $\int_a^b f(x) \, dx$ as the line integral of a vector field along an oriented arc.

Answer: Let I^+ be the interval $I = [a, b]$ oriented from a to b. Then, in the notation of 6.1.15, $\int_a^b f(x) \, dx = \int_{I^+} f \, dx$.

6.2   Integration around simple closed curves

A simple closed curve in $\mathbb{R}^n$ is a closed loop without any self-intersections. It differs from a simple arc only in that it has no end points. In this section we adapt the methods of Sections 2.6 and 6.1 to define the orientation of simple closed curves and to study the integrals associated with them.

6.2.1 Definition. *A subset C of $\mathbb{R}^n$ is called a* simple closed curve *if there is a piecewise C^1 path $\alpha: [a, a + h] \subseteq \mathbb{R} \to \mathbb{R}^n$, whose image is C, such that $\alpha(a) = \alpha(a + h)$ and α is 1–1 on $[a, a + h[$. The* *function* α *is then called a* simple parametrization *of C based at* $\alpha(a)$ in C.

6.2.2 Example.   The circle $x^2 + y^2 = 1$ is a simple closed curve in $\mathbb{R}^2$.
   The simple parametrization of the circle $\alpha(t) = (\cos t, \sin t)$, $t \in [0, 2\pi]$, is based at $(1, 0)$. As t increases from 0 to 2π, the point $\alpha(t)$ completes a single counterclockwise trip around the circle. See Fig. 6.3.

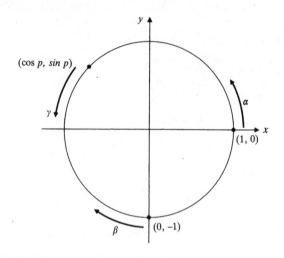

Fig. 6.3

The simple parametrization $\beta(t) = (\sin 2t, \cos 2t)$, $t \in [-\frac{1}{2}\pi, \frac{1}{2}\pi]$ traces the circle clockwise and is based at $(0, -1)$.
   For any fixed $p \in \mathbb{R}$, $\gamma(t) = (\cos(p + t), \sin(p + t))$, $t \in [0, 2\pi]$ is a simple parametrization of the circle associated with a counterclockwise direction. In this case the parametrization is based at $(\cos p, \sin p)$.

6.2.3 Example.   The function

$$\alpha(t) = ((2 + \cos 3t)\cos 2t, (2 + \cos 3t)\sin 2t, \sin 3t), \qquad t \in [0, 2\pi]$$

is a simple parametrization, based at $(3, 0, 0)$, of the *trefoil knot* illustrated in Fig. 6.4. Its projection onto the x, y plane is the image of the path $\alpha_{1,2}(t) = r(t)(\cos 2t, \sin 2t)$, $t \in [0, 2\pi]$, where $r(t) = 2 + \cos 3t$.

As in the case of simple arcs, a simple closed curve may have corners.

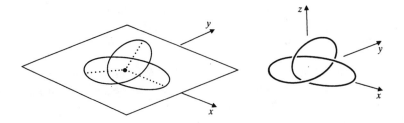

Fig. 6.4 Trefoil knot and projection onto x, y plane

6.2.4 *Example.* The astroid $x^{2/3} + y^{2/3} = 1$ is a simple closed curve in $\mathbb{R}^2$, simply parametrized by

$$\alpha(t) = (\cos^3 t, \sin^3 t), \qquad t \in [0, 2\pi].$$

See Fig. 6.5.

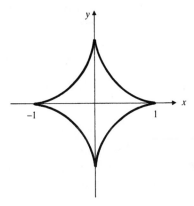

Fig. 6.5 Astroid $x^{2/3} + y^{2/3} = 1$

6.2.5 *Example.* The function

$$\alpha(t) = (\cos t, \sin t, |t|), \qquad t \in [-\pi, \pi]$$

is a simple parametrization of a simple closed curve in $\mathbb{R}^3$ based at $(-1, 0, \pi)$ and lying on the cylinder $x^2 + y^2 = 1$.

In defining the orientation of a simple closed curve C we shall need to know how different parametrizations of C are related. One awkward feature is that different parametrizations may be based at

different points of C, as we have seen in Example 6.2.2. It is to deal
with this problem that we introduce the periodic extension of a
simple parametrization.

6.2.6 Definition.   *Let* $\alpha:[a, a + h] \subseteq \mathbb{R} \rightarrow \mathbb{R}^n$ *be a simple
parametrization of a simple closed curve C. The* periodic extension
of α *is defined to be the function* $\alpha_*:\mathbb{R} \rightarrow \mathbb{R}^n$, *where*

$$\alpha_*(t + mh) = \alpha(t) \qquad \textit{for all } t \in [a, a + h] \textit{ and all } m \in \mathbb{Z}.$$

Since $\alpha(a) = \alpha(a + h)$ and α is continuous, the function α_* is well
defined and continuous. It has period h. It copies the effect of α on
each interval $[a + mh, a + mh + h]$ in $\mathbb{R}$.

The following lemma shows how α_* copies the effect of α on *any*
interval of length h in $\mathbb{R}$.

6.2.7 Lemma.   *Let* $\alpha:[a, a + h] \subseteq \mathbb{R} \rightarrow \mathbb{R}^n$ *be a simple parametrization of a simple closed curve C in* $\mathbb{R}^n$, *and let* $\alpha_*:\mathbb{R} \rightarrow \mathbb{R}^n$ *be
the periodic extension of* α. *Then for any* $p \in \mathbb{R}$, *the function* α_*
restricted to $[p, p + h]$ *is a simple parametrization of C.*

Proof.   Since $\alpha_*(p) = \alpha_*(p + h)$ and α_* is piecewise C^1, we need
only prove that α_* is 1–1 on $[p, p + h[$. Suppose that
$\alpha_*(u) = \alpha_*(v)$. Choose $n, m \in \mathbb{Z}$ such that $u - nh$ and $v - mh$ lie in
$[a, a + h[$. Then

$$\alpha(u - nh) = \alpha_*(u) = \alpha_*(v) = \alpha(v - mh).$$

Now α is 1–1 on $[a, a + h[$. Therefore $u - nh = v - mh$, and so
$u - v$ is an integral multiple of h. Hence, if u and v both lie in
$[p, p + h[$ then $u = v$, which proves that α_* is 1–1 on $[p, p + h[$.

6.2.8 Example.   The circle $x^2 + y^2 = 1$ in $\mathbb{R}^2$ is simply parametrized by
$\alpha:[0, 2\pi] \subseteq \mathbb{R} \rightarrow \mathbb{R}^2$ given by $\alpha(t) = (\cos t, \sin t)$, $t \in [0, 2\pi]$. The periodic
extension $\alpha_*:\mathbb{R} \rightarrow \mathbb{R}^2$ is given by $\alpha_*(u) = (\cos u, \sin u)$, $u \in \mathbb{R}$. For any
fixed $p \in \mathbb{R}$, the function α_* restricted to $[p, p + 2\pi]$ is a simple
parametrization of C. It is based at $(\cos p, \sin p)$, and is illustrated as γ in
Fig. 6.3.

6.2.9 Example.   Let $\alpha:[-\pi, \pi] \subseteq \mathbb{R} \rightarrow \mathbb{R}^2$ be given by $\alpha(t) =$
$(\cos t, \sin t, |t|)$, $t \in [-\pi, \pi]$. The periodic extension $\alpha_*:\mathbb{R} \rightarrow \mathbb{R}^2$ of α is
defined for $u \in \mathbb{R}$ by

$$\alpha_*(u) = (\cos u, \sin u, |u - 2m_u\pi|)$$

where $m_u \in \mathbb{Z}$ is so chosen that $-\pi \leqslant u - 2m_u\pi < \pi$.

We shall compare simple parametrizations of a simple closed curve by comparing their periodic extensions.

6.2.10 Theorem.   *Let $\alpha : [a, a + h] \subseteq \mathbb{R} \to \mathbb{R}^n$ and $\beta : [b, b + k] \subseteq \mathbb{R} \to \mathbb{R}^n$ be simple parametrizations of a simple closed curve C in $\mathbb{R}^n$, and let α_* and β_* be their periodic extensions. Then there exists a strictly monotonic continuous function ϕ_* from $\mathbb{R}$ onto $\mathbb{R}$ such that $\beta_* = \alpha_* \circ \phi_*$. Moreover, if α and β are piecewise smooth then ϕ_* is piecewise smooth.*

Proof.   Put $\alpha(a) = \mathbf{r} \in C$. Then there exists a unique $c \in [b, b + k[$ such that $\beta(c) = \mathbf{r}$. Let $\gamma : [c, c + k] \subseteq \mathbb{R} \to \mathbb{R}^n$ be the restriction of β_* to $[c, c + k]$. By, Lemma 6.2.7, γ is a simple parametrization of C. Since $\gamma(c) = \mathbf{r} = \alpha(a)$, both α and γ are based at $\mathbf{r}$.

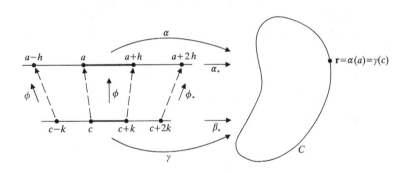

Fig. 6.6

Let γ_0 and α_0 denote respectively the functions γ restricted to the open interval $]c, c + k[$ and the function α restricted to $]a, a + h[$. Both γ_0 and β_0 are 1–1 and their image is $C \backslash \{\mathbf{r}\}$. Therefore the function $\phi_0 = \alpha_0^{-1} \circ \gamma_0$ is 1–1 from $]c, c + k[$ onto $]a, a + h[$. The method used in part (i) of the proof of Theorem 2.6.23 shows that ϕ_0 is continuous as a function from $]c, c + k[$ into the compact interval $[a, a + h]$. Therefore ϕ_0 is strictly monotonic.

Since the image of ϕ_0 is $]a, a + h[$ we can extend ϕ_0 to a continuous strictly monotonic function ϕ from $[c, c + k]$ onto $[a, a + h]$ by taking $\phi(c) = a$ and $\phi(c + k) = a + h$ if ϕ_0 is increasing, and $\phi(c) = a + h$, $\phi(c + k) = a$ if ϕ_0 is decreasing. Since $\gamma_0 = \alpha_0 \circ \phi_0$ on $]c, c + k[$, it is clear that $\gamma = \alpha \circ \phi$ on $[c, c + k]$.

Passing to the periodic extensions γ_*, α_* and β_*, we construct a strictly monotonic continuous function ϕ_* from $\mathbb{R}$ on to $\mathbb{R}$ such that $\beta_* = \alpha_* \circ \phi_*$ as follows:

[i] if ϕ is increasing then define

$$\phi_*(t + mk) = \phi(t) + mk \qquad \text{for all } t \in [c, c+k] \text{ and all } m \in \mathbb{Z}.$$

[ii] if ϕ is decreasing then define

$$\phi_*(t + mk) = \phi(t) - mk \qquad \text{for all } t \in [c, c+k] \text{ and all } m \in \mathbb{Z}.$$

The desired properties of ϕ_* can be verified with reference to Fig. 6.6, where the case ϕ_* increasing is illustrated.

Finally, the piecewise smoothness condition is proved by the method of part (ii) of Theorem 2.6.23.

6.2.11 *Example.* Consider the two simple parametrizations of the circle $x^2 + y^2 = 1$ in $\mathbb{R}^2$ given by $\alpha(t) = (\cos t, \sin t)$, $t \in [0, 2\pi]$ and $\beta(t) = (\sin 2t, \cos 2t)$, $t \in [-\frac{1}{2}\pi, \frac{1}{2}\pi]$. The effect is sketched in Fig. 6.3.

Following through the argument of Theorem 6.2.10 we find that $h = 2\pi$, $k = \pi$, that α is based at $\alpha(0) = (1, 0)$, and that $\beta(c) = (1, 0)$ when $c = \frac{1}{4}\pi$. This leads to $\gamma : [\frac{1}{4}\pi, \frac{5}{4}\pi] \subseteq \mathbb{R} \to \mathbb{R}^2$ given by $\gamma(u) = (\sin 2u, \cos 2u)$. We find the monotonic continuous function ϕ from $[\frac{1}{4}\pi, \frac{5}{4}\pi]$ onto $[0, 2\pi]$ such that $\gamma = \alpha \circ \phi$ by solving

$$(\sin 2u, \cos 2u) = (\cos \phi(u), \sin \phi(u)), \qquad u \in [\tfrac{1}{4}\pi, \tfrac{5}{4}\pi].$$

This gives

$$\phi(u) = \tfrac{5}{2}\pi - 2u, \qquad u \in [\tfrac{1}{4}\pi, \tfrac{5}{4}\pi].$$

Notice that ϕ is monotonic decreasing.

Finally, the function ϕ_* such that $\alpha_* = \beta_* \circ \phi_*$ will be found to satisfy

$$\phi_*(u) = \tfrac{5}{2}\pi - 2u, \qquad u \in \mathbb{R}.$$

The function ϕ_* of Theorem 6.2.10 is not unique. In the above example we could equally well choose

$$\phi_*(u) = \tfrac{1}{2}\pi - 2u, \qquad u \in \mathbb{R}.$$

Referring to Fig. 6.6, this effectively corresponds to a periodic shift of the image of ϕ_* (the domain of α_*) through $h = 2\pi$. The possibilities for ϕ_* are related as follows.

6.2.12 *Theorem.* *Let* $\alpha : [a, a+h] \subseteq \mathbb{R} \to \mathbb{R}^n$ *and* $\beta : [b, b+k] \subseteq \mathbb{R} \to \mathbb{R}^n$ *be simple parametrizations of a simple closed curve* C *in* $\mathbb{R}^n$.

If $\phi_* : \mathbb{R} \to \mathbb{R}$ *and* $\psi_* : \mathbb{R} \to \mathbb{R}$ *are continuous functions such that*

6.2.13 $$\beta_* = \alpha_* \circ \phi_* \qquad and \qquad \beta_* = \alpha_* \circ \psi_*$$

then there exists $m \in \mathbb{Z}$ *such that*

6.2.14 $$\phi_*(u) = \psi_*(u) + mh, \qquad for\ all\ u \in \mathbb{R}.$$

In particular ϕ_* *is strictly increasing (decreasing) if and only if* ψ_* *is strictly increasing (decreasing).*

Proof. By 6.2.13, $\alpha_*(\phi_*(u)) = \alpha_*(\psi_*(u))$ for each $u \in \mathbb{R}$. Therefore, by the definition of α_*, corresponding to each $u \in \mathbb{R}$ there is $m_u \in \mathbb{Z}$ such that

$$\phi_*(u) - \psi_*(u) = m_u h.$$

Now $h > 0$ and so the integer m_u depends continuously upon u. Therefore m_u is constant, and 6.2.14 follows. The rest of the proof is left as exercise.

In view of Theorems 6.2.10 and Theorem 6.2.12 the following relationship between parametrizations of a simple closed curve C is well defined.

6.2.15 Definition. *Let* α *and* β *be two simple parametrizations of a simple closed curve and let* ϕ_* *be a strictly monotonic continuous function from* $\mathbb{R}$ *onto* $\mathbb{R}$ *such that* $\beta_* = \alpha_* \circ \phi_*$. *Then* α *and* β *have the same (opposite) orientation if* ϕ_* *is strictly increasing (decreasing).*

6.2.16 Example. The parametrizations α and β of the unit circle considered in Example 6.2.11 have opposite orientation since ϕ_* is strictly decreasing. Notice that $\alpha(t)$ traces the circle counterclockwise and $\beta(t)$ traces it clockwise.

6.2.17 Example. The simple parametrization α and the inverse parametrization α^- of a simple closed curve have opposite orientation (Exercise 6.2.7).

The relationship of having the same orientation is an equivalence relation on the set of all simple parametrizations of a simple closed curve. There are two equivalence classes—the one containing a parametrization α is denoted by $[\alpha]$.

6.2.18. Definition.   *Let C be a simple closed curve and let α be a simple parametrization of C.*

[i] *The pair* $\{C, [\alpha]\}$ *is called an* oriented simple closed curve. *It will be denoted either by* C^+ *or by* C^-. *If* $C^+ = \{C, [\alpha]\}$ *then* $C^- = \{C, [\alpha^-]\}$ *and vice versa.*

[ii] *Let* $C^+ = \{C, [\alpha]\}$. *Any parametrization in* $[\alpha]$ *is called a* parametrization *of* C^+.

There are therefore, as expected, two orientations of a simple closed curve. The simple parametrizations of the circle $x^2 + y^2 = 1$ in $\mathbb{R}^2$ for example divide into two classes, those that trace the circle clockwise and those that trace the circle counterclockwise.

The following theorems lead to the definition of the integrals of scalar and vector fields around simple closed curves. As in the case of simple arcs the definitions depend upon the integrals being independent of parametrization of the curve.

6.2.19 Theorem.   *Let* $\alpha : [a, a + h] \subseteq \mathbb{R} \to \mathbb{R}^n$ *and* $\beta : [b, b + k] \subseteq \mathbb{R} \to \mathbb{R}^n$ *be simple parametrizations of a simple closed curve C in* $\mathbb{R}^n$ *with the same orientation, and let S be a subset of* $\mathbb{R}^n$ *containing C.*

[i] *For any continuous scalar field* $f : S \subseteq \mathbb{R}^n \to \mathbb{R}$

$$\int_\alpha f \, ds = \int_\beta f \, ds.$$

[ii] *For any continuous vector field* $F : S \subseteq \mathbb{R}^n \to \mathbb{R}^n$

$$\int_\alpha F \cdot d\alpha = \int_\beta F \cdot d\beta.$$

Proof.   We prove the formula for the vector field, leaving the case of the scalar field to the reader.

Find a strictly increasing continuous function $\phi_* : \mathbb{R} \to \mathbb{R}$ such that $\beta_* = \alpha_* \circ \phi_*$.

Let $\phi_*(c) = a$; then $\phi_*(c + k) = a + h$ and taking $\phi_*(c + \frac{1}{2}k) = p$ we have $a < p < a + h$.

The paths $\alpha \mid [a, p]$ and $\beta_* \mid [c, c + \frac{1}{2}k]$ are simple parametrizations of the same simple arc C_1 in S. Since they have the same orientation, by Theorem 6.1.3

6.2.20 $\displaystyle\int_a^p F(\alpha(t)) \cdot \alpha'(t) \, dt = \int_c^{c+1/2k} F(\beta_*(u)) \cdot \beta'_*(u) \, du.$

Similarly,

6.2.21 $$\int_p^{a+h} F(\alpha(t)) \cdot \alpha'(t) \, \mathrm{d}t = \int_{c+1/2k}^{c+k} F(\beta_*(u)) \cdot \beta_*'(u) \, \mathrm{d}u.$$

Adding 6.2.20 and 6.2.21 we find that

$$\int_\alpha F \cdot \mathrm{d}\alpha = \int_\beta F \cdot \mathrm{d}\beta.$$

6.2.22 Corollary.   *Let α and β be simple parametrizations of a simple closed curve C in $\mathbb{R}^n$ with opposite orientations, and let S be a subset of $\mathbb{R}^n$ containing C.*
   [i] *For any continuous scalar field $f : S \subseteq \mathbb{R}^n \to \mathbb{R}$*

$$\int_\alpha f \, \mathrm{d}s = \int_\beta f \, \mathrm{d}s.$$

   [ii] *For any continuous vector field $F : S \subseteq \mathbb{R}^n \to \mathbb{R}^n$*

$$\int_\alpha F \cdot \mathrm{d}\mathbf{r} = -\int_\beta F \cdot \mathrm{d}\mathbf{r}.$$

Proof.   Apply Theorem 6.2.19 to the parametrizations α and β^-.

Theorem 6.2.19 and its corollary show that the following definitions are not ambiguous.

6.2.23 Definition.   *Let $f : S \subseteq \mathbb{R}^n \to \mathbb{R}$ be a continuous scalar field, and let C be a simple closed curve in S. The* integral of f over C *is defined and denoted by*

$$\int_C f \, \mathrm{d}s = \int_\alpha f \, \mathrm{d}s$$

where α is any simple parametrization of C.

6.2.24 Definition.   *Let $F : S \subseteq \mathbb{R}^n \to \mathbb{R}^n$ be a continuous vector field, and let C^+ be an oriented simple closed curve in S. The* integral of F around C^+ *is defined and denoted by*

$$\int_{C^+} F \cdot \mathrm{d}\mathbf{r} = \int_\alpha F \cdot \mathrm{d}\alpha$$

where α is any simple parametrization of C^+.

Corollary 6.2.22(ii) implies that

$$\int_{C^-} F \cdot d\mathbf{r} = -\int_{C^+} F \cdot d\mathbf{r}$$

The integral of a vector field F around an oriented simple closed curve C^+ is often denoted by

$$\oint_C F \cdot d\mathbf{r},$$

the circle through the integral sign indicating that the curve C is closed. As in the case of the integral of F along an oriented simple arc in $\mathbb{R}^n$, the expression $F \cdot d\mathbf{r}$ can be written $F_1 \, dx_1 + \cdots + F_n \, dx_n$, as we have done in Exercises 6.2.

6.2.25 Example. The points $A = (0, 2)$, $B = (-1, 0)$, $C = (0, -2)$, $D = (1, 0)$ lie at the corners of a diamond shape consisting of the straight edges AB, BC, CD and DA. Let C^+ be the oriented simple closed curve consisting of these edges, with counterclockwise orientation. See Fig. 6.7.

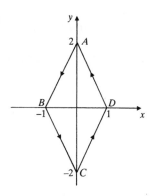

Fig. 6.7

If F is a continuous force field in $\mathbb{R}^2$ then the work done by F on a particle as it moves once around the diamond in a counterclockwise sense is given by $\int_{C^+} F \cdot d\mathbf{r}$. Let us calculate this for the case

$$F(x, y) = (y, -x) \qquad (x, y) \in \mathbb{R}^2.$$

We consider the oriented simple closed curve C^+ as consisting of four oriented simple arcs with appropriate simple parametrizations as follows:

[i] C_1^+ from A to B, parametrized by $\alpha : [0, 1] \subseteq \mathbb{R} \to \mathbb{R}^2$, where $\alpha(t) = (-t, 2 - 2t)$, $t \in [0, 1]$;

[ii] C_2^+ from B to C, parametrized by, $\beta : [0, 1] \subseteq \mathbb{R} \to \mathbb{R}^2$, where $\beta(t) = (-1 + t, -2t)$, $t \in [0, 1]$;

[iii] C_3^+ from C to D, parametrized by $\gamma : [0, 1] \subseteq \mathbb{R} \to \mathbb{R}^2$, where $\gamma(t) = (t, -2 + 2t)$, $t \in [0, 1]$;

[iv] C_4^+ from D to A, parametrized by $\mu : [0, 1] \subseteq \mathbb{R} \to \mathbb{R}^2$, where $\mu(t) = (1 - t, 2t)$, $t \in [0, 1]$.

Hence

$$\int_{C^+} F \cdot d\mathbf{r} = \int_\alpha F \cdot d\alpha + \int_\beta F \cdot d\beta + \int_\gamma F \cdot d\gamma + \int_\mu F \cdot d\mu$$

$$= \int_0^1 -2 \, dt + \int_0^1 -2 \, dt + \int_0^1 -2 \, dt + \int_0^1 -2 \, dt = -8.$$

Similarly, by considering the inverse paths of α, β, γ, and μ we obtain

$$\int_{C^-} F \cdot d\mathbf{r} = 8.$$

As in the case of a simple arc, the integral of scalar field may be used to define the length of a simple closed curve.

6.2.26 Definition. *Let C be a simple closed curve in $\mathbb{R}^n$. The length of C is defined by $l(C) = \int_C 1 \, ds$.*

6.2.27 Example. *Let C be the ellipse $x^2/a^2 + y^2/b^2 = 1$ in $\mathbb{R}^2$ where $a \geq b > 0$. The length of C may be found by using the simple parametrization $\alpha(t) = (a \cos t, b \sin t)$, $t \in [0, 2\pi]$. We obtain*

$$l(C) = l(\alpha) = 4a \int_0^{\pi/2} \sqrt{(1 - k^2 \cos^2 t)} \, dt,$$

where $k^2 = (a^2 - b^2)/a^2$.

Exercises 6.2

1.   Calculate the line integral $\int_{C^+} F \cdot d\mathbf{r}$ where $F(x, y) = (2xy - x^2, x + y^2)$ and
(a) C^+ is the rectangle with corners $(0, 0)$ $(1, 0)$, $(1, 2)$, $(0, 2)$ oriented counterclockwise;
(b) C^+ is the triangle with corners $(0, 0)$, $(1, 0)$, $(0, 2)$ oriented counterclockwise;

(c) C^+ is the closed curve formed by the parabolas $y = x^2$ and $y^2 = x$, $0 \leqslant x \leqslant 1$, $0 \leqslant y \leqslant 1$, with clockwise orientation.

Answers: (a) 0; (b) 1/3; (c) $-1/30$.

2.   Let $f : \mathbb{R}^2 \to \mathbb{R}$ be the scalar field defined by $f(x, y) = 2xy$. Calculate the integrals $\int_C f \, ds$ for the simple closed curves C of Exercise 1.

Answers: (a) 6; (b) $4\sqrt{5}/3$; (c) $(25\sqrt{5} + 1)/30$.

3.   Calculate $\int_{C^+} -y \, dx + x \, dy$, where
(a) C^+ is the unit circle $x^2 + y^2 = 1$ with clockwise orientation;
(b) C^+ is the circle $x^2 + y^2 = c^2$, $c > 0$, with counterclockwise orientation;
(c) C^+ is the counterclockwise oriented ellipse $\frac{1}{2}x^2 + y^2 = 1$;
(d) C^+ is the clockwise oriented unit square, vertices $(0, 0)$, $(0, 1)$, $(1, 1)$, $(1, 0)$.

Answers: (a) -2π; (b) $2\pi c^2$; (c) 4π; (d) -2.

4.   A particle moves in $\mathbb{R}^2$ in a force field given by

$$F(x, y) = \left(\frac{-y}{x^2 + y^2}, \frac{x}{x^2 + y^2} \right), \qquad (x, y) \neq (0, 0).$$

Calculate the work done on the particle by F as the particle
(a)   performs one counterclockwise revolution of the unit circle $x^2 + y^2 = 1$;
(b)   performs one clockwise revolution of the circle $x^2 + y^2 = c^2$;
(c)   performs one clockwise revolution of the circle $(x - 2)^2 + y^2 = 1$.

Answers: compute $\int_{C^+} F \cdot d\mathbf{r}$ for appropriately chosen oriented simple closed curves C^+. (a) 2π; (b) -2π; (c) 0.

5.   Calculate $\int_{C^+} y \, dx + z \, dy + x \, dz$, where C^+ is the ellipse in $\mathbb{R}^3$ $\{(x, y, z) \in \mathbb{R}^3 \mid x^2 + y^2 = 1, \, x + y + z = 1\}$ with orientation $(0, 1, 0) \to (-1, 0, 2) \to (1, 0, 0)$ (counterclockwise orientation viewed from 'above' the ellipse on the z axis).

Answer: -3π. An appropriate simple parametrization of the oriented ellipse is $\alpha(t) = (-\sin t, \cos t, 1 + \sin t - \cos t)$, $t \in [0, 2\pi]$.

6.   Let $f : S \subseteq \mathbb{R}^n \to \mathbb{R}^n$ be a continuous vector field, and let C^+ be an oriented simple closed curve in S. Define the *mean component of F around* C^+ by

$$\frac{1}{l(C)} \int_{C^+} F \cdot d\mathbf{r},$$

where $l(C)$ is the length of C. Find the mean component of the following fields around the circle C^+ centre the origin, radius c, oriented counterclockwise.

(a) $F(x, y) = (-y, x)$;

(b) $F(x, y) = \left(\dfrac{y}{x^2 + y^2}, \dfrac{-x}{x^2 + y^2}\right)$;

(c) $F(x, y) = (y, x)$.

Answers: (a) c; (b) $-1/c$; (c) 0 (sketch the field!).

7. Let $\alpha : [a, a + h] \subseteq \mathbb{R} \to \mathbb{R}^n$ be a simple parametrization of a simple closed curve C. Prove that the inverse parametrization $\alpha^- : [a, a + h] \subseteq \mathbb{R} \to \mathbb{R}^n$ of C defined by

$$\alpha^-(u) = \alpha(2a + h - u), \qquad u \in [a, a + h]$$

has opposite orientation to that of α.

Hint: find the (unique) continuous function ϕ from $[a, a + h]$ onto $[a, a + h]$ such that $\alpha^- = \alpha \circ \phi$, and show that ϕ is strictly decreasing.

10. Let α and β be the simple parametrizations of the ellipse $x^2/a^2 + y^2/b^2 = 1$ given by

$$\alpha(t) = (a \cos t, b \sin t), \qquad t \in [0, 2\pi],$$
$$\beta(t) = (a \sin 2t, -b \cos 2t), \qquad t \in [\pi/3, 4\pi/3]$$

and let α_* and β_* be their periodic extensions. Find a strictly monotonic continuous function ϕ_* from $\mathbb{R}$ onto $\mathbb{R}$ such that $\beta_* = \alpha_* \circ \phi_*$.

Answer: $\phi_*(u) = \tfrac{1}{4}\pi + 2u + m\pi$ for any choice of $m \in \mathbb{Z}$.

6.3 Rotation. Irrotational fields

An important application of the line integral of a vector field around a simple closed curve arises in fluid dynamics. The motion of a fluid (a stream of water, for example) is described by its velocity field F, where $F(\mathbf{r})$ is the velocity of the fluid at the point with position vector $\mathbf{r}$. Imagine a small cork moving with the fluid. Then, besides moving downstream, the cork may also rotate about a vertical axis. The rotation is pronounced close to the outlet of an emptying bath, for example. It occurs less dramatically in a stream as illustrated in Fig. 6.8 where the velocity of the fluid is rather greater near one bank than near the other. It does not occur at all in an evenly flowing stream.

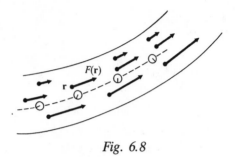

Fig. 6.8

Let $F : S \subseteq \mathbb{R}^2 \to \mathbb{R}^2$ be a continuous vector field modelling the *velocity* on the surface of a fluid. Let C^+ be an oriented simple closed curve in S. The line integral $\int_{C^+} F \cdot d\mathbf{r}$ can be interpreted as the *integral of the tangential component of F around C^+*. If C is the simple closed curve where the cork breaks the surface of the liquid we would expect the cork to rotate in the direction of the orientation of C^+ if $\int_{C^+} F \cdot d\mathbf{r}$ is positive, and in the opposite direction (or not at all) if this integral is negative (or zero). The integral $\int_{C^+} F \cdot d\mathbf{r}$ is called the *circulation of F around C^+*.

Suppose that the curve C is a circle, centre O, radius c. We obtain an absolute measure of the cork's rate of rotation (about a vertical axis through O) by calculating the mean angular velocity about O of the fluid around the circle C^+. Using the relation $v = c\omega$ between the tangential component of velocity v at a point P on C and the angular velocity ω of P about O, we obtain

$$\text{mean angular velocity around } C^+ = \frac{1}{l(C)} \int_{C^+} \left(\frac{F}{c} \right) \cdot d\mathbf{r}$$

$$= \frac{1}{2\pi c^2} \int_{C^+} F \cdot d\mathbf{r}.$$

Notice that πc^2 is the area enclosed by the circle C.

Generalizing, let C^+ be an arbitrary oriented simple closed curve in S. The integral

$$\frac{1}{\text{area enclosed by } C} \int_{C^+} F \cdot d\mathbf{r}$$

is called the *rotation of F around C^+*. It provides a measure in radians per unit time of twice the net rate of turning of the fluid around C^+.

6.3.1 *Example.* Let $F : \mathbb{R}^2 \to \mathbb{R}^2$ be defined by

$$F(x, y) = (1, \tfrac{1}{2}x^2) \qquad (x, y) \in \mathbb{R}^2.$$

Let C^+ be the circle in $\mathbb{R}^2$ centred at (a, b), radius c, with counterclockwise orientation. Using the natural parametrization of C^+ given by

$$(x(t), y(t)) = (a + c \cos t, b + c \sin t), \qquad t \in [0, 2\pi]$$

the circulation of F around C^+ is found to be

$$\int_{C^+} F \cdot d\mathbf{r} = \int_0^{2\pi} (1, \tfrac{1}{2}(a + c \cos t)^2) \cdot (-c \sin t, c \cos t) \, dt$$
$$= ac^2\pi.$$

Hence the rotation of F around C^+ is a. A circular cork centred at (a, b) and moving in the velocity field F would experience a mean angular velocity $2a$ about its centre, the turning effect being clockwise or counterclockwise according as $a < 0$ or $a > 0$. See Fig. 6.9.

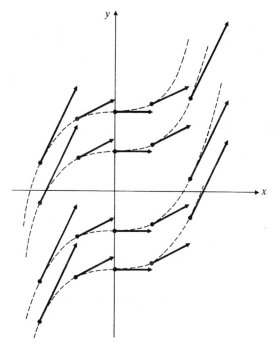

Fig. 6.9 Velocity field $F(x, y) = (1, \tfrac{1}{2}x^2)$

6.3.2 *Example.* Let $F:\mathbb{R}^2\backslash\{0\}\to\mathbb{R}^2$ be defined by

$$F(x, y) = \left(\frac{-y}{x^2 + y^2}, \frac{x}{x^2 + y^2}\right), \qquad (x, y) \neq (0, 0).$$

[i] Let C_1^+ be the circle centred at the origin, radius c, with counterclockwise orientation. We find that

6.3.3 $$\int_{C_1^+} F \cdot d\mathbf{r} = \int_0^{2\pi} 1\, dt = 2\pi.$$

Hence the rotation of F around C_1^+ is $2/c^2$. A sketch of the field F confirms the positive (counterclockwise) rotation. (See Fig. 5.14, page 308.) The rotation increases unboundedly with decreasing radius c. There is a whirlpool effect at the origin.

[ii] Let C_2 be the circle centred at $(a, b) \neq (0, 0)$ and radius $c < \sqrt{(a^2 + b^2)}$. So C_2 does not enclose the origin. In this case we find that, for either orientation of C_2,

6.3.4 $$\int_{C_2^+} F \cdot d\mathbf{r} = 0.$$

Hence the rotation of F around C_2^+ is zero. Referring again to Fig. 5.14, the apparent counterclockwise rotation of the field about the origin is balanced on the circle C_2 by an apparent clockwise effect produced by the diminution of speed with increasing distance from the origin.

The integral 6.3.4 can be established directly from the following theorem.

6.3.5 *Theorem.* *Let $F:D \subseteq \mathbb{R}^2 \to \mathbb{R}^2$ be a conservative vector field defined on an open subset D of $\mathbb{R}^2$. Then the circulation of F around any oriented simple closed curve C^+ in D is zero.*

Proof.   Immediate from Theorem 5.5.21.

6.3.6 *Example.* Returning to Example 6.3.2, the field F given there is conservative on $\mathbb{R}^2\backslash H$, where H is the half-line $\{t(a, b) \in \mathbb{R}^2 \mid t \leqslant 0\}$. See Exercise 5.5.8. This establishes 6.3.4. On the other hand, F is not conservative on $\mathbb{R}^2\backslash\{0\}$. Therefore we cannot use Theorem 6.3.5 to obtain information about the rotation of F around simple closed curves enclosing the origin.

Let $F:D \subseteq \mathbb{R}^2 \to \mathbb{R}^2$ be a continuous velocity field defined on an open subset D of $\mathbb{R}^2$. Fix a point $\mathbf{p} \in D$. We define the *rotation of F*

at **p** by a limiting process. Let C_ε^+ be the circle centred at **p**, radius ε, with counterclockwise orientation, and let S_ε be the circular disc enclosed by C_ε. Then the rotation of F at **p** is defined by

6.3.7 $\qquad (\text{rot } F)(\mathbf{p}) = \lim_{\varepsilon \to 0} (\text{rotation of } F \text{ around } C_\varepsilon^+)$

$$= \lim_{\varepsilon \to 0} \frac{1}{\text{area } S_\varepsilon} \int_{C_\varepsilon^+} F \cdot d\mathbf{r}.$$

provided this limit exists. The number $(\text{rot } F)(\mathbf{p})$ is positive, negative or zero according as the turning effect at **p** is counterclockwise, clockwise or zero.

6.3.8 Example. In Example 6.3.1, where $F(x, y) = (1, \frac{1}{2}x^2)$,

$$(\text{rot } F)(a, b) = a.$$

In Example 6.3.2, where $F(x, y) = (-y/(x^2 + y^2), x/(x^2 + y^2))$, $(x, y) \neq (0, 0)$,

$$(\text{rot } F)(x, y) = 0 \qquad \text{for all } (x, y) \in \mathbb{R}^2 \backslash \{\mathbf{0}\}.$$

This follows from 6.3.4.

6.3.9 Definition. *Let $F: D \subseteq \mathbb{R}^2 \to \mathbb{R}^2$ be a continuous (velocity) field defined on an open subset D of $\mathbb{R}^2$. the field F is said to be irrotational if*

$$(\text{rot } F)(\mathbf{p}) = 0 \qquad \text{for all } \mathbf{p} \in D.$$

It follows from Theorem 6.3.5 that a conservative field $F: D \subseteq \mathbb{R}^2 \to \mathbb{R}^2$, defined on an open subset D of $\mathbb{R}^2$, is irrotational. Example 6.3.8 shows the converse to be false: the field $F: \mathbb{R}^2 \backslash \{\mathbf{0}\} \to \mathbb{R}^2$ given by

6.3.10 $\quad F(x, y) = \left(\dfrac{-y}{x^2 + y^2}, \dfrac{x}{x^2 + y^2} \right), \qquad (x, y) \neq (0, 0)$

is irrotational, but it is not conservative (Example 5.5.28). It is true, however, that if $F: D \subseteq \mathbb{R}^2 \to \mathbb{R}^2$ is a C^1 irrotational field then it is *locally conservative* in the sense that to each $\mathbf{p} \in D$ there corresponds a neighbourhood of **p** in which F is conservative. We cannot expect a more general converse, since the property of a field being irrotational concerns its local structure (individual points and their neighbourhoods), whereas a field being conservative concerns its global structure: it depends on the work done by the field along various simple arcs joining points in its domain.

In this section we have introduced the basic ideas relating to rotational flow in two-dimensional fluids. A large part of the remainder of this book is devoted to developing these ideas. In Chapter 7 we shall consider two-dimensional flow in terms of Green's Theorem. The main result of Chapter 9 (Stokes' Theorem) has important applications to three-dimensional fluid flow. We shall use it to assign to each point **p** in a C^1 velocity field $F : D \subseteq \mathbb{R}^3 \to \mathbb{R}^3$ an absolute measure of the rotational tendency there. This is achieved by the vector-valued function curl $F : D \subseteq \mathbb{R}^3 \to \mathbb{R}^3$, which generalizes the (scalar-valued) function rot F introduced in this section.

Exercises 6.3.

1. Let $F : \mathbb{R}^2 \to \mathbb{R}^2$ be defined by $F(x, y) = \frac{1}{4}(-y, x)$. (The field is illustrated in Fig. 5.2, page 282.) Find the rotation of F around the oriented simple curve C^+, where
    (a) C^+ is the circle $x^2 + y^2 = c^2$ oriented clockwise;
    (b) C^+ is the circle centred at (a, b), radius c, oriented counterclockwise.

Answers: (a) $-\frac{1}{2}$; (b) $\frac{1}{2}$.

2. Repeat Exercise 1 for the field $F(x, y) = (-y(x^2 + y^2), x(x^2 + y^2))$.

Answers: (a) $-2c^2$; (b) $4a^2 + 4b^2 + 2c^2$.

3. For the field $F(x, y) = (-y(x^2 + y^2), x(x^2 + y^2))$, find (rot F)(**p**) at **p** $= (a, b) \in \mathbb{R}^2$.

Answer: $4a^2 + 4b^2$.

4. Let $F : D \subseteq \mathbb{R}^2 \to \mathbb{R}^2$ be a velocity field. A curve in D whose tangent at each of its points **p** is in the direction of $F(\mathbf{p})$ is called a *stream line* or a *field line of* the field F.
    Find the stream lines of the field $F(x, y) = (1, \frac{1}{2}x^2)$. (The field is sketched in Fig. 6.9.)

Answer: suppose that α parametrizes a stream line C of F. Then

$$\alpha'(t) = (\alpha_1'(t), \alpha_2'(t)) = F(\alpha_1(t), \alpha_2(t)) = (1, \tfrac{1}{2}(\alpha_1(t))^2).$$

Integrating each coordinate equation in turn, deduce that the equation of C is $y = \frac{1}{6}x^3 + k$ for some constant k.

5. (a) Sketch the following velocity fields, indicating some stream lines.

Prove that the fields are irrotational.

(i)   The constant field $F:\mathbb{R}^2 \to \mathbb{R}^2$ given by $F(x, y) = (a, b)$, for all $(x, y) \in \mathbb{R}^2$;

(ii)   The field $F(x, y) = (\frac{1}{4}x, \frac{1}{4})$, $(x, y) \in \mathbb{R}^2$.

(iii) The field $F:\mathbb{R}^2\backslash\{\mathbf{0}\} \to \mathbb{R}^2$ given by

$$F(\mathbf{r}) = \frac{1}{r}\mathbf{r}, \qquad \mathbf{r} = (x, y) \in \mathbb{R}^2\backslash\{\mathbf{0}\},$$

where $r = \|\mathbf{r}\|$.

Answer: the fields are all conservative and therefore irrotational.

(b) Let $g:\mathbb{R} \to \mathbb{R}$ be a continuous real-valued function. Prove that the field $F:\mathbb{R}^2 \to \mathbb{R}^2$ defined by

$$F(\mathbf{r}) = g(r)\mathbf{r}, \qquad \mathbf{r} \in \mathbb{R}^2$$

is irrotational.

Answer: $F = \operatorname{grad} f$ where $f(\mathbf{r}) = \int rg(r)\,\mathrm{d}r$. So F is conservative.

6. Show that the field $F:\mathbb{R}^2\backslash\{\mathbf{0}\} \to \mathbb{R}^2$ defined by

$$F(x, y) = \frac{e^x}{x^2 + y^2}(x \sin y - y \cos y, \, x \cos y + y \sin y),$$

$$(x, y) \neq (0, 0),$$

is irrotational but not conservative.

Answer: to prove that F is not conservative consider the oriented unit circle C^+ parametrized by $\alpha(t) = (\cos t, \sin t)$, $t \in [0, 2\pi]$. Then

$$\int_{C^+} F \cdot \mathrm{d}\mathbf{r} = \int_0^{2\pi} e^{\cos t} \cos(\sin t)\,\mathrm{d}t > 0.$$

The apparrently mammoth task of proving that F is irrotational will reduce to a simple calculation, once we have Green's Theorem. See Exercise 7.6.2.

Readers familiar with Complex Variable theory will appreciate the origin of the above exercise. The integral $\int_{C^+} F \cdot \mathrm{d}\mathbf{r}$ is the imaginary part of the contour integral of e^z/z around C^+. The conclusions of the exercise follow from the observation that

$$\int_{C^+} \frac{e^z}{z}\,\mathrm{d}z = \begin{cases} 2\pi i & \text{if } C \text{ encloses the origin,} \\ 0 & \text{otherwise.} \end{cases}$$

Similarly, the field 6.3.10 considered in the text arises from the complex function $1/z$.

7

Double integrals in $\mathbb{R}^2$

7.1 Integral over a rectangle

Let R be the compact rectangle $[a, b] \times [c, d] = \{(x, y) \in \mathbb{R}^2 \mid a \leqslant x \leqslant b, c \leqslant y \leqslant d\}$ in $\mathbb{R}^2$ and let $f : R \subseteq \mathbb{R}^2 \to \mathbb{R}$ be a bounded function. Since f is bounded, there exists (a lower bound) $m \in \mathbb{R}$ and (an upper bound) $M \in \mathbb{R}$ such that

$$m \leqslant f(x, y) \leqslant M, \qquad \text{for all } (x, y) \in R.$$

A partition of $[a, b]$

$$a = x_0 < x_1 < \cdots < x_{k-1} < x_k = b$$

and a partition of $[c, d]$

$$c = y_0 < y_1 < \cdots < y_{l-1} < y_l = d$$

lead to a partition $\mathcal{P}$ of R into kl subrectangles of the form

$$R_{ij} = \{(x, y) \in \mathbb{R}^2 \mid x_{i-1} \leqslant x \leqslant x_i, y_{j-1} \leqslant y \leqslant y_j\}$$

where $i = 1, \ldots, k$ and $j = 1, \ldots, l$. The area of the subrectangle R_{ij} (which is shaded in Fig. 7.1(i)) is

$$A_{ij}(x_i - x_{i-1})(y_j - y_{j-1}) > 0.$$

Since f is bounded on R it is also bounded on each subrectangle. For each i and j let m_{ij} and M_{ij} be respectively the greatest lower bound and the least upper bound of f on R_{ij}. For every choice of $\mathbf{p}_{ij} \in R_{ij}$ we have

7.1.1 $$m_{ij}A_{ij} \leqslant f(\mathbf{p}_{ij})A_{ij} \leqslant M_{ij}A_{ij}.$$

When f is positive on R_{ij}, the left-hand side and right-hand side of 7.1.1 are respectively lower and upper approximations to the volume above R_{ij} bounded by the graph $z = f(x, y)$ (see Fig. 7.1(ii)). This suggests that for any bounded function f on R and any partition $\mathcal{P}$ of R we define

7.1.2 $\quad L(f, \mathcal{P}) = \sum_{i,j} m_{ij}A_{ij}, \qquad$ the lower Riemann sum

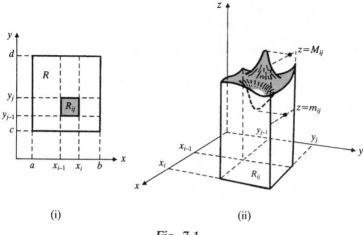

(i)   (ii)

Fig. 7.1

and

7.1.3 $\qquad U(f, \mathcal{P}) = \sum_{i,j} M_{ij} A_{ij}, \qquad$ the upper Riemann sum

of f corresponding to the partition $\mathcal{P}$.

It follows from 7.1.1 that

7.1.4 $\qquad L(f, \mathcal{P}) \leq \sum_{i,j} f(\mathbf{p}_{ij})(x_i - x_{i-1})(y_j - y_{j-1}) \leq U(f, \mathcal{P}).$

7.1.5 *Example.* Let $R = [-1, 0] \times [-1, 0] \subseteq \mathbb{R}^2$ and let $f : R \subseteq \mathbb{R}^2 \to \mathbb{R}$ be defined by $f(x, y) = -x$, $(x, y) \in R$. The graph of f is that part of the plane $z = -x$ sketched in Fig. 7.2.

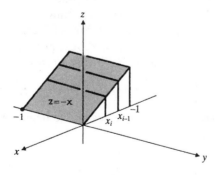

Fig. 7.2

Let V be the volume of the wedge between R and the graph $z = -x$. We show that V can be approximated as closely as we like both by an upper Riemann sum and by a lower Riemann sum.

Given any $\varepsilon > 0$ let $\mathscr{P}$ be the partition of R into rectangles determined by $y_0 = -1$, $y_1 = 0$ and $-1 = x_0 < x_1 < \cdots < x_{k-1} < x_k = 0$ where $x_i - x_{i-1} < \varepsilon$ for all $i = 1, \ldots, k$. For each i, $m_{i1} = -x_i$ and $M_{i1} = -x_{i-1}$. Hence

7.1.6 $$L(f, \mathscr{P}) = \sum_i - x_i(x_i - x_{i-1})(y_1 - y_0) = \sum_i - x_i(x_i - x_{i-1})$$

and

7.1.7 $$U(f, \mathscr{P}) = \sum_i - x_{i-1}(x_i - x_{i-1})(y_1 - y_0) = \sum_i - x_{i-1}(x_i - x_{i-1}).$$

Consideration of Fig. 7.2 shows that

$$L(f, \mathscr{P}) < V < U(f, \mathscr{P}),$$

whereas 7.1.6 and 7.1.7 imply that

$$U(f, \mathscr{P}) - L(f, \mathscr{P}) = \sum_i (x_i - x_{i-1})(x_i - x_{i-1}) < \varepsilon \sum_i (x_i - x_{i-1}) = \varepsilon.$$

Hence both $L(f, \mathscr{P})$ and $U(f, \mathscr{P})$ are within ε of V.

7.1.8 Example.   Let $R = [0, 1] \times [0, 1] \subseteq \mathbb{R}^2$ and let $g : R \subseteq \mathbb{R}^2 \to \mathbb{R}$ be defined by

$$g(x, y) = \begin{cases} 1 & \text{if } x \text{ and } y \text{ are both rational} \\ 0 & \text{otherwise.} \end{cases}$$

Then for any partition $\mathscr{P}$ of R, $L(g, \mathscr{P}) = 0$ and $U(g, \mathscr{P}) = 1$. Hence 7.1.4 is satisfied but in contrast to Example 7.1.5 there is no real number which can be approximated arbitrarily closely both by lower sums and by upper sums.

We say that a partition $\mathscr{P}^*$ of R is a *refinement* of a partition $\mathscr{P}$ if every subrectangle of $\mathscr{P}^*$ is a subset of a subrectangle of $\mathscr{P}$. Suppose now that $\mathscr{P}_1$ and $\mathscr{P}_2$ are partitions of R and that $\mathscr{P}^*$ is the refinement of both $\mathscr{P}_1$ and $\mathscr{P}_2$ obtained by superimposing $\mathscr{P}_1$ on top of $\mathscr{P}_2$. It is not difficult to show, following on from 7.1.4, that

$$L(f, \mathscr{P}_1) \leqslant L(f, \mathscr{P}^*) \leqslant U(f, \mathscr{P}^*) \leqslant U(f, \mathscr{P}_2).$$

The set of lower sums therefore lies on the negative side of the set of upper sums.

7.1.9 Definition.   *A bounded function $f : R \subseteq \mathbb{R}^2 \to \mathbb{R}$ is said to be* (Riemann) integrable *over the compact rectangle R if the least upper* bound of all lower Riemann sums is equal to the greatest lower bound of all upper Riemann sums. *The common value is then called the* double integral of *f* over *R or simply the* integral of *f* over *R*. *It is denoted by*

$$\iint_R f \, \mathrm{d}A \qquad or \ by \qquad \iint_R f(x, y) \, \mathrm{d}x \, \mathrm{d}y.$$

Other common notations are

$$\iint_R f \qquad or \qquad \iint_R f(x, y) \, \mathrm{d}(x, y).$$

7.1.10 Theorem. *A bounded function* $f : R \subseteq \mathbb{R}^2 \to \mathbb{R}$ *is integrable over the compact rectangle R if and only if to each* $\varepsilon > 0$ *there correspond partitions* $\mathscr{P}_1$ *and* $\mathscr{P}_2$ *of R such that*

$$U(f, \mathscr{P}_1) - L(f, \mathscr{P}_2) < \varepsilon,$$

or equivalently, to each $\varepsilon > 0$ *there corresponds a partition* $\mathscr{P}$ *of R such that*

$$U(f, \mathscr{P}) - L(f, \mathscr{P}) < \varepsilon.$$

Proof. Exercise.

7.1.11 Example. The function *f* of Example 7.1.5 is integrable. The function *g* of Example 7.1.8 is not integrable.

7.1.12 Theorem. *Let* $f : R \subseteq \mathbb{R}^2 \to \mathbb{R}$ *and* $g : R \subseteq \mathbb{R}^2 \to \mathbb{R}$ *be integrable bounded functions over the compact rectangle R, and let c be a real number. Then* $f + g$ *and cf are integrable over R and*

$$\iint_R (f + g) \, \mathrm{d}A = \iint_R f \, \mathrm{d}A + \iint_R g \, \mathrm{d}A \qquad and \qquad \iint_R cf \, \mathrm{d}A = c \iint_R f \, \mathrm{d}A$$

Proof. Exercise.

When *f* is a positive continuous function on *R* its integral corresponds to the volume in $\mathbb{R}^3$ above *R* in the *x*, *y* plane and below the graph $z = f(x, y)$.

7.1.13 Example. Considering the function f of Example 7.1.5 we have

$$\iint_R -x \, dx \, dy = V = \tfrac{1}{2}.$$

7.1.14 Theorem. *Let $f:R \subseteq \mathbb{R}^2 \to \mathbb{R}$ be a continuous function on a compact rectangle $R = [a, b] \times [c, d]$ in $\mathbb{R}^2$. Then f is integrable over R.*

Proof. The compactness of R implies that the continuity of f is uniform, by Theorem 4.2.12. to each $\varepsilon > 0$, therefore, there corresponds $\delta > 0$ such that

7.1.15 $|f(\mathbf{v}) - f(\mathbf{w})| < \varepsilon$ whenever $\mathbf{v}, \mathbf{w} \in R$ and $\|\mathbf{v} - \mathbf{w}\| < \delta$.

Let $\mathscr{P}$ be a partition of R such that the diagonal length of each subrectangle R_{ij} of $\mathscr{P}$ is less than δ. Then from 7.1.15,

$$M_{ij} - m_{ij} \leqslant \varepsilon,$$

and, from 7.1.2 and 7.1.3,

$$U(f, \mathscr{P}) - L(f, \mathscr{P}) = \sum_{i,j} (M_{ij} - m_{ij})(x_i - x_{i-1})(y_j - y_{j-1})$$
$$\leqslant \varepsilon(b - a)(d - c).$$

Since $\varepsilon(b - a)(d - c)$ can be chosen arbitrarily small, it follows, by Theorem 7.1.11, that f is integrable over R.

Exercises 7.1

1. Let $R = [0, 1] \times [0, 1] \subseteq \mathbb{R}^2$ and let $f:R \subseteq \mathbb{R}^2 \to \mathbb{R}$ be the continuous function defined by $f(x, y) = xy$, $(x, y) \in R$. For a fixed integer n let $\mathscr{P}$ be the partition of R into n^2 subsquares obtained by partitioning the edges of R into n equal pieces. Show that

$$U(f, \mathscr{P}) = \sum_{i=1}^{n} \sum_{j=1}^{n} \frac{i}{n} \frac{j}{n} \frac{1}{n^2} = \frac{(\tfrac{1}{2}n(n + 1))^2}{n^4},$$

$$L(f, \mathscr{P}) = \sum_{i=1}^{n} \sum_{j=1}^{n} \frac{i-1}{n} \frac{j-1}{n} \frac{1}{n^2} = \frac{(\tfrac{1}{2}(n - 1)n)^2}{n^4}$$

Deduce that

$$\iint_R f \, dA = \tfrac{1}{4}.$$

2. Let $R = [-1, 0] \times [-1, 0] \subseteq \mathbb{R}^2$. Evaluate by the method of Exercise 1
    the integral $\iint_R -x \, dx \, dy$.

Answer: $\frac{1}{2}$, as was shown geometrically in the text.

3. Evaluate (a) $\iint_R x \, dx \, dy$, (b) $\iint_R y \, dx \, dy$, where $R = [0, 1] \times [1, 3] \subseteq \mathbb{R}^2$

Answer: (a) 1; (b) 4.

7.2   Null sets in $\mathbb{R}^2$

There are two questions that arise naturally following the definition
of the double integral in Section 7.1. Firstly, what conditions on a
bounded function, more general than continuity, are sufficient to
ensure that the double integral exists? Secondly, is there a
technique for evaluating the integral which does not involve
returning to first principles? This second question will be considered
in Section 7.3.

We give one answer to the first question in Theorem 7.2.12. The
null subsets needed in the statement of the theorem are defined in
terms of the partitions of $\mathbb{R}^2$ which we now describe.

Two finite sequences of numbers $x_0 < x_1 < \cdots < x_k$ and
$y_0 < y_1 < \cdots < y_l$ lead to two families of grid lines $x = x_i$
$(i = 0, \ldots, k)$ and $y = y_j$ $(j = 0, \ldots, l)$ in $\mathbb{R}^2$. These lines lead to a
partition $\mathscr{P}$ of $\mathbb{R}^2$ into $(k + 2)(l + 2)$ closed rectangles (see Fig. 7.3).

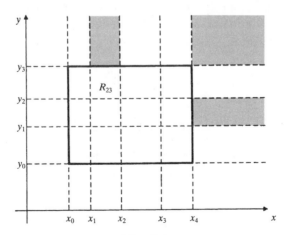

Fig. 7.3

Some of these rectangles are unbounded (the ones shaded in Fig. 7.3 for example). However, kl of them are bounded and therefore compact—each of these being of the form

$$R_{ij} = \{(x, y) \in \mathbb{R}^2 \mid x_{i-1} \leqslant x \leqslant x_i, \ y_{j-1} \leqslant y \leqslant y_j\},$$
$$i = 1, \ldots, k \quad \text{and} \quad j = 1, \ldots, l.$$

7.2.1 Example. The compact rectangles in the above partition of $\mathbb{R}^2$ form a partition of $[x_0, x_k] \times [y_0, y_l]$.

Conversely, any partition of a compact rectangle $[a, b] \times [c, d]$ in $\mathbb{R}^2$ has a natural extension to a partition of $\mathbb{R}^2$.

7.2.2 Example. Let $S_1, \ldots, S_m$ be compact rectangles in $\mathbb{R}^2$, each of whose sides is parallel to one of the coordinate axes. A partition of $\mathbb{R}^2$ can be formed by extending the edges of S_i for each i.

7.2.3 Definition. *If all the grid lines of a partition $\mathcal{P}$ of $\mathbb{R}^2$ are also grid lines of a partition $\mathcal{P}^*$ then $\mathcal{P}^*$ is said to be a refinement of $\mathcal{P}$. Equivalently $\mathcal{P}^*$ refines $\mathcal{P}$ if every rectangle in $\mathcal{P}^*$ is a subset of a rectangle in $\mathcal{P}$.*

7.2.4 Example. Let $\mathcal{P}_1$ and $\mathcal{P}_2$ be partitions of $\mathbb{R}^2$ and let $\mathcal{P}^*$ be obtained by using all the grid lines of both $\mathcal{P}_1$ and $\mathcal{P}_2$. Then $\mathcal{P}^*$ is a refinement of both $\mathcal{P}_1$ and $\mathcal{P}_2$.

Let $\mathcal{P}$ be a partition of $\mathbb{R}^2$. For any non-empty subset A of $\mathbb{R}^2$, there are some rectangles in $\mathcal{P}$ which have points in common with A and some which do not. In Fig. 7.4 those rectangles of a partition

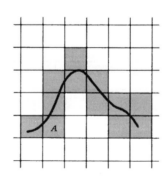

Fig. 7.4

that do meet a simple arc A have been shaded. The following definition is concerned with the total area (possibly infinite) of those rectangles that meet the subset.

7.2.5 Definition.   *Let $\mathcal{P}$ be a partition of $\mathbb{R}^2$. For any subset A of $\mathbb{R}^2$, the sum of the areas of those rectangles of $\mathcal{P}$ whose intersection with A is non-empty is called the* contact *of $\mathcal{P}$ with A and is denoted by $\kappa(A, \mathcal{P})$.*

Notice that $\kappa(A, \mathcal{P}) > 0$ when A is not empty. It is clear that for any bounded set A, a partition of $\mathbb{R}^2$ can be chosen having finite contact with A.

7.2.6 Example.   Let $R = [0, 1] \times [0, 1] \subseteq \mathbb{R}^2$, and let A be the simple closed curve forming the edge of R. For a given natural number n, let $\mathcal{P}_n$ be the partition of $\mathbb{R}^2$ determined by the grid lines

$$x = \frac{r-1}{n} \quad \text{and} \quad y = \frac{r-1}{n}, \quad \text{for } r = 0, 1, \ldots, n+2.$$

See Fig. 7.5. The rectangles which meet A are shaded. The total area of the

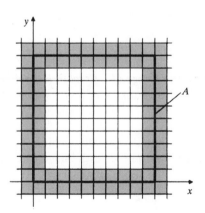

Fig. 7.5

compact rectangles in $\mathcal{P}_n$ is $(1 + 2/n)^2$ and the total area of these compact rectangles that do not meet A is $(1 - 2/n)^2$. Hence

7.2.7
$$\kappa(A, \mathcal{P}_n) = \left(1 + \frac{2}{n}\right)^2 - \left(1 - \frac{2}{n}\right)^2 = \frac{8}{n}.$$

Notice that in Example 7.2.6, by making n large enough we can, by 7.2.7, make $\kappa(A, \mathscr{P}_n)$ as small as we like. This means that A is a null set in the following sense.

7.2.8 Definition.   *A set* $N \subseteq \mathbb{R}^2$ *is said to be a* null set *in* $\mathbb{R}^2$ *if to each* $\varepsilon > 0$ *there corresponds a partition* $\mathscr{P}$ *of* $\mathbb{R}^2$ *such that* $\kappa(N, \mathscr{P}) < \varepsilon$.

In considering properties of null sets it will be helpful to know the following theorems concerning contact.

7.2.9 Theorem.   *For any two subsets A and B of* $\mathbb{R}^2$ *and any partition* $\mathscr{P}$ *of* $\mathbb{R}^2$,

$$\kappa(A, \mathscr{P}) \leqslant \kappa(A \cup B, \mathscr{P}) \leqslant \kappa(A, \mathscr{P}) + \kappa(B, \mathscr{P}).$$

Proof.   Any rectangle of $\mathscr{P}$ which meets A also meets $A \cup B$. Any rectangle of $\mathscr{P}$ which meets $A \cup B$ also meets A or B or both.

7.2.10 Theorem.   *Let A be a subset of* $\mathbb{R}^2$, *and let $\mathscr{P}^*$ be a refinement of a partition $\mathscr{P}$ of* $\mathbb{R}^2$. *Then* $\kappa(A, \mathscr{P}^*) \leqslant \kappa(A, \mathscr{P})$.

Proof.   Exercise.

The following theorem provides important examples of null sets.

7.2.11 Theorem.   *The image of a C^1 path* $\alpha : [a, b] \subseteq \mathbb{R} \to \mathbb{R}^2$ *is a null set in* $\mathbb{R}^2$. *In particular simple arcs and simple closed curves in* $\mathbb{R}^2$ *are null sets.*

Proof.   We omit the details. The idea behind the proof is as follows. Let α be a C^1 path in $\mathbb{R}^2$ with image C. Since α' is continuous the path α has finite length. Hence it is possible to find a partition of $\mathbb{R}^2$ whose contact with C is as small as we please.

The C^1 condition on α cannot be dropped in Theorem 7.2.11. As we remarked in Section 2.6 there exist continuous functions defined on a compact interval $[a, b]$ whose image in $\mathbb{R}^2$ fills a square and therefore is not a null set.

The significance of null sets lies in the following important result answering the first of the two questions asked at the beginning of the section. The theorem generalizes Theorem 7.1.14.

7.2.12 Theorem. *Let* $R = [a, b] \times [c, d]$ *be a compact rectangle in* $\mathbb{R}^2$, *and let* $f : R \subseteq \mathbb{R}^2 \to \mathbb{R}$ *be a bounded function. If the set of points at which f is discontinuous forms a null set in* $\mathbb{R}^2$ *then f is integrable over R.*

Proof. Assume that

$$N = \{(x, y) \subseteq \mathbb{R}^2 \mid (x, y) \in R, f \text{ is not continuous at } (x, y)\}$$

is a null set. Corresponding to any $\varepsilon > 0$ there is a partition $\mathscr{P}$ of $\mathbb{R}^2$ such that

7.2.13 $$\kappa(N, \mathscr{P}) < \varepsilon.$$

We may assume, by Theorem 7.2.10, that the grid lines of $\mathscr{P}$ include $x = a$, $x = b$ and $y = c$, $y = d$. The partition $\mathscr{P}$ is therefore the extension of a partition $\mathscr{Q}$ of the rectangle R.

Let $Q_1, \ldots, Q_m$ be the subrectangles in $\mathscr{Q}$ that meet N and let $R_1, \ldots, R_n$ be the subrectangles in $\mathscr{Q}$ that do not meet N.

The function f is continuous on the compact subset $K = \bigcup_1^n R_j$ of $\mathbb{R}^2$ and hence it is uniformly continuous on K. Therefore there exists $\delta > 0$ such that

7.2.14 $\quad |f(\mathbf{v})| - |f(\mathbf{w})| < \varepsilon$ whenever $\|\mathbf{v} - \mathbf{w}\| < \delta,$ $\qquad \mathbf{v}, \mathbf{w} \in K.$

Let $\mathscr{Q}^*$ be a refinement of $\mathscr{Q}$ such that the diagonal length of each subrectangle in $\mathscr{Q}^*$ is less than δ. We consider the expression

7.2.15 $$U(f, \mathscr{Q}^*) - L(f, \mathscr{Q}^*).$$

The area of K is less than or equal to $(b - a)(d - c)$ and the variation of f on each subrectangle in $\mathscr{Q}^*$ which is a subset of K is, by 7.2.14, less than ε. Hence the contribution to expression 7.2.15 from the subrectangles in $\mathscr{Q}^*$ which are subsets of K is less than $\varepsilon(b - a)(d - c)$.

Similarly the area of $J = \bigcup_1^n Q_i$ is, by 7.2.13, less than ε since $\mathscr{Q}^*$ is a refinement of $\mathscr{Q}$. the variation of f on J is less than or equal to $M - m$ where m and M are respectively a lower and upper bound of f on R. Hence the contribution to expression 7.2.15 from subrectangles in $\mathscr{Q}^*$ which are subsets of J is less than $\varepsilon(M - m)$. Therefore

7.2.16 $\quad U(f, \mathscr{Q}^*) - L(f, \mathscr{Q}^*) < \varepsilon(b - a)(d - c) + \varepsilon(M - m).$

Since, by taking $\varepsilon > 0$ small enough, the right-hand side of 7.2.16

can be made as small as we please, the result follows from Theorem 7.1.10.

7.2.17 *Example.* Let A be the annulus $\{(x, y) \in \mathbb{R}^2 \mid 1 \leqslant x^2 + y^2 < 4\}$ in $\mathbb{R}^2$ and let $R = [-3, 3] \times [-3, 3] \subseteq \mathbb{R}^2$. The function $f : R \subseteq \mathbb{R}^2 \to \mathbb{R}$ defined by

$$f(x, y) = \begin{cases} x^2 + \sin y, & (x, y) \in A, \\ 0, & (x, y) \notin A \end{cases}$$

is bounded and continuous everywhere on R except along the two circles that form the boundary of A. These circles, being simple closed curves, form a null set in $\mathbb{R}^2$ and so f is integrable over R.

7.2.18 *Theorem.* *Let N be a null subset of a compact rectangle R in $\mathbb{R}^2$.*

   [i] *If $f : R \subseteq \mathbb{R}^2 \to \mathbb{R}$ is a bounded function with the property that $f(\mathbf{p}) = 0$ for all $\mathbf{p} \in R \backslash N$, then f is integrable over R and $\iint\limits_R f \, \mathrm{d}A = 0$.*

   [ii] *Let $g : R \subseteq \mathbb{R}^2 \to \mathbb{R}$ be integrable over R and let $h : R \subseteq \mathbb{R}^2 \to \mathbb{R}$ be a bounded function such that $g(\mathbf{p}) = h(\mathbf{p})$ for all $\mathbf{p} \in R \backslash N$. Then h is integrable over R and $\iint\limits_R h \, \mathrm{d}A = \iint\limits_R g \, \mathrm{d}A$.*

Proof. [*i*] Let M and m be respectively upper and lower bounds of a function $f : R \subseteq \mathbb{R}^2 \to \mathbb{R}$. For any $\varepsilon > 0$ find a partition $\mathscr{P}$ of $\mathbb{R}^2$ such that $\mu = \kappa(N, \mathscr{P}) < \varepsilon$. We can assume that the edges of R are grid lines of $\mathscr{P}$. Let $\mathscr{Q}$ be the partition of R determined by $\mathscr{P}$.

If $f(\mathbf{p}) = 0$ for all $\mathbf{p} \in R \backslash N$ then $U(f, \mathscr{Q}) \leqslant M\mu$ and $L(f, \mathscr{Q}) \geqslant m\mu$. Therefore

$$U(f, \mathscr{Q}) - L(f, \mathscr{Q}) \leqslant (M - m)\mu < (M - m)\varepsilon.$$

Hence f is integrable, by Theorem 7.1.10. Also

$$m\mu \leqslant L(f, \mathscr{Q}) \leqslant \iint\limits_R f \, \mathrm{d}A \leqslant U(f, \mathscr{Q}) \leqslant M\mu.$$

Since $0 < \mu < \varepsilon$, therefore $\iint\limits_R f \, \mathrm{d}A = 0$.

   [*ii*] Let $f = h - g$. Then by part (i) f is integrable. Since $h = f + g$,

$$\iint\limits_R h \, \mathrm{d}A = \iint\limits_R f \, \mathrm{d}A + \iint\limits_R g \, \mathrm{d}A = \iint\limits_R g \, \mathrm{d}A.$$

In the final example of this section we show that a function may be integrable over R even though its discontinuities form a set which is not null.

7.2.19 *Example.* Let R be the compact rectangle $[0, 1] \times [0, 1]$ in $\mathbb{R}^2$ and define the subset S of R by

$$S = \{(x, y) \in \mathbb{R}^2 \mid (x, y) \in R, x \text{ and } y \text{ both rational}\}.$$

Let $\mathcal{P}$ be any partition of $\mathbb{R}^2$. Any subrectangle of $\mathcal{P}$ that meets R also meets S. Since the area of R is 1, it follows that $\kappa(S, \mathcal{P}) \geq 1$. Hence S is not a null set in $\mathbb{R}^2$.

Each point of S can be expressed in standard form $(p/q, r/s)$ where p, q, r and s are non-negative integers and the two fractions are in their lowest terms.

Define a function $h : R \subseteq \mathbb{R}^2 \to \mathbb{R}^2$ as follows

$$h(x, y) = \begin{cases} \dfrac{1}{qs} & \text{when } (x, y) = \left(\dfrac{p}{q}, \dfrac{r}{s}\right) \in S, \text{ in standard form,} \\ 0 & \text{when } (x, y) \in R \backslash S. \end{cases}$$

Clearly, for any given $(a, b) \in S$, the function h takes the value 0 as close as we like to (a, b). Therefore h is not continuous at (a, b). Therefore, since S is not a null set, the set of discontinuities of h is not a null set.

To prove that h is integrable over R we consider the difference $U(h, \mathcal{Q}) - L(h, \mathcal{Q})$ for different partitions $\mathcal{Q}$ of R. From the definition of the function h, $L(h, \mathcal{Q}) = 0$ for all $\mathcal{Q}$. Therefore if the integral exists at all, it is 0.

In order to estimate $U(h, \mathcal{Q})$ we shall need to know that

7.2.20 $$0 \leq h(x, y) \leq 1,$$

Corresponding to any ε with $1 > \varepsilon > 0$ consider the set $T = \{(x, y) \in \mathbb{R}^2 \mid (x, y) \in R, h(x, y) \geq \varepsilon\}$. The definition of h implies that T is a finite subset of S. Suppose that T contains n elements. Find a partition $\mathcal{Q}$ of R such that the edge of each subrectangle in $\mathcal{Q}$ has length less than ε/n. The area of each subrectangle is therefore less than ε^2/n^2.

The contribution to $U(h, \mathcal{Q})$ from those subrectangles of $\mathcal{Q}$ which do not meet T is, by definition of T, less that ε (remember the area of R is 1). Similarly, since each point of T lies in at most 4 subrectangles, the total area of the subrectangles of $\mathcal{Q}$ which do meet T is less than $4n(\varepsilon^2/n^2)$. Hence their contribution to $U(h, \mathcal{Q})$ is less than $4n(\varepsilon^2/n^2)$, by 7.2.20. Therefore

$$U(h, \mathcal{Q}) < \varepsilon + 4n(\varepsilon^2/n^2) < 5\varepsilon.$$

Hence

$$U(h, \mathcal{Q}) - L(h, \mathcal{Q}) < 5\varepsilon$$

and therefore h is integrable over R, and $\iint\limits_R h \, dA = 0$.

Theorem 7.2.12 gives an answer to the first of our two questions about the integral of a bounded function over a rectangle. The full answer is given in Exercise 7.2.2(b).

We turn to the second question, concerning a technique for evaluating the integral, in the next section.

Exercises 7.2

1. Let A be the image of the discontinuous function $\phi : [0, 1] \subseteq \mathbb{R} \to \mathbb{R}^2$ defined by $\phi(0) = (0, 0)$ and

    $$\phi(t) = \left(t, \sin\frac{1}{t} \right), \qquad 0 < t \leq 1.$$

    Is A a null set?

    Answer: yes. Notice that for any $0 < \varepsilon < 1$ the function ϕ restricted to $[\varepsilon, 1]$ is C^1.

2. (a) Prove that a subset A of $\mathbb{R}^2$ is a null set if and only if to any $\varepsilon > 0$ there correspond a finite number of compact rectangles R_i, $i = 1, \ldots, n$, of total area not greater than ε such that $A \subseteq \bigcup_1^n R_i$.

    Hint: show that it is sufficient to consider rectangles R_i whose sides are parallel to the axes. Given the existence of such rectangles R_i, $i = 1, \ldots, n$, construct a partition $\mathcal{P}$ such that $\kappa(A, \mathcal{P}) \leq \varepsilon$.
    (b) Let R be the compact rectangle $[0, 1] \times [0, 1]$ in $\mathbb{R}^2$ and let

    $$S = \{(x, y) \in \mathbb{R}^2 \mid (x, y) \in R, \ x \text{ and } y \text{ both rational}\}.$$

    In Example 7.2.19 we saw that S is not a null set. Show, however, that to any $\varepsilon > 0$ there correspond an *infinite* number of compact rectangles $R_i \subseteq R$, $i \in \mathbb{N}$, of total area not exceeding ε, such that $S \subseteq \bigcup_1^\infty R_i$.

    Hint: the set S is countable. Assume an enumeration, and include the first point of S in a rectangle R_1 of area $\varepsilon/2$, the second in R_2 of area $\varepsilon/4, \ldots$, the nth in R_n of area $\varepsilon/2^n, \ldots$. In view of the above property, the set S is said to have *measure zero*.
    A bounded function $f : R \subseteq \mathbb{R}^2 \to \mathbb{R}$ is integrable if and only if the set of discontinuities of f has measure zero.

3. Let $h : R \subseteq \mathbb{R}^2 \to \mathbb{R}$ be the function defined in Example 7.2.19. Prove

that h is continuous at $(a, b) \in R$ if and only if at least one of a, b is irrational.

Hint: given $\varepsilon > 0$, there are only a finite number of rational pairs (x, y) such that $h(x, y) \geq \varepsilon$. Consider a neighbourhood of (a, b) that excludes such rational pairs.

4. Let $R = [0, 1] \times [0, 1] \subseteq \mathbb{R}^2$, let $\mathscr{P}$ be a partition of $\mathbb{R}^2$, and let $S = \{(x, y) \in \mathbb{R}^2 \mid (x, y) \in R, x \text{ and } y \text{ both rational}\}$. In Example 7.2.19 we saw that $\kappa(S, \mathscr{P}) \geq 1$. Show that also $\kappa(R \backslash S, \mathscr{P}) \geq 1$.

7.3   Repeated integrals

Let $f : R \subseteq \mathbb{R}^2 \to \mathbb{R}$ be a positive continuous function on the compact rectangle $R = [a, b] \times [c, d]$ in $\mathbb{R}^2$. By Theorem 7.1.14 the double integral of f over R exists. It measures the volume in $\mathbb{R}^3$ bounded by R in the (x, y)-plane and the graph $z = f(x, y)$. See Fig. 7.6(i).

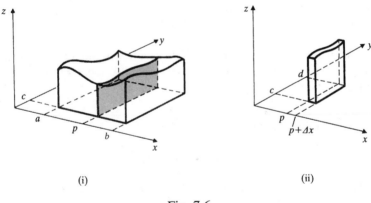

(i)   (ii)

Fig. 7.6

The plane $x = p$ cuts this volume for each $a \leq p \leq b$. The area of the associated cross-section (shaded in Fig. 7.6(i)) is

7.3.1
$$F(p) = \int_c^d f(p, y) \, \mathrm{d}y.$$

Since this integral exists for every p, 7.3.1 defines a function $F : [a, b] \subseteq \mathbb{R} \to \mathbb{R}$.

7.3.2 *Example*.   Consider the function $f : R \subseteq \mathbb{R}^2 \to \mathbb{R}$ defined by

$$f(x, y) = xy, \qquad 0 \leqslant x \leqslant 1, \, 0 \leqslant y \leqslant 1$$

where R is the rectangle $[0, 1] \times [0, 1]$ in $\mathbb{R}^2$.

The associated function $F : [a, b] \subseteq \mathbb{R} \to \mathbb{R}$ defined in 7.3.1 is given by

$$F(x) = \int_0^1 xy \, \mathrm{d}y = [\tfrac{1}{2}xy^2]_{y=0}^{y=1} = \tfrac{1}{2}x.$$

In Fig. 7.6(ii) we have sketched a thin slice of the volume which lies between the planes $x = p$ and $x = p + \Delta x$. The volume of this thin slice is approximately $F(p) \, \Delta x$. This suggests that we have the following expression for the total volume

7.3.3 $$\iint\limits_R f(x, y) \, \mathrm{d}x \, \mathrm{d}y = \int_a^b F(x) \, \mathrm{d}x = \int_a^b \left\{ \int_c^d f(x, y) \, \mathrm{d}y \right\} \mathrm{d}x.$$

The integral on the right-hand side of 7.3.3 is called a *repeated integral*. It is more conveniently denoted by

7.3.4 $$\int_a^b \mathrm{d}x \int_c^d f(x, y) \, \mathrm{d}y.$$

Notice that if 7.3.3 is true then it gives a technique for evaluating a double integral which involves performing two elementary integrations.

7.3.5 *Example*.   Following on from Example 7.3.2,

$$\int_0^1 \mathrm{d}x \int_0^1 xy \, \mathrm{d}y = \int_0^1 \tfrac{1}{2}x \, \mathrm{d}x = [\tfrac{1}{4}x^2]_0^1 = \tfrac{1}{4}.$$

This conclusion is in accordance with 7.3.3 since we have seen in Exercise 7.1.1 that the integral of f over R is also $\tfrac{1}{4}$.

We could repeat the above argument in terms of cross-sections of the volume cut by planes of the form $y = q$, $c \leqslant q \leqslant d$. We would expect that

7.3.6 $$\iint\limits_R f(x, y) \, \mathrm{d}x \, \mathrm{d}y = \int_c^d \left\{ \int_a^b f(x, y) \, \mathrm{d}x \right\} \mathrm{d}y.$$

The integral of the right-hand side of 7.3.6 is again a repeated integral. It is denoted by

7.3.7
$$\int_c^d \mathrm{d}y \int_a^b f(x, y)\, \mathrm{d}x.$$

Comparison of 7.3.3 and 7.3.6 suggests that for well enough behaved functions $f : R \subseteq \mathbb{R}^2 \to \mathbb{R}$,

7.3.8
$$\iint_R f(x, y)\, \mathrm{d}x\, \mathrm{d}y = \int_a^b \mathrm{d}x \int_c^d f(x, y)\, \mathrm{d}y = \int_c^d \mathrm{d}y \int_a^b f(x, y)\, \mathrm{d}x.$$

We shall establish conditions under which 7.3.8 holds later in this section (Theorem 7.3.17). Before we do that, however, it is instructive to contrast the following examples.

7.3.9 Example.

$$\int_0^1 \mathrm{d}x \int_1^3 (x^2 + y)\, \mathrm{d}y = \int_0^1 [x^2 y + \tfrac{1}{2} y]_{y=1}^{y=3}\, \mathrm{d}x = \int_0^1 (2x^2 + 4)\, \mathrm{d}x = \tfrac{14}{3},$$

$$\int_1^3 \mathrm{d}y \int_0^1 (x^2 + y)\, \mathrm{d}x = \int_1^3 (\tfrac{1}{3} + y)\, \mathrm{d}y = \tfrac{14}{3}.$$

7.3.10 Example. Define $f : R \subseteq \mathbb{R}^2 \to \mathbb{R}$ by

$$f(x, y) = \begin{cases} 2y & \text{if } x \text{ irrational,} \\ 1 & \text{if } x \text{ rational,} \end{cases}$$

where $R = [0, 1] \times [0, 1] \subseteq \mathbb{R}^2$. Then

$$\int_0^1 f(x, y)\, \mathrm{d}y = \begin{cases} \displaystyle\int_0^1 2y\, \mathrm{d}y = 1 & \text{if } x \text{ irrational,} \\ \displaystyle\int_0^1 1\, \mathrm{d}y = 1 & \text{if } x \text{ rational.} \end{cases}$$

Hence

$$\int_0^1 \mathrm{d}x \int_0^1 f(x, y)\, \mathrm{d}y = \int_0^1 \mathrm{d}x = 1.$$

On the other hand, for any fixed y, the integral of $f(x, y)$ with respect to x over $[0, 1]$ does not exist. Hence

$$\int_0^1 \mathrm{d}y \int_0^1 f(x, y)\, \mathrm{d}x \text{ does not exist.}$$

We turn now to the general theorems which provide the required technique for evaluating a double integral. The notation given in 7.3.4 and 7.3.7 which was motivated by considering positive continuous functions will be adopted generally whenever the repeated integral exists. We use the notation developed in Section 7.1 in relation to a partition of a compact rectangle.

7.3.11 Theorem. *Let R be the compact rectangle $[a, b] \times [c, d]$ in $\mathbb{R}^2$ and let $f : R \subseteq \mathbb{R}^2 \to \mathbb{R}$ be a bounded function which is integrable over R. Suppose that for each fixed $x \in [a, b]$ the integral*

$$F(x) = \int_c^d f(x, y) \, dy$$

exists. Then the function $F : [a, b] \subseteq \mathbb{R} \to \mathbb{R}$ is integrable and

$$\iint_R f(x, y) \, dx \, dy = \int_a^b dx \int_c^d f(x, y) \, dy.$$

Proof. We first prove that the function F is bounded on $[a, b]$. Let m and M be respectively lower and upper bounds for f on R. Then for each given $p \in [a, b]$ and all $y \in [c, d]$,

$$m \leq f(p, y) \leq M.$$

Hence, by Theorem 1.6.19,

$$m(d - c) \leq \int_c^d f(p, y) \, dy \leq M(d - c).$$

Therefore $m(d - c) \leq F(p) \leq M(d - c)$ and so F is bounded. It follows that the upper Riemann sum $U(F, \Pi)$ and lower Riemann sum $L(F, \Pi)$ of F corresponding to any partition Π of $[a, b]$ exist.

Choose any $\varepsilon > 0$. Since f is integrable over R there exists a partition $\mathscr{P}$ of R such that

7.3.12 $U(f, \mathscr{P}) - L(f, \mathscr{P}) < \varepsilon.$

Suppose that $\mathscr{P}$ derives from a partition $a = x_0 < \cdots < x_{k-1} < x_k = b$ of $[a, b]$ and a partition $c = y_0 < \cdots < y_{l-1} < y_l = d$ of $[c, d]$, and denote the partition of $[a, b]$ by Π.

For each $i = 1, \ldots, k$ choose a point $p_i \in [x_{i-1}, x_i]$. Then for all $j = 1, \ldots, l$ and all y with $y_{j-1} \leq y \leq y_j$, we have in the notation of Section 7.1

$$m_{ij} \leq f(p_i, y) \leq M_{ij}.$$

Hence , by Theorem 1.6.19,

7.3.13 $$m_{ij}(y_j - y_{j-1}) \leqslant \int_{y_{j-1}}^{y_j} f(p_i, y) \, \mathrm{d}y \leqslant M_{ij}(y_j - y_{j-1}).$$

Multiplying the inequalities 7.3.13 by $(x_i - x_{i-1})$ and summing over all i and j we obtain

7.3.14 $$L(f, \mathscr{P}) \leqslant \sum_i F(p_i)(x_i - x_{i-1}) \leqslant U(f, \mathscr{P}).$$

But the inequality 7.3.14 is true for all choices of $p_i \in [x_{i-1}, x_i]$ and so

7.3.15 $$L(f, \mathscr{P}) \leqslant L(F, \Pi) \leqslant U(F, \Pi) \leqslant U(f, \mathscr{P}).$$

Hence, from 7.3.12,

7.3.16 $$U(F, \Pi) - L(F, \Pi) < \varepsilon.$$

Since Π corresponds to the arbitrary $\varepsilon > 0$, 7.3.16 implies that f is integrable over $[a, b]$ and 7.3.15 in turn implies that

$$\int_a^b F(x) \, \mathrm{d}x = \iint_R f(x, y) \, \mathrm{d}x \, \mathrm{d}y,$$

and the proof is complete.

Clearly the above argument can also be used (with the roles of x and y interchanged) to prove the corresponding result concerning a bounded integrable function $f : R \subseteq \mathbb{R}^2 \to \mathbb{R}$ which has the property that

$$\int_a^b f(x, y) \, \mathrm{d}x$$

exists for each $y \in [c, d]$. The reader is recommended to make the necessary alterations to the theorem and its proof.

The two results together give us the following theorem.

7.3.17 Theorem. (*Fubini's Theorem*)   *Let $f : R \subseteq \mathbb{R}^2 \to \mathbb{R}$ be a bounded function which is integrable over a compact rectangle R in $\mathbb{R}^2$. If f is integrable as a function of x for each y and integrable as a function of y for each x then the two repeated integrals exist and are both equal to the double integral of f over R.*

The following theorem is an important special case.

7.3.18 Theorem.   *Let* $f:R \subseteq \mathbb{R}^2 \to \mathbb{R}$ *be continuous on the compact rectangle* R, *where* $R = [a, b] \times [c, d]$. *Then* f *is integrable over* R *and*

$$\iint\limits_R f(x, y) \, \mathrm{d}x \, \mathrm{d}y = \int_a^b \mathrm{d}x \int_c^d f(x, y) \, \mathrm{d}y = \int_c^d \mathrm{d}y \int_a^b f(x, y) \, \mathrm{d}x.$$

Proof.   The continuous function f satisfies the hypothesis of Theorem 7.3.17.

7.3.19 Example.   The function $f:[-1, 0] \times [-1, 0] \subseteq \mathbb{R}^2 \to \mathbb{R}$ defined by $f(x, y) = -x$, $(x, y) \in [-1, 0] \times [-1, 0] = R$ is continuous. hence

$$\iint\limits_R -x \, \mathrm{d}x \, \mathrm{d}y = \int_{-1}^0 \mathrm{d}x \int_{-1}^0 -x \, \mathrm{d}y = \int_{-1}^0 -x \, \mathrm{d}x = \tfrac{1}{2}.$$

The double integral is the volume of the wedge sketched in Fig. 7.2 and considered in Example 7.1.5.

Example 7.3.10 shows that care is needed in general when reversing the order of integration in a repeated integral. Furthermore there exist functions which are integrable but which have the property that neither repeated integral exists. See Exercise 7.3.3.

Exercises 7.3

1.   Let R be the rectangle $[-1, 2] \times [-1, 1]$. For each of the following continuous functions $f:R \subseteq \mathbb{R}^2 \to \mathbb{R}$ evaluate the double integral $\iint\limits_R f \, \mathrm{d}A$ in two ways: (a) as the repeated integral

$\int_{-1}^1 \mathrm{d}y \int_{-1}^2 f(x, y) \, \mathrm{d}x$; (b) as the repeated integral $\int_{-1}^2 \mathrm{d}x \int_{-1}^1 f(x, y) \, \mathrm{d}y$.
   (i) $f(x, y) = xy(x + y)$;
   (ii) $f(x, y) = x^2 \sin y$;
   (iii) $f(x, y) = |x + y|$.

Hint for (iii): $\int_{-1}^2 |x + y| \, \mathrm{d}x = \int_{-1}^{-y} -(x + y) \, \mathrm{d}x + \int_{-y}^2 (x + y) \, \mathrm{d}x.$

Answers: (i) 1; (ii) 0; (iii) 17/3.

2.   Prove that the function f of Example 7.3.10 is not integrable over the square $R = [0, 1] \times [0, 1]$.

Hint: prove that for any partition $\mathscr{P}$ of R, $L(f, \mathscr{P}) < 3/4$ and $U(f, \mathscr{P}) > 5/4$.

3.   Define the function $f:R \subseteq \mathbb{R}^2 \to \mathbb{R}$ over the unit square $R = [0, 1] \times [0, 1]$

by the rule

$$f(x, y) = \begin{cases} 1 & \text{when } x = \tfrac{1}{2}, y \text{ rational } or \ y = \tfrac{1}{2}, x \text{ rational} \\ 0 & \text{otherwise.} \end{cases}$$

Prove that f is integrable over R, but that neither of the repeated integrals $\int_0^1 dy \int_0^1 f(x, y)\, dx$ and $\int_0^1 dx \int_0^1 f(x, y)\, dy$ exists.

Hint: integrability follows from Theorem 7.2.12.

4. Consider the integrable function $h : R \subseteq \mathbb{R}^2 \to \mathbb{R}$ defined in Example 7.2.19. Do the repeated integrals

$$\int_0^1 dy \int_0^1 h(x, y)\, dx \qquad \text{and} \qquad \int_0^1 dx \int_0^1 h(x, y)\, dy$$

exist?

Hint: in the first case fix a rational number y in $[0, 1]$ and show that $\int_0^1 h(x, y)\, dx$ exists by considering lower and upper Riemann sums of h.

Answer: both repeated integrals exist and are zero, since $\int_0^1 h(x, b)\, dx = \int_0^1 h(a, y)\, dy = 0$ for all a and b in $[0, 1]$.

5. Let $f : R \subseteq \mathbb{R}^2 \to \mathbb{R}$ be defined over the rectangle $R = [a, b] \times [c, d]$. Prove that if the double integral $\iint_R f(x, y)\, dx\, dy$ exists then the two repeated integrals

$$\int_a^b dx \int_c^d f(x, y)\, dy \qquad \text{and} \qquad \int_c^d dy \int_a^b f(x, y)\, dx$$

cannot exist without being equal.

7.4   Integrals over general subsets of $\mathbb{R}^2$

The integral of a bounded real-valued function over a bounded subset S of $\mathbb{R}^2$ is defined in terms of the integral of a related function over a compact rectangle containing S.

7.4.1 Definition. *Let* $f : S \subseteq \mathbb{R}^2 \to \mathbb{R}$ *be a function defined on a bounded subset S of $\mathbb{R}^2$. Corresponding to any compact rectangle R such that $S \subseteq R$ the associated function* $f^* : R \subseteq \mathbb{R}^2 \to \mathbb{R}$ *is defined to be the extension of f given by*

$$f^*(\mathbf{p}) = \begin{cases} f(\mathbf{p}) & \text{if} \quad \mathbf{p} \in S, \\ 0 & \text{if} \quad \mathbf{p} \in R \backslash S. \end{cases}$$

7.4.2 Definition.   *Let* $f: S \subseteq \mathbb{R}^2 \to \mathbb{R}$ *be a bounded function defined on a bounded subset S of* $\mathbb{R}^2$ *and let R be a compact rectangle such that* $S \subseteq R$. *The function f is said to be* integrable *over S if the associated function* $f^*: R \subseteq \mathbb{R}^2 \to \mathbb{R}$ *is integrable over R. The* (double) integral *of f over S is then defined to be the integral of* f^* *over R.*

We adopt a natural extension of the notation in Definition 7.1.9 and denote the integral of *f* over *S* by

$$\iint\limits_S f \, dA \qquad \text{or by} \qquad \iint\limits_S f(x, y) \, dx \, dy.$$

The integral is also denoted by

$$\iint\limits_S f \qquad \text{or by} \qquad \iint\limits_S f(x, y) d(x, y).$$

The effect of Definition 7.4.2 is that

$$\iint\limits_S f \, dA = \iint\limits_R f^* \, dA.$$

The criterion for integrability and the value of the integral given in Definition 7.4.2 are independent of the choice of the compact rectangle *R* containing *S*.

7.4.3 Example.   Let *S* be the unit disc $\{(x, y) \in \mathbb{R}^2 \mid x^2 + y^2 \leqslant 1\}$ in $\mathbb{R}^2$ and let $f: S \subseteq \mathbb{R}^2 \to \mathbb{R}$ be the constant function given by $f(\mathbf{p}) = 1$ for all $\mathbf{p} \in S$. The disc is a subset of the compact rectangle $R = [-1, 1] \times [-1, 1]$. The associated function $f^*: R \subseteq \mathbb{R}^2 \to \mathbb{R}$ given by

$$f^*(\mathbf{p}) = 1 \text{ if } \mathbf{p} \in S \qquad \text{and} \qquad f^*(\mathbf{p}) = 0 \text{ if } \mathbf{p} \in R \backslash S$$

is continuous everywhere in *R* except at the points on the boundary of the disc. The boundary is the circle $x^2 + y^2 = 1$ in $\mathbb{R}^2$. The circle, being a simple closed curve, is (by Theorem 7.2.11) a null set. Therefore, by Theorem 7.2.12, f^* is integrable over *R* and hence *f* is integrable over *S*.

We can evaluate the integral by appealing to Theorem 7.3.11 since, for each $x \in [-1, 1]$, $f^*(x, y)$ is integrable as a function of *y*. Indeed, in the notation adopted there

$$F(x) = \int_{-1}^{1} f^*(x, y) \, dy = \int_{-1}^{-\sqrt{(1-x^2)}} 0 \, dy + \int_{-\sqrt{(1-x^2)}}^{\sqrt{(1-x^2)}} 1 \, dy + \int_{\sqrt{(1-x^2)}}^{1} 0 \, dy$$

$$= 2\sqrt{(1 - x^2)}.$$

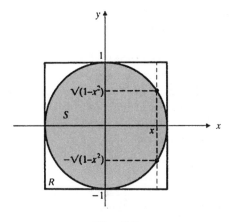

Fig. 7.7

Hence

$$\iint\limits_{S} f(x, y)\, \mathrm{d}x\, \mathrm{d}y = \iint\limits_{R} f^*(x, y)\, \mathrm{d}x\, \mathrm{d}y = \int_{-1}^{1} \mathrm{d}x \int_{-1}^{1} f^*(x, y)\, \mathrm{d}y$$

$$= \int_{-1}^{1} F(x)\, \mathrm{d}x = \int_{-1}^{1} 2\sqrt{(1 - x^2)}\, \mathrm{d}x = \pi.$$

As expected, the integral of the constant function 1 over the unit disc is just the area of the disc.

The above example illustrates the following formal definition of the area of a subset of $\mathbb{R}^2$.

7.4.4 Definition. *Let S be a bounded subset of* $\mathbb{R}^2$. *The* area *of S is defined to be* $\iint\limits_{S} 1\, \mathrm{d}A$, *if that integral exists.*

7.4.5 Example. All null sets in $\mathbb{R}^2$ have zero area. See Theorem 7.2.18(i).

7.4.6 Example. If the area of a bounded set S in $\mathbb{R}^2$ exists then its value is

$$\inf \{\kappa(S,\, \mathscr{P}) \,|\, \text{partitions } \mathscr{P} \text{ of } \mathbb{R}^2\}.$$

This expression, whether S has area or not, is often called the *outer content* of S.

We shall obtain presently (Theorem 7.4.13) a necessary and

sufficient condition for a bounded subset of $\mathbb{R}^2$ to have a meaningful area. We need the following definition.

7.4.7 Definition.   *Let S be a subset of $\mathbb{R}^n$. The point $\mathbf{p} \in \mathbb{R}^n$ is said to be a* boundary point *of S if for each $\varepsilon > 0$ the neighbourhood $N(\mathbf{p}, \varepsilon)$ meets both S and $\mathbb{R}^n \backslash S$. The set of all such boundary points is called the* boundary *of S and is denoted by ∂S.*

7.4.8 *Example*.   It is often useful to know that

7.4.9 $$\partial S = \bar{S} \backslash \text{Int } S.$$

Proof.   First $\mathbf{p} \in \bar{S}$ if and only if, for each $\varepsilon > 0$, $N(\mathbf{p}, \varepsilon)$ meets S (Theorem 3.2.28), and second $\mathbf{p} \in \text{Int } S$ if and only if there exists $\varepsilon > 0$ such that $N(\mathbf{p}, \varepsilon)$ does not meet $\mathbb{R}^2 \backslash S$. this proves 7.4.9.

It follows from 7.4.9, since $\text{Int } S \subseteq \bar{S}$, that

7.4.10 $$S \cup \partial S = \bar{S}.$$

7.4.11 *Example*.   Let $\phi : \mathbb{R} \to \mathbb{R}$ be a continuous function and let

$$S = \{(x, y) \in \mathbb{R}^2 \mid y \leqslant \phi(x)\}.$$

The set S and the graph G of ϕ are sketched in Fig. 7.8. We show that $\partial S = G$.

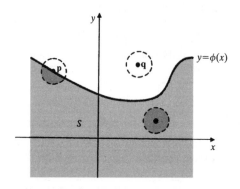

Fig. 7.8

Consider the function $f : \mathbb{R}^2 \to \mathbb{R}$ defined by

$$f(x, y) = \phi(x) - y, \qquad (x, y) \in \mathbb{R}^2.$$

Since ϕ is continuous, the function f is also continuous. Furthermore

$$f^{-1}(0) = G, f^{-1}(]-\infty, 0[) = \mathbb{R}^2\backslash S \quad \text{and} \quad f^{-1}(]0, \infty[) = S\backslash G,$$

and the sets $\mathbb{R}^2\backslash S$ and $S\backslash G$ are open sets in $\mathbb{R}^2$ (Theorem 4.2.10).

First, every point of G lies in ∂S. For let $\mathbf{p} = (t, \phi(t)) \in G$. Then given any $\varepsilon > 0$ the neighbourhood $N(\mathbf{p}, \varepsilon)$ contains the points $\mathbf{p} \in S$ and $(t, \phi(t) + \varepsilon/2) \in \mathbb{R}^2\backslash S$. Hence $\mathbf{p} \in \partial S$.

Second, no point of $\mathbb{R}^2\backslash S$ lies in ∂S. For consider $\mathbf{q} \in \mathbb{R}^2\backslash S$. Since $\mathbb{R}^2\backslash S$ is an open subset of $\mathbb{R}^2$, there exists a neighbourhood of $\mathbf{q}$ lying in $\mathbb{R}^2\backslash S$. This neighbourhood does not meet S and so $\mathbf{q} \notin \partial S$.

Third, by a similar argument, no point of $S\backslash G$ lies in ∂S. Therefore, since $\mathbb{R}^2 = G \cup (\mathbb{R}^2\backslash S) \cup (S\backslash G)$, the boundary of S is G.

The condition of continuity in this example cannot be relaxed. See Exercise 7.4.1.

7.4.12 *Example.* The annulus $S = \{(x, y) \in \mathbb{R}^2 \mid 1 \leqslant x^2 + y^2 < 4\}$ is a bounded set; it lies for example in the rectangle $[-3, 3] \times [-2, 2]$. The boundary of S consists of two concentric circles, $x^2 + y^2 = 1$, which belongs to S, and $x^2 + y^2 = 4$, which does not. Notice that the boundary of S is not in one piece.

The following theorem provides an important class of integrable functions.

7.4.13 *Theorem.* *Let S be a bounded subset of $\mathbb{R}^2$ with null boundary, and let $f : S \subseteq \mathbb{R}^2 \rightarrow \mathbb{R}$ be a bounded function which is continuous everywhere except on a null set in $\mathbb{R}^2$. Then f is integrable over S.*

Proof. Let N be the set of those discontinuities of f that lie in the interior of S. Then N is a null set and therefore $N \cup \partial S$ is also a null set. Let R be a compact rectangle such that $\bar{S} \subseteq \text{Int } R$, and let $f^* : R \subseteq \mathbb{R}^2 \rightarrow \mathbb{R}$ be the extension of f to R which takes the value zero outside S. Then f^* is continuous on $(\text{Int } S)\backslash N$ and on $R\backslash\bar{S}$. Therefore the discontinuities of f^* lie in $N \cup (\bar{S}\backslash\text{Int } S) = N \cup \partial S$, by 7.4.9. Since this is a null set, it follows by Theorem 7.2.12 that f^* is integrable over R. Therefore f is integrable over S.

The following theorem concerns the existence of area of a bounded subset of $\mathbb{R}^2$.

7.4.14 *Theorem.* *Let S be a bounded subset of $\mathbb{R}^2$. Then the integral $\iint\limits_S 1 \, dA$ exists if and only if ∂S is a null set.*

Proof.   If ∂S is a null set then $\iint\limits_{S} 1 \, dA$ exists by Theorem 7.4.13.

Suppose conversely that ∂S is not a null set. Let R be a compact rectangle in $\mathbb{R}^2$ such that $\bar{S} \subseteq \text{Int } R$. Then there exists an $\varepsilon > 0$ such that

7.4.15 $\kappa(\partial S, \mathcal{Q}) > \varepsilon$ for every partition $\mathcal{Q}$ of R.

Define $f : R \subseteq \mathbb{R}^2 \to \mathbb{R}$ to be the function such that $f(\mathbf{p}) = 1$ if $\mathbf{p} \in S$ and $f(\mathbf{p}) = 0$ if $p \in R \backslash S$. (This is the extension of the constant function 1 on S to take the value zero outside S.)

Let $\mathcal{P}$ be any partition of R. We shall prove that there exists a refinement $\mathcal{P}^*$ of $\mathcal{P}$ such that

7.4.16 $$U(f, \mathcal{P}) - L(f, \mathcal{P}) \geqslant \tfrac{1}{2}\kappa(\partial S, \mathcal{P}^*)$$

It then follows from 7.4.16, 7.4.15 and Theorem 7.1.10 that f is not integrable over R and hence that $\iint\limits_{S} 1 \, dA$ does not exist.

It remains to prove 7.4.16 for a suitable refinement $\mathcal{P}^*$ of $\mathcal{P}$. We begin by refining $\mathcal{P}$ through the addition of two grid lines between each pair of adjacent horizontal and vertical grid lines of $\mathcal{P}$ as shown in Fig. 7.9(i), where the new lines are dotted.

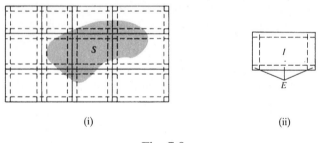

(i)   (ii)

Fig. 7.9

Notice that each rectangle of the refinement of $\mathcal{P}$ is either an *edge rectangle* (that is, it meets the edge of a rectangle of $\mathcal{P}$) or an *interior rectangle* (that is, it lies in the interior of a rectangle of $\mathcal{P}$). In fact, each rectangle of $\mathcal{P}$ is partitioned into one interior rectangle and eight edge rectangles (Fig. 7.9(ii)). Now choose $\mathcal{P}^*$ to be such a refinement of $\mathcal{P}$ for which *the sum of the areas of all edge rectangles is less than* $U(f, \mathcal{P}) - L(f, \mathcal{P})$.

Let $E_1, \ldots, E_n$ be the edge rectangles of $\mathscr{P}^*$ that meet ∂S. Then

7.4.17 $$\sum_1^n \text{area } E_j \leqslant U(f, \mathscr{P}) - L(f, \mathscr{P}).$$

Similarly let $I_1, \ldots, I_k$ be the interior rectangles of $\mathscr{P}^*$ that meet ∂S, and let $B_1, \ldots, B_k$ be the corresponding rectangles of $\mathscr{P}$ in whose interior they lie. Then each B_r has a point of ∂S in its interior and so contains points of both S and $R\backslash S$. Hence $f(B_r) = \{0, 1\}$, and therefore

7.4.18 $$\sum_1^k \text{area } I_r \leqslant \sum_1^k \text{area } B_r \leqslant U(f, \mathscr{P}) - L(f, \mathscr{P}).$$

Since

7.4.19 $$\kappa(\partial S, \mathscr{P}^*) = \sum_1^n \text{area } E_j + \sum_1^k \text{area } I_r$$

the inequality 7.4.16 follows from 7.4.17 and 7.4.18. This completes the proof.

In most applications, functions are integrated over bounded subsets of $\mathbb{R}^2$ whose boundaries consist of a finite number of simple closed curves. By Theorem 7.2.11, such boundaries are null sets. Part (i) of the following example shows, however, that in general arguments we have to be careful of the boundary.

7.4.20 *Example.* Let $S = \{(x, y) \in \mathbb{R}^2 \mid 0 \leqslant x \leqslant 1, 0 \leqslant y \leqslant 1,$ both x and y rational$\}$. Then $\partial S = [0, 1] \times [0, 1]$, which is not a null set.

[i] The function $f : S \subseteq \mathbb{R}^2 \to \mathbb{R}$ given by, $f(x, y) = 1$ for all $(x, y) \in S$ is continuous on S but is not integrable over S. See Example 7.1.8.

[ii] The function $g : S \subseteq \mathbb{R}^2 \to \mathbb{R}$ given by $g(x, y) = 0$ for all $(x, y) \in S$ is integrable over S.

[iii] The function $h : S \subseteq \mathbb{R}^2 \to \mathbb{R}$ given by

$$h(x, y) = \frac{1}{qs} \text{ whenever } (x, y) = \left(\frac{p}{q}, \frac{r}{s}\right) \in S, \text{ in standard form,}$$

is not continuous anywhere in S and yet is integrable over S. See Example 7.2.19.

We must be careful, even when the domain of integration is a bounded open set, for such a set can have a non-null boundary. See Exercise 7.4.3.

The following two theorems describe particularly important cases, where the double integral of f over S can be evaluated in terms of repeated single integrals.

7.4.21 Theorem.   *Let* $\phi:[a, b] \subseteq \mathbb{R} \to \mathbb{R}$ *and* $\psi:[a, b] \subseteq \mathbb{R} \to \mathbb{R}$ *be continuous functions such that* $\phi(x) < \psi(x)$ *for all* $x \in]a, b[$. *Let*

$$S = \{(x, y) \in \mathbb{R}^2 \mid a \leq x \leq b, \ \phi(x) \leq y \leq \psi(x)\}$$

and suppose that the boundary of S *is a simple closed curve. Then any continuous function* $f:S \subseteq \mathbb{R}^2 \to \mathbb{R}$ *is integrable over* S, *and*

$$\iint_S f(x, y)\, dx\, dy = \int_a^b dx \int_{\phi(x)}^{\psi(x)} f(x, y)\, dy.$$

Proof.   Consider the region S sketched in Fig. 7.10. Let

$$c = \inf\{\phi(x) \mid a \leq x \leq b\} \quad \text{and} \quad d = \sup\{\psi(x) \mid a \leq x \leq b\}.$$

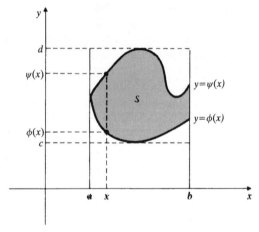

Fig. 7.10 An x-simple region

Then S is a subset of the compact rectangle $R = [a, b] \times [c, d]$. By Theorem 7.2.11 the boundary of S is a null set. Furthermore, f is bounded on S. Therefore, by Theorem 7.4.13 f is integrable over S.

To complete the proof we apply Theorem 7.3.11 to evaluate the integral of f^* over R. We need to know that for each $x \in [a, b]$ the function $f^*(x, y)$ is integrable as a function of y. Now $f^*(x, y) = 0$ if $c \leqslant y < \phi(x)$, $f^*(x, y) = f(x, y)$ if $\phi(x) \leqslant y \leqslant \psi(x)$ and $f^*(x, y) = 0$ if $\psi(x) < y \leqslant d$. Furthermore, for each $x \in [a, b]$, $f(x, y)$ is continuous as a function of $y \in [\phi(x), \psi(x)]$. Therefore $f^*(x, y)$ is integrable as a function of y and

$$\int_c^d f^*(x, y) \, dy = \int_c^{\phi(x)} 0 \, dy + \int_{\phi(x)}^{\psi(x)} f(x, y) \, dy + \int_{\psi(x)}^d 0 \, dy$$

$$= \int_{\phi(x)}^{\psi(x)} f(x, y) \, dy.$$

Hence, by Theorem 7.3.11,

$$\iint_S f(x, y) \, dx \, dy = \iint_R f^*(x, y) \, dx \, dy = \int_a^b dx \int_{\phi(x)}^{\psi(x)} f(x, y) \, dy,$$

and the proof is complete.

A set S of the type defined in Theorem 7.4.21 is called an *x-simple region* (in $\mathbb{R}^2$). See Fig. 7.10

The following theorem is similar to Theorem 7.4.21 and has much the same proof except that the roles of x and y are interchanged.

7.4.22 Theorem. *Let* $\phi : [c, d] \subseteq \mathbb{R} \to \mathbb{R}$ *and* $\psi : [c, d] \subseteq \mathbb{R} \to \mathbb{R}$ *be continuous functions such that* $\phi(y) < \psi(y)$ *for all* $y \in \,]c, d[$. *Let*

$$S = \{(x, y) \in \mathbb{R}^2 \mid c \leqslant y \leqslant d, \; \phi(y) \leqslant x \leqslant \psi(y)\}$$

and suppose that the boundary of S is a simple closed curve. Then any continuous function $f : S \subseteq \mathbb{R}^2 \to \mathbb{R}$ *is integrable over S and*

$$\iint_S f(x, y) \, dx \, dy = \int_c^d dy \int_{\phi(y)}^{\beta(y)} f(x, y) \, dx.$$

Proof.   Exercise.

A set S of the type defined in Theorem 7.4.22 is called a *y-simple region* (in $\mathbb{R}^2$). See Fig. 7.11(i).

The region sketched in Fig. 7.11(ii) is both *x*-simple and *y*-simple. In general, a bounded subset S of $\mathbb{R}^2$ will be neither *x*-simple nor *y*-simple. It may however be possible to integrate a function $f : S \subseteq \mathbb{R}^2 \to \mathbb{R}$ over such a set by subdividing S into subsets which are of one or other of the two types. The integral over S can then be found by adding up the integrals of f over these subsets. As an illustration of this remark consider the following example.

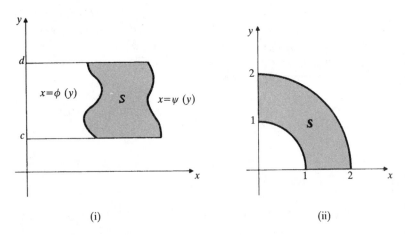

Fig. 7.11 (i) A y-simple region; (ii) An x-simple and y-simple region

7.4.23 Example.   Find the area of the region S enclosed by the parabolas $y^2 = x + 3$ and $y^2 = -2x + 6$ (Fig. 7.12). The parabolas intersect where $x + 3 = -2x + 6$. Hence their points of intersection are $(1, -2)$ and $(1, 2)$. We subdivide S into two x-simple regions A and B by the line $x = 1$, as indicated in Fig. 7.12.

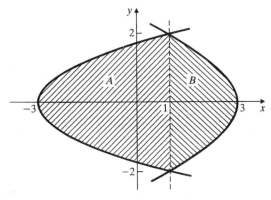

Fig. 7.12

We obtain

$$\text{Area of } S = \iint_A 1 \, dx \, dy + \iint_B 1 \, dx \, dy$$

$$= \int_{-3}^{1} dx \int_{-\sqrt{(x+3)}}^{\sqrt{(x+3)}} 1 \, dy + \int_{1}^{3} dx \int_{-\sqrt{(-2x+6)}}^{\sqrt{(-2x+6)}} 1 \, dy = \tfrac{32}{3} + \tfrac{16}{3} = 16.$$

Alternatively, and more simply, we may regard S as a single y-simple region, and obtain

$$\text{Area of } S = \int_{-2}^{2} dy \int_{y^2-3}^{3-1/2y^2} 1 \, dx = 16.$$

We end this section by recording some useful technical results.

7.4.24 Theorem.   *Let* $f: S \subseteq \mathbb{R}^2 \to \mathbb{R}$ *and* $g: S \subseteq \mathbb{R}^2 \to \mathbb{R}$ *be bounded functions which are integrable over the bounded set* S, *and let* c *be a real number. Then* $f + g$ *and* cf *are integrable over* S, *and*

$$\iint_S (f+g) \, dA = \iint_S f \, dA + \iint_S g \, dA \quad \text{and} \quad \iint_S cf \, dA = c \iint_S f \, dA.$$

Proof.   Apply Definition 7.4.2 to Theorem 7.1.12.

7.4.25. Theorem.   *Let* $f: S \subseteq \mathbb{R}^2 \to \mathbb{R}$ *and* $g: S \subseteq \mathbb{R}^2 \to \mathbb{R}$ *be bounded functions which are integrable over the bounded set* S *and such that* $f(x, y) \leqslant g(x, y)$ *for all* $(x, y) \in S$. *then*

$$\iint_S f \, dA \leqslant \iint_S g \, dA.$$

Proof.   Exercise.

7.4.26 Corollary. *Integral Mean-Value Theorem.*   *Let* S *be a compact subset of* $\mathbb{R}^2$ *with null boundary, and let* $f: S \subseteq \mathbb{R}^2 \to \mathbb{R}$ *be a continuous function. Then* **[i]:**

$$m(\text{area } S) \leqslant \iint_S f \, dA \leqslant M(\text{area } S)$$

where m *and* M *are respectively the greatest lower bound and the least upper bound of f on S, and* **[ii]**: *if S is path connected there exists* $(a, b) \in S$ *such that*

$$\iint_S f \, dA = f(a, b)(\text{area } S).$$

Proof.   Since f is a continuous function on a compact set, both m and M are finite. Furthermore, if S is path connected, then f attains all values between m and M inclusively. Part [i] now follows by applying Theorem 7.4.25 to the inequality $m \leqslant f(x, y) \leqslant M$ for all $(x, y) \in S$. Part [ii] follows from [i].

7.4.27 Theorem.   *Let S and T be disjoint bounded subsets of* $\mathbb{R}^2$ *and let* $f : S \cup T \subseteq \mathbb{R}^2 \to \mathbb{R}$ *be bounded. If f is integrable over S and if f is integrable over T then f is integrable over* $S \cup T$ *and*

$$\iint_{S \cup T} f \, dA = \iint_S f \, dA + \iint_T f \, dA.$$

Proof.   Let R be a compact rectangle such that $S \cup T \subseteq R$. For any subset $B \subseteq S \cap T$ let $f_B : R \subseteq \mathbb{R}^2 \to \mathbb{R}$ be defined by

$$f_B(\mathbf{p}) = \begin{cases} f(\mathbf{p}) & \text{if} \quad \mathbf{p} \in B \\ 0 & \text{if} \quad \mathbf{p} \in R \backslash B. \end{cases}$$

Then, by Definition 7.4.2,

$$\iint_S f \, dA + \iint_T f \, dA = \iint_R f_S \, dA + \iint_R f_T \, dA$$

$$= \iint_R (f_S + f_T) \, dA \qquad \text{by Theorem 7.4.24}$$

$$= \iint_R f_{S \cup T} \, dA \qquad \text{since } S \cap T \text{ is empty}$$

$$= \iint_{S \cup T} f \, dA \qquad \text{by Definition 7.4.2.}$$

The next result demonstrates the irrelevance of null subsets in domains of integration.

7.4.28 Theorem.   *Let S be a bounded subset of* $\mathbb{R}^2$, *let* $N \subseteq S$ *be a null set, and let* $f : S \subseteq \mathbb{R}^2 \to \mathbb{R}$ *and* $g : S \subseteq \mathbb{R}^2 \to \mathbb{R}$ *be bounded functions which agree on* $S \backslash N$.

[i]   *The function f is integrable over S if and only if f is integrable over* $S \backslash N$, *and*

$$\iint_S f \, dA = \iint_{S \backslash N} f \, dA$$

when the integrals exist.

[ii]   *If f is integrable over S then g is integrable over S, and*

$$\iint_S f \, dA = \iint_S g \, dA.$$

Proof.   (*i*)   Let R be a compact rectangle in $\mathbb{R}^2$ such that $S \subseteq R$. Assume that f is integrable over S. Then, in the notation of the proof of Theorem 7.4.27

$$\iint_S f \, dA = \iint_R f_S \, dA.$$

By Theorem 7.2.18[ii]

$$\iint_R f_S \, dA = \iint_R f_{S \backslash N} \, dA,$$

since the functions f_S and $f_{S \backslash N}$ disagree on R only on the null set N. The result follows. The converse is proved similarly.

[*ii*]   If f is integrable over S then

$$\iint_S f \, dA = \iint_{S \backslash N} f \, dA = \iint_{S \backslash N} g \, dA = \iint_S g \, dA.$$

Exercises 7.4

1.   (a)   Let $\phi : \mathbb{R} \to \mathbb{R}$ be the discontinuous function defined by

$$\phi(x) = \begin{cases} 1 & \text{when} \quad x \leqslant 2 \\ 0 & \text{when} \quad x > 2. \end{cases}$$

Define the subset S of $\mathbb{R}^2$ by

7.4.29 $S = \{(x, y) \in \mathbb{R}^2 \,|\, y \leqslant \phi(x)\}.$

Prove that ∂S is the union of the graph of ϕ and the set
$\{(2, y) \in \mathbb{R}^2 \mid 0 \leqslant y \leqslant 1\}$. Compare with Example 7.4.11.
(b)   Let $\phi : \mathbb{R} \to \mathbb{R}$ be defined by $\phi(0) = 0$ and $\phi(x) = \sin(1/x)$, $x \neq 0$,
and let S be given by 7.4.29. Find ∂S.

Answer: ∂S is the union of the graph of ϕ and the set $\{(0, y) \in \mathbb{R}^2 \mid -1 \leqslant y \leqslant 1\}$.

2.   Let R be a compact rectangle in $\mathbb{R}^2$ and let S be a subset of $\mathbb{R}^2$. Prove
that if R meets both S and $R \backslash S$ then R meets the boundary ∂S of S.

Hint: suppose $\mathbf{p} \in R \cap S$ and $\mathbf{q} \in R \backslash S$. Define the line segment L_x by

$$L_x = \{\mathbf{p} + t(\mathbf{q} - \mathbf{p}). \, 0 \leqslant t \leqslant x\}.$$

then $L_x \subseteq R$, $0 \leqslant x \leqslant 1$. Also $L_0 = \{\mathbf{p}\} \subseteq S$, and L_1 meets $R \backslash S$, so
$L_1 \nsubseteq S$. Define

$$c = \sup \{x \in [0, 1] \mid L_x \subseteq S\}.$$

Show that $\mathbf{p} + c(\mathbf{q} - \mathbf{p}) \in (\partial S) \cap R$.

3.   Let R be the unit square $[0, 1] \times [0, 1]$ in $\mathbb{R}^2$. Let $\mathbf{p}_1, \ldots, \mathbf{p}_k, \ldots,$ be
an enumeration of the points in Int R with rational coordinates. For
each k let S_k be an open rectangle in R containing $\mathbf{p}_k$ and of area
$\leqslant 1/2^{k+1}$. Put $S = \bigcup_k S_k$. It is easy to verify that (i) S is a bounded open
subset of $\mathbb{R}^2$; (ii) $\sum_k$ area $S_k \leqslant \frac{1}{2}$; (iii) $\bar{S} = R$. Prove that ∂S, the
boundary of S, is not a null set.

Proof: let $\mathcal{D}$ be any partition of R, and let $R_1, \ldots, R_n$ be the rectangles in
$\mathcal{D}$. Since $\bar{S} = R$, every rectangle R_i meets S. Let $R_1, \ldots, R_m$ be the
rectangles such that $R_j \subseteq S$, $j = 1, \ldots, m$. Then $\sum_1^m$ area $R_j \leqslant \frac{1}{2}$. Every
other rectangle R_i, $i = m + 1, \ldots, n$, meets both S and $R \backslash S$, and
therefore meets ∂S (Exercise 2). Hence $\kappa(\partial S, \mathcal{D}) \geqslant \frac{1}{2}$.

4.   Sketch the elliptical disc

$$S = \{(x, y) \in \mathbb{R}^2 \mid x^2/a^2 + y^2/b^2 \leqslant 1\},$$

where a and b are positive constants.
    Find continuous functions $\phi : [-a, a] \subseteq \mathbb{R} \to \mathbb{R}$ and $\psi : [-a, a] \subseteq \mathbb{R} \to \mathbb{R}$ such that

$$S = \{(x, y) \in \mathbb{R}^2 \mid -a \leqslant x \leqslant a, \, \phi(x) \leqslant y \leqslant \psi(x)\}.$$

Compute $\iint\limits_S f(x, y) \, dx \, dy$ as a repeated integral for the cases

(a) $f(x, y) = 1$, $(x, y) \in S$; (b) $f(x, y) = x^2$, $(x, y) \in S$. The first
integral gives the area of the disc. The second integral measures the
second moment of area of the disc about the y-axis.

Answers: $\phi(x, y) = -b\sqrt{(1 - x^2/a^2)}$, $\psi(x, y) = b\sqrt{(1 - x^2/a^2)}$. (a) πab;
   (b) $\frac{1}{4}\pi ba^3$.

5. By computing a suitable repeated integral, find
    (a) the area enclosed by the parabolas $y = x^2$ and $y = 2 - x^2$;
    (b) the area bounded by the parabola $y^2 = x$ and the line $x = 1$;
    (c) the area enclosed by the parabola $y^2 = 4x$ and the line $x + y = 3$.
    (This is done most simply by a single application of Theorem 7.4.22.
    As an alternative, perform two applications of Theorem 7.4.21,
    leading to $\int_0^1 4\sqrt{x}\, dx + \int_1^9 (3 - x + 2\sqrt{x})\, dx$.)

Answers: (a) 8/3; (b) 4/3; (c) 64/3.

6. Let S be a subset of $\mathbb{R}^2$ with area $\mathcal{A} = \iint\limits_S 1 \, dA$. Let $f : S \subseteq \mathbb{R}^2 \to \mathbb{R}$ be
    integrable over S. Provided $\mathcal{A} \neq 0$, the *mean value of f over S* is
    defined to be the number

    $$\bar{f} = \frac{1}{\mathcal{A}} \iint\limits_S f \, dA.$$

    In particular, the mean values of $f(x, y) = x$ and $f(x, y) = y$ over S give
    the coordinates $(\bar{x}, \bar{y})$ of the *centroid* of S. Find the centroids of the
    regions $S \subseteq \mathbb{R}^2$ whose areas were calculated in Exercise 5(a), (b), (c).

Answers: (a) $(0, 1)$; (b) $(3/5, 0)$; (c) $(17/5, -2)$.

7. (a) Given that f is continuous over $[0, 1] \times [0, 1] \subseteq \mathbb{R}^2$, prove that

    $$\int_0^1 dx \int_x^1 f(x, y) \, dy = \int_0^1 dy \int_0^y f(x, y) \, dx.$$

Hint: Show that the repeated integrals are both equal to $\iint\limits_S f \, dA$ where S is
a suitable triangle in $\mathbb{R}^2$.

    (b) Evaluate $\int_0^1 dx \int_x^1 e^{y^2} \, dy$.

Answer: $\frac{1}{2}(e - 1)$.

8. Sketch the region of integration of

    (a) $\displaystyle\int_0^1 dx \int_x^{2x} f(x, y) \, dy$;   (b) $\displaystyle\int_{-1}^1 dx \int_0^{|x|} f(x, y) \, dy$.

    Deduce that

    (a) $\displaystyle\int_0^1 dx \int_x^{2x} f(x, y) \, dy = \int_0^1 dy \int_{1/2y}^y f(x, y) \, dx + \int_1^2 dy \int_{1/2y}^1 f(x, y) \, dx$;

(b) $\displaystyle\int_{-1}^{1} \mathrm{d}x \int_{0}^{|x|} f(x, y)\, \mathrm{d}y = \int_{0}^{1} \mathrm{d}y \int_{y}^{1} (f(x, y) + f(-x, y))\, \mathrm{d}x.$

Hint for (a): sketch the subset of $\mathbb{R}^2$ enclosed by the lines $x = 0$, $x = 1$, $y = x$ and $y = 2x$.

9. (a) Show that the set S (part of a circular annulus) sketched in Fig. 7.11(ii) is both x-simple and y-simple.

Proof: it is x-simple because $S = \{(x, y) \subseteq \mathbb{R}^2 \mid 0 \leq x \leq 2,\ \phi(x) \leq y \leq \psi(x)\}$, where

$$\phi(x) = \begin{cases} \sqrt{(1 - x^2)} & \text{when} \quad 0 \leq x \leq 1, \\ 0 & \text{when} \quad 1 \leq x \leq 2. \end{cases}$$

and $\psi(x) = \sqrt{(4 - x^2)}$, $0 \leq x \leq 2$. It follows similarly that S is a y-simple region.

(b) Prove that the elliptical region $S = \{(x, y) \in \mathbb{R}^2 \mid x^2/a^2 + y^2/b^2 \leq 1\}$ is both x-simple and y-simple.

10. Check that the x-simple region sketched in Fig. 7.10 is not y-simple, but that it can be subdivided into two y-simple regions.

11. Sketch the annulus $S = \{(x, y) \in \mathbb{R}^2 \mid 1 \leq x^2 + y^2 \leq 4\}$. Show that S can be subdivided into four x-simple regions. Applying Theorem 7.4.21, evaluate the integral $\iint_{S} x^2 y^2\, \mathrm{d}x\, \mathrm{d}y$.

Answer: $21\pi/8$.

12. Evaluate the integral $\iint_{S} x^2 y\, \mathrm{d}x\, \mathrm{d}y$, where S is the annulus

$$\{(x, y) \in \mathbb{R}^2 \mid 1 \leq x^2 + y^2 \leq 4\}.$$

Answer: 0. The result follows directly from the observation that the function $f(x, y) = x^2 y$ is skew-symmetric about the x-axis, since $f(x, -y) = -f(x, y)$, and symmetric about the y-axis, since $f(-x, y) = f(x, y)$.

13. Let $f : S \subseteq \mathbb{R}^2 \to \mathbb{R}$ be integrable over S. Show that f may not be integrable over a subset A in S.

Hint: let $S = [0, 1] \times [0, 1]$, let $f(\mathbf{x}) = 1$ for all $\mathbf{x} \in S$ and let $A = \{(x, y) \in \mathbb{R}^2 \mid (x, y) \in S,\ x \text{ and } y \text{ both rational}\}$.

14. Prove the following generalization of Theorem 7.4.27. Let S and T be bounded sets in $\mathbb{R}^2$ which intersect in a null set, and let $f : S \cup T \subseteq \mathbb{R}^2 \to \mathbb{R}$ be bounded. If f is integrable over S and also over T then f is integrable over $S \cup T$, and

$$\iint_{S \cup T} f\, \mathrm{d}A = \iint_{S} f\, \mathrm{d}A + \iint_{T} f\, \mathrm{d}A.$$

Hint: put $Q = S \backslash (S \cap T)$. Then Q and T are disjoint subsets of $\mathbb{R}^2$. By Theorem 7.4.28 f is integrable over Q. Apply Theorem 7.4.27 to f over $Q \cup T$.

15. Let S be a bounded set in $\mathbb{R}^2$ with null boundary. Let $f : \bar{S} \subseteq \mathbb{R}^2 \to \mathbb{R}$ be a bounded function. Prove that f is integrable over S if and only if f is integrable over $\bar{S}$, and, if the integrals exist, that

$$\iint_S f \, dA = \iint_{\bar{S}} f \, dA.$$

7.5   Green's Theorem

We have seen in Section 6.2 that there are two orientations associated with any simple closed curve. It will be important in this section to be able to distinguish between these orientations without reference to a particular parametrization. The orientations of the unit circle $x^2 + y^2 = 1$ are, for example, counterclockwise (parametrized by $(\cos t, \sin t)$, $t \in [0, 2\pi]$) or clockwise (parametrized by $(\cos t, -\sin t)$, $t \in [0, 2\pi]$). In general, however, such a description is not straightforward. For example, the counterclockwise oriented circle in Fig. 7.13 may be distorted through the plane into either of the two other oriented curves in the figure. In both cases the counterclockwise sense is not so easy to spot.

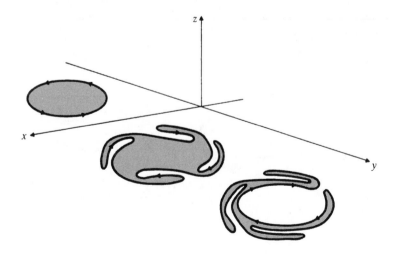

Fig. 7.13

378   *Double integrals in $\mathbb{R}^2$*

In tackling this problem we shall assume without proof the surprisingly deep theorem, the *Jordan Curve Theorem,* which tells us in particular that a simple closed curve in $\mathbb{R}^2$ separates the plane into two subsets, the *inside* and the *outside,* of which it is the common boundary. That is, we shall assume that if C is a simple closed curve in $\mathbb{R}^2$ then $\mathbb{R}^2 \backslash C$ is the union of two disjoint open sets U and V such that $\partial U = \partial V = C$, and U (the inside) is bounded and V (the outside) is unbounded. Figure 7.13, where the insides of the curves are shaded, indicates some of the difficulties of a general argument. For further details see Apostol, *Mathematical Analysis,* Chapter 8.

One intuitive approach to describing an orientation of C is to sketch $\mathbb{R}^3$ with a right handed set of axes—as in Fig. 7.13—and then to identify $(x, y) \in C$ with $(x, y, 0)$ in the sketch. In this way we obtain a sketch of C in the x, y plane in $\mathbb{R}^3$. If we now imagine ourselves walking around C on the z-positive side of the plane—following a given orientation—then the inside of C will lie always to our left or always to our right. In the case of the three oriented curves sketched in Fig. 7.13 for example, the insides lie to the left. On the other hand if the circle were oriented clockwise, the inside would lie to the right.

7.5.1 Definition.   *Let C^+ be an oriented simple closed curve in $\mathbb{R}^2$ with inside $U \subseteq \mathbb{R}^2$. Then C^+ is said to have* counterclockwise *orientation if, on traversing C^+ in the above sense, the set U lies to the left. If U lies to the right, then the orientation is said to be* clockwise.

7.5.2 Example.   All three curves in Fig. 7.13 have counterclockwise orientation.

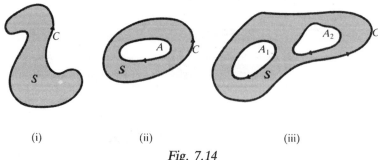

(i)   (ii)   (iii)

Fig. 7.14

In this section we shall be concerned with integrals over certain subsets S of $\mathbb{R}^2$ whose boundary consists of a number of simple closed curves. See Fig. 7.14.

7.5.3 Definition. *Let C be a simple closed curve in $\mathbb{R}^2$, and let $A_1, \ldots, A_k$ be disjoint simple closed curves lying inside C such that for all $i \neq j$, A_i and A_j are outside each other. The subset $D \subseteq \mathbb{R}^2$ of points lying inside C and outside each of $A_1, \ldots, A_k$ is called an* open region *in $\mathbb{R}^2$. The set $S \subseteq \mathbb{R}^2$,*

$$S = \bar{D} = D \cup C \cup A_1 \cup \cdots \cup A_k$$

is called a compact region.

7.5.4 Example. In Fig. 7.14(i) the compact region S is $\bar{U} = U \cup C$ where U is the inside of the simple closed curve C. Two other compact regions are illustrated in Fig. 7.14(ii) and (iii).

Green's Theorem in the Plane, one of the classical theorems of vector analysis, equates the line integral of a C^1 vector field around the boundary of a compact region and the (double) integral of a related continuous scalar field over the region. The theorem runs as follows.

7.5.5 Green's Theorem in the Plane. *Let S be a compact region in $\mathbb{R}^2$ and let $F : S \subseteq \mathbb{R}^2 \to \mathbb{R}^2$ be a C^1 vector field. Then*

$$\int_{\partial S^+} F \cdot d\mathbf{r} = \iint_S \left(\frac{\partial F_2}{\partial x} - \frac{\partial F_1}{dy} \right) dA$$

where F_1 and F_2 are the coordinate functions of F and ∂S^+ is the boundary of S positively oriented in the following sense.

7.5.6 Definition. *Let the compact region $S \subseteq \mathbb{R}^2$ be formed from simple closed curves $C, A_1, \ldots, A_k$, as in Definition 7.5.3. Then the positively oriented boundary ∂S^+ of S is*

$$\partial S = C \cup A_1 \cup \cdots \cup A_k$$

together with an orientation of the simple closed curves $C, A_1, \ldots, A_k$ such that traversing any one of them in the direction of its orientation keeps the region S on the left. This involves orienting C counterclockwise and $A_1, \ldots, A_k$ clockwise.

7.5.7 *Example*.   In Fig. 7.14 the positively oriented boundaries of the shaded regions S are indicated. Notice that in (ii), for example, the inner edge A is oriented clockwise and the outer edge C counterclockwise.

7.5.8 *Example*.   Let $U \subseteq \mathbb{R}^2$ be the inside of a simple closed curve C in $\mathbb{R}^2$. Then $S = \bar{U} = U \cup C$ is a compact region and the positively oriented boundary of S is C with counterclockwise orientation. See for example Figs. 7.13 and 7.14(i).

We can use Green's Theorem to obtain a definition of the counterclockwise orientation of a simple closed curve C which is independent of a physical model of the curve. For let C^+ denote the curve with *either* of its two possible orientations. Then by Green's Theorem, with $F(x, y) = (0, x)$,

$$\int_{C^+} (0, x) \cdot \mathrm{d}\mathbf{r} = \pm \iint_S 1 \, \mathrm{d}A = \pm \text{area } S$$

where S is the compact region whose interior is the inside of C. We could therefore *define* C^+ as having counterclockwise orientation if the integral $\int_{C^+} (0, x) \cdot \mathrm{d}\mathbf{r}$ is positive. In this case the value of the line integral is the area of the inside of C.

We shall not prove Green's Theorem in its full generality. In this section we consider a number of special cases.

We begin by considering Green's Theorem for x-simple regions. Let

$$S = \{(x, y) \in \mathbb{R}^2 \mid a \leqslant x \leqslant b, \ \phi(x) \leqslant y \leqslant \psi(x)\}$$

be an x-simple region. Then S is a compact region, and its boundary is the simple closed curve made up of four simple curves B_1, B_2, B_3,

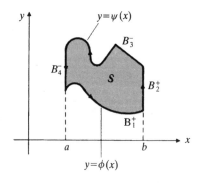

Fig. 7.15

B_4 as illustrated in Fig. 7.15. Assign to these curves orientations as follows:

$B_1^+ = \{(x, y) \in \mathbb{R}^2 \mid y = \phi(x)\}$ oriented from $(a, \phi(a))$ to $(b, \phi(b))$

$B_2^+ = \{(x, y) \in \mathbb{R}^2 \mid x = b, \phi(b) \leqslant y \leqslant \psi(b)\}$

oriented from $(b, \phi(b))$ *to* $(b, \psi(b))$

$B_3^+ = \{(x, y) \in \mathbb{R}^2 \mid y = \psi(x)\}$ oriented from $(a, \psi(a))$ to $(b, \psi(b))$

$B_4^+ = \{(x, y) \in \mathbb{R}^2 \mid x = a, \phi(a) \leqslant y \leqslant \psi(a)\}$

oriented from $(a, \phi(a))$ to $(a, \psi(a))$.

Then the positively oriented boundary ∂S^+ of S is formed from B_1^+, B_2^+, B_3^-, B_4^-.

For any continuous vector field $F : S \subseteq \mathbb{R}^2 \to \mathbb{R}^2$,

7.5.9 $\displaystyle\int_{\partial S^+} F \cdot d\mathbf{r} = \int_{B_1^+} F \cdot d\mathbf{r} + \int_{B_2^+} F \cdot d\mathbf{r} + \int_{B_3^-} F \cdot d\mathbf{r} + \int_{B_4^-} F \cdot d\mathbf{r}.$

Suppose now that F is a C^1 vector field on S of the form

$$F(x, y) = (F_1(x, y), 0), \qquad (x, y) \in S$$

where $F_1 : S \subseteq \mathbb{R}^2 \to \mathbb{R}$ has continuous partial derivatives.

Let $\alpha : [p, q] \subseteq \mathbb{R} \to \mathbb{R}^2$ be a C^1 parametrization of B_1^+. Then

$$\begin{aligned}
\int_{B_1^+} F \cdot d\mathbf{r} &= \int_p^q F_1(\alpha(t)) \alpha_1'(t) \, dt \\
&= \int_p^q F_1(\alpha_1(t), \phi(\alpha_1(t))) \alpha_1'(t) \, dt \\
&= \int_a^b F_1(x, \phi(x)) \, dx.
\end{aligned}$$

Similarly

$$\int_{B_3^+} F \cdot d\mathbf{r} = \int_a^b F_1(x, \psi(x)) \, dx.$$

On the other hand, since F is orthogonal to B_2^+ and B_4^+,

$$\int_{B_2^+} F \cdot d\mathbf{r} = \int_{B_4^+} F \cdot d\mathbf{r} = 0.$$

Hence, from 7.5.9,

$$\int_{\partial S^+} (F_1, 0) \cdot d\mathbf{r} = \int_a^b F_1(x, \phi(x)) \, dx - \int_a^b F_1(x, \psi(x)) \, dx$$

and therefore, since $\partial F_1/\partial y$ is continuous on S,

7.5.10 $$\int_{\partial S^+} (F_1, 0) \cdot d\mathbf{r} = - \int_a^b dx \int_{\phi(x)}^{\psi(x)} \frac{\partial F_1}{\partial y} (x, y) \, dy.$$

We therefore have the following theorem.

7.5.11 Theorem. *Let S be an x-simple region in $\mathbb{R}^2$. Then for any C^1 function $F_1 : S \subseteq \mathbb{R}^2 \to \mathbb{R}$*

$$\int_{\partial S^+} (F_1, 0) \cdot d\mathbf{r} = - \iint_S \frac{\partial F_1}{\partial y} \, dA$$

where ∂S^+ is the positively oriented boundary of S.

Proof. The theorem follows from 7.5.10 and Theorem 7.4.21.

Theorem 7.5.11 is matched by a similar result in which the roles of x and y are interchanged. It is clear that a y-simple region is also a compact region.

7.5.12 Theorem. *Let S be a y-simple region in $\mathbb{R}^2$. Then for any C^1 function $F_2 : S \subseteq \mathbb{R}^2 \to \mathbb{R}$*

$$\int_{\partial S^+} (0, F_2) \cdot d\mathbf{r} = \iint_S \frac{\partial F_2}{\partial x} \, dA$$

where ∂S^+ is the positively oriented boundary of S.

Proof. Let $\phi : [c, d] \subseteq \mathbb{R} \to \mathbb{R}$ and $\psi : [c, d] \subseteq R \to R$ be piecewise C^1 functions such that $\phi(y) < \psi(y)$ for all $y \in]c, d[$ and let

$$S = \{(x, y) \in \mathbb{R}^2 \mid c \leq y \leq d, \, \phi(y) \leq x \leq \psi(y)\}.$$

An argument similar to the one used for x-simple regions

established that

$$\int_{\partial S^+} (0, F_2) \cdot d\mathbf{r} = \int_c^d F_2(\psi(y), y) \, dy - \int_c^d F_2(\phi(y), y) \, dy$$

$$= \int_c^d dy \int_{\phi(y)}^{\psi(y)} \frac{\partial F_2}{\partial x}(x, y) \, dx.$$

The result follows from Theorem 7.4.22.

We obtain an important special case of Green's Theorem by combining Theorems 7.5.11 and 7.5.12.

7.5.13 Definition. *A subset of $\mathbb{R}^2$ is said to be a simple region if it is both x-simple and y-simple.*

An ellipse together with its inside is a simple region, as are the sets sketched in Figures 7.11(ii) and 7.12. A simple region is clearly a compact region.

7.5.14 Green's Theorem for simple regions. *Let $S \subseteq \mathbb{R}^2$ be a simple region and let $F : S \subseteq \mathbb{R}^2 \to \mathbb{R}^2$ be a C^1 vector field. Then*

7.5.15
$$\int_{\partial S^+} F \cdot d\mathbf{r} = \iint_S \left(\frac{\partial F_2}{\partial x} - \frac{\partial F_1}{\partial y} \right) dA$$

where ∂S^+ is the positively oriented boundary of S.

Proof. For all $(x, y) \in S$

$$F(x, y) = (F_1(x, y), F_2(x, y)) = (F_1(x, y), 0) + (0, F_2(x, y)).$$

Hence

$$\int_{\partial S^+} F \cdot d\mathbf{r} = \int_{\partial S^+} (F_1, 0) \cdot d\mathbf{r} + \int_{\partial S^+} (0, F_2) \cdot d\mathbf{r}.$$

Theorem 7.5.14 now follows from Theorems 7.5.11 and 7.5.12 since both coordinate functions F_1 and F_2 have continuous partial derivatives.

Expression 7.5.15 is commonly written in the form

7.5.16
$$\int_{\partial S^+} (F_1 \, dx + F_2 \, dy) = \iint_S \left(\frac{\partial F_2}{\partial x} - \frac{\partial F_1}{\partial y} \right) dx \, dy.$$

7.5.17 *Example.* Let $R = [-1, 0] \times [-1, 0] \subseteq \mathbb{R}^2$ and let $F : R \to \mathbb{R}^2 \to \mathbb{R}^2$ be given by

$$F(x, y) = (xy, 0), \qquad (x, y) \in R.$$

We have seen in Example 7.3.19 that

$$\iint_R \left(\frac{\partial F_2}{\partial x} - \frac{\partial F_1}{\partial y} \right) dA = \iint_R -x \, dx \, dy = \tfrac{1}{2}.$$

On the other hand it is not difficult to see that

$$\int_{\partial R^+} F \cdot d\mathbf{r} = \int_\alpha F \cdot d\mathbf{r} = \int_0^1 (1 - t, 0) \cdot (-1, 0) \, dt = \tfrac{1}{2}$$

where $\alpha : [0, 1] \subseteq \mathbb{R} \to \mathbb{R}^2$ is the parametrization of the edge of R from $(-1, -1)$ to $(0, -1)$ given by $\alpha(t) = (t - 1, -1)$, $t \in [0, 1]$. The line integrals along the other edges are all zero.

As in the argument following Example 7.5.8, we can find the area of a compact region S by evaluating the line integral of a suitable vector field around its positively oriented boundary ∂S^+. For the case where ∂S is a simple closed curve, if $\alpha : [c, d] \subseteq \mathbb{R} \to \mathbb{R}^2$ is a simple, counterclockwise parametrization of ∂S then

7.5.18   Area $S = \displaystyle\int_\alpha (0, x) \cdot d\alpha = \int_\alpha (-y, 0) \cdot d\alpha = \tfrac{1}{2} \int_\alpha (-y, x) \cdot d\alpha.$

This follows immediately from Theorem 7.5.5 and has as a consequenece the following result.

7.5.19 *Theorem.* *Let* $\alpha : [c, d] \subseteq \mathbb{R} \to \mathbb{R}^2$ *be a simple counterclockwise parametrization of the boundary* ∂S *of a simple region* S, *and let*

$$\alpha(t) = (x(t), y(t)), \qquad t \in [c, d].$$

Then

$$\text{Area } S = \int_c^d x(t) y'(t) \, dt = - \int_c^d x'(t) y(t) \, dt$$

$$= \tfrac{1}{2} \int_c^d (x(t) y'(t) - x'(t) y(t)) \, dt.$$

Proof.   Exercise.

7.5.20 *Example.* The ellipse

$$\frac{x^2}{a^2}+\frac{y^2}{b^2}=1$$

has counterclockwise parametrization $\alpha:[0, 2\pi]\subseteq\mathbb{R}\to\mathbb{R}^2$, where

$$\alpha(t) = (a\cos t, b\sin t), \qquad t \in [0, 2\pi].$$

The area of the inside of the ellipse is given by any of the three expressions

$$\int_0^{2\pi} ab\cos^2 t\,\mathrm{d}t, \ -\int_0^{2\pi}(-ab\sin^2 t)\,\mathrm{d}t \quad \text{and} \quad \tfrac{1}{2}ab\int_0^{2\pi}(\cos^2 t + \sin^2 t)\,\mathrm{d}t.$$

They are all equal to πab.

Suppose now that V and W are simple regions with $V \subseteq \text{Int } W$. Then

$$S = W \backslash \text{Int } V$$

is a compact region whose boundary is $\partial W \cup \partial V$. See Fig. 7.16. Let

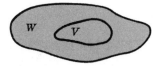 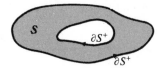

Fig. 7.16

∂V^+ and ∂W^+ be the counterclockwise oriented boundaries of V and W respectively. Then the positively oriented boundary of S is

7.5.21 $$\partial S^+ = \partial W^+ \cup \partial V^-.$$

For any C^1 vector field $F: W \subseteq \mathbb{R}^2 \to \mathbb{R}^2$,

$$\iint_S \left(\frac{\partial F_2}{\partial x} - \frac{\partial F_1}{\partial y}\right)\mathrm{d}A = \iint_W \left(\frac{\partial F_2}{\partial x} - \frac{\partial F_1}{\partial y}\right)\mathrm{d}A - \iint_{\text{Int } V} \left(\frac{\partial F_2}{\partial x} - \frac{\partial F_1}{\partial y}\right)\mathrm{d}A$$

$$= \iint_W \left(\frac{\partial F_2}{\partial x} - \frac{\partial F_1}{\partial y}\right)\mathrm{d}A - \iint_V \left(\frac{\partial F_2}{\partial x} - \frac{\partial F_1}{\partial y}\right)\mathrm{d}A$$

$$= \int_{\partial W^+} F \cdot \mathrm{d}\mathbf{r} - \int_{\partial V^+} F \cdot \mathrm{d}\mathbf{r}, \quad \text{by Theorem 7.5.14}$$

$$= \int_{\partial W^+} F \cdot \mathrm{d}\mathbf{r} + \int_{\partial V^-} F \cdot \mathrm{d}\mathbf{r}.$$

That is, by 7.5.21,

7.5.22
$$\iint\limits_{S} \left(\frac{\partial F_2}{\partial x} - \frac{\partial F_1}{\partial y} \right) \mathrm{d}A = \int_{\partial S^+} F \cdot \mathrm{d}\mathbf{r}.$$

This establishes Green's Theorem for 'simple regions with a simple hole'. An easy generalization leads to Green's Theorem for simple regions with a finite number of simple holes.

7.5.23 *Example.* Let S be the annulus $\{(x, y) \in \mathbb{R}^2 \mid 1 \leqslant x \leqslant 4\}$. Then ∂S^+, the positively oriented boundary of S is the union of the counterclockwise oriented circle $x^2 + y^2 = 4$ and the clockwise oriented circle $x^2 + y^2 = 1$. Let $F: \mathbb{R}^2 \to \mathbb{R}^2$ be the vector field

$$F(x, y) = (-y, x), \qquad (x, y) \in \mathbb{R}^2.$$

The line integral around a counterclockwise oriented circle centre O in $\mathbb{R}^2$, radius c, is $2\pi c^2$ (Exercise 6.2.3). Hence

$$\int_{\partial S^+} F \cdot \mathrm{d}\mathbf{r} = 2\pi(2^2) - 2\pi(1^2) = 6\pi.$$

On the other hand, since the area of S is 3π,

$$\iint\limits_{S} \left(\frac{\partial F_2}{\partial x} - \frac{\partial F_1}{\partial y} \right) \mathrm{d}A = \iint\limits_{S} 2 \, \mathrm{d}A = 6\pi$$

which confirms 7.5.22.

The argument leading to a proof of more general forms of Green's Theorem is more difficult. However, it is often possible to find particular techniques for particular cases. For example, it may be possible to divide S, the region over which we wish to integrate, into a finite number of simple regions by using a finite number of simple closed curves. See Fig. 7.17.

Let the positively oriented boundary of S be ∂S^+ and let the simple regions into which S is subdivided be $S_1, \ldots, S_k$ with positively (counterclockwise) oriented boundaries $\partial S_1^+, \ldots, \partial S_k^+$ respectively.

Notice that if a simple arc lies in just *one* of $\partial S_1, \ldots, \partial S_k$, say ∂S_r, then it also lies in ∂S and is oriented in the same way in both ∂S_r^+ and ∂S^+. this is because if S_r lies on the left on traversing the curve then so does S.

On the other hand, if a simple arc lies in both ∂S_i and ∂S_j $(i \neq j)$ then it is given opposite orientations in ∂S_i^+ and ∂S_j^+. See Fig. 7.17.

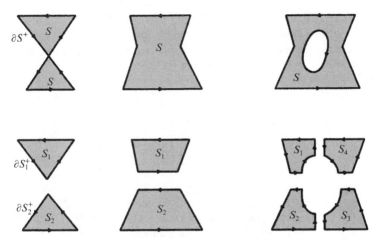

Fig. 7.17 Piecewise simple regions and their decomposition into simple regions

In this case the line integrals of a vector field along that part of ∂S_i^+ and along that part of ∂S_j^+ are equal in magnitude and of opposite sign.

We shall call compact regions that can be subdivided in this way, *piecewise simple regions*.

7.5.24 Green's Theorem for piecewise simple regions. *Let S be a piecewise simple region, and let ∂S^+ be its positively oriented boundary. Then for any C^1 vector field $F : S \subseteq \mathbb{R}^2 \to \mathbb{R}^2$,*

7.5.25
$$\int_{\partial S^+} F \cdot \mathrm{d}r = \iint_S \left(\frac{\partial F_2}{\partial x} - \frac{\partial F_1}{\partial y} \right) \mathrm{d}A.$$

Proof. With the above notation, since the line integrals cancel along common boundaries,

7.5.26
$$\int_{\partial S_1^+} F \cdot \mathrm{d}r + \cdots + \int_{\partial S_k^+} F \cdot \mathrm{d}r = \int_{\partial S^+} F \cdot \mathrm{d}r.$$

Expression 7.5.25 follows by applying Theorem 7.5.14 to each term on the left-hand side of 7.5.26.

In the next section we use Green's Theorem to say something more about irrotational and conservative fields. We also give alternative forms of the theorem, in order to show how it is related to the other classical theorems of vector analysis—Stokes' Theorem and Gauss' Theorem—which are concerned with integration of vector fields over surfaces and over volumes.

Exercises 7.5.

1. Verify Green's Theorem in the plane

$$\int_{\partial S^+} F \cdot d\mathbf{r} = \iint_S \left(\frac{\partial F_2}{\partial x} - \frac{\partial F_1}{\partial y} \right) dA$$

where $F(x, y) = (2xy - x^2, x + y^2)$ and
(a) S is the rectangular region with corners $(0, 0), (0, 2), (1, 2), (1, 0)$;
(b) S is the triangular region with corners $(0, 0), (0, 2), (1, 0)$;
(c) S is the region enclosed by the parabolas $y = x^2$ and $y^2 = x$.

Answers: both integrals are equal to (a) 0; (b) 1/3; (c) 1/30.

2. Verify Green's Theorem in the plane where S is the annulus $\{(x, y) \in \mathbb{R}^2 \mid a^2 \leqslant x^2 + y^2 \leqslant b^2\}$ and

(a) $F(x, y) = \left(\dfrac{-y}{\sqrt{(x^2 + y^2)}}, \dfrac{x}{\sqrt{(x^2 + y^2)}} \right)$;

(b) $F(x, y) = \left(\dfrac{-y}{x^2 + y^2}, \dfrac{x}{x^2 + y^2} \right)$;

(c) $F(x, y) = \left(\dfrac{x}{x^2 + y^2}, \dfrac{y}{x^2 + y^2} \right)$.

Answers: (a) $2\pi(b - a)$; (b) 0; (c) 0.

3. Use Green's Theorem to show that the line integral

$$\int_{C^+} \frac{-y}{x^2 + y^2} dx + \frac{x}{x^2 + y^2} dy$$

takes the value
(a) 0, if C^+ is any oriented simple closed curve in $\mathbb{R}^2 \backslash \{(0, 0)\}$ that does not enclose the origin;
(b) 2π, if C^+ is any counterclockwise oriented simple closed curve around the origin.

Hint: put

$$F(x, y) = \left(\frac{-y}{x^2 + y^2}, \frac{x}{x^2 + y^2} \right), \qquad (x, y) \neq (0, 0).$$

To prove (a) apply Green's Theorem to the region $S \subseteq \mathbb{R}^2$ enclosed by

C. To prove (b) apply Green's Theorem to an annulus whose boundary is *C* and a non-intersecting *circle* centre the origin. Compare Exercise 2(b).

4. Prove that

$$\int_{C^+} \frac{x}{x^2+y^2}\,dx + \frac{y}{x^2+y^2}\,dy = 0,$$

where C^+ is any oriented simple closed curve in $\mathbb{R}^2 \backslash \{(0,0)\}$.

5. Let C^+ be an oriented simple arc in $\mathbb{R}^2$ running from $(0,0)$ to (a,b). Prove, using Green's Theorem that the line integral

$$\int_{C^+} (x+y^2)\,dx + 2xy\,dy$$

is independent of the choice of C^+.

Hint: consider a second oriented simple arc B^+ from $(0,0)$ to (a,b). Prove that $\int_{C^+} + \int_{B^-} = 0$. Compare Exercise 5.5.7.

6. Sketch the astroid $x^{2/3} + y^{2/3} = a^{2/3}$ and use Theorem 7.5.19 to calculate the area of the region enclosed by it.

Answer: $3\pi a^2/8$. A suitable simple parametrization of the astroid is $\alpha(t) = (a\cos^3 t, a\sin^3 t)$, $t \in [0, 2\pi]$.

7. Prove that the area of a simple region S in $\mathbb{R}^2$ is given by each of the three line integrals

$$\text{Area of } S = \int_{\partial S^+} x\,dy = \int_{\partial S^+} -y\,dx = \tfrac{1}{2}\int_{\partial S^+} x\,dy - y\,dx,$$

where ∂S^+ is the positively oriented boundary of S.

8. Find the area of one loop of the figure 8 curve $x = \sin t$, $y = \sin 2t$.

Answer: 4/3.

9. Find the area bounded by one arc of the cycloid $x = a(t - \sin t)$, $y = a(1 - \cos t)$, $a > 0$, $t \in [0, 2\pi]$ and the *x*-axis.

Answer: $3\pi a^2$.

7.6 Rot, curl, and div

In Section 6.3 we justified calling the line integral of a continuous vector field $F: D \subseteq \mathbb{R}^2 \to \mathbb{R}^2$ around an oriented simple closed curve C^+ the *circulation* of F around C^+. We then defined the rotation of

F around C^+ and the rotation of F at a point (a, b) in D. Suppose now that F is a C^1 vector field. Green's Theorem allows us to find an alternative expression for $(\text{rot } F)(a, b)$.

Let S_ε be the closed disc with centre (a, b) and radius $\varepsilon > 0$, where ε is chosen small enough for S_ε to lie in D. From its definition in 6.3.7 and from Green's Theorem 7.5.14, the rotation of F around ∂S_ε^+ is given by

$$\frac{1}{\text{area } S_\varepsilon} \int_{\partial S_\varepsilon^+} F \cdot d\mathbf{r} = \frac{1}{\text{area } S_\varepsilon} \iint_{S_\varepsilon} \left(\frac{\partial F_2}{\partial x} - \frac{\partial F_1}{\partial y} \right) dA$$

where ∂S_ε^+ is the boundary circle with counterclockwise orientation. There therefore exists, by the Integral Mean-Value Theorem 7.4.26, a point $(a + h_\varepsilon, b + k_\varepsilon) \in S_\varepsilon$ such that the rotation of F around ∂S_ε^+ is

$$\frac{\partial F_2}{\partial x}(a + h_\varepsilon, b + k_\varepsilon) - \frac{\partial F_1}{\partial y}(a + h_\varepsilon, b + k_\varepsilon).$$

Since $(\text{rot } F)(a, b)$ is the limit of the rotation around S_ε^+ as ε tends to 0, and since F_1 and F_2 have continuous partial derivatives,

7.6.1 $$(\text{rot } F)(a, b) = \frac{\partial F_2}{\partial x}(a, b) - \frac{\partial F_1}{\partial y}(a, b).$$

This useful result, which can of course be taken as a definition of rot F, leads to the following theorem.

7.6.2 Theorem.   *A C^1 vector field $F : D \subseteq \mathbb{R}^2 \to \mathbb{R}^2$ defined on an open set D in $\mathbb{R}^2$ is irrotational if and only if* $\dfrac{\partial F_2}{\partial x} = \dfrac{\partial F_1}{\partial y}$.

Proof.   Immediate from 7.6.1.

The condition of Theorem 7.6.2 is also necessary and sufficient for a field to be conservative provided we suitably restrict the domain of F.

7.6.3 Definition.   *A subset $D \subseteq \mathbb{R}^2$ is* simply connected *if it is path connected and if whenever a simple closed curve lies in D then so does its inside.*

(i)   (ii)   (iii)

Fig. 7.18 (i), (ii) Simply connected, (iii) not simply connected

7.6.4 *Example.*

  [i] The rectangle and the Y-shape sketched in Fig. 7.18 are simply connected, but the annulus is not.

  [ii] The open subset $D = \mathbb{R}^2 \backslash \{\mathbf{0}\}$ of $\mathbb{R}^2$ is not simply connected. The unit circle $x^2 + y^2 = 1$ lies in D but, since $\mathbf{0} \notin D$, its inside does not.

  [iii] Let $A \subseteq \mathbb{R}^2$ be the half-axis

$$A = \{(x, 0) \in \mathbb{R}^2 \mid x \leq 0\}.$$

Then $\mathbb{R}^2 \backslash A$ is simply connected.

The following theorem generalizes the result of Exercise 5.5.12. We shall assume the general form of Green's Theorem given in 7.5.5.

7.6.5 *Theorem.*   *A C^1 vector field $F : D \subseteq \mathbb{R}^2 \to \mathbb{R}^2$ defined on an open, simply connected subset D of $\mathbb{R}^2$ is conservative if and only if $\partial F_2 / \partial x = \partial F_1 / \partial y$.*

Proof.   First, if F is conservative then $F = \operatorname{grad} f$ for some C^2 scalar field $f : D \subseteq \mathbb{R}^2 \to \mathbb{R}$. By Theorem 3.10.4 the mixed partial derivatives $\partial^2 f / \partial x \, \partial y$ and $\partial^2 f / \partial y \, \partial x$ are equal. It follows that $\partial F_2 / \partial x = \partial F_1 / \partial y$.

  Second, suppose that $\partial F_2 / \partial x = \partial F_1 / \partial y$. Let C be a simple closed curve in D. Then since D is simply connected, the inside U of C also lies in D. Put $S = U \cup C$ and let C^+ have counterclockwise orientation. Green's Theorem implies that

$$\int_{C^+} F \cdot d\mathbf{r} = \iint_S \left(\frac{\partial F_2}{\partial x} - \frac{\partial F_1}{\partial y} \right) dA = 0.$$

Since this is true for all simple closed curves in D, Theorem 5.5.21 implies that F is conservative.

7.6.6 *Example.* Consider the function $F : \mathbb{R}^2 \backslash \{\mathbf{0}\} \to \mathbb{R}^2$ defined by

$$F(x, y) = \left(\frac{-y}{x^2 + y^2}, \frac{x}{x^2 + y^2} \right), \qquad (x, y) \neq (0, 0).$$

Then

$$\frac{\partial F_2}{\partial x} = \frac{\partial F_1}{\partial y} = \frac{y^2 - x^2}{(x^2 + y^2)^2}, \qquad (x, y) \neq (0, 0).$$

Hence F is irrotational. However, on the domain $\mathbb{R}^2 \backslash \{\mathbf{0}\}$, F is *not* conservative. For example, the line integral of F counterclockwise around the unit circle centre $(0, 0)$ is 2π. See also Example 5.5.28.

However, if the domain of F is limited to the simply connected set $\mathbb{R}^2 \backslash A$ where $A = \{(x, 0) \in \mathbb{R}^2 \,|\, x \leqslant 0\}$, then F is conservative. Indeed $F = \mathrm{grad}\, f$ where $f : \mathbb{R}^2 \backslash A \subseteq \mathbb{R}^2 \to \mathbb{R}$ is given in Exercise 5.5.8.

Expression 7.6.1 allows us to rephrase the conclusion of Green's Theorem as

7.6.7
$$\int_{\partial S^+} F \cdot \mathbf{dr} = \iint_S \mathrm{rot}\, F \, \mathrm{d}A.$$

We can obtain a physical interpretation of 7.6.7 if, as in Section 6.3, we think of f as representing the flow of a fluid in the (x, y) plane. Remember that $(\mathrm{rot}\, F)(x, y)$ is positive, negative or zero according to whether the flow near (x, y) is counterclockwise, clockwise or uniform. The identity 7.6.7 tells us that the circulation $\int_{\partial S^+} F \cdot \mathbf{dr}$ around the positively oriented boundary of S is equal to the aggregate of local rotational effects on S. It may of course be zero despite the presence of local clockwise and counterclockwise rotations.

7.6.8 *Example.* Consider the velocity field $F : \mathbb{R}^2 \to \mathbb{R}^2$ defined by

$$F(x, y) = (1, \tfrac{1}{2}x^2), \qquad (x, y) \in \mathbb{R}^2.$$

Then

$$(\mathrm{rot}\, F)(x, y) = x, \qquad (x, y) \in \mathbb{R}^2.$$

Let S be a circular disc, centre $(0, b)$, radius c, where b and $c > 0$ are arbitrary. Then

$$\int_{\partial S^+} F \cdot \mathbf{dr} = \iint_S \mathrm{rot}\, F \, \mathrm{d}A = \iint_S x \, \mathrm{d}x \, \mathrm{d}y = 0.$$

The vanishing of the circulation of F around ∂S^+ is made plausible by reference to Fig. 6.9 (page 337), where the velocity field F is sketched.

We can associate a vector field $F : D \subseteq \mathbb{R}^2 \to \mathbb{R}^2$ with a vector field $F^* : D \times \mathbb{R}^3 \to \mathbb{R}^3$ defined by

7.6.9 $F^*(x, y, z) = (F_1(x, y), F_2(x, y), 0), \qquad (x, y) \in D.$

Suppose that F is differentiable. Then so is F^*, and by Definition 4.10.1.

7.6.10 $(\operatorname{curl} F^*)(x, y, z) = \left(\dfrac{\partial F_2}{\partial x} - \dfrac{\partial F_1}{\partial y} \right)(x, y)\mathbf{k} = (\operatorname{rot} F)(x, y)\mathbf{k},$

where $\mathbf{k} = (0, 0, 1)$ is the unit vector in the z direction.

Expression 7.6.7 then becomes a statement of Green's Theorem in the form

7.6.11 $$\int_{\partial S^+} F \cdot d\mathbf{r} = \iint_S \operatorname{curl} F^* \cdot \mathbf{k} \, dA.$$

This is the form which generalizes to Stokes' Theorem.

A section of the field F^* with the x, y plane is illustrated in Fig. 7.19. Notice that $(\operatorname{curl} F^*)(\mathbf{p})$ points in the direction of $\mathbf{k}$ if the effective flow around $\mathbf{p}$ is counterclockwise as viewed from 'above' the x, y plane; it points in the direction $-\mathbf{k}$ if the effective flow around the point is clockwise. In Fig. 7.19 the numbers $(\operatorname{rot} F)(\mathbf{p})$ and $(\operatorname{rot} F)(\mathbf{q})$ can be estimated by considering the circulation around small counterclockwise oriented circles in the x, y plane

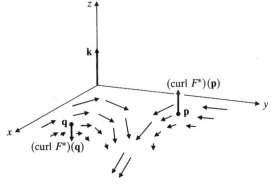

Fig. 7.19

surrounding **p** and **q** respectively. Here (rot F)(**p**) is positive and (rot F)(**q**) is negative.

A picture similar to Fig. 7.19 illustrates the section of F^* with any plane parallel to the x, y plane.

It follows from 7.6.10 that rot $F = 0$ if and only if curl $F^* = \mathbf{0}$. So the field F in $\mathbb{R}^2$ is irrotational (Definition 6.3.9) if and only if the associated field F^* in $\mathbb{R}^3$ is irrotational (Definition 4.10.8).

A second application of Green's Theorem to the fluid flow F in a compact region S relates the net fluid flow across the boundary ∂S of S and the net expansion or contraction of the fluid within S. For this application we suppose that the boundary of S is smooth, that is, it is composed of a finite number of smooth simple closed curves. As usual, let ∂S^+ be the positively oriented boundary of S, and let $T : \partial S \subseteq \mathbb{R}^2 \rightarrow \mathbb{R}^2$ be the unit tangent field to ∂S^+. For any $\mathbf{r} \in \partial S$ we define the *unit normal to ∂S at $\mathbf{r}$ pointing away from S* to be the vector

7.6.12 $$N(\mathbf{r}) = (T_2(\mathbf{r}), -T_1(\mathbf{r})),$$

where $T(\mathbf{r}) = (T_1(\mathbf{r}), T_2(\mathbf{r}))$. So $N(\mathbf{r})$ is obtained by a clockwise rotation through $\frac{1}{2}\pi$ of the unit tangent $T(\mathbf{r})$. It is the unit normal vector to ∂S that points to the right as one follows the oriented boundary ∂S^+. See Fig. 7.20.

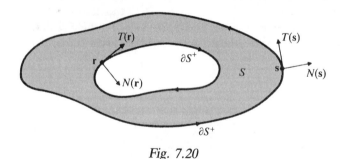

Fig. 7.20

The following theorem provides the relationship referred to above. Remember that for any C^1 vector field $F : S \subseteq \mathbb{R}^2 \rightarrow \mathbb{R}^2$

$$\text{div } F = \frac{\partial F_1}{\partial x} + \frac{\partial F_2}{\partial y}.$$

7.6.13 Divergence Theorem in the Plane. *Let S be a piecewise simple region in $\mathbb{R}^2$ with smooth boundary ∂S. For any C^1 vector field $F : S \subseteq \mathbb{R}^2 \to \mathbb{R}^2$*

7.6.14 $$\int_{\partial S} F \cdot N \, \mathrm{d}s = \iint_S \operatorname{div} F \, \mathrm{d}A$$

where $N(\mathbf{r})$ is the unit normal to ∂S at $\mathbf{r} \in \partial S$ pointing away from S.

Proof. [*i*] Consider first the case where S is a simple region. Let $\alpha : [a, b] \subseteq \mathbb{R} \to \mathbb{R}^2$ be a simple (smooth) parametrization of ∂S^+, the positively oriented boundary of S. The unit tangent to ∂S^+ at $\alpha(t)$ is given by

$$T(\alpha(t)) = (\alpha_1'(t), \, \alpha_2'(t))/\|\alpha'(t)\|,$$

and the unit normal to ∂S at $\alpha(t)$ pointing away from S is

$$N(\alpha(t)) = (\alpha_2'(t), \, -\alpha_1'(t))/\|\alpha'(t)\|.$$

Define a vector field $G : S \subseteq \mathbb{R}^2 \to \mathbb{R}^2$ by

$$G(x, y) = (-F_2(x, y), F_1(x, y)), \qquad (x, y) \in S.$$

Then G is a C^1 field and

7.6.15 $\quad F(\alpha(t)) \cdot N(\alpha(t)) = G(\alpha(t)) \cdot T(\alpha(t)), \qquad t \in [a, b].$

See Fig. 7.21.

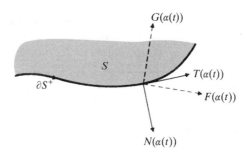

Fig. 7.21

Hence

$$\int_{\partial S} F \cdot N \, \mathrm{d}s = \int_a^b F(\alpha(t)) \cdot N(\alpha(t)) \, \|\alpha'(t)\| \, \mathrm{d}t$$

$$= \int_a^b G(\alpha(t)) \cdot T(\alpha(t)) \, \|\alpha'(t)\| \, \mathrm{d}t \qquad \text{by 7.6.15}$$

$$= \int_a^b G(\alpha(t)) \cdot \alpha'(t) \, \mathrm{d}t$$

$$= \int_{\partial S^+} G \cdot \mathrm{d}\mathbf{r} = \iint_S \left(\frac{\partial G_2}{\partial x} - \frac{\partial G_1}{\partial y} \right) \mathrm{d}A \quad \text{by Green's Theorem}$$

$$= \iint_S \left(\frac{\partial F_1}{\partial x} + \frac{\partial F_2}{\partial y} \right) \mathrm{d}A = \iint_S \mathrm{div}\, F \, \mathrm{d}A.$$

[*ii*] Suppose now that S is a piecewise simple region that can be cut up into a finite number of simple regions $S_1, \ldots, S_k$ by a finite number of simple curves. Where two of these simple regions S_i and S_j meet along a simple curve C the associated path integrals satisfy

$$\int_C F \cdot N_i \, \mathrm{d}s + \int_C F \cdot N_j \, \mathrm{d}s = 0,$$

since at any point $\mathbf{p} \in C$, $N_i(\mathbf{p})$ (the unit normal to ∂S_i at $\mathbf{p}$ pointing away from S_i) is in the opposite direction to $N_j(\mathbf{p})$ (the unit normal to ∂S_j at $\mathbf{p}$ pointing away from S_j).

Hence 7.6.14 follows by adding the corresponding expressions relating to each of the simple regions $S_1, \ldots, S_k$.

In the proof of Theorem 7.6.13 the condition that the boundary of S be smooth can be relaxed to piecewise smoothness. In that case there may be a finite number of points $\mathbf{r} \in \partial S$ on which the normal vector $N(\mathbf{r})$ cannot be defined. However, relation 7.6.14 still applies when suitably interpreted.

7.6.16 *Example.*   Let S be the simple region $\{(x, y) \in \mathbb{R}^2 \mid |2x| + |y| \leq 2\}$ sketched in Fig. 7.22 and let $F: \mathbb{R}^2 \to \mathbb{R}^2$ be the C^1 vector field defined by

$$F(x, y) = (y, -x), \qquad (x, y) \in \mathbb{R}^2.$$

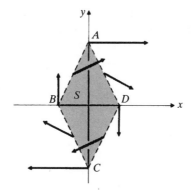

Fig. 7.22 Vector field $F(x, y) = (y, -x)$ on boundary of diamond S

Then div $F = 0$, and, by 7.6.14,

$$\int_{\partial S} F \cdot N \, ds = \iint_S \operatorname{div} F \, dA = 0.$$

The component $F \cdot N$ is sometimes positive (when $F(x, y)$ is directed away from S) and sometimes negative (when $F(x, y)$ is directed into S). These contributions are balanced in the integral of $F \cdot N$ over ∂S.

Incidentally, by Green's Theorem

$$\int_{\partial S^+} F \cdot d\mathbf{r} = \iint_S \operatorname{rot} F \, dA = \iint_S -2 \, dA = -8,$$

since the area of S is 4. This calculation agrees with that in Example 6.2.25.

Concerning the physical application of Theorem 7.6.13, if we again regard F as the velocity field of a fluid flow in a compact region S, then $(\operatorname{div} F)(\mathbf{p})$, $\mathbf{p} \in S$, measures the rate of expansion per unit area of the fluid at $\mathbf{p}$ (see Section 4.11). Hence $\iint_S \operatorname{div} F \, dA$ aggregates the rate of expansion of the fluid in S. From physical considerations, this rate of expansion must equal the rate at which fluid flows out of S across the boundary ∂S, and this is given by the line integral $\int_{\partial S} F \cdot N \, ds$. The equality is confirmed in 7.6.14. The integral $\int_{\partial S} F \cdot N \, ds$ is called the *outward flux of F across S*.

Expression 7.6.14 is the form of Green's Theorem which generalizes to Gauss' Divergence Theorem.

It is interesting to compare Theorem 7.6.13 with the following form of Green's Theorem for a piecewise simple region S with smooth boundary. As above, let $T : \partial S \subseteq \mathbb{R}^2 \to \mathbb{R}^2$ be the unit tangent field to ∂S^+. From Green's Theorem, written in the form 7.6.7, we obtain

7.6.17
$$\int_{\partial S} F \cdot T \, \mathrm{d}s = \iint_S \operatorname{rot} F \, \mathrm{d}A.$$

Compare this with 7.6.14

$$\int_{\partial S} F \cdot N \, \mathrm{d}s = \iint_S \operatorname{div} F \, \mathrm{d}A.$$

These formulae when interpreted in terms of fluid flow tell us that the integrals over the boundary of S of the tangential and normal components of F are equal respectively to the aggregate of the local rotational and expansional effects on S.

We can use Theorem 7.6.13 to obtain an expression for the divergence of a vector field which parallels the definition of its rotation given in 6.3.7. Let $F : \mathbb{R}^2 \to \mathbb{R}^2$ be a C^1 vector field and let (a, b) be a point of $\mathbb{R}^2$. Then for any closed disc S_ε, centre (a, b) and radius $\varepsilon > 0$,

$$\frac{1}{\operatorname{area} S_\varepsilon} \int_{\partial S_\varepsilon} F \cdot N \, \mathrm{d}s = \frac{1}{\operatorname{area} S_\varepsilon} \iint_{S_\varepsilon} \operatorname{div} F \, \mathrm{d}A \qquad \text{by 7.6.14}$$

$$= (\operatorname{div} F)(a + h_\varepsilon, b + k_\varepsilon),$$

for some point $(a + h_\varepsilon, b + k_\varepsilon) \in S_\varepsilon$, by the Integral Mean-Value Theorem. Since div F is continuous we obtain

7.6.18
$$(\operatorname{div} F)(a, b) = \lim_{\varepsilon \to 0} \frac{1}{\operatorname{area} S_\varepsilon} \int_{\partial S_\varepsilon} F \cdot N \, \mathrm{d}s.$$

By comparison, 7.6.17 leads to the following expression for rot F:

7.6.19
$$(\operatorname{rot} F)(a, b) = \lim_{\varepsilon \to 0} \frac{1}{\operatorname{area} S_\varepsilon} \int_{\partial S_\varepsilon} F \cdot T \, \mathrm{d}s.$$

Exercises 7.6

1. Let $F : \mathbb{R}^2 \backslash \{\mathbf{0}\} \to \mathbb{R}^2$ be the C^1 vector field defined by

$$F(x, y) = \left(\frac{x}{x^2 + y^2}, \frac{y}{x^2 + y^2} \right), \qquad (x, y) \neq (0, 0).$$

Show that F is irrotational. Is the field F conservative?

Answer: yes. Note that Theorem 7.6.5 cannot be used to prove this, since $\mathbb{R}^2\backslash\{\mathbf{0}\}$ is not simply connected. A direct calculation shows that $F = \operatorname{grad} f$, where $f(x, y) = \frac{1}{2}\ln(x^2 + y^2)$, $(x, y) \neq (0, 0)$. See also Exercise 7.5.4.

2. Verify by applying Theorem 7.6.2 that the field $F:\mathbb{R}^2\backslash\{\mathbf{0}\} \to \mathbb{R}^2$ defined by

$$F(x, y) = \frac{e^x}{x^2 + y^2}(x \sin y - y \cos y, x \cos y + y \sin y), \qquad (x, y) \neq (0, 0)$$

is irrotational. (The field F is not conservative. See Exercise 6.3.6.)

3. Let $F:\mathbb{R}^2 \to \mathbb{R}^2$ be the C^1 vector field defined by

$$F(x, y) = (y^2, x^2), \qquad (x, y) \in \mathbb{R}^2.$$

Calculate $(\operatorname{rot} F)(x, y)$. Verify Green's Theorem 7.6.7

$$\int_{\partial S^+} F \cdot d\mathbf{r} = \iint_S \operatorname{rot} F \, dA$$

where S is the circular disc centre (a, b), radius k, by showing that each integral takes the value $2\pi k^2(a - b)$. Sketch the vector field F.

Note: the field lines of F are $y^3 = x^3 + c$, where c is a constant.

4. Let D be an open set in $\mathbb{R}^2$. Show that the vector field $F:D \subseteq \mathbb{R}^2 \to \mathbb{R}^2$ is conservative if and only if the associated field $F^*:D \times \mathbb{R} \subseteq \mathbb{R}^3 \to \mathbb{R}^3$ (defined in 7.6.9) is conservative.

Hint: suppose $F^* = \operatorname{grad} f^*$ for some scalar field $f^*:D \times \mathbb{R} \subseteq \mathbb{R}^3 \to \mathbb{R}$. Then $\partial f^*/\partial z = 0$, so there exists $f:D \subseteq \mathbb{R}^2 \to \mathbb{R}^2$ such that $f^*(x, y, z) = f(x, y)$ for all $(x, y) \in D$, $z \in \mathbb{R}$. Deduce that $F = \operatorname{grad} f$.

5. Let $Z = \{(0, 0, z)\}$, $z \in \mathbb{R}$ (the z-axis in $\mathbb{R}^3$). The vector field $F^*:\mathbb{R}^3\backslash Z \subseteq \mathbb{R}^3 \to \mathbb{R}^3$ defined by

$$F^*(x, y, z) = \left(\frac{-y}{x^2 + y^2}, \frac{x}{x^2 + y^2}, 0\right), \qquad (x, y) \neq (0, 0)$$

is irrotational (Example 4.10.7). Show that F^* is not conservative. Show, however, that F^* is conservative if its domain is limited to the set $\mathbb{R}^3\backslash H$, where H is the half-plane (containing Z) given by $\{(x, 0, z) \in \mathbb{R}^3 \mid x \leq 0\}$.

Hint: Exercise 4 and Example 7.6.6.

6. In Example 7.6.16, verify directly that $\int_{\partial S} F \cdot N \, ds = 0$.

Answer: parametrize the directed edge AB (Fig. 7.22) by $\alpha(t) = (-t, 2 - 2t)$, $t \in [0, 1]$. Then $\alpha'(t) = (-1, -2)$, $N(t) = (-2, 1)$, $F(\alpha(t)) = (2 - 2t, t)$, and $(F \cdot N)(\alpha(t)) = 5t - 4$. Hence

$$\int_{AB} F \cdot N \, ds = \int_0^1 (F \cdot N)(\alpha(t)) \, \|\alpha'(t)\| \, dt = -3\sqrt{5}/2.$$

The corresponding path integrals over BC, CD, DA are respectively $3\sqrt{5}/2$, $-3\sqrt{5}/2$, and $3\sqrt{5}/2$.

7. Verify the Divergence Theorem in the Plane

$$\int_{\partial S} F \cdot N \, ds = \iint_S \text{div } F \, dA$$

where $F : \mathbb{R}^2 \to \mathbb{R}^2$ is the C^1 vector field defined by

$$F(x, y) = (\tfrac{1}{4}x, \tfrac{1}{4}), \qquad (x, y) \in \mathbb{R}^2$$

(and which is sketched in Fig. 4.18), and S is
   (a) the unit square with corners $A(0, 0)$, $B(1, 0)$, $C(1, 1)$, $D(0, 1)$;
   (b) the diamond with corners $P(0, 1)$, $Q(-1, 0)$, $R(0, -1)$, $T(1, 0)$;
   (c) the circular region $(x - 1)^2 + y^2 \le 1$.

Answers: (a) the integrals of $F \cdot N$ over AB, BC, CD, DA are $-\tfrac{1}{4}, \tfrac{1}{4}, \tfrac{1}{4}, 0$;
   (b) the integrals of $F \cdot N$ over PQ, QR, RT, TP are $\tfrac{3}{8}, -\tfrac{1}{8}, -\tfrac{1}{8}, \tfrac{3}{8}$;
   (c) parametrize S by $\alpha(t) = (1 + \cos t, \sin t)$, $t \in [0, 2\pi]$.
   Then $T(\alpha(t)) = (-\sin t, \cos t)$ and $N(\alpha(t)) = (\cos t, \sin t)$.

8. Verify the Divergence Theorem in the Plane where $F(x, y) = (2xy - x^2, x + y^2)$, and S is the rectangular region with corners $(0, 0)$, $(0, 2)$, $(1, 2)$, $(1, 0)$.

9. Verify the Divergence Theorem in the Plane where S is the annulus $0 < a^2 \le x^2 + y^2 \le b^2$ and

   (a) $F(x, y) = \left(\dfrac{-y}{x^2 + y^2}, \dfrac{x}{x^2 + y^2}\right)$;   (b) $F(x, y) = \left(\dfrac{x}{x^2 + y^2}, \dfrac{y}{x^2 + y^2}\right)$.

   Interpret your results relative to a sketch of the fields.

7.7   Change of variables in double integrals

The reader will be familiar with the change of variable formula of elementary calculus

7.7.1 $$\int_a^b f(\alpha(t))\alpha'(t) \, dt = \int_{\alpha(a)}^{\alpha(b)} f(x) \, dx.$$

The formula relates two integrals through the change of variable $x = \alpha(t)$. For example, the substitution $x = \alpha(t) = 5 - t^2$ can be used to evaluate

7.7.2 $$\int_{-1}^{2} 2t\sqrt{(5 - t^2)}\,\mathrm{d}t = -\int_{4}^{1} \sqrt{x}\,\mathrm{d}x = 14/3.$$

The integrals in 7.7.1 depend on an *orientation* of the intervals of integration. For example, to evaluate the right-hand integral we must integrate 'from $\alpha(a)$ to $\alpha(b)$', where, possibly, $\alpha(a) > \alpha(b)$.

The effect of orientation can be illustrated with the integrals 7.7.2 and the substitution $x = 5 - t^2$. As t increases from -1 to 0 to 1 to 2 the corresponding values of x pass from 4 to 5 to 4 to 1. Considered in this way, the substitution $x = 5 - t^2$ in the left-hand integral of 7.7.2 leads to three pieces

$$\int_{4}^{5} -\sqrt{x}\,\mathrm{d}x + \int_{5}^{4} -\sqrt{x}\,\mathrm{d}x + \int_{4}^{1} -\sqrt{x}\,\mathrm{d}x.$$

The 'doubling back' and consequent cancelling of the first two (oriented) integrals happens because the continuous path α given by $\alpha(t) = 5 - t^2$ is not 1–1 on the interval $[-1, 2]$.

In this section we shall obtain a change of variables formula for double integrals. However, the double integrals we shall consider are integrals over *unoriented* subsets of $\mathbb{R}^2$. Recall that the double integral $\iint_{R} f(x, y)\,\mathrm{d}x\,\mathrm{d}y$ of a function $f : R \subseteq \mathbb{R}^2 \to \mathbb{R}$ over a rectangle R is defined in terms of Riemann sums associated with partitions of R without reference to an orientation of the subset R of $\mathbb{R}^2$. In the same spirit we may interpret the single integral $\int_{a}^{b} g(x)\,\mathrm{d}x$, where $a < b$, as the integral of a function $g : I \subseteq \mathbb{R} \to \mathbb{R}$ over the subset $I = [a, b]$ of $\mathbb{R}$, and denote it by

$$\int_{I} g\,\mathrm{d}s \qquad \text{or} \qquad \int_{I} g(x)\,\mathrm{d}x.$$

The following special case of formula 7.7.1 is the basis for generalization to double integrals.

7.7.3 Theorem.   *Let $\alpha : I = [a, b] \subseteq \mathbb{R} \to \mathbb{R}$ be a 1–1 C^1 path. Then for any continuous function $f : \alpha(I) \subseteq \mathbb{R} \to \mathbb{R}$*

7.7.4 $$\int_{\alpha(I)} f(x)\,\mathrm{d}x = \int_{I} f(\alpha(t))\,|\alpha'(t)|\,\mathrm{d}t.$$

402   *Double integrals in* $\mathbb{R}^2$

Proof. The conditions on α ensure that α is strictly monotonic on $[a, b]$ and that $\alpha([a, b])$ is an interval with end points $\alpha(a)$ and $\alpha(b)$. If α is strictly increasing then $|\alpha'(t)| = \alpha'(t)$ for all $t \in [a, b]$ and $\alpha(I) = [\alpha(a), \alpha(b)]$. If α is strictly decreasing then $|\alpha'(t)| = -\alpha'(t)$ for all $t \in [a, b]$ and $\alpha(I) = [\alpha(b), \alpha(a)]$. In both cases 7.7.4 follows from 7.7.1.

With a view to generalizing Theorem 7.7.3 to double integrals let us consider the integrals in 7.7.4 in terms of their approximating Riemann sums. Since α is strictly monotonic on the interval $I = [a, b]$, there is a 1–1 correspondence between the partitions of I and those of the interval $\alpha(I)$. In particular, a subinterval $[t, t + \Delta t]$ in I of length $\Delta t > 0$ is mapped by α to a subinterval in $\alpha(I)$ of length $|\alpha(t + \Delta t) - \alpha(t)|$. By the Mean-Value Theorem there exists $p \in]t, t + \Delta t[$ such that

$$|\alpha(t + \Delta t) - \alpha(t)| = |\alpha'(p)|\, \Delta t.$$

By the assumed continuity of α' we conclude that for small values of $\Delta t > 0$ the number $|\alpha'(t)|$ measures the magnification of the interval $[t, t + \Delta t]$ under the mapping α.

The following argument suggests that, in a similar way, the magnification factor of a small rectangle in $\mathbb{R}^2$ under a C^1 mapping $G : \mathbb{R}^2 \to \mathbb{R}^2$ is the absolute value of $\det J_{G,(u,v)}$, where (u, v) is a point of the rectangle.

Let R be a rectangle in $\mathbb{R}^2$ with corners $\mathbf{a} = (u, v)$, $\mathbf{b} = (u + \Delta u, v)$, $\mathbf{c} = (u + \Delta u, v + \Delta v)$ and $\mathbf{d} = (u, v + \Delta v)$. See Fig. 7.23. The area of R is $\Delta u\, \Delta v$.

For any C^1 function $G : \mathbb{R}^2 \to \mathbb{R}^2$ the area of $S = G(R)$ is approximately the area of the parallelogram with adjacent sides $\mathbf{b}' - \mathbf{a}'$ and $\mathbf{d}' - \mathbf{a}'$, where $\mathbf{a}'$, $\mathbf{b}'$, $\mathbf{c}'$, and $\mathbf{d}'$ are respectively the

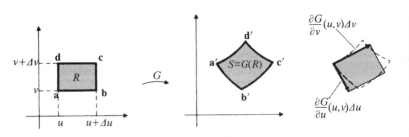

Fig. 7.23

images under G of **a**, **b**, **c**, and **d** (see Fig. 7.23). We observe that

$$\lim_{\Delta u \to 0} \frac{\mathbf{b}' - \mathbf{a}'}{\Delta u} = \lim_{\Delta u \to 0} \frac{G(u + \Delta u, v) - G(u, v)}{\Delta u} = \frac{\partial G}{\partial u}(u, v)$$

and

$$\lim_{\Delta v \to 0} \frac{\mathbf{d}' - \mathbf{a}'}{\Delta v} = \lim_{\Delta v \to 0} \frac{G(u, v + \Delta v) - G(u, v)}{\Delta v} = \frac{\partial G}{\partial v}(u, v).$$

Therefore we have the approximations

$$\mathbf{b}' - \mathbf{a}' \approx \frac{\partial G}{\partial u}(u, v) \, \Delta u \qquad \text{and} \qquad \mathbf{d}' - \mathbf{a}' \approx \frac{\partial G}{\partial v}(u, v) \, \Delta v,$$

and so the area of S is approximated by the area of the parallelogram with adjacent sides

$$\left(\frac{\partial G_1}{\partial u}(u, v) \, \Delta u, \frac{\partial G_2}{\partial u}(u, v) \, \Delta u \right) \quad \text{and}$$

$$\left(\frac{\partial G_1}{\partial v}(u, v) \, \Delta v, \frac{\partial G_2}{\partial v}(u, v) \, \Delta v \right).$$

By Theorem 1.2.23, the area of this parallelogram is equal to the *absolute value* of

$$\det \begin{bmatrix} \dfrac{\partial G_1}{\partial u}(u, v) \, \Delta u & \dfrac{\partial G_1}{\partial v}(u, v) \, \Delta v \\[2ex] \dfrac{\partial G_2}{\partial u}(u, v) \, \Delta u & \dfrac{\partial G_2}{\partial v}(u, v) \, \Delta v \end{bmatrix} = \det J_{G, (u, v)} \, \Delta u \, \Delta v.$$

Now let A be a subset of $\mathbb{R}^2$. A function $G : A \subseteq \mathbb{R}^2 \to \mathbb{R}^2$ may reasonably be thought of as representing a change of variables

$$x = G_1(u, v), \qquad y = G_2(u, v)$$

relating $(u, v) \in A$ and $(x, y) \in G(A)$ if, for example, G is 1–1 and both G and G^{-1} are C^1 functions. We would then expect G to transform a rectangular grid covering A into a grid covering $G(A)$ as illustrated in Fig. 7.24.

Let $f : G(A) \subseteq \mathbb{R}^2 \to \mathbb{R}$ be a continuous function. The Change of Variables Theorem for double integrals gives conditions under which $\iint_{G(A)} f(x, y) \, \mathrm{d}x \, \mathrm{d}y$ can be expressed as a double integral over A.

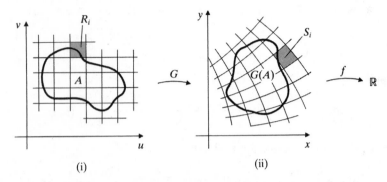

Fig. 7.24 (i) $\mathbb{R}^2 : u, v$ plane; (ii) $\mathbb{R}^2 : x, y$ plane

We can arrive informally at the appropriate identity by manipulating Riemann sums corresponding to a partition $\mathscr{P}$ of $\mathbb{R}^2$ as follows. Let $R_1, \ldots, R_n$ be the compact rectangles of $\mathscr{P}$ that meet A, and for each $i = 1, \ldots, n$ choose $\mathbf{r}_i \in R_i \cap A$. Let $G(\mathbf{r}_i) = \mathbf{s}_i$ and let $G(R_i) = S_i$. Then we have the following sequence of approximations for the integral of f over $G(A)$.

$$\iint\limits_{G(A)} f(x, y) \, dx \, dy \approx \sum_i f(\mathbf{s}_i)(\text{area } S_i)$$

$$\approx \sum_i f(G(\mathbf{r}_i)) \left| \det J_{G, \mathbf{r}_i} \right| (\text{area } R_i)$$

$$\approx \iint\limits_A f(G(u, v)) \left| \det J_{G, (u, v)} \right| du \, dv.$$

It is surprising how much care is needed in justifying these steps in a formal proof.

7.7.5 *Theorem (Change of variables).*   Let $G : K \subseteq \mathbb{R}^2 \to \mathbb{R}^2$ be a C^1 function defined on a compact set K in $\mathbb{R}^2$, and let $D \subseteq K$ be an open subset of $\mathbb{R}^2$ such that

[i] $K \backslash D$ is a null set,
[ii] G is 1–1 on D,
[iii] $\det J_{G, (u, v)} \neq 0$ for all $(u, v) \in D$.

Then, for any bounded function $f : G(K) \subseteq \mathbb{R}^2 \to \mathbb{R}$ *which is continuous on* $G(D)$

7.7.6 $$\iint\limits_{G(K)} f(x, y) \, dx \, dy = \iint\limits_K f(G(u, v)) \left| \det J_{G, (u, v)} \right| du \, dv.$$

There are many versions of this theorem. It is often easiest to apply in the form that we have stated it. Notice in particular that G need not be 1–1 on the whole of K.

In traditional terms, Theorem 7.7.5 states that under the change of variables

$$x = G_1(u, v), \qquad y = G_2(u, v)$$

from Cartesian x, y coordinates to curvilinear u, v coordinates, the integral of f over $G(K)$ is transformed according to the rule

$$\iint_{G(K)} f(x, y) \, dx \, dy = \iint_K \tilde{f}(u, v) \left| \frac{\partial(x, y)}{\partial(u, v)} \right| \, du \, dv,$$

where $\tilde{f}(u, v) = f(G_1(u, v), G_2(u, v))$, and

$$\frac{\partial(x, y)}{\partial(u, v)} = \det \begin{bmatrix} \dfrac{\partial x}{\partial u}(u, v) & \dfrac{\partial x}{\partial v}(u, v) \\ \dfrac{\partial y}{\partial u}(u, v) & \dfrac{\partial y}{\partial v}(u, v) \end{bmatrix} = \det J_{G, (u, v)}.$$

We end this section with some applications of the theorem and with a brief outline of a method that can be used to prove it. However, a formal proof is beyond the scope of this book.

7.7.7 Example. Let S be the annulus $S = \{(x, y) \in \mathbb{R}^2 \mid 1 \leqslant x^2 + y^2 \leqslant 4\}$ in $\mathbb{R}^2$. The integral

$$\iint_S x^2 y^2 \, dx \, dy$$

can be evaluated as the sum of four integrals over x-simple regions in the x, y plane (Exercise 7.4.11). The following method is much simpler. Consider the transformation to polar coordinates

$$(x, y) = G(r, \theta) = (r \cos \theta, r \sin \theta), \qquad (r, \theta) \in \mathbb{R}^2.$$

Define $K = \{(r, \theta) \in \mathbb{R}^2 \mid 1 \leqslant r \leqslant 2, \ 0 \leqslant \theta \leqslant 2\pi\}$. Then $S = G(K)$. See Fig. 7.25. Let $f : \mathbb{R}^2 \to \mathbb{R}$ be defined by

$$f(x, y) = x^2 y^2, \qquad (x, y) \in \mathbb{R}^2.$$

Then the conditions of Theorem 7.7.5 are satisfied (where D is the open rectangle $]1, 2[\times]0, 2[$, and $K \backslash D$ its null boundary). In particular

$$\det J_{G, (r, \theta)} = \frac{\partial(x, y)}{\partial(r, \theta)}(r, \theta) = \det \begin{bmatrix} \cos \theta & -r \sin \theta \\ \sin \theta & r \cos \theta \end{bmatrix} = r$$

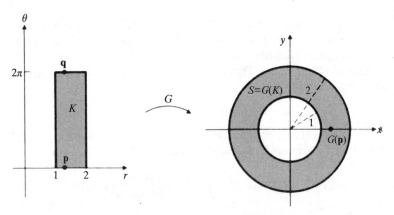

Fig. 7.25 Rectangle K and image under polar transformation
$$G(r,\ \theta) = (r\cos\theta,\ r\sin\theta)$$

which is positive for all $(r,\ \theta) \in D$. Hence, by 7.7.6,

$$\iint\limits_{S} x^2 y^2 \, \mathrm{d}x \, \mathrm{d}y = \iint\limits_{K} (r^2\cos^2\theta)(r^2\sin^2\theta)r \, \mathrm{d}r \, \mathrm{d}\theta$$

$$= \int_1^2 \mathrm{d}r \int_0^2 r^5 \cos^2\theta \sin^2\theta \, \mathrm{d}\theta.$$

Now

$$\int_0^{2\pi} \cos^2\theta \sin^2\theta \, \mathrm{d}\theta = \tfrac{1}{4}\int_0^{2\pi} \sin^2 2\theta \, \mathrm{d}\theta = \tfrac{1}{4}\pi.$$

Hence

$$\iint\limits_{S} x^2 y^2 \, \mathrm{d}x \, \mathrm{d}y = \tfrac{1}{4}\pi \int_1^2 r^5 \, \mathrm{d}r = \frac{21\pi}{8}.$$

Notice that the transformation G takes the point $\mathbf{p}$ on the
boundary of K (Fig. 7.25) to an interior point of $G(K)$. However, it
takes points in D, the interior of K, to points in the interior of
$G(K)$. The function G is 1–1 on D but not 1–1 on K: the point $\mathbf{q}$ in
Fig. 7.25 has the same image as $\mathbf{p}$.

The above example illustrates the following general result.

7.7.8 Theorem.   Let H be the half-plane $H = \{(r,\ \theta) \in \mathbb{R}^2 \mid r \geqslant 0\}$
and let $G : H \subseteq \mathbb{R}^2 \to \mathbb{R}^2$ *be the polar coordinate transformation*

given by

$$(x, y) = G(r, \theta) = (r \cos \theta, r \sin \theta), \qquad r \geq 0.$$

Let D be an open bounded subset of H whose boundary is a null set, and let $K = \bar{D}$. If G is 1–1 on D and if $f: G(K) \subseteq \mathbb{R}^2 \to \mathbb{R}$ is continuous then

7.7.9
$$\iint\limits_{G(K)} f(x, y) \, dx \, dy = \iint\limits_{K} f(r \cos \theta, r \sin \theta) r \, dr \, d\theta.$$

Proof. The conditions of Theorem 7.7.5 are satisfied.

Theorem 7.7.8 is useful when one is required to integrate a function f over a bounded set S of the x, y plane which is contained within two polar angles $\theta = \alpha$ and $\theta = \beta$ (see Fig. 7.26). The corresponding r, θ integral can then be evaluated as a repeated integral by finding K in the r, θ plane such that $S = G(K)$. With ψ_1

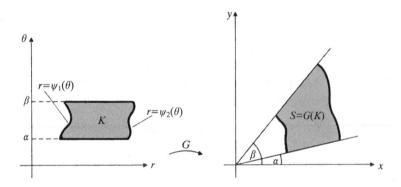

Fig. 7.26 Polar coordinate transformation
$$G(r, \theta) = (r \cos \theta, r \sin \theta)$$

and ψ_2 as shown in the figure, we then have

$$\iint\limits_{G(K)} f(x, y) \, dx \, dy = \int_\alpha^\beta d\theta \int_{\psi_1(\theta)}^{\psi_2(\theta)} f(r \cos \theta, r \sin \theta) r \, dr.$$

7.7.10 *Example.* Find the area in the x, y plane of the set S bounded by the x-axis and the lemniscate which is given in polar coordinates by

7.7.11
$$r^2 = a^2 \cos 2\theta, \qquad 0 \leq \theta \leq \tfrac{1}{4}\pi.$$

The lemniscate consists of the points $(r \cos \theta, r \sin \theta)$ in the x, y plane subject to the condition 7.7.11. See Fig. 7.27.

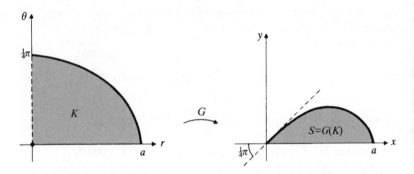

Fig. 7.27

The set S is the image under the polar transformation G of the set K sketched in Fig. 7.27, where K is chosen so that G is 1–1 on its interior D. The conditions of Theorem 7.7.8 are satisfied. Hence

$$\text{Area of } S = \iint\limits_{G(K)} 1 \, dx \, dy = \iint\limits_K r \, dr \, d\theta = \int_0^{1/4\pi} d\theta \int_0^{a\sqrt{\cos 2\theta}} r \, dr = \tfrac{1}{4}a^2.$$

Our final example illustrates a change of variables from Cartesian to parabolic coordinates.

7.7.12 *Example.* *Parabolic Coordinates.* Consider the coordinate transformation $(x, y) = G(u, v)$ given by

$$x = u^2 - v^2, \qquad y = uv.$$

Then

$$\det J_{G(u, v)} = \det \begin{bmatrix} 2u & -2v \\ v & u \end{bmatrix} = 2(u^2 + v^2),$$

which is non-zero except at the origin.

Let U be the subset of the u, v plane consisting of the interior of the rectangle $A(0, 2)$, $B(1, 2)$, $C(1, -2)$, $D(0, -2)$, and let $K = \bar{U}$. See Fig. 7.28. Then G is 1–1 on U and $G(K)$ is closed and bounded by the parabolas

$$x = \tfrac{1}{4}y^2 - 4 \qquad \text{and} \qquad x = 1 - y^2.$$

For example, the edge of K

$$AB = \{(u, 2) \in \mathbb{R}^2 \mid 0 \leq u \leq 1\}$$

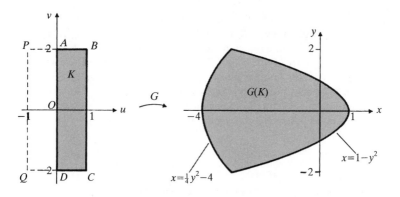

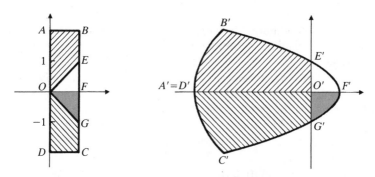

Fig. 7.28 Parabolic coordinate transformation
$$G(u, v) = (u^2 - v^2, uv)$$

is mapped by G to the boundary curve of $G(K)$

$$A'B' = \{(x, y) \in \mathbb{R}^2 \mid x = u^2 - 4, y = 2u, 0 \leqslant u \leqslant 1\},$$

and this is a piece of the parabola $x = \frac{1}{4}y^2 - 4$. Note that G is not 1–1 on K, for G takes both OA and OD onto part of the negative x-axis $y = 0$, $0 \geqslant x \geqslant -4$.

Let us apply Theorem 7.7.5 to evaluate the integral

7.7.13
$$I = \iint\limits_{G(K)} y^2 \, dx \, dy.$$

The conditions of the theorem are satisfied. hence, by 7.7.6,

$$I = \iint\limits_{K} 2u^2v^2(u^2 + v^2) \, du \, dv = 2\int_0^1 du \int_{-2}^2 (u^4v^2 + u^2v^4) \, dv.$$

The repeated integral is easily evaluated, and we obtain $I = 32/3$.

It is interesting to note that K and the larger rectangle $K^* = PBCQ$ (see Fig. 7.28) have the same image under G. It would be a mistake, however, to claim the equality of integrals 7.7.6 with K^* replacing K. The reason is that G is not 1–1 on an open set $U^* \subseteq K^*$, where $K^*\backslash U^*$ is null. So Theorem 7.7.5 does not apply. (See also Exercise 7.7.12.)

The integral 7.7.13 can also be evaluated without a change of variables by considering the set $G(K)$ as a y-simple region. We obtain

$$I = \int_{-2}^{2} y^2 \, dx \int_{1/4y^2-4}^{1-y^2} dy = 32/3.$$

In the present example this is clearly easier than a change to parabolic coordinates. The reader should always bear in mind the two possibilities of a direct integration and a change of variables.

We end this section with a brief note on a method of proof of the Change of Variables Theorem 7.7.5. The result is established in three stages.

[*i*]   First prove the theorem for a function that affects only one of the variables—such a function is said to be *primitive*. This case is a consequence of the single integral result.

[*ii*]   Next consider the theorem for a function which is a composition of primitive functions. Then show that under the hypotheses of Theorem 7.7.5 the function $G : D \subseteq \mathbb{R}^2 \to \mathbb{R}^2$ is *locally* a composition of primitive functions, that is, corresponding to any $(u, v) \in D$ there is a neighbourhood U of (u, v) on which G is a composition of primitive functions.

[*iii*]   Finally prove Theorem 7.7.5 by finding open sets $U_1, \ldots, U_n$ in D which 'almost' cover K and on each of which G is the composition of primitive functions. The integral on the left hand side of 7.7.6 is approximated by a sum of integrals taken over $G(U_i)$. To achieve this reduction one uses the important concept of a *partition of unity*.

This is the approach given, for example, in Rudin, *Principles of Mathematical Analysis* and in Spivak, *Calculus on Manifolds*.

Exercises 7.7

1.   Consider the coordinate transformation $(x, y) = G(u, v)$ given by

$$G(u, v) = (au, bv), \qquad (u, v) \in \mathbb{R}^2$$

where a and b are positive constants.

(a)   Prove that G is 1–1 on $\mathbb{R}^2$.
(b)   Show that G maps the circular disc $K = \{(u, v) \in \mathbb{R}^2 \,|\, u^2 + v^2 \leqslant 1\}$ in the u, v plane onto the elliptical disc $G(K) = \{(x, y) \in \mathbb{R}^2 \,|\, x^2/a^2 + y^2/b^2 \leqslant 1\}$ in the x, y plane.
(c)   Calculate $\det J_{G,(u,v)}$.
(d)   Calculate, using Theorem 7.7.5, the area of the elliptical disc $G(K)$. (Assume known that the area of K is π.)

Answer: πab.

2.   Consider the parabolic coordinate transformation $(x, y) = G(u, v)$, given by

$$x = u^2 - v^2, \qquad y = 2uv \qquad (u, v) \in \mathbb{R}^2.$$

Let k be the compact square in the u, v plane with corners $A(0, 0)$, $B(0, 2)$, $C(2, 2)$, $D(2, 0)$.
(a)   Prove that G is 1–1 on K.
(b)   Sketch the set $S = G(K)$ in the x, y plane.
(c)   Compute the area of S

    (i)   by integrating $\iint_S 1 \, dx \, dy$ directly over S;

    (ii)   by applying the Change of Variables Theorem 7.7.5.

Answer: 128/3.

3.   Three pairs (x, y), (u, v), (w, s) of coordinates of the plane are related by

$$(x, y) = G(u, v) \qquad \text{and} \qquad (u, v) = H(w, s),$$

where $H : D \subseteq \mathbb{R}^2 \to \mathbb{R}^2$ and $G : H(D) \subseteq \mathbb{R}^2 \to \mathbb{R}^2$ are 1–1 C^1 functions defined on open sets D and $H(D)$. Obtain the Chain Rule

$$\frac{\partial(x, y)}{\partial(w, s)}(\mathbf{p}) = \frac{\partial(x, y)}{\partial(u, v)}(H(\mathbf{p})) \frac{\partial(u, v)}{\partial(w, s)}(\mathbf{p}) \qquad \mathbf{p} \in D.$$

In particular, if the function H is inverse to G then the coordinates (x, y) and (u, v) are functionally related by

$$(x, y) = G(u, v) \qquad \text{and} \qquad (u, v) = H(x, y).$$

Deduce in this case that

7.7.14 $\qquad \dfrac{\partial(x, y)}{\partial(u, v)}(H(\mathbf{p})) \dfrac{\partial(u, v)}{\partial(x, y)}(\mathbf{p}) = 1, \qquad \mathbf{p} \in D.$

The rule 7.7.14 is often presented in the abbreviated form

$$\frac{\partial(x, y)}{\partial(u, v)} \frac{\partial(u, v)}{\partial(x, y)} = 1.$$

4. The *hyperbolic coordinates* u, v are implicitly defined in the x, y half-plane $\{(x, y) \in \mathbb{R}^2 \mid x > 0\}$ by

$$u = x^2 - y^2, \qquad v = 2xy.$$

Justify the name by sketching a selection of curves $u = $ constant and $v = $ constant.
    Prove that

$$\frac{\partial(x, y)}{\partial(u, v)}(u, v) = \frac{1}{4\sqrt{(u^2 + v^2)}}.$$

Hint: consider $\partial(u, v)/\partial(x, y)$ and apply the result of Exercise 3.

5. Sketch the set S in the first quadrant of the x, y plane bounded by the curves $x^2 - y^2 = 1$, $x^2 - y^2 = 9$, $xy = 2$, and $xy = 4$. Compute

$$\iint\limits_S (x^2 + y^2) \, dx \, dy$$

by changing to hyperbolic coordinates, given implicitly by

$$x^2 - y^2 = u, \qquad 2xy = v,$$

and applying Theorem 7.7.5.

Hint: use the result of Exercise 4. The alternative method of subdividing S into x-simple regions and applying Theorem 7.4.21 leads to complicated calculations.

Answer: 8.

6. Show that

$$\iint\limits_S x^2 y^2 \, dx \, dy = 1/360$$

where S is the triangular region bounded by the lines $y = 0$, $y = x$ and $y = 1 - x$
    (a) by computing a repeated integral;
    (b) by the change of variables $x = u + v$, $y = u - v$ and then computing a repeated integral. (In the second case the repeated integral takes the simple form $\int_0^{1/2} dv \int_0^v f(u, v) \, du$.)

7. Evaluate $\displaystyle\iint\limits_S \frac{1}{\sqrt{(x^2 + y^2)}} \, dx \, dy$ where S is the annulus $\{(x, y) \in \mathbb{R}^2 \mid a^2 \leqslant x^2 + y^2 \leqslant b^2\}$ by transforming to polar coordinates.

Answer: $2\pi(b - a)$.

8. Find the area in the first quadrant of the x, y plane bounded by the

x-axis, the curve $r = a(1 + \cos\theta)$ (given in polar coordinates) and the circle $r = a$.

Answer: $a^2(\pi + 8)/8$.

9. Compute $\iint_S x \, dx \, dy$, where S is the region defined by $x \geq 0$, $y \geq 0$,

$x^2 + y^2 \geq 1$ and $x^2 + y^2 \leq 4$. (See Fig. 7.11(ii) where S is sketched.) What are the coordinates of the centroid of S?

Answers: 7/3; (28/9, 28/9). Transform to polar coordinates.

10. Find the area bounded by the lemniscate $(x^2 + y^2)^2 = 2a^2(x^2 - y^2)$ by changing to polar coordinates.

Answer: $2a^2$.

11. (a) Compute the integral

$$\iint_{S(a)} e^{-(x^2+y^2)} \, dx \, dy,$$

where $S(a)$ is the circular disc bounded by the circle $x^2 + y^2 = a^2$. Show that

$$\lim_{a \to \infty} \iint_{S(a)} e^{-(x^2+y^2)} \, dx \, dy = \pi.$$

(b) Let $T(b)$, $b > 0$, be the square in $\mathbb{R}^2$ with corners $(\pm b, \pm b)$. Show that

$$\iint_{T(b)} e^{-(x^2+y^2)} \, dx \, dy = \left(\int_{-b}^{b} e^{-t^2} \, dt \right)^2$$

Hint: apply theorem 7.3.18.

(c) Prove that

$$\lim_{a \to \infty} \iint_{S(a)} e^{-(x^2+y^2)} \, dx \, dy = \lim_{b \to \infty} \iint_{T(b)} e^{-(x^2+y^2)} \, dx \, dy.$$

Deduce that

$$\int_{-\infty}^{\infty} e^{-t^2} \, dt = \sqrt{\pi}.$$

12. Compute the integral of $2u^2v^2(u^2 + v^2)$ over the rectangle $K^* = PBCQ$ described in Example 7.7.12. Compare your result with that of Example 7.7.12.

Answer: 64/3.

7.8   An application. Orientation of simple closed curves

Let D be a simply connected open set in $\mathbb{R}^2$. The Change of Variables Theorem enables us to explore further the significance of the Jacobian of a 1–1 C^1 function $G : D \subseteq \mathbb{R}^2 \to \mathbb{R}^2$. Let C^+ be a counterclockwise oriented simple closed curve in D, and let $\alpha : [a, b] \subseteq \mathbb{R} \to \mathbb{R}^2$ be a simple parametrization of C^+. Then $G \circ \alpha$ is a simple parametrization of the simple closed curve $G(C)$. We shall show that $G \circ \alpha$ *traces $G(C)$ counterclockwise or clockwise whenever* $\det J_{G,(u,v)}$ *is respectively positive or negative for all* $(u, v) \in D$. Hence the sign of $\det J_{G,(u,v)}$ determines whether G preserves or reverses the orientation of closed curves in D.

Let T be the region interior to C (so $\partial T = C$), and let S be the region interior to $G(C)$. Suppose that $\det J_{G,(u,v)} > 0$ for all $(u, v) \in D$. By the Change of Variables Theorem,

$$\text{Area } S = \iint_S 1 \, dx \, dy = \iint_T \det J_{G,(u,v)} \, du \, dv$$

$$= \iint_T \left(\frac{\partial G_1}{\partial u} \frac{\partial G_2}{\partial v} - \frac{\partial G_1}{\partial v} \frac{\partial G_2}{\partial u} \right) dA$$

$$= \iint_T \left(\frac{\partial}{\partial u} \left(G_1 \frac{\partial G_2}{\partial v} \right) - \frac{\partial}{\partial v} \left(G_1 \frac{\partial G_2}{\partial u} \right) \right) dA$$

$$= \int_{C^+} G_1 \frac{\partial G_2}{\partial u} \, du + G_1 \frac{\partial G_2}{\partial v} \, dv \qquad \text{by Green's Theorem}$$

$$= \int_{C^+} G_1 \operatorname{grad} G_2 \cdot d\mathbf{r}$$

$$= \int_a^b G_1(\alpha(t))(\operatorname{grad} G_2)(\alpha(t)) \cdot \alpha'(t) \, dt,$$

$$\text{since } \alpha \text{ parametrizes } C^+$$

$$= \int_a^b (G_1 \circ \alpha)(t)(G_2 \circ \alpha)'(t) \, dt \qquad \text{by Theorem 3.8.7}$$

$$= \int_a^b \beta_1(t) \beta_2'(t) \, dt \qquad \text{where } \beta = G \circ \alpha$$

$$= \int_\beta (0, x) \cdot d\beta$$

$$= \begin{cases} \text{Area } S & \text{if } G(C) \text{ is traced counterclockwise by } G \circ \alpha, \\ -\text{Area } S & \text{if } G(C) \text{ is traced clockwise by } G \circ \alpha \end{cases}$$

by 7.5.18. Therefore $G(C)$ is traced counterclockwise by $G \circ \alpha$.

Similarly a negative determinant $\det J_{G, (u, v)}$ for all $(u, v) \in D$ implies that G reverses orientations.

Exercises 7.8

1. Let a and b be non-zero constants. Verify that the function $G : \mathbb{R}^2 \to \mathbb{R}^2$ defined by

$$G(u, v) = (au, bv) \qquad (u, v) \in \mathbb{R}^2$$

is 1–1 and C^1 on $\mathbb{R}^2$, and that

$$\det J_{G, (u, v)} = ab \qquad \text{for all } (u, v) \in \mathbb{R}^2.$$

Let C^+ be the counterclockwise oriented unit circle centred at the origin of $\mathbb{R}^2$. Sketch the simple closed curve $G(C)$, and find the orientation of $G(C^+)$ for the cases (a) $a = 2$, $b = 2$; (b) $a = 2$, $b = -1$.

Answers: (a) circle $x^2 + y^2 = 4$ oriented counterclockwise; (b) ellipse $\frac{1}{4}x^2 + y^2 = 1$ oriented clockwise.

2. The transformation of the plane

$$G(u, v) = \left(\frac{u}{u^2 + v^2}, \frac{v}{u^2 + v^2} \right), \qquad (u, v) \neq (0, 0)$$

is called an *inversion* of the plane with respect to the unit circle (centre the origin). Note that in polar coordinates the point $(r \cos \theta, r \sin \theta)$, $r \neq 0$, is mapped to $\left(\frac{1}{r} \cos \theta, \frac{1}{r} \sin \theta \right)$. In particular, points on the unit circle are fixed, points inside the unit circle are mapped to points outside the unit circle and vice versa.

(a) Show that G transforms a circle not passing through the origin into a circle.

Hint: put $x = u/(u^2 + v^2)$, $y = v/(u^2 + v^2)$. Then $u = x/(x^2 + y^2)$, $v = y/(x^2 + y^2)$. Now consider the image under G of the circle $(u - a)^2 + (v - b)^2 = c^2$.

(b) Prove that

$$\det J_{G, (u, v)} = -\frac{1}{(u^2 + v^2)^2} < 0, \qquad (u, v) \neq (0, 0).$$

(c) Sketch the circle C with equation $(u - 2)^2 + v^2 = 1$ and its image

$G(C)$. Verify that C with counterclockwise orientation is mapped to $G(C)$ with clockwise orientation.

Hint: parametrize C by $\alpha(t) = (2 + \cos t, \sin t)$, $t \in [0, 2\pi]$, for example.
(d)   Show that G transforms the counterclockwise oriented circle A^+ of radius a, centre the origin, into the counterclockwise oriented circle $G(A^+)$ of radius $1/a$, centre the origin. Why does this not contradict the results of this section?

Answer: The region of $\mathbb{R}^2$ interior to A contains the origin, which lies outside the domain of G. Note that $\mathbb{R}^2 \backslash \{(0, 0)\}$ is not simply connected.

8

Surfaces in $\mathbb{R}^3$

Much of the work of the last chapter was concerned with integration over subsets of $\mathbb{R}^2$. We are now in a position to show that the same theory can be developed for integration over subsets, not of the plane, but of curved surfaces in $\mathbb{R}^3$.

The theory of integration over subsets of $\mathbb{R}^2$ relies heavily upon the geometry of $\mathbb{R}^2$. In this chapter we establish the corresponding geometry of surfaces in $\mathbb{R}^3$. In particular we study subsets of $\mathbb{R}^3$, called *smooth simple surfaces,* which are curvilinear copies in $\mathbb{R}^3$ of the simple regions in $\mathbb{R}^2$.

The most important theorem that we proved in Chapter 7 was Green's Theorem 7.5.14 which equates the line integral of a vector valued function around the boundary of a simple region and the integral of a related real valued function over the region itself. The work of this chapter will enable us, in Chapter 9, to define integrals over smooth simple surfaces and then to prove Stokes' Theorem (the curvilinear form of Green's Theorem) which equates the line integral of a function around the edge of a smooth simple surface and the integral of a related function over the surface itself.

8.1 Smooth simple surfaces

In Chapters 5 and 6 we developed a theory of integration of real-valued and vector-valued functions over curves in $\mathbb{R}^n$. The basic results involved smooth simple curves, that is, subsets of $\mathbb{R}^n$ which are the image of 1–1, smooth functions of the form $\alpha : I \subseteq \mathbb{R} \to \mathbb{R}^n$, where I is a compact interval.

Analogously, the theory of integration of real-valued and vector-valued functions over surfaces in $\mathbb{R}^3$ involves smooth simple surfaces, that is, subsets of $\mathbb{R}^3$ which are the image of 1–1 functions of the form $\rho : S \subseteq \mathbb{R}^2 \to \mathbb{R}^3$, where S is a simple region and ρ is smooth in the sense of Definition 4.3.24.

The smoothness condition involves extending ρ on (the compact set) S to a smooth function defined on an open domain containing

S. This can be done in many ways. It can be proved, however, that the partial derivatives of any two such extensions agree at all points of *S*, including the boundary ∂S. Consequently we define the partial derivatives of the smooth function ρ at a point **p** on ∂S to be equal to the partial derivatives at **p** of *any* smooth extension of ρ.

8.1.1 Definition.   *A subset K of* $\mathbb{R}^3$ *is called a* (smooth) *simple surface if there is a simple region S in* $\mathbb{R}^2$ *and a* 1–1 (*smooth*) C^1 *function* $\rho : S \subseteq \mathbb{R}^2 \to \mathbb{R}^3$ *such that* $\rho(S) = K$. *The function* ρ *is then called a* simple parametrization *of K.*

Let *K* be a smooth simple surface, simply parametrized by $\rho : S \subseteq \mathbb{R}^2 \to \mathbb{R}^3$. Denoting coordinates in $\mathbb{R}^2$ by (u, v) and coordinates in $\mathbb{R}^3$ by (x, y, z), the image $\rho(S) = K$ is described parametrically by

$$x = \rho_1(u, v), \qquad y = \rho_2(u, v), \qquad z = \rho_3(u, v)$$

for all $(u, v) \in S$. The smoothness condition on ρ means that the Jacobian matrix of ρ at **p** in *S*

$$J_{\rho, \mathbf{p}} = \begin{bmatrix} \dfrac{\partial \rho_1}{\partial u}(\mathbf{p}) & \dfrac{\partial \rho_1}{\partial v}(\mathbf{p}) \\[2mm] \dfrac{\partial \rho_2}{\partial u}(\mathbf{p}) & \dfrac{\partial \rho_2}{\partial v}(\mathbf{p}) \\[2mm] \dfrac{\partial \rho_3}{\partial u}(\mathbf{p}) & \dfrac{\partial \rho_3}{\partial v}(\mathbf{p}) \end{bmatrix}$$

has rank 2. In other words, *the vectors* $(\partial \rho / \partial u)(\mathbf{p})$ *and* $(\partial \rho / \partial v)(\mathbf{p})$ *are linearly independent.*

Just as the smoothness of a path implies that its image has no sharp corners, so smoothness of the simple parametrization $\rho : S \subseteq \mathbb{R}^2 \to \mathbb{R}^3$ implies that the simple surface $\rho(S)$ has no sharp peaks or ridges. We shall justify this statement presently.

In the following example the edge of a smooth simple surface *K* is taken to be $\rho(\partial S)$ where $\rho : S \subseteq \mathbb{R}^2 \to \mathbb{R}^3$ is a simple parametrization of *K*. It can be shown that the edge is independent of which simple parametrization is chosen.

8.1.2 Example.   Let *S* be a simple region in $\mathbb{R}^2$ and let $\rho : S \subseteq \mathbb{R}^2 \to \mathbb{R}^3$ be the 1–1 C^1 function defined by

$$\rho(u, v) = (u, v, u^2 + v^2), \qquad (u, v) \in S.$$

Then at any point (a, b) in S

$$\frac{\partial \rho}{\partial u}(a, b) = (1, 0, 2a) \qquad \text{and} \qquad \frac{\partial \rho}{\partial v}(a, b) = (0, 1, 2b).$$

These vectors are linearly independent for all choices of a, b. Therefore ρ is smooth on S for any choice of simple region S, and so ρ is a simple parametrization of the smooth simple surface $K = \rho(S)$. For example, if S is the circular disc $\{(u, v) \in \mathbb{R}^2 \mid u^2 + v^2 \leq 1\}$ then K is the parabolic bowl $z = x^2 + y^2 \leq 1$ with circular edge $z = x^2 + y^2 = 1$. On the other hand, if S is the square $\{(u, v) \in \mathbb{R}^2 \mid |u| \leq 1, |v| \leq 1\}$ then K is the parabolic bowl $z = x^2 + y^2$ whose edge constitutes the four parabolic curves $z = 1 + y^2$, $|y| \leq 1$, $x = \pm 1$ and $z = x^2 + 1$, $|x| \leq 1$, $y = \pm 1$. See Fig. 8.1. Notice that the four corners in the edge correspond to the four corners of the square S.

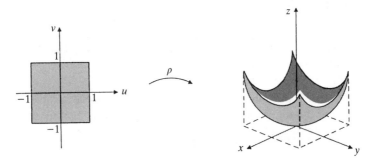

Fig. 8.1

8.1.3 *Example.* Consider the 'northern' hemisphere, radius $c > 0$, centre the origin,

$$K = \{(x, y, z) \in \mathbb{R}^3 \mid x^2 + y^2 + z^2 = c^2, z \geq 0\}.$$

Is K a (smooth) simple surface? Two 'natural' parametrizations of K suggest themselves.

[i] The *graph* parametrization

$$\gamma(x, y) = (x, y, \sqrt{(c^2 - x^2 - y^2)}), \qquad x^2 + y^2 \leq c^2,$$

where we think of K as the graph $z = g(x, y)$ of the function

$g(x, y) = \sqrt{(c^2 - x^2 - y^2)}$, defined on the simple region (disc radius c) $S = \{(x, y) \in \mathbb{R}^2 \,|\, x^2 + y^2 \leqslant c^2\}$. The function $\gamma : S \subseteq \mathbb{R}^2 \to \mathbb{R}^3$ is 1–1 but not C^1; in fact γ is not even differentiable at points on the boundary of S. Therefore γ does not answer our question about the nature of K.

  [ii]   The *spherical coordinate* parametrization $\sigma : T \subseteq \mathbb{R}^2 \to \mathbb{R}^3$, where $T = \{(\phi, \theta) \in \mathbb{R}^2 \,|\, 0 \leqslant \phi \leqslant \frac{1}{2}\pi, \, 0 \leqslant \theta \leqslant 2\pi\}$ and

$$\sigma(\phi, \theta) = (c \sin \phi \cos \theta, \, c \sin \phi \sin \theta, \, c \cos \phi), \qquad (\phi, \theta) \in T.$$

See Example 1.5.1. Then T is a simple region (a rectangle) in $\mathbb{R}^2$ and σ is clearly C^1. However, σ is not 1–1 on T, so σ does not answer our question either.

Despite these two failures at achieving a simple parametrization of K of either type, it turns out that a hemisphere is a smooth simple surface. A simple parametrization is obtained by *stereographic projection*. See Exercise 8.1.2.

8.1.4 *Example.* Let S be a simple region in $\mathbb{R}^2$ and let $g : S \subseteq \mathbb{R}^2 \to \mathbb{R}$ be a C^1 function. Then the graph of g is a smooth simple surface. It has the simple parametrization $\gamma : S \subseteq \mathbb{R}^2 \to \mathbb{R}^3$, where

$$\gamma(x, y) = (x, y, g(x, y)), \qquad (x, y) \in S,$$

for clearly γ is 1–1 and its partial derivatives $\partial \gamma / \partial x$ and $\partial \gamma / \partial y$ are linearly independent.

We can now establish a link between smooth simple surfaces and our earlier definition of a smooth surface in $\mathbb{R}^3$ as a level set of a smooth function $f : E \subseteq \mathbb{R}^3 \to \mathbb{R}$ (Exercise 3.8.10). Suppose that

$$K = \{(x, y, z) \in E \,|\, f(x, y, z) = c\}.$$

Consider a point $\mathbf{p} = (a, b, c)$ in K. Since f is smooth, $(\mathrm{grad}\, f)(\mathbf{p}) \neq \mathbf{0}$. Suppose that $(\partial f / \partial z)(\mathbf{p}) \neq 0$. Then, by the Implicit Function Theorem 4.8.1 applied to $F(x, y, z) = f(x, y, z) - c$, there is a C^1 function $g : N \subseteq \mathbb{R}^2 \to \mathbb{R}$ defined on a neighbourhood N of (a, b) in $\mathbb{R}^2$ such that the points of K close to $\mathbf{p}$ are described by the graph $z = g(x, y)$, $(x, y) \in N$. The alternative conditions $(\partial f / \partial y)(\mathbf{p}) \neq 0$, or $(\partial f / \partial x)(\mathbf{p}) \neq 0$ lead to similar descriptions $y = g(x, z)$ or $x = g(y, z)$. It follows from Example 8.1.4 that K is made up of overlapping smooth simple surfaces.

8.1.5 *Example.* The sphere $x^2 + y^2 + z^2 = c^2$, $c > 0$, is a level set of the smooth function $f : \mathbb{R}^3 \backslash \{\mathbf{0}\} \to \mathbb{R}$ given by $f(x, y, z) = x^2 + y^2 + z^2$. The sphere is not a simple surface. (A simple surface must have a non-trivial edge.) However, it is made up of two smooth simple surfaces, the southern and the northern hemispheres, centre the origin, radius c.

Put briefly, we have shown above that a level set of a smooth function is *locally* a smooth simple surface. Conversely, a smooth simple surface is locally a level set of a smooth function. For let K be a smooth simple surface simply parametrized by $\rho : S \subseteq \mathbb{R}^2 \to \mathbb{R}^3$. Consider a point $\mathbf{p}$ in S. Since the Jacobian matrix $J_{\rho, \mathbf{p}}$ has rank 2, the determinant

$$\frac{\partial(\rho_i, \rho_j)}{\partial(u, v)}(\mathbf{p}) = \det \begin{bmatrix} \dfrac{\partial \rho_i}{\partial u}(\mathbf{p}) & \dfrac{\partial \rho_i}{\partial v}(\mathbf{p}) \\[2ex] \dfrac{\partial \rho_j}{\partial u}(\mathbf{p}) & \dfrac{\partial \rho_j}{\partial v}(\mathbf{p}) \end{bmatrix}$$

is non-zero for some $i, j = 1, 2, 3$ and $i \neq j$. Suppose by way of illustration that this is so when $i = 1$, $j = 3$. Consider the parametric description of K

$$x = \rho_1(u, v) \qquad y = \rho_2(u, v), \qquad z = \rho_3(u, v), \qquad (u, v) \in S.$$

By the Inverse Function Theorem 4.6.7 the equations

$$x = \rho_1(u, v), \qquad z = \rho_3(u, v)$$

can be solved uniquely for u and v with (u, v) sufficiently close to $\mathbf{p}$, and values of (x, z) sufficiently close to $(\rho_1(\mathbf{p}), \rho_3(\mathbf{p}))$. Moreover the solutions $u = g(x, z)$, $v = h(x, z)$ are C^1. It follows that in a suitable neighbourhood of $\rho(\mathbf{p})$ the surface K is given by the equation

$$y = \rho_2(g(x, z), h(x, z)).$$

So in this neighbourhood K is a level set of the smooth function $f(x, y, z) = y - \rho_2(g(x, z), h(x, z))$.

Whereas level sets often give a convenient *definition* of a surface, the appropriate tool for *integration* over surfaces is the smooth simple surface. Since a smooth simple surface K in $\mathbb{R}^3$ is a copy of a simple region S in $\mathbb{R}^2$ (through a 1–1 smooth function $\rho : S \subseteq \mathbb{R}^2 \to \mathbb{R}^3$) it is possible to relate integrals over K to integrals over S, and these we have learnt to deal with in the last chapter.

Partial justification for calling a parametrization $\rho : S \subseteq \mathbb{R}^2 \to \mathbb{R}^3$ smooth when the partial derivatives $\partial \rho / \partial u$ and $\partial \rho / \partial v$ are linearly independent is given by the following theorem. But first we give an example of what can happen when a C^1 function $\sigma : S \subseteq \mathbb{R}^2 \to \mathbb{R}^3$ is not smooth.

8.1.6 *Example.* Consider the 'square cone' $z = |x| + |y|$ in $\mathbb{R}^3$ and the C^1 parametrization of it $\sigma : \mathbb{R}^2 \to \mathbb{R}^3$ defined by

$$\sigma(u, v) = (u^3, v^3, |u|^3 + |v|^3) \qquad (u, v) \in \mathbb{R}^2.$$

See Fig. 8.2. Then, for all $\mathbf{p} = (u, v) \in \mathbb{R}^2$,

$$\frac{\partial \sigma}{\partial u}(\mathbf{p}) = (3u^2, 0, 3u\,|u|) \qquad \text{and} \qquad \frac{\partial \sigma}{\partial v}(\mathbf{p}) = (0, 3v^2, 3v\,|v|).$$

The function σ is smooth everywhere except along the axes $u = 0$ and $v = 0$. These two axes are taken by σ to the ridges in the cone.

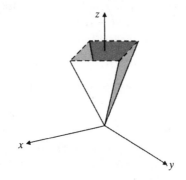

Fig. 8.2 Square cone $z = |x| + |y|$

8.1.7 *Theorem.* *Let D be an open subset of $\mathbb{R}^2$, and let $\sigma : D \subseteq \mathbb{R}^2 \to \mathbb{R}^n$, $n \geqslant 2$, be a C^1 function. Then σ is smooth at $\mathbf{p} \in D$ if and only if, for each smooth path $\alpha : [-1, 1] \subseteq \mathbb{R} \to \mathbb{R}^2$ in D such that $\alpha(0) = \mathbf{p}$, the path $\sigma \circ \alpha : [-1, 1] \subseteq \mathbb{R} \to \mathbb{R}^n$ in $\sigma(D)$ is smooth at $0 \in \mathbb{R}$.*

Proof. Assume first that σ is smooth at $\mathbf{p} \in D$. Let $\alpha : [-1, 1] \subseteq \mathbb{R} \to \mathbb{R}^2$ be a smooth path in D such that $\alpha(0) = \mathbf{p}$. By the Chain Rule, Exercise 4.4.8,

$$(\sigma \circ \alpha)'(0) = \alpha_1'(0) \frac{\partial \sigma}{\partial u}(\mathbf{p}) + \alpha_2'(0) \frac{\partial \sigma}{\partial v}(\mathbf{p}).$$

Since $(\partial\sigma/\partial u)(\mathbf{p})$ and $(\partial\sigma/\partial v)(\mathbf{p})$ are linearly independent, and $\alpha'(0) \neq \mathbf{0}$ it follows that $(\sigma \circ \alpha)'(0) \neq \mathbf{0}$, and so $\sigma \circ \alpha$ is smooth at 0.

Conversely, suppose that σ is not smooth at $\mathbf{p} \in D$. Then there exist $(a, b) \neq (0, 0)$ such that

$$a \frac{\partial\sigma}{\partial u}(\mathbf{p}) + b \frac{\partial\sigma}{\partial v}(\mathbf{p}) = \mathbf{0}.$$

Consider the smooth path in D given by $\alpha(t) = \mathbf{p} + t(a, b)$, $t \in [-k, k]$ (for suitably small k). Then $(\sigma \circ \alpha)'(0) = \mathbf{0}$ and so $\sigma \circ \alpha : [-k, k] \subseteq \mathbb{R} \to \mathbb{R}^3$ is not smooth at $0 \in \mathbb{R}$. By a change of scale we can redefine α on the interval $[-1, 1]$, and the proof is complete.

Theorem 8.1.7 tells us that a function $\sigma : D \subseteq \mathbb{R}^2 \to \mathbb{R}^n$ is smooth if and only if it transforms all smooth paths in D into smooth paths in $\sigma(D)$.

As in the case of smooth simple arcs, a smooth simple surface may be the image of non-smooth functions as well as smooth functions. Example 8.1.3 provides an illustration. Here is another.

8.1.8 *Example.* The parabolic bowl $z = x^2 + y^2 \leq 1$ considered in Example 8.1.2 as the image of a smooth function, can also be parametrized by the C^1 function $\sigma : T \subseteq \mathbb{R}^2 \to \mathbb{R}^3$ given by

$$\sigma(u, v) = (u^3, v^3, u^6 + v^6), \qquad (u, v) \in T.$$

where T is the simple region $\{(u, v) \in \mathbb{R}^2 \mid u^6 + v^6 \leq 1\}$. In this case

$$\frac{\partial\sigma}{\partial u}(u, v) = (3u^2, 0, 6u^5) \qquad \text{and} \qquad \frac{\partial\sigma}{\partial v}(u, v) = (0, 3v^2, 6v^5).$$

Therefore the function σ is not smooth at points on the u-axis and on the v-axis.

In contrast to the parabolic bowl which admits both smooth and non-smooth parametrizations, there are no smooth parametrizations of the square cone. The presence of the vertex of the cone rules out the possibility of a smooth parametrization, as does the presence of the ridges.

The next theorem shows how two simple parametrizations of a smooth simple surface are related.

8.1.9 *Definition.* *Let K be a smooth simple surface in $\mathbb{R}^3$ and let*

$\rho : S \subseteq \mathbb{R}^2 \to \mathbb{R}^3$ *and* $\eta : T \subseteq \mathbb{R}^2 \to \mathbb{R}^3$ *be simple parametrizations of K. The parametrization* ρ *is said to be* equivalent *to* η *if there is a smooth function* ϕ *from S onto T such that* $\rho = n \circ \phi$.

The smoothness condition on $\phi : S \subseteq \mathbb{R}^2 \to \mathbb{R}^2$ means that the 2×2 Jacobian matrix $J_{\phi, \mathbf{p}}$ is non-singular at all points $\mathbf{p} \in S$. Since simple parametrizations are 1–1, the function ϕ referred to in Definition 8.1.9 must also be 1–1.

8.1.10 *Theorem.* *Let K be a smooth simple surface in* $\mathbb{R}^3$ *and let* $\rho : S \subseteq \mathbb{R}^2 \to \mathbb{R}^3$ *and* $\eta : T \subseteq \mathbb{R}^2 \to \mathbb{R}^3$ *be simple parametrizations of K. Then* ρ *is equivalent to* η.

The proof of Theorem 8.1.10 which we omit, is based on the Inverse Function Theorem.

The reader will notice similarities in the properties of smooth simple arcs in $\mathbb{R}^n$ and simple surfaces in $\mathbb{R}^3$. The resemblance begins with their definitions. Both are images of a 1–1 smooth function defined on a compact set, an interval in $\mathbb{R}$ for simple arcs, and a simple region in $\mathbb{R}^2$ for simple surfaces. The concept of equivalence is similarly defined, and now we have the result that in both cases there is just one equivalence class of 1–1 smooth simple parametrizations.

Further similarities between smooth simple arcs and smooth simple surfaces arise in the discussion of orientation, as we shall see in the next section.

8.1.11 *Definition.* *Let K be a smooth simple surface in* $\mathbb{R}^3$ *and let* $\rho : S \subseteq \mathbb{R}^2 \to \mathbb{R}^3$ *be a simple parametrization of K.*

[i] *The* edge *of K is the simple closed curve* $\rho(\partial S)$. *It is denoted by* ∂K.

[ii] *The* inside *of K is the set* $\rho(\text{Int } S)$. *It is denoted by* K^0.

It can be shown that the edge and inside of K do not depend upon the parametrization chosen.

8.1.12 *Example.* The smooth simple surface (parabolic bowl) $z = x^2 + y^2$, $z \leq 1$ in $\mathbb{R}^3$ has the following two simple parametrizations.

[i] $\rho : S \subseteq \mathbb{R}^2 \to \mathbb{R}^3$, where S is the circular disc $\{(u, v) \in \mathbb{R}^2 \mid u^2 + v^2 \leq 1\}$ and

$$\rho(u, v) = (u, v, u^2 + v^2), \qquad (u, v) \in S.$$

[ii] $\eta : T \subseteq \mathbb{R}^2 \to \mathbb{R}^3$, where T is the elliptical disc $\{(s, t) \in \mathbb{R}^2 \mid 4s^2 + t^2 \leqslant 1\}$ and

$$\eta(s, t) = (2s, -t, 4s^2 + t^2), \qquad (s, t) \in T.$$

The 1–1 smooth function ϕ from S onto T such that $\rho = \eta \circ \phi$ is found as follows. Consider $(u, v) \in S$ and put $\phi(u, v) = (s, t) \in T$. Then

$$\rho(u, v) = \eta(\phi(u, v)) = \eta(s, t),$$

that is,

$$(u, v, u^2 + v^2) = (2s, -t, 4s^2 + t^2), \qquad (u, v) \in S.$$

Solving for s, t, we obtain $s = \tfrac{1}{2}u$, $t = -v$, and so

$$\phi(u, v) = (\tfrac{1}{2}u, -v), \qquad (u, v) \in S.$$

Notice that ϕ maps Int S onto Int T and ∂S onto ∂T. The edge of the bowl is

$$\rho(\partial S) = \eta(\partial T) = \{(x, y, z) \in \mathbb{R}^3 \mid x^2 + y^2 = 1, \ z = 1\}.$$

Finally in this section we introduce a generalization of a smooth simple surface.

8.1.13 Definition. *A subset K of $\mathbb{R}^3$ is said to be a* (smooth) *piecewise simple surface if there is a piecewise simple region S in $\mathbb{R}^2$ and a 1–1* (smooth) *C^1 function $\rho : S \subseteq \mathbb{R}^2 \to \mathbb{R}^3$ such that $\rho(S) = K$. The function ρ is called a* simple parametrization *of K.*

A simple surface is of course also piecewise simple. The compact cylinder $x^2 + y^2 = 1$, $|z| \leqslant 1$ is a piecewise simple surface that is not simple (Exercises 8.1.3 and 8.1.4).

The results of this section can be extended without difficulty to apply to a smooth piecewise simple surface K. In particular, any two simple parametrizations of K are equivalent in the sense of Definition 8.1.9. The edge of K is unambiguously defined to be the set $\rho(\partial S)$ where $\rho : S \subseteq \mathbb{R}^2 \to \mathbb{R}^3$ is any simple parametrization of K. The inside of K is the set $K^0 = \rho(\text{Int } S)$.

Exercises 8.1

1. Sketch roughly the following smooth simple surfaces K and find a simple parametrization $\rho : S \subseteq \mathbb{R}^2 \to \mathbb{R}^3$ of K.
    (a) The spherical cap $x^2 + y^2 + z^2 = 1$, $\tfrac{1}{2} \leqslant z \leqslant 1$.
    (b) The spherical cap $x^2 + y^2 + z^2 = 1$, $-1 \leqslant x \leqslant -\tfrac{1}{3}$.

(c)   The intersection of the spherical caps (a) and (b).
(d)   The rectangular surface $y = 3$, $|x| \leq 1$, $-1 \leq z \leq 2$.

Answers: (a) $\rho(u, v) = (u, v, \sqrt{(1 - u^2 - v^2)})$, $u^2 + v^2 \leq \frac{3}{4}$.
  (b) $\rho(s, t) = (-\sqrt{(1 - s^2 - t^2)}, s, t)$, $s^2 + t^2 \leq \frac{8}{9}$.
  (c) $\rho(u, v) = (u, v, \sqrt{(1 - u^2 - v^2)})$, $u^2 + v^2 \leq \frac{3}{4}$ and $-\frac{1}{2}\sqrt{3} \leq u \leq -\frac{1}{3}$.
  (d) $\rho(u, v) = (u, 3, v)$, $|u| \leq 1$, $-1 \leq v \leq 2$.

2.   Let K be the 'southern hemisphere' of the unit sphere in $\mathbb{R}^3$, centre the origin. Verify that the simple parametrization of K obtained through stereographic projection from $(0, 0, 1)$ (Fig. 8.3) is defined implicitly by

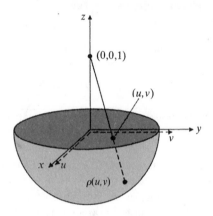

Fig. 8.3 Stereographic projection

$\rho(u, v) = (x, y, z)$, $u^2 + v^2 \leq 1$, where

$$u = \frac{x}{1 - z}, \qquad v = \frac{y}{1 - z}, \qquad z = -\sqrt{(1 - x^2 - y^2)}.$$

Deduce the explicit formula

$$\rho(u, v) = \frac{1}{1 + u^2 + v^2}(2u, 2v, u^2 + v^2 - 1), \qquad u^2 + v^2 \leq 1.$$

and verify that ρ is a smooth parametrization of K.

3.   Let K be the compact cylinder $x^2 + y^2 = 1$, $|z| \leq 1$. Show that K is a piecewise simple surface.

Proof: let S be the annulus $\{(u, v) \in \mathbb{R}^2 \mid 1 \leq u^2 + v^2 \leq 9\}$. Define

$\rho : S \subseteq \mathbb{R}^2 \to \mathbb{R}^3$ by

$$\rho(u, v) = (u/\sqrt{(u^2 + v^2)},\ v/\sqrt{(u^2 + v^2)},\ \sqrt{(u^2 + v^2)} - 2),$$

$$(u, v) \in S.$$

Then ρ is a simple parametrization of the cylinder.

4. Show that the cylinder K of Exercise 3 is not a simple surface.

Hint: assume a simple parametrization $\rho : S \subseteq \mathbb{R}^2 \to \mathbb{R}^3$ of K. Let C and C^* be the simple closed curves $x^2 + y^2 = 1,\ z = 1$, and $x^2 + y^2 = 1,\ z = -1$ in K. Then, $D = \rho^{-1}(C)$ and $D^* = \rho^{-1}(C^*)$ are disjoint (closed) Jordan curves in S. Then $K \backslash (C \cup C^*)$ is path connected, but $S \backslash (D \cup D^*)$ is not path connected.

5. Show that the sphere $x^2 + y^2 + z^2 = c^2$ is not a piecewise simple surface.

Hint: assume a simple parametrization $\rho : S \subseteq \mathbb{R}^2 \to \mathbb{R}^3$ of the sphere K. Then $C = \rho(\partial S)$ consists of a finite number of simple closed curves in K. The set $S \backslash \partial S$ is path connected, but $K \backslash C$ is not.

6. Let K be the semicircular sheet $x^2 + y^2 = k^2,\ x \geqslant 0,\ |z| \leqslant 1$. Verify that K has the following simple parametrizations $\rho : S \subseteq \mathbb{R}^2 \to \mathbb{R}^3$ and $\eta : T \subseteq \mathbb{R}^2 \to \mathbb{R}^3$:
    (a) $\rho(u, v) = (k \cos u,\ k \sin u,\ v),$ $-\tfrac{1}{2}\pi \leqslant u \leqslant \tfrac{1}{2}\pi,$ $-1 \leqslant v \leqslant 1,$
    (b) $\eta(s, t) = (k \sin s,\ k \cos s,\ -t)$ $0 \leqslant s \leqslant \pi,$ $-1 \leqslant t \leqslant 1.$
    Find the 1–1 continuous function ϕ from S onto T such that $\rho = \eta \circ \phi$, and verify that ϕ is smooth.

Hint: solve for $s,\ t$ the equation $\rho(u, v) = \eta(s, t)$; then $(s, t) = \phi(u, v)$.

Answer: $\phi(u, v) = (\tfrac{1}{2}\pi - u,\ -v).$

The following exercises illustrate Theorem 8.1.7 for the cases $n = 3$ and $n = 2$.

7. (a) Define the C^1 function $\sigma : \mathbb{R}^2 \to \mathbb{R}^3$ by

$$\sigma(u, v) = (u^2 - v^2,\ uv,\ 0), \qquad (u, v) \in \mathbb{R}^2.$$

Show that σ is not smooth on $\mathbb{R}^2$, but that it is smooth on any open set D in $\mathbb{R}^2$ that does not contain the origin.
    (b) Verify that the C^1 path $\alpha : [-\pi, \tfrac{1}{2}\pi] \subseteq \mathbb{R} \to \mathbb{R}^2$ defined by

$$\alpha(t) = (\cos t,\ \sin t), \qquad -\pi \leqslant t \leqslant \tfrac{1}{2}\pi$$

is smooth, and that the C^1 path $\alpha^* = \sigma \circ \alpha : [-\pi, \tfrac{1}{2}\pi] \subseteq \mathbb{R} \to \mathbb{R}^3$ is also smooth. Describe the image C of α and the image $\sigma(C)$ of α^*.
    (c) Consider similarly the C^1 path (through the origin of $\mathbb{R}^2$) $\beta : [-1, 1] \subseteq \mathbb{R} \to \mathbb{R}^2$ defined by

$$\beta(t) = (t, t^2), \qquad -1 \leqslant t \leqslant 1.$$

Verify that β is smooth, but that $\beta^* = \sigma \circ \beta$ is not smooth. Sketch the image of β and the image of β^*.

Answers: (a) $\partial\sigma/\partial u$ and $\partial\sigma/\partial v$ are linearly independent except at the origin. (b) The path α traces three-quarters of the unit circle $u^2 + v^2 = 1$. The path α^* traces the ellipse $x^2 + 4y^2 = 1$, $z = 0$ one and a half times. Note that α is 1–1, but α^* is not. (c) The image of β^* has a cusp at the origin.

8. Define the C^1 function $G : \mathbb{R}^2 \to \mathbb{R}^2$ by

$$G(u, v) = (u^2 + v^2, u + v), \qquad (u, v) \in \mathbb{R}^2.$$

Show that G is not smooth on any open set D in the u, v plane that meets the line $u = v$.

Let $\alpha : [0, 2] \subseteq \mathbb{R} \to \mathbb{R}^2$ be the C^1 path given by

$$\alpha(t) = \begin{cases} (t, 2 - t) & \text{when } 0 \leqslant t \leqslant 1 \\ (t, 1/t) & \text{when } 1 \leqslant t \leqslant 2. \end{cases}$$

Show that α is smooth but that $\alpha^* = G \circ \alpha$ is not smooth. Sketch the image of α and the image of α^*.

Note: the function G is 1–1 on the half-plane $u \geqslant v$ and on the half-plane $u \leqslant v$. It maps both these half-planes, through the transformation $x = u^2 + v^2$, $y = u + v$, onto part of the x, y plane defined by $x \geqslant \frac{1}{2}y^2$. Sketch the image of α in the u, v plane and image of α^* in the x, y plane.

This example illustrates the importance of smoothness in Change of variables problems (Section 7.7). Note that condition (iii) of the Change of variables Theorem 7.7.5 is a smoothness condition on the function G.

8.2   Tangents and normals

Let K be a simple surface. Suppose that $\lambda : [a, b] \subseteq \mathbb{R} \to \mathbb{R}^3$ is a C^1 path whose image lies in K and that $\lambda(c) = \mathbf{q}$. Then $\lambda'(c)$ is a tangent vector to the curve $\lambda([a, b])$ at $\mathbf{q}$ and could reasonably be regarded also as a tangent vector to K at $\mathbf{q}$.

8.2.1 Definition.   *Let K be a simple surface and let $\mathbf{q}$ be a point of K^0. Then $\mathbf{v} \in \mathbb{R}^3$ is a* tangent vector *to K at $\mathbf{q}$ if there is a C^1 path $\lambda : [a, b] \subseteq \mathbb{R} \to \mathbb{R}^3$, whose image lies in K, such that, for some $c \in [a, b]$, $\lambda(c) = \mathbf{q}$ and $\lambda'(c) = \mathbf{v}$. The set of all tangent vectors to K at $\mathbf{q}$ is denoted by $T_{\mathbf{q}}(K)$.*

Now let $\rho : S \subseteq \mathbb{R}^2 \to \mathbb{R}^3$ be a simple parametrization of a smooth simple surface K, let **p** be an interior point of S, and put $\rho(\mathbf{p}) = \mathbf{q} \in K$. Then the two paths $\alpha : [-d, d] \subseteq \mathbb{R} \to \mathbb{R}^2$ and $\beta : [-d, d] \subseteq \mathbb{R} \to \mathbb{R}^2$ given by

$$\alpha(t) = \mathbf{p} + t(1, 0) \quad \text{and} \quad \beta(t) = \mathbf{p} + t(0, 1) \quad \text{for all } t \in [-d, d]$$

both pass through $\alpha(0) = \beta(0) = \mathbf{p}$ and, provided $d > 0$ is small enough, lie in S. Hence $\rho \circ \alpha$ and $\rho \circ \beta$ are C^1 paths through $(\rho \circ \alpha)(0) = (\rho \circ \beta)(0) = \mathbf{q}$ in K. By the Chain Rule

8.2.2 $(\rho \circ \alpha)'(0) = \dfrac{\partial \rho}{\partial u}(\mathbf{p}) \quad \text{and} \quad (\rho \circ \beta)'(0) = \dfrac{\partial \rho}{\partial v}(\mathbf{p}).$

These two vectors therefore, are tangential to K at **q**. See Fig. 8.4.

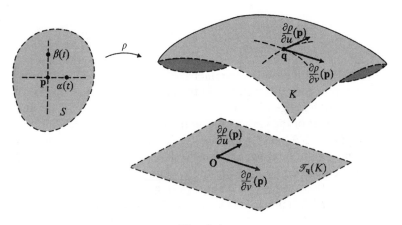

Fig. 8.4

They are linearly independent, since ρ is smooth.

Since **p** is an interior point of S, for any real numbers k and l the path $k\alpha + l\beta : [-d, d] \subseteq \mathbb{R} \to \mathbb{R}^2$ through **p** lies in S, provided $d > 0$ is small enough. It follows that every linear combination of the vectors $(\partial \rho / \partial u)(\mathbf{p})$ and $(\partial \rho / \partial v)(\mathbf{p})$ is tangential to K at **q**. Therefore $T_{\mathbf{q}}(K) \supseteq \text{span} \{(\partial \rho / \partial u)(\mathbf{p}), (\partial \rho / \partial v)(\mathbf{p})\}$.

Consider next an arbitrary C^1 path $\gamma : [a, b] \subseteq \mathbb{R} \to \mathbb{R}^2$ through **p** in S, and suppose that $\gamma(c) = \mathbf{p}$. If $\gamma'(c) = (k, l)$ then

$$\gamma'(c) = k(1, 0) + l(0, 1) = k\alpha'(0) + l\beta'(0).$$

It follows by the Chain Rule

$$(\rho \circ \gamma)'(c) = (\text{grad } \rho)(\mathbf{p}) \cdot \gamma'(c) = k(\rho \circ \alpha)'(0) + (\rho \circ \beta)'(0),$$

and by 8.2.2 that

$$(\rho \circ \gamma)'(c) = k \frac{\partial \rho}{\partial u}(\mathbf{p}) + l \frac{\partial \rho}{\partial v}(\mathbf{p}) \in \text{span} \left\{ \frac{\partial \rho}{\partial u}(\mathbf{p}), \frac{\partial \rho}{\partial v}(\mathbf{p}) \right\}.$$

It can be shown that *any* C^1 path in K through $\mathbf{q} \in K$ is the image under ρ of a C^1 path in S through $\mathbf{p} \in S$, so we have the following result.

8.2.3 Theorem.   *Let K be a smooth simple surface and let $\mathbf{q}$ lie in K^0. For any simple parametrization $\rho : S \subseteq \mathbb{R}^2 \to \mathbb{R}^3$ of K, if $\rho(\mathbf{p}) = \mathbf{q}$ then*

8.2.4 $$T_{\mathbf{q}}(K) = \text{span} \left\{ \frac{\partial \rho}{\partial u}(\mathbf{p}), \frac{\partial \rho}{\partial v}(\mathbf{p}) \right\}.$$

Theorem 8.2.3 tells us that the set $T_{\mathbf{q}}(k)$ is a plane and that it is independent of the simple parametrization ρ of K used to define it. The theorem concerns points $\mathbf{q}$ in the inside K^0 of K. For points $\mathbf{q} = \rho(\mathbf{p})$ on the edge of K we define $T_{\mathbf{q}}(K)$ by 8.2.4. The definition is independent of the choice of simple parametrization ρ of K.

The plane $T_{\mathbf{q}}(K)$ passes through $\mathbf{0} \in \mathbb{R}^3$ and may not even meet K, let alone contain $\mathbf{q}$. However, by translating $T_{\mathbf{q}}(K)$ in $\mathbb{R}^3$ through $\mathbf{q}$ we obtain a plane that is tangential to K at $\mathbf{q}$. The set $\mathbf{q} + T_{\mathbf{q}}(K)$ is naturally called the *tangent plane to K at $\mathbf{q}$*.

Since $T_{\mathbf{q}}(K)$ is a plane in $\mathbb{R}^3$, the set of all vectors in $\mathbb{R}^3$ which are orthogonal to $T_{\mathbf{q}}(K)$ form a one-dimensional subspace. Any vector in this line is orthogonal at $\mathbf{q}$ to every curve in K through $\mathbf{q}$. Such a vector, if non-zero, is called a *normal vector to K at $\mathbf{q}$*.

Again, let $\rho : S \subseteq \mathbb{R}^2 \to \mathbb{R}^3$ be a simple parametrization of K and let $\rho(\mathbf{p}) = \mathbf{q}$. Then the cross product

8.2.5 $$\nu_\rho(\mathbf{p}) = \frac{\partial \rho}{\partial u}(\mathbf{p}) \times \frac{\partial \rho}{\partial v}(\mathbf{p})$$

is a normal vector to K at $\mathbf{q}$.

The tangent plane $\mathbf{q} + T_{\mathbf{q}}(K)$ has equation

8.2.6 $$\nu_\rho(\mathbf{p}) \cdot ((x, y, z) - \mathbf{q}) = 0.$$

The equation is independent of the choice of simple parametrization

ρ of K, since if η is another simple parametrization of K with $\eta(\mathbf{w}) = \mathbf{q}$ then $\nu_\rho(\mathbf{p})$ and $\nu_\eta(\mathbf{w})$ are both orthogonal to $T_\mathbf{q}(K)$ and therefore linearly dependent. The precise relationship between them is as follows.

8.2.7 Theorem. *Let K be a smooth simple surface in $\mathbb{R}^3$, let $\rho : S \subseteq \mathbb{R}^2 \to \mathbb{R}^3$ and $\eta : T \subseteq \mathbb{R}^2 \to \mathbb{R}^3$ be simple parametrizations of K, and let ϕ be the smooth function from S onto T such that $\rho = \eta \circ \phi$. If $\mathbf{p} \in S$ and $\mathbf{w} \in T$ are such that $\rho(\mathbf{p}) = \eta(\mathbf{w})$ then*

8.2.8
$$\nu_\rho(\mathbf{p}) = \det J_{\phi,\mathbf{p}} \nu_\eta(\mathbf{w}).$$

Proof.   Parametrizing points in S by (u, v) and points in T by (s, t), the relation $\rho = \eta \circ \phi$ implies that $(\rho_2, \rho_3) = (\eta_2, \eta_3) \circ \phi$ and (Exercise 4.4.9)

$$\frac{\partial(\rho_2, \rho_3)}{\partial(u, v)}(\mathbf{p}) = \frac{\partial(\eta_2, \eta_3)}{\partial(s, t)}(\mathbf{w}) \frac{\partial(\phi_1, \phi_2)}{\partial(u, v)}(\mathbf{p}),$$

with similar expresions for $\partial(\rho_3, \rho_1)/\partial(u, v)$ and $\partial(\rho_1, \rho_2)/\partial(u, v)$.
   Now

8.2.9
$$\nu_\rho(\mathbf{p}) = \frac{\partial \rho}{\partial u}(\mathbf{p}) \times \frac{\partial \rho}{\partial v}(\mathbf{p})$$
$$= \left(\frac{\partial(\rho_2, \rho_3)}{\partial(u, v)}(\mathbf{p}), \frac{\partial(\rho_3, \rho_1)}{\partial(u, v)}(\mathbf{p}), \frac{\partial(\rho_1, \rho_2)}{\partial(u, v)}(\mathbf{p}) \right).$$

Therefore

$$\nu_\rho(\mathbf{p}) = \nu_\eta(\mathbf{w}) \frac{\partial(\phi_1, \phi_2)}{\partial(u, v)}(\mathbf{p}),$$

which is 8.2.8.

8.2.10 Example.   Let K be the parabolic bowl $z = x^2 + y^2 \leqslant 1$.
   [i]   Consider the simple parametrization $\rho : S \subseteq \mathbb{R}^2 \to \mathbb{R}^3$ of K where

$$\rho(u, v) = (u, v, u^2 + v^2), \qquad (u, v) \in S$$

and S is the circular disc $\{(u, v) \in \mathbb{R}^2 \mid u^2 + v^2 \leqslant 1\}$. Let $\mathbf{p} = (a, b) \in S$. Then $\mathbf{q} = \rho(\mathbf{p}) = (a, b, a^2 + b^2)$ and

$$\frac{\partial \rho}{\partial u}(\mathbf{p}) = (1, 0, 2a) \qquad \text{and} \qquad \frac{\partial \rho}{\partial v}(\mathbf{p}) = (0, 1, 2b).$$

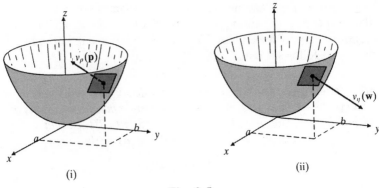

Fig. 8.5

By the cross product formula 1.2.20

$$v_\rho(\mathbf{p}) = \frac{\partial \rho}{\partial u}(\mathbf{p}) \times \frac{\partial \rho}{\partial v}(\mathbf{p}) = (-2a, -2b, 1).$$

See Fig. 8.5. The equation of the tangent plane to K at $\mathbf{q} = \rho(\mathbf{p})$ is, by 8.2.6,

$$(-2a, -2b, 1) \cdot (x - a, y - b, z - a^2 - b^2) = 0$$

that is,

8.2.11 $$2ax + 2by - z = a^2 + b^2.$$

[ii]   Contrast ρ with the alternative simple parametrization $\eta : T \subseteq \mathbb{R}^2 \to \mathbb{R}^3$ of K given by

$$\eta(s, t) = (2s, -t, 4s^2 + t^2), \qquad (s, t) \in T$$

where T is the elliptical disc $\{(s, t) \in \mathbb{R}^2 \mid 4s^2 + t^2 \leq 1\}$. Taking $\mathbf{w} = (\tfrac{1}{2}a, -b)$ we have $\mathbf{w} \in T$ and $\eta(\mathbf{w}) = (a, b, a^2 + b^2) = \mathbf{q}$. Since

$$\frac{\partial \eta}{\partial s}(\mathbf{w}) = (2, 0, 4a) \qquad \text{and} \qquad \frac{\partial \eta}{\partial t}(\mathbf{w}) = (0, -1, -2b)$$

we obtain

$$v_\eta(\mathbf{w}) = \frac{\partial \eta}{\partial s}(\mathbf{w}) \times \frac{\partial \eta}{\partial t}(\mathbf{w}) = (4a, 4b, -2) = -2v_\rho(\mathbf{p}).$$

The 1–1 smooth function ϕ from S onto T such that $\rho = \eta \circ \phi$ is given by (Example 8.1.12)

$$\phi(u, v) = (\tfrac{1}{2}u, -v) \qquad (u, v) \in S.$$

Hence

$$J_{\phi,(u,v)} = \begin{bmatrix} \frac{1}{2} & 0 \\ 0 & -1 \end{bmatrix} \qquad (u, v) \in S,$$

and therefore $\det J_{\phi,(u,v)} = -\frac{1}{2}$. This confirms 8.2.8. The reader can verify that if the parametrization η is used to calculate the tangent plane to K at $\mathbf{q}$, then again equation 8.2.11 is obtained.

Exercises 8.2

1. For each of the following smooth simple surfaces K verify that the given functions $\rho : S \subseteq \mathbb{R}^2 \to \mathbb{R}^3$ and $\eta : T \subseteq \mathbb{R}^2 \to \mathbb{R}^3$ are simple parametrizations of K. For the given point $\mathbf{q} \in K$
   (a) find $\mathbf{p} \in S$ and $\mathbf{w} \in T$ such that $\rho(\mathbf{p}) = \eta(\mathbf{w}) = \mathbf{q}$;
   (b) calculate $v_\rho(\mathbf{p})$ and $v_\eta(\mathbf{w})$, check that they are linearly dependent, and verify formula 8.2.8;
   (c) find the equation of the tangent plane to K at $\mathbf{q}$;
   (d) illustrate your answers with a sketch.
      (i) The *parabolic bowl* $z = x^2 + y^2$, $|z| \leq 10$, $\mathbf{q} = (1, 2, 5)$,

$$\rho(u, v) = (u, v, u^2 + v^2), \qquad\qquad u^2 + v^2 \leq 10,$$
$$\eta(s, t) = (s + t, s - t, 2s^2 + 2t^2), \qquad\qquad s^2 + t^2 \leq 5.$$

      (ii) The *spherical cap* $x^2 + y^2 + z^2 = 1$, $\frac{1}{2} \leq z \leq 1$, $\mathbf{q} = (\frac{1}{3}, \frac{2}{3}, \frac{2}{3})$,

$$\rho(u, v) = (u, v, \sqrt{(1 - u^2 - v^2)}), \qquad\qquad u^2 + v^2 \leq \tfrac{3}{4},$$
$$\eta(s, t) = (2s, -t, \sqrt{(1 - 4s^2 - t^2)}), \qquad\qquad 4s^2 + t^2 \leq \tfrac{3}{4}.$$

      (iii) The *(triangular) plane sheet* $x + y + z = 1$, $x \geq 0$, $y \geq 0$, $z \geq 0$, $\mathbf{q} = (\frac{1}{2}, \frac{1}{4}, \frac{1}{4})$,

$$\rho(u, v) = (u, v, 1 - u - v),$$
$$\eta(s, t) = (s + t, 1 - 2s, s - t).$$

Answers: (i) $v_\rho(1, 2) = (-2, -4, 1)$ pointing 'inwards' at $\mathbf{q}$, $v_\eta(\frac{3}{2}, -\frac{1}{2}) = (4, 8, -2)$ pointing 'outwards'. Tangent plane $-2x - 4y + z = -5$.
$\phi(u, v) = \eta^{-1} \circ \rho(u, v) = (\frac{1}{2}u + \frac{1}{2}v, \frac{1}{2}u - \frac{1}{2}v)$. (ii) $v_\rho(\frac{1}{3}, \frac{2}{3}) = (\frac{1}{2}, 1, 1)$ pointing 'outwards', $v_\eta(\frac{1}{6}, -\frac{2}{3}) = (-1, -2, -2)$ pointing 'inwards' $\phi(u, v) = (\frac{1}{2}u, -v)$. (iii) $v_\rho(u, v) = (1, 1, 1)$, $v_\eta(s, t) = (2, 2, 2)$.

8.3 Orientation of smooth simple surfaces

In Sections 2.6 and 6.1 we discussed the two orientations of a smooth simple arc C. There are three equivalent ways of distinguishing between them—first by ordering the end points of C, second by giving an equivalence class of simple parametrizations of C, and third by giving a continuous unit tangent vector field to C. Essentially an orientation of C assigns an order to each pair of points lying within C consistent with the ordering of an interval.

The orientation of a smooth simple surface K is open to much the same treatment. It involves the orientation of the simple closed curves that lie within K.

When the simple surface is a plane simple region S lying in the x, y plane, we can describe the orientation of an oriented simple closed curve C in S as being either clockwise or counterclockwise. The curve C is oriented *counterclockwise* if, while walking around the curve on the positive z side of the plane in the direction of orientation, the inside of C remains on the left. (See the discussion in Section 7.5.) Using this idea we can assign an orientation to a *simple region* in $\mathbb{R}^2$.

8.3.1 Definition.   *A simple region S in $\mathbb{R}^2$ together with a counterclockwise orientation of all simple closed curves in S is said to be a* positively oriented *simple region. We denote it by S^+.*

Suppose now that K is a smooth simple surface in $\mathbb{R}^3$ and that $\rho: S \subseteq \mathbb{R}^2 \to \mathbb{R}^3$ is a simple parametrization of K. Then any simple closed curve in K is the image under ρ of a simple closed curve in S. The counterclockwise orientation of this curve in S leads, through ρ, to an orientation of the curve in K. We can picture this by thinking of S^+ as a flexible mat with counterclockwise oriented simple closed curves drawn in it. The function ρ tells us how to stretch the mat over K. The oriented curves in S then become oriented curves in K.

In order to make this more precise it will be helpful to introduce some notation. Remember that we have denoted an oriented simple closed curve by $(C, [\alpha])$ where C is the curve and $[\alpha]$ is the equivalence class of simple parametrizations of C having the same orientation as α.

8.3.2 Definition.   *Let $(C, [\alpha])$ be an oriented simple closed curve in $\mathbb{R}^2$ and let $f: C \subseteq \mathbb{R}^2 \to \mathbb{R}^n$ be a 1–1 C^1 function. Then $f(C, [\alpha])$ is defined to be the simple closed curve $f(C)$ with orientation determined by $f \circ \alpha$. that is,*

$$f(C, [\alpha]) = (f(C), [f \circ \alpha]).$$

We leave it to the reader to justify this definition by showing that if α and β are equivalent parametrizations of C then $f \circ \alpha$ and $f \circ \beta$ are equivalent parametrizations of $f(C)$.

8.3.3 Example. Let K be the parabolic bowl $z = x^2 + y^2$, $z \leq 4$, in $\mathbb{R}^3$ and let $\rho : S \subseteq \mathbb{R}^2 \to \mathbb{R}^3$ be the simple parametrization of K given by

$$\rho(u, v) = (u, v, u^2 + v^2), \qquad (u, v) \in S,$$

where S is the disc $u^2 + v^2 \leq 4$ in $\mathbb{R}^2$.

Let C be the circle $u^2 + v^2 = 4$ in S and let $\alpha : [0, 2\pi] \subseteq \mathbb{R} \to \mathbb{R}^2$ be the (counterclockwise) parametrization of C given by

$$\alpha(t) = (2 \cos t, 2 \sin t), \qquad t \in [0, 2\pi].$$

Then $\rho(C, [\alpha])$ is the simple closed curve

$$\rho(C) = \{(x, y, z) \in \mathbb{R}^3 \mid x^2 + y^2 = 4, z = 4\}$$

with orientation determined by

8.3.4 $\qquad (\rho \circ \alpha)(t) = (2 \cos t, 2 \sin t, 4), \qquad t \in [0, 2\pi].$

The oriented simple closed curves $C^+ = (C, [\alpha])$ and $\rho(C^+)$ are sketched in Fig. 8.6, as are two other (counterclockwise) oriented curves in S and their images under ρ.

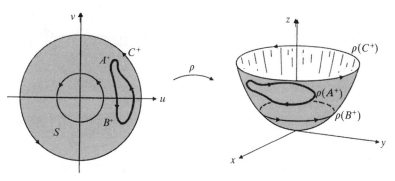

Fig. 8.6

Confining attention to *counterclockwise* oriented simple closed curves in a simple region S, we can use a simple parametrization $\rho : S \subseteq \mathbb{R}^2 \to \mathbb{R}^3$ of a simple surface K in $\mathbb{R}^3$ to determine an orientation of all simple closed curves in K.

8.3.5 *Definition.*   *An* oriented simple surface *is a simple surface K whose simple closed curves are oriented through the above procedure. The parametrization* ρ, *and any other simple parametrization which leads to the same orientation, is called a* simple parametrization *of the oriented surface.*

Informally, an oriented simple surface is the image of a positively oriented simple region S^+ under a simple parametrization ρ. It is conveniently denoted by $\rho(S^+)$.

8.3.6 *Example.*   **[i]**   The simple parametrization ρ of the parabolic bowl K considered in Example 8.3.3 leads to an orientation of K. See Fig. 8.6.

[ii]   In contrast, consider the simple parametrization $\eta : T \subseteq \mathbb{R}^2 \rightarrow \mathbb{R}^3$ of the parabolic bowl given by

$$\eta(r, s) = (2r, -2s, 4(r^2 + s^2)), \qquad (r, s) \in T,$$

where T is the disc $r^2 + s^2 \leqslant 1$ in $\mathbb{R}^2$.

The effect of η on several counterclockwise oriented curves in T is illustrated in Fig. 8.7. In particular let Z^+ be the circle $r^2 + s^2 = 1$ in T,

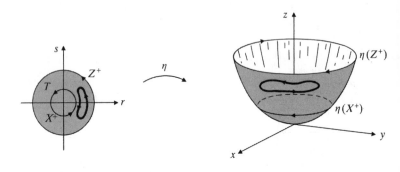

Fig. 8.7

parametrized (counterclockwise) by $\alpha : [0, 2\pi] \subseteq \mathbb{R} \rightarrow \mathbb{R}^2$ given by $\alpha(t) = (\cos t, \sin t)$, $t \in [0, 2\pi]$. Then $\alpha(Z^+)$ is the simple closed curve

$$\alpha(Z) = \{(x, y, z) \in \mathbb{R}^2 \mid x^2 + y^2 = 4,\ z = 4\}$$

with orientation determined by

$$(\eta \circ \alpha)(t) = (2 \cos t, -2 \sin t, 4), \qquad t \in [0, 2\pi].$$

Notice (see Example 8.3.3) that $\rho(C) = \eta(Z)$, that both C^+ and Z^+ are oriented counterclockwise, but that $\rho(C^+)$ and $\eta(Z^+)$ have opposite orientations. Therefore the simple parametrizations ρ and η give different orientations to K. If S and T are thought of as flexible mats with counterclockwise oriented curves drawn in them, then ρ simply lifts S up to fit over K whereas η first turns T over and then lifts it up over K. The sets $\rho(S)$ and $\eta(T)$ are both equal to K, but $\rho(S^+)$ and $\eta(T^+)$ are different oriented surfaces.

8.3.7 Definition. *Let $\rho : S \subseteq \mathbb{R}^2 \to \mathbb{R}^3$ and $\eta : T \subseteq \mathbb{R}^2 \to \mathbb{R}^3$ be simple parametrizations of a simple surface K in $\mathbb{R}^3$. Then ρ and η are said to have the same (opposite) orientation if they determine the same (opposite) orientation of K, that is, if $\rho(C^+)$ and $\eta(Z^+)$ have the same (opposite) orientation whenever C^+ and Z^+ are counterclockwise oriented simple closed curves in S and T respectively such that $\rho(C) = \eta(Z)$.*

8.3.8 Theorem. *Let $\rho : S \subseteq \mathbb{R}^2 \to \mathbb{R}^3$ and $\eta : T \subseteq \mathbb{R}^2 \to \mathbb{R}^3$ be simple parametrizations of a smooth simple surface K in $\mathbb{R}^3$, and let $\phi : S \subseteq \mathbb{R}^2 \to \mathbb{R}^2$ be the 1–1 smooth function from S onto T such that $\rho = \eta \circ \phi$. then ρ and η determine the same orientation of K if*

8.3.9 $$\det J_{\phi,(u,v)} > 0 \qquad \text{for all } (u, v) \in S,$$

and they determine opposite orientations of K if

8.3.10 $$\det J_{\phi,(u,v)} < 0 \qquad \text{for all } (u, v) \in S.$$

Proof. Let C^+ and Z^+ be counterclockwise oriented simple closed curves in S and T respectively such that $\rho(C) = \eta(Z)$. Since S is simply connected, the argument of Section 7.8 implies that if condition 8.3.9 is true then $\phi(C^+) = Z^+$. But, by Definition 8.3.2, $\phi(C^+) = Z^+$ if and only if $\eta(\phi(C^+)) = \eta(Z^+)$, that is, $\rho(C^+) = \eta(Z^+)$. So in this case ρ and η determine the same orientation of K.

On the other hand if condition 8.3.10 holds then $\phi(C^+) = Z^-$ and ρ and η determine opposite orientations of K.

We say that the simple parametrizations ρ and η of the smooth simple surface are *properly equivalent* if they determine the same orientation. Compare this with the definition of proper equivalence of parametrizations of curves (Section 2.6).

8.3.11 Example. We have seen in Examples 8.3.3 and 8.3.6 two parametrizations ρ and η of the parabolic bowl which do not determine the same orientation of the bowl. The appropriate function $\phi : S \subseteq \mathbb{R}^2 \to \mathbb{R}^2$ such that $\rho = \eta \circ \phi$ is given by

$$\phi(u, v) = (\eta^{-1} \circ \rho)(u, v) = (\tfrac{1}{2}u, -\tfrac{1}{2}v), \qquad (u, v) \in S.$$

Notice that $\det J_{\phi, (u, v)} = -\tfrac{1}{4}$ is negative for all $(u, v) \in S$.

The following example shows that every simple surface has at least two orientations.

8.3.12 Example. Let $\rho : S \subseteq \mathbb{R}^2 \to \mathbb{R}^3$ be a simple parametrization of a simple surface K and let S^* be the simple region (reflection of S in the horizontal axis)

$$S^* = \{(r, s) \in \mathbb{R}^2 \mid (r, -s) \in S\}.$$

Then $\mu : S^* \subseteq \mathbb{R}^2 \to \mathbb{R}^3$ defined by

8.3.13 $\qquad\qquad \mu(r, s) = \rho(r, -s) \qquad$ for all $(r, s) \in S^*$

is also a simple parametrization of K. But

$$\phi(u, v) = (\mu^{-1} \circ \rho)(u, v) = (u, -v), \qquad \text{for all } (u, v) \in S,$$

and therefore, since $\det J_{\phi, (u, v)} = -1$ throughout S, ρ and μ give different orientation to K. The parametrization μ defined by 8.3.13 is conveniently denoted by ρ^- and called the *opposite parametrization to* ρ. (Compare the definition of inverse path, 2.7.19.) The opposite parametrization ρ^- should not be confused with the inverse function ρ^{-1}.

8.3.14 Example. The two parametrizations ρ and η of the parabolic bowl (8.3.3 and 8.3.6) have opposite orientation since by Example 8.3.11

$$\det J_{\phi, (u, v)} = -\tfrac{1}{4} \qquad \text{for all } (u, v) \in S.$$

compare Figs. 8.6 and 8.7

8.3.15 Example. Let K be a simple surface in $\mathbb{R}^3$ and let $\rho : S \subseteq \mathbb{R}^2 \to \mathbb{R}^3$ be a simple parametrization of K. Then ρ and ρ^- have opposite orientation.

We can now prove that every smooth simple surface has precisely two orientations. (In contrast, a more general surface may have more than two orientations. See Exercise 8.5.7.)

8.3.16 Theorem. *Let K be a smooth simple surface in $\mathbb{R}^3$. Then K has two, and only two, orientations.*

Proof. Let $\rho : S \subseteq \mathbb{R}^2 \to \mathbb{R}^3$ be a simple parametrization of K. Then ρ and ρ^- determine two different orientations of K.

Let $\eta : T \subseteq \mathbb{R}^2 \to \mathbb{R}^3$ be another simple parametrization of K and let $\phi : S \subseteq \mathbb{R}^2 \to \mathbb{R}^2$ be the 1–1 smooth function from S onto T such that $\rho = \eta \circ \phi$. Then η either has the same orientation as ρ or the same orientation as ρ^-. For, if not, then both conditions 8.3.9 and 8.3.10 fail, and that contradicts the fact that ϕ is smooth on the path connected set S.

An equivalence relation is established on the set of all simple parametrizations of a smooth simple surface K by regarding two as being equivalent if they have the same orientation. (This is the equivalence relation based on the concept of proper equivalence.) It follows from Theorem 8.3.16 that there are two equivalence classes. If one of them is $[\rho]$ then the other is $[\rho^-]$. The following notation for an oriented surface underlines this point.

8.3.17 Definition. *Let K be a smooth simple surface with an orientation determined by a simple parametrization $\rho : S \subseteq \mathbb{R}^2 \to \mathbb{R}^3$. Then the oriented simple surface $\rho(S^+)$ is denoted by $(K, [\rho])$.*

The two orientations of K are therefore $(K, [\rho])$ and $(K, [\rho^-])$. In moving from one to the other, the orientation of all simple closed curves in K is reversed.

In a second way of specifying an orientation of a smooth simple surface K we use the fact (Theorem 8.2.7) that if $\rho : S \subseteq \mathbb{R}^2 \to \mathbb{R}^3$ and $\eta : T \subseteq \mathbb{R}^2 \to \mathbb{R}^3$ are simple parametrizations in K then whenever $\rho(\mathbf{p}) = \eta(\mathbf{w}) = \mathbf{q}$, the normal vectors $v_\rho(\mathbf{p})$ and $v_\eta(\mathbf{w})$ are related by

8.3.18 $$v_\rho(\mathbf{p}) = (\det J_{\phi,\mathbf{p}})(v_\eta(\mathbf{w}))$$

where $\rho = \eta \circ \phi$.

Notice that the normal vectors are in the same (opposite) direction if and only if ρ and η have the same (opposite) orientation. Thus the direction of the normal vector identifies the class of parametrizations.

8.3.19 Definition. *Let K be a smooth simple surface. A function $N : K \subseteq \mathbb{R}^2 \to \mathbb{R}^3$ is called a* unit normal field *on K if, for all $\mathbf{q} \in K$, $N(\mathbf{q})$ is a unit normal vector to K at $\mathbf{q}$.*

Every simple parametrization of a smooth simple surface leads to a continuous unit normal vector field.

8.3.20 Theorem. *Let* $\rho : S \subseteq \mathbb{R}^2 \to \mathbb{R}^3$ *and* $\eta : T \subseteq \mathbb{R}^2 \to \mathbb{R}^3$ *be simple parametrizations of a simple surface* K *in* $\mathbb{R}^3$.
   [i]   *The function* $N_\rho : K \subseteq \mathbb{R}^3 \to \mathbb{R}^3$ *defined by*

8.3.21 $$N_\rho(\mathbf{q}) = \frac{v_\rho(\rho^{-1}(\mathbf{q}))}{\|v_\rho(\rho^{-1}(\mathbf{q}))\|} \qquad \text{for all } \mathbf{q} \in K$$

is a continuous unit normal vector field on K;
   [ii] $N_\rho = N_\eta$ *if and only if* ρ *and* η *have the same orientation*;
   [iii] $N_\rho = -N_\eta$ *if and only if* ρ *and* η *have opposite orientations*.

Proof.   Immediate from 8.3.18 and Theorem 8.3.8.

8.3.22 Example.   Let K be the parabolic bowl $z = x^2 + y^2$, $z \leqslant 4$, in $\mathbb{R}^3$. The simple parametrizations ρ and η of K discussed in Examples 8.3.3 and 8.3.6 lead to unit normal vectors, for all $(x, y, z) \in K$,

$$N_\rho(x, y, z) = \frac{v_\rho(x, y)}{\|v_\rho(x, y)\|} = \frac{(-2x, -2y, 1)}{\|(-2x, -2y, 1)\|}$$

$$= \frac{1}{\sqrt{(4x^2 + 4y^2 + 1)}} (-2x, -2y, 1),$$

and

$$N_\eta(x, y, z) = \frac{v_\eta(\tfrac{1}{2}x, \tfrac{1}{2}y)}{\|v_\eta(\tfrac{1}{2}x, \tfrac{1}{2}y)\|} = \frac{(8x, 8y, -4)}{\|(8x, 8y, -4)\|}$$

$$= \frac{1}{\sqrt{(4x^2 + 4y^2 + 1)}} (2x, 2y, -1).$$

Notice that ρ and η have opposite orientations and $N_\eta(x, y, z) = -N_\rho(x, y, z)$.

8.3.23 Theorem.   *There are just two continuous unit normal vector fields on a smooth simple surface.*

Proof.   Let N be a continuous unit normal vector field on a simple surface K, and let N^* be another.

For each point $\mathbf{q} \in K$, either $N^*(\mathbf{q}) = N(\mathbf{q})$ or $N^*(\mathbf{q}) = -N(\mathbf{q})$. In the first case the dot product $N^*(\mathbf{q}) \cdot N(\mathbf{q})$ is 1 and in the second case it is -1.

Consider the continuous scalar field $\delta : K \subseteq \mathbb{R}^3 \to \mathbb{R}$ defined by

$$\delta(\mathbf{r}) = N^*(\mathbf{r}) \cdot N(\mathbf{r}), \qquad \text{for all } \mathbf{r} \in K.$$

We know that $\delta(K) \subseteq \{-1, 1\}$. But K, and therefore $\delta(K)$, is path connected. Hence either $\delta(K) = \{1\}$ or $\delta(K) = \{-1\}$. In the first case $N^* = N$ and in the second $N^* = -N$.

The two continuous unit normal fields on a smooth simple surface K therefore match the two equivalence classes $[\rho]$ and $[\rho^-]$ of simple parametrizations of K through 8.3.21. Picking one of the fields determines an orientation and vice versa. We therefore have another notation for an oriented simple surface:

$$(K, N),$$

where N is the continuous unit normal vector field determining the orientation of K. If correspondingly $(K, N) = (K, [\rho])$ then $N = N_\rho$ in the notation 8.3.21. We can picture how the vector field N on K is related to ρ by thinking of S as a flexible mat in the u, v plane in $\mathbb{R}^3$ with a multitude of spikes sticking up out of S in the positive w direction. The parametrization ρ effectively tells us how to stretch this mat over K. The vector field N then points in the direction of the spikes. See Fig. 8.8.

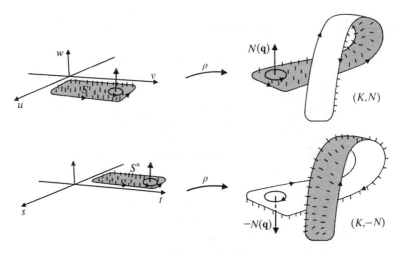

Fig. 8.8

The figure also shows how, for each $\mathbf{q} \in K$, the vector $N(\mathbf{q})$ is related to the orientation of a small simple closed curve in K around $\mathbf{q}$. A 'right-hand' screw turned in the direction of orientation will move in the direction of the normal. Notice that this is how the counterclockwise orientation is related to the positive w direction in the domain of ρ.

The description (K, N) of an oriented smooth simple surface is without reference to a parametrization. If ρ is a parametrization such that $(K, N) = (K, [\rho])$ then $(K, [\rho^-]) = (K, -N)$, which is the oppositely oriented smooth simple surface.

An orientation of a smooth simple surface K is determined by any one of many different properties. For example, naming a particular simple parametrization or specifying a unit normal to K at some *one* point in K determines the orientation completely. Equally, naming a single oriented simple closed curve in K would do. This is usually done by giving an orientation to the edge E of K. If the simple parametrizations of K divide into two classes $[\rho]$ and $[\rho^-]$ then all parametrizations in $[\rho]$ determine one orientation, say E^+, of E, and all parametrizations in $[\rho^-]$ determine the opposite orientation E^-.

8.3.24 Definition.   *The oriented edge E^+ of the oriented smooth simple surface $(K, [\rho]) = (K, N)$ is denoted by $\partial(K, [\rho])$ or by $\partial(K, N)$.*

Clearly if $\rho : S \subseteq \mathbb{R}^2 \to \mathbb{R}^3$ is a simple parametrization of (K, N) then $\partial(K, N) = \rho(\partial S^+)$.

8.3.25 Remark.   If the orientation of a smooth simple surface K is determined by assigning an orientation to its edge E, then the corresponding unit normal vector field N on K is such that when walking around E in the direction of its orientation *with normal vectors $N(\mathbf{q})$ from heel to head*, the surface K lies on the left. See Fig. 8.8. This is precisely how the counterclockwise orientation of ∂S is related to S in the u, v plane when normal vectors are taken in the positive w direction.

Exercises 8.3

1.   Find a simple parametrization of the following oriented smooth simple surfaces (K, N). Sketch the surface K, its associated continuous unit normal vector field N, and its oriented edge $\partial(K, N)$.
     (a)   The spherical cap $x^2 + y^2 + z^2 = 1$, $\frac{1}{2} \leq z \leq 1$ with unit normal field N pointing away from the centre (origin);

(b)   the same spherical cap with field N pointing towards the centre;
(c)   the rectangular surface $y = 3$, $|x| \leq 1$, $-1 \leq z \leq 2$ with field N
pointing in the positive y direction;
(d)   the plane surface $K = \{(x, y, 0) \in \mathbb{R}^3 \mid (x, y) \in S\}$ where S is a
simple region in $\mathbb{R}^2$, with field N pointing in the negative z direction.

Answers: (a) $\rho(u, v) = (u, v, \sqrt{(1 - u^2 - v^2)})$, $u^2 + v^2 \leq \frac{3}{4}$; (b) the opposite
parametrization $\rho^-(r, s) = (r, -s, \sqrt{(1 - r^2 - s^2)})$, $r^2 + s^2 \leq \frac{3}{4}$;
(c) $\mu(r, s) = (r, 3, -s)$, $|r| \leq 1$, $-2 \leq s \leq 1$; (d) $\rho(u, v) = (u, -v, 0)$,
$(u, -v) \in S$.

2.   Let K be the parabolic cap

$$z = 4 - x^2 - y^2, \qquad z \geq 0.$$

Assign an orientation to K by orienting its edge E: $x^2 + y^2 = 4$, $z = 0$,
through the parametrization $\alpha(t) = (2 \cos t, -2 \sin t, 0)$, $0 \leq t \leq 2\pi$.
    Sketch the surface K and its oriented edge E^+, and find a simple
parametrization ρ of this oriented surface. Sketch also the associated
continuous unit normal vector field N such that $(K, [\rho]) = (K, N)$.

Answer: $\rho(u, v) = (u, -v, 4 - u^2 - v^2)$, $u^2 + v^2 \leq 4$, for example.
$N(x, y, 4 - x^2 - y^2) = (1 + 4x^2 + 4y^2)^{-1/2}(-2x, 2y, -1)$, pointing
'inwards'.

3.   Verify that $\det J_{\phi, (u, v)}$ is negative for all $(u, v) \in \mathbb{R}^2$, where
(a) $\phi(u, v) = (u, -v)$, (b) $\phi(u, v) = (-u, v)$, (c) $\phi(u, v) = (v, u)$.
Show that, in each case, if $\eta : T \subseteq \mathbb{R}^2 \to \mathbb{R}^3$ is a simple parametrization
of an oriented smooth simple surface (K, N), then $\eta \circ \phi : \phi^{-1}(T) \subseteq$
$\mathbb{R}^2 \to \mathbb{R}^3$ is a simple parametrization of $(K, -N)$.

4.   Let (K, N) be the hyperbolic paraboloid $z = x^2 - y^2$, $-1 \leq x \leq 2$,
$0 \leq y \leq 1$, with orientation determined by the simple parametrization

$$\eta(r, s) = (r, s, r^2 - s^2) \qquad -1 \leq r \leq 2, 0 \leq s \leq 1.$$

Obtain, from Exercise 3, three parametrizations of $(K, -N)$.

Answers: (a) $\rho(u, v) = (u, -v, u^2 - v^2)$, $-1 \leq u \leq 2$, $-1 \leq v \leq 0$;
(b) $\rho(u, v) = (-u, v, u^2 - v^2)$, $-2 \leq u \leq 1$, $0 \leq v \leq 1$;
(c) $\rho(u, v) = (v, u, v^2 - u^2)$, $0 \leq u \leq 1$, $-1 \leq v \leq 2$.

8.4   Orientation of smooth piecewise simple surfaces

As in the case of simple regions, the orientation of a piecewise
simple region in $\mathbb{R}^2$ involves the orientation of simple closed curves
that lie within it. However, some care is needed.

8.4.1 *Example.* Let S be the piecewise simple region (annulus) in $\mathbb{R}^2$

$$S = \{(u, v) \in \mathbb{R}^2 \mid \tfrac{1}{2} \leqslant \sqrt{(u^2 + v^2)} \leqslant 2\}$$

and let $\phi : S \subseteq \mathbb{R}^2 \to \mathbb{R}^2$ be the 1–1 smooth function which firstly inverts S onto itself with respect to the circle $u^2 + v^2 = 1$ (see Exercise 7.8.2) and then reflects S in the u-axis. That is,

$$\phi(u, v) = \left(\frac{u}{u^2 + v^2}, \frac{-v}{u^2 + v^2} \right), \qquad (u, v) \in S.$$

Then ϕ reverses the orientation of some simple closed curves in S and preserves the orientation of others (Fig. 8.9). For example, ϕ maps the unit

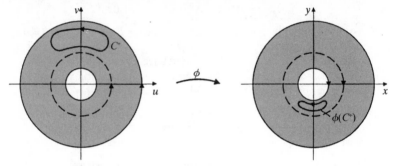

Fig. 8.9

circle $u^2 + v^2 = 1$ oriented counterclockwise onto the same circle oriented clockwise. However, since

$$\det J_{\phi, (u, v)} = \frac{1}{(u^2 + v^2)^2}, \qquad (u, v) \in S$$

the function ϕ preserves the orientation of all simple closed curves whose *inside* is a subset of S.

The following definition ensures that there are just two possible orientations of a piecewise simple region in $\mathbb{R}^2$.

8.4.2 *Definition.* *A piecewise simple region S in $\mathbb{R}^2$ is positively oriented by assigning a counterclockwise orientation to the boundary of every simple closed curve whose inside lies in S. We denote it by S^+.*

The theory of oriented smooth piecewise simple surfaces now follows the same path as that of oriented smooth simple surfaces.

We again find two equivalence classes of parametrizations $[\rho]$ and $[\rho^-]$ and correspondingly two continuous unit normal vector fields N and $-N$ to a smooth piecewise simple surface K, and K has two possible orientations. If one of them is $(K, [\rho]) = (K, N)$ then the other is $(K, [\rho^-]) = (K, -N)$.

The orientation of a smooth piecewise simple surface K is determined by the orientation assigned to the edge of any one simple surface lying in K. It is not, however, determined by the orientation of every simple closed curve in K (see Exercise 8.4.2).

The orientation of K is also determined by the orientation of its edge which it inherits from the boundary of the oriented piecewise simple region S^+ used to define it.

8.4.3 *Example.* Consider the section of the parabolic bowl $z = x^2 + y^2$, $1 \leq z \leq 4$, with inward facing normals. The oriented surface (K, N) is sketched in Fig. 8.10. Let S be the piecewise simple region

$$S = \{(u, v) \in \mathbb{R}^2 \mid 1 \leq u^2 + v^2 \leq 4\},$$

lying between the circles $u^2 + v^2 = 4$ and $u^2 + v^2 = 1$ denoted respectively by C and A. Let $\rho : S \subseteq \mathbb{R}^2 \to \mathbb{R}^3$ be defined by

$$\rho(u, v) = (u, v, u^2 + v^2), \qquad (u, v) \in S.$$

Then ρ is a simple parametrization of (K, N).

The edge of (K, N) is also sketched in Fig. 8.10. It consists of two circles $\rho(C)$ and $\rho(A)$ oriented as shown. Notice that walking along any part of the edge in the direction of orientation, with normal vectors $N(\mathbf{q})$ from heel to head, keeps the surface K on the left.

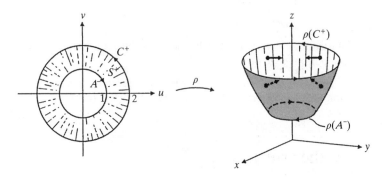

Fig. 8.10

Exercises 8.4

1. Sketch the cylinder K (smooth piecewise simple surface) defined by
    $x^2 + y^2 = 1$, $\frac{1}{2} \leqslant z \leqslant 2$.

    Verify that the simple parametrizations of K

    $$\rho(u, v) = \left\{ \frac{u}{\sqrt{(u^2 + v^2)}}, \frac{v}{\sqrt{(u^2 + v^2)}}, \sqrt{(u^2 + v^2)} \right\},$$

    $$\frac{1}{2} \leqslant \sqrt{(u^2 + v^2)} \leqslant 2$$

    and

    $$\eta(u, v) = \left(\frac{u}{\sqrt{(u^2 + v^2)}}, \frac{-v}{\sqrt{(u^2 + v^2)}}, \frac{1}{\sqrt{(u^2 + v^2)}} \right),$$

    $$\frac{1}{2} \leqslant \sqrt{(u^2 + v^2)} \leqslant 2$$

    determine the same orientation of K, which is given by the continuous
    unit vector field N on K of inward facing normals.

    Hint: the parametrizations ρ and η are related by $\eta = \rho \circ \phi$, where
    $\phi : S \subseteq \mathbb{R}^2 \to \mathbb{R}^2$ is the function defined in Example 8.4.1. In particular,
    $\det J_{\phi, (u, v)} > 0$ for all $(u, v) \in S$.

2. Let (K, N) be the oriented cylinder considered in Exercise 1. Let C be
    the simple closed curve $x^2 + y^2 = 1$, $z = 1$ in K. Show that the
    parametrizations ρ and η assign opposite orientations to C.

    Hint: note again the connection with Example 8.4.1. The functions ρ and η
    both take S^+ to (K, N), but they map the unit circle $u^2 + v^2 = 1$ in S,
    oriented counterclockwise, to C^+ and C^- respectively.

3. Again let K be the cylinder $x^2 + y^2 = 1$, $\frac{1}{2} \leqslant z \leqslant 2$. Find a
    parametrization μ of the oriented surface K, the orientation of K being
    determined by orienting its 'top' edge $x^2 + y^2 = 1$, $z = 2$ by the path
    $\alpha : [0, 2\pi] \subseteq \mathbb{R} \to \mathbb{R}^3$, where

    $$\alpha(t) = (\cos t, -\sin t, 2), \qquad 0 \leqslant t \leqslant 2\pi.$$

    Illustrate by means of a sketch, and mark the appropriate orientation
    of the bottom edge of K.

    Answer: choose $\mu = \rho^-$, for example, where ρ is given in Exercise 1. The
    bottom edge is correctly oriented by the path $\beta(t) = (\cos t, \sin t, \frac{1}{2})$,
    $0 \leqslant t \leqslant 2\pi$.

8.5   Other surfaces

The surfaces we have considered so far in this chapter are not
sufficiently general to include many familiar surfaces, such as the

sphere, cube, cone and Möbius strip. For the purpose of integration it is convenient to express them as a union of smooth simple surfaces with non-intersecting insides, which are suitably sown together along their edges. The following general definition of a surface formally describes the construction.

8.5.1 Definition. *Let $S_1, \ldots, S_k$ be simple regions in $\mathbb{R}^2$ whose interiors are pairwise disjoint. Let $S = S_1 \cup \cdots \cup S_k$, let $B = \partial S_1 \cup \cdots \cup \partial S_k$, and let*
$$U = \operatorname{Int} S_1 \cup \cdots \cup \operatorname{Int} S_k.$$
For any C^1 function $\rho : S \subseteq \mathbb{R}^2 \to \mathbb{R}^3$ such that
   [i] $\rho : U \subseteq \mathbb{R}^2 \to \mathbb{R}^3$ *is 1–1 and smooth,*
   [ii] $\rho(U) \cap \rho(B) = \phi$,
the set $K = \rho(S)$ is called a surface *in $\mathbb{R}^3$. The function ρ is a* parametrization *of K.*

Notice that the parametrization ρ is not required to be 1–1 on B. Notice also that $\rho(U)$ is a disjoint union of the insides of smooth simple surfaces which form almost the whole of K in the sense that $\rho(B)$ is a finite union of curves in the surface. Tangent planes and normal lines can therefore be drawn 'almost everywhere' in the surface K.

8.5.2 Example. A cube in $\mathbb{R}^3$ is a surface by this definition. Its six faces are smooth simple surfaces. They can be simply parametrized on non-overlapping simple regions $S_1, \ldots, S_6$ in $\mathbb{R}^2$. The resulting parametrization ρ of the cube satisfies the conditions of Definition 8.5.1.

It is often possible to find a convenient parametrization of a surface using just one simple region S.

8.5.3 Example. The sphere $x^2 + y^2 + z^2 = c^2$ can be seen to satisfy Definition 8.5.1 by regarding it as two hemispheres sown together. Each hemisphere is a simple surface simply parametrized by a stereographic projection. For the purpose of integrating over the sphere it is much more convenient to use the C^1 spherical coordinate parametrization $\rho : S \subseteq \mathbb{R}^2 \to \mathbb{R}^3$ given by
$$\rho(\phi, \theta) = (c \sin \phi \cos \theta, c \sin \phi \sin \theta, c \cos \phi), \qquad (\phi, \theta) \in S,$$
where S is the simple (rectangular) region $S = [0, \pi] \times [0, 2\pi]$ in $\mathbb{R}^2$. Note that ρ is 1–1 and smooth on $\operatorname{Int} S$, but that ρ is not 1–1 on S. The set $\rho(\partial S)$ is the semicircle $x^2 + z^2 = 1$, $x \geq 0$, $y = 0$.

8.5.4 *Example.* The cone K defined by $z = \sqrt{(x^2 + y^2)}$, $0 \leqslant z \leqslant h$ is a surface. Using the C^1 parametrization of K

$$\rho(u, v) = (v \cos u, v \sin u, v), \qquad (u, v) \in S$$

on the simple (rectangular) region $S = [0, 2\pi] \times [0, h]$ in $\mathbb{R}^2$, the 'bad' points $(u, 0)$ where ρ is not 1–1 appear on the boundary of S. The function ρ is 1–1 and smooth on Int S.

8.5.5 *Example.* Let K be the cylinder $x^2 + y^2 = k^2$, $|z| \leqslant 1$. This is a piecewise simple surface, since it is the image of an annulus under a 1–1 smooth function. The following more natural parametrization of K displays it as a surface satisfying Definition 8.5.1. Let S be the rectangular strip

8.5.6 $S = \{(\theta, t) \in \mathbb{R}^2 \mid 0 \leqslant \theta \leqslant \pi, \ -1 \leqslant t \leqslant 1\}.$

The C^1 function $\rho : S \subseteq \mathbb{R}^2 \to \mathbb{R}^3$ defined by

8.5.7 $\rho(\theta, t) = (k \cos 2\theta, k \sin 2\theta, t) \qquad (\theta, t) \in S$

transforms S into the cylinder K (Fig. 8.11). Notice that ρ is 1–1 and smooth on Int S but that it is not 1–1 on S.

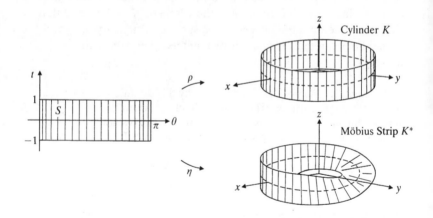

Fig. 8.11

We can illustrate the effect of ρ by imagining S to be a strip of paper and joining its ends together to form the cylinder. An interesting variation is obtained if we give one end of the strip S a single twist before making the join. The result is a *Möbius strip*. Define the C^1 function $\eta : S \subseteq \mathbb{R}^2 \to \mathbb{R}^3$ by

8.5.8 $\eta(\theta, t) = ((k - t \sin \theta) \cos 2\theta, (k - t \sin \theta) \sin 2\theta, t \cos \theta).$

Then $K^* = \eta(S)$ is a Möbius strip (Fig. 8.11). Notice that, for $-1 \leqslant t \leqslant 1$,

$$\eta(0, t) = (k, 0, t) \quad \text{and} \quad \eta(\pi, t) = (k, 0, -t),$$

which confirms the twist in the strip. Notice also that the points $\eta(\tfrac{1}{2}\pi, t) = (-k + t, 0, 0)$ lie in the x-axis. The curve running along the middle of the strip $t = 0$ is the circle $x^2 + y^2 = k^2$ in the x, y plane.

Since $\eta(0, t) = \eta(\pi, -t)$, the function η is not 1–1 on S. However, it is 1–1 and smooth on Int S, and so K^* is a surface by Definition 8.5.1.

Let K be a smooth piecewise simple surface, simply parametrized by $\rho : S \subseteq \mathbb{R}^2 \to \mathbb{R}^3$. In Section 8.4 we defined the oriented surface $(K, [\rho])$ to be the image under ρ of the positively oriented region S^+ in $\mathbb{R}^2$. Suppose that S can be subdivided into the simple regions $S_1, \ldots, S_k$, with non-overlapping interiors. Then whenever the boundaries ∂S_i and ∂S_j overlap in a simple arc C, the positively oriented curves ∂S_i^+ and ∂S_j^+ give opposite orientations of the arc C. (See Fig. 7.17 p. 387.) Consequently the oriented simple closed curves $\rho(\partial S_i^+)$ and $\rho(\partial S_j^+)$ in $(K, [\rho])$ give opposite orientations to the simple arc $\rho(C)$ in K. We use this property to define an orientable surface.

8.5.9 Definition.   Let K be a surface in $\mathbb{R}^3$. Then K is said to be orientable *if it has a parametrization* $\rho : S = S_1 \cup \cdots \cup S_k \subseteq \mathbb{R}^2 \to \mathbb{R}^3$ *satisfying the conditions of Definition 8.5.1, such that*

[iii]   *no simple arc in K is traced more than twice by ρ acting on* $\partial S_1^+ \cdots \partial S_k^+$,

[iv]   *a simple arc A in K which is traced twice is traced once in each direction.*

If the parametrization ρ satisfies these conditions we say that K is oriented by ρ, *and the oriented surface is denoted by* $(K, [\rho])$.

In Definition 8.5.9, as usual ∂S_i^+ denoted the positively (counterclockwise) oriented boundary of S_i in $\mathbb{R}^2$.

It is clear that smooth (piecewise) simple surfaces satisfy the condition of orientability.

The parametrization ρ of an oriented surface $(K, [\rho])$ leads to a continuous unit normal vector field N_ρ by the formula

8.5.10 $$N_\rho(\mathbf{q}) = \frac{v_\rho(\rho^{-1}(\mathbf{q}))}{\|v_\rho(\rho^{-1}(\mathbf{q}))\|},$$

where the domain of N_ρ is restricted to the subset $\rho(\text{Int } S_1 \cup \cdots \cup$

Int S_k) of K. The vector field N_ρ identifies the orientation of K: the relationship between the direction of $N_\rho(\mathbf{q})$ and the orientation of a small simple closed curve in K around $\mathbf{q}$ is again as illustrated in Fig. 8.8. We again have the notation (K, N) for the oriented surface $(K, [\rho])$, where $N = N_\rho$.

8.5.11 *Example.* The sphere is an orientable surface. Consider for example the sphere K, radius $c > 0$, defined by $x^2 + y^2 + z^2 = c^2$. The K is parametrized by $\rho : S \subseteq \mathbb{R}^2 \to \mathbb{R}^3$ where $S = [0, \pi] \times [0, 2\pi]$ and

$$\rho(\phi, \theta) = (c \sin \phi \cos \theta, c \sin \phi \sin \theta, c \cos \phi), \qquad (\phi, \theta) \in S.$$

The boundary ∂S is mapped by ρ onto the line of longitude A in K defined by $\theta = 0$ (Fig. 8.12). Consider ρ acting on the positively oriented boundary ∂S^+. Starting from the origin $\phi = 0$, $\theta = 0$ and moving counterclockwise around ∂S the effect of ρ is to (i) trace A from the north pole to the south pole, (ii) stay at the south pole (when $\phi = \pi$ and θ increased from 0 to 2π), (iii) trace A from the south pole to the north pole, (iv) stay at the south pole. So by Definition 8.5.9 the sphere has been oriented by the parametrization ρ.

An elementary calculation shows that for all $(\phi, \theta) \in S$

$$v_\rho(\phi, \theta) = c^2(\sin^2 \phi \cos \theta, \sin^2 \phi \sin \theta, \sin \phi \cos \phi)$$

and that

$$\|v_\rho(\phi, \theta)\| = c^2 \sin \phi.$$

In particular, $v_\rho(\tfrac{1}{2}\pi, 0) = c^2(1, 0, 0)$, and this is a normal vector to K at the

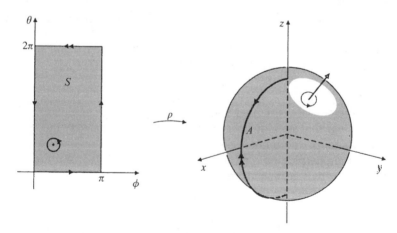

Fig. 8.12

point $\rho(\tfrac{1}{2}\pi, 0) = (1, 0, 0)$. So the parametrization ρ leads to the continuous unit normal vector field N_ρ of *outward facing* unit normals. Since ν_ρ is continuous on S and non-zero provided only that $\phi \neq 0$ or π, the field N_ρ is defined by 8.5.10 at all points on the sphere K excluding the north pole and the south pole. Its definition can be continuously extended to incorporate these points into its domain.

A sphere has just two orientations. It is said to be *positively oriented* by the continuous field of outward facing normals (pointing away from the centre of the sphere). If the parametrization ρ leads to the positive orientation then ρ^- leads to the orientation by inward facing normals.

8.5.12 *Example.* The cone and the cylinder are orientable surfaces. The parametrizations given in Examples 8.5.4 and 8.5.5 satisfy the conditions of Definition 8.5.9. There are again just two orientations. The corresponding continuous unit normal vector fields can be defined at all points on the cylinder and at all points on the cone excluding its vertex.

8.5.13 *Example.* The Möbius strip (Example 8.5.5) is a non-orientable surface. In constructing the Möbius strip from a rectangular strip S, the single twist given to one end of S before making the join ensures that the parts of ∂S^+ which are joined together have the same orientation along the join. See Fig. 8.13.

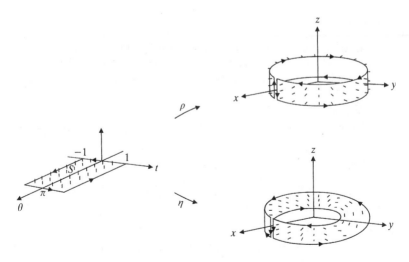

Fig. 8.13

Imagine the strip S as a spiked mat, the spikes (unit normals) sticking up out of S. Then by giving S the single twist before making the join, inevitably the spikes will display a reversal in direction (discontinuity) at the join. So, although the Möbius strip is a *locally* smooth surface (it has no peaks or ridges) it is not possible to define a continuous unit normal vector on it. In this connection see also Exercise 8.5.5.

8.5.14 *Example.* The cube is an orientable surface. The parametrizations of its six faces have to be carefully chosen to satisfy the definition of orientability. There are again just two orientations, the positive orientation by outward facing normals, illustrated in Fig. 8.14, and the orientation by inward facing normals.

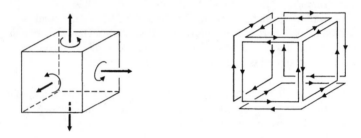

Fig. 8.14 Positively oriented cube

Exercises 8.5

1. For each of the following surfaces K verify that the given function $\rho : S \subseteq \mathbb{R}^2 \rightarrow \mathbb{R}^3$ is a parametrization of K. Let $\mathbf{q} = \rho(\mathbf{p})$, where $\mathbf{p} \in \text{Int } S$. Calculate $\nu_\rho(\mathbf{p})$ and hence find the equation of the tangent plane to K at $\mathbf{q}$.
    (a)   The *sphere* $x^2 + y^2 + z^2 = c^2$, $c > 0$,

    $$\rho(\phi, \theta) = (c \sin \phi \cos \theta, c \sin \phi \sin \theta, c \cos \phi), \qquad (\phi, \theta) \in S,$$

    where $S = [0, \pi] \times [0, 2\pi]$.
    (b)   The *circular cone* $z = \sqrt{(x^2 + y^2)}$, $z \leq 1$,

    $$\rho(u, v) = (v \cos u, v \sin u, v) \qquad (u, v) \in S$$

    where $S = [0, 2\pi] \times [0, 1]$.
    (c)   The *cylinder* $x^2 + y^2 = k^2$, $|z| \leq 1$,

    $$\rho(\theta, t) = (k \cos 2\theta, k \sin 2\theta, t), \qquad (\theta, t) \in S$$

    where $S = [0, \pi] \times [-1, 1]$.

(d) The *Möbius strip*

$$\rho(\theta, t) = ((k - t \sin \theta) \cos 2\theta, (k - t \sin \theta) \sin 2\theta, t \cos \theta)$$

$(\theta, t) \in S = [0, \pi] \times [-1, 1]$.

(e) One loop of the *figure 8 sheet* (compare Example 2.6.5)

$$\rho(u, v) = (\sin u, \sin 2u, v), \qquad (u, v) \in S$$

where $S = [0, \pi] \times [-1, 1]$.

Answers: (a) $v_\rho(\phi, \theta) = c^2 \sin \phi (\sin \phi \cos \theta, \sin \phi \sin \theta, \cos \phi) = c \sin \phi \rho(\phi, \theta)$.

Equation of tangent plane at $\rho(\phi, \theta)$ is

$$x \sin \phi \cos \theta + y \sin \phi \sin \theta + z \cos \phi = c.$$

(b) $v_\rho(u, v) = (v \cos u, v \sin u, -v)$. (c) $v_\rho(\theta, t) = (2k \cos 2\theta, 2k \sin 2\theta, 0)$. (d) $v_\rho(\theta, t) = t(-\sin 2\theta, \cos 2\theta, 0) + 2(k - t \sin \theta)(\cos \theta \cos 2\theta, \cos \theta \sin 2\theta, \sin \theta)$. (e) $v_\rho(u, v) = (2 \cos 2u, -\cos u, 0)$.

2. The parametrizations $\rho : S \subseteq \mathbb{R}^2 \to \mathbb{R}^3$ of the surfaces K in Exercise 1 are not 1–1 smooth on the boundary of S. Comment on the possibility of fitting tangent planes to K at points on $\rho(\partial S)$.

Answers: (a) $\rho(\phi, 0) = \rho(\phi, 2\pi) = (c \sin \phi, 0, c \cos \phi)$ and $v_\rho(\phi, 0) = v_\rho(\phi, 2\pi)$. Normals agree. (b) $\rho(0, v) = \rho(2\pi, v)$ and $v_\rho(0, v) = v_\rho(2\pi, v) = (v, 0, -v)$. At vertex of cone, $v = 0$, $v_\rho(u, 0) = (0, 0, 0)$. Tangent plane not defined at $(0, 0, 0)$. (c) $\rho(0, t) = \rho(\pi, t)$ and $v_\rho(0, t) = v_\rho(\pi, t)$. Normals agree. (d) $\rho(0, t) = \rho(\pi, -t) = (k, 0, t)$ but $v_\rho(0, t) = -v_\rho(\pi, -t)$. Tangent plane defined at $(k, 0, t)$. (e) $\rho(0, v) = \rho(\pi, v) = (0, 0, v)$ but $v_\rho(0, v) = (2, -1, 0)$ and $v_\rho(\pi, v) = (2, 1, 0)$. Tangent plane not defined at $(0, 0, v)$.

3. Let S be the compact square in $\mathbb{R}^2$

$$S = \{(x, y) \in \mathbb{R}^2 \mid -\pi \leqslant x \leqslant \pi, -\pi \leqslant y \leqslant \pi\}$$

and let r and R be positive constants with $r < R$. The image of the function $\rho : S \subseteq \mathbb{R}^2 \to \mathbb{R}^3$ defined for all $(u, v) \in S$ by

$$\rho(u, v) = ((R + r \cos v) \cos u, (R + r \cos v) \sin u, r \sin v)$$

is the *torus* K sketched in Fig. 8.15. It is shaped like a ring centred at the origin and of circular cross-section r, where R is the distance from the origin to the centre of a cross-section.

Verify that K is an orientable surface, and identity the unit normal vector field N_ρ. Show also that the torus is implicitly defined by the equation

$$(\sqrt{(x^2 + y^2)} - R)^2 + z^2 = r^2.$$

Answer: (K, N_ρ) is (positively) oriented by outward facing normals.

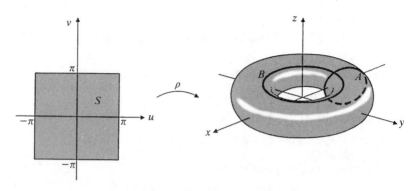

Fig. 8.15

4. Let K be a surface. Prove that if there exists a *continuous* unit normal vector field $N: K \subseteq \mathbb{R}^3 \to \mathbb{R}^3$ then K is orientable.

5. Find a parametrization of the sphere $x^2 + y^2 + z^2 = c^2$ which determines the orientation by inward facing unit normals.

Hint: consider ρ^-, where ρ is the spherical coordinate parametrization of Example 8.5.11.

6. This exercise illustrates that a surface need not be path connected. Show that the hyperboloid of two sheets K defined by $z^2 = 1 + x^2 + y^2$, $|z| \leq 2$ is a surface by Definition 8.5.1.

Answer: the idea is to find disjoint regions S_1 and S_2 in $\mathbb{R}^2$ and a parametrization $\rho: S = S_1 \cup S_2 \subseteq \mathbb{R}^2 \to \mathbb{R}^3$ of K such that $\rho(S_1)$ and $\rho(S_2)$ are respectively the sheets $z = \sqrt{(1 + x^2 + y^2)} \leq 2$ and $z = -\sqrt{(1 + x^2 + y^2)} \geq -2$. This can be achieved, for example, by taking S_1 and S_2 to be the compact discs in $\mathbb{R}^2$ of radius $\sqrt{3}$ and centres $(2, 0)$ and $(-2, 0)$ respectively, and defining ρ on S_1 by

$$\rho(u, v) = (u - 2, v, \sqrt{(1 + (u - 2)^2 + v^2)}), \qquad 0 \leq (u - 2)^2 + v^2 \leq 3$$

and ρ on S_2 by

$$\rho(u, v) = (u + 2, v, -\sqrt{(1 + (u + 2)^2 + v^2)}), \qquad 0 \leq (u + 2)^2 + v^2 \leq 3.$$

7. Let K be the hyperboloid of two sheets $z^2 = 1 + x^2 + y^2$, $|z| \leq 2$, considered in Exercise 6. Obtain four different orientations of K.

Answer: in the notation of Exercise 6, four inequivalent classes of parametrizations of K are given by (a) ρ on S, (b) ρ^- on S, (c) ρ on S_1 and ρ^- on S_2, and (d) ρ^- on S_1 and ρ on S_2.

8. Find a parametrization of the cylinder $x^2 + y^2 = k^2$, $|z| \leq 1$, which orients it by inward facing normals.

Integration over surfaces

9.1 Surface area

The area of a simple region S in $\mathbb{R}^2$ can be estimated by chopping S up into small rectangles and adding up their area. This estimate is a Riemann sum of the integral $\iint_S 1 \, dA$, which, if it exists, is defined to be the area of S.

We shall use the same technique in this section to define the area of a smooth simple surface and related subsets of $\mathbb{R}^3$. Let $\rho : S \subseteq \mathbb{R}^2 \to \mathbb{R}^3$ be a simple parametrization of a smooth simple surface K. Let $\{R_{ij}\}$ be a grid of rectangles covering S. The first step towards a reasonable definition of the area of K is to find a reasonable estimate for the area of $\rho(R_{ij})$ for each small compact rectangle R_{ij}. We shall find that the sum of these estimates is a Riemann sum of a certain integral over S. The area of K will be defined to be that integral.

For any rectangle R in S, the set $\rho(R)$ is a distorted copy of R lying in K. Provided that R is small this distortion can be estimated by using the differentiability of ρ. Suppose that $\mathbf{p} = (p_1, p_2) \in S$ is a vertex of R and that $R = [p_1, p_1 + h_1] \times [p_2, p_2 + h_2] \subseteq S$.

For all small $\mathbf{k} \in \mathbb{R}^2$, since ρ is differentiable,

9.1.1 $\rho(\mathbf{p} + \mathbf{k}) \approx \rho(\mathbf{p}) + L_{\rho, \mathbf{p}}(\mathbf{k}) = \rho(\mathbf{p}) + k_1 \dfrac{\partial \rho}{\partial u}(\mathbf{p}) + k_2 \dfrac{\partial \rho}{\partial v}(\mathbf{p}).$

For a fixed $\mathbf{p} \in \mathbb{R}^2$, let $\bar{\rho} : \mathbb{R}^2 \to \mathbb{R}^2$ be the affine function defined by

9.1.2 $\bar{\rho}(\mathbf{p} + \mathbf{k}) = \rho(\mathbf{p}) + k_1 \dfrac{\partial \rho}{\partial u}(\mathbf{p}) + k_2 \dfrac{\partial \rho}{\partial v}(\mathbf{p}), \qquad \mathbf{k} \in \mathbb{R}^2.$

The image of $\bar{\rho}$ is the tangent plane $\rho(\mathbf{p}) + T_{\rho(\mathbf{p})}(K)$ and $\bar{\rho}(R)$ is a parallelogram in this tangent plane with vertices

$$\bar{\rho}(p_1, p_2) = \rho(\mathbf{p}), \qquad \bar{\rho}(p_1 + h_1, p_2) = \rho(\mathbf{p}) + h_1 \dfrac{\partial \rho}{\partial u}(\mathbf{p}),$$

$$\bar{\rho}(p_1, p_2 + h_2) = \rho(\mathbf{p}) + h_2 \frac{\partial \rho}{\partial v}(\mathbf{p}),$$

$$\bar{\rho}(p_1 + h_1, p_2 + h_2) = \rho(\mathbf{p}) + h_1 \frac{\partial \rho}{\partial u}(\mathbf{p}) + h_2 \frac{\partial \rho}{\partial v}(\mathbf{p}).$$

See Fig. 9.1.

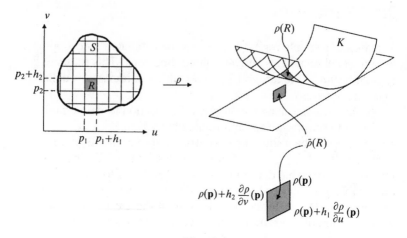

Fig. 9.1

Expression 9.1.1 tells us that $\rho(\mathbf{p} + \mathbf{k}) \approx \bar{\rho}(\mathbf{p} + \mathbf{k})$ for small $\mathbf{k}$, and therefore, when R is a small rectangle, the area of $\bar{\rho}(R)$ is a good approximation to the area of $\rho(R)$. But (Theorem 1.2.22(iv)) the area of the parallelogram with the given vertices is

$$\text{area } \bar{\rho}(R) = \left\| \frac{\partial \rho}{\partial u}(\mathbf{p}) \times \frac{\partial \rho}{\partial v}(\mathbf{p}) \right\| |h_1 h_2| = \|v_\rho(\mathbf{p})\| \text{ area } R.$$

Hence

9.1.3 $\text{area } \rho(R) \approx \|v_\rho(\mathbf{p})\| \text{ area } R.$

Suppose now that $\{R_{ij}\}$ is a grid of small rectangles covering S where the bottom left-hand vertex of R_{ij} is $\mathbf{p}_{ij}$. We have the following sequence of approximations for the area of $K = \rho(S)$:

$$\text{area } K = \text{area } \rho(S) \approx \sum_{i,j} \text{area } \rho(R_{ij}) \approx \sum_{i,j} \|v_\rho(\mathbf{p}_{ij})\| \text{ area } R_{ij}.$$

The last expression is, however, a Riemann sum of the integral

$$\iint_S \|v_\rho(u, v)\| \, du \, dv.$$

We now define formally the area of a smooth simple surface in terms of a simple parametrization. We then prove that the value does not depend upon the parametrization which is chosen.

9.1.4 Definition. *Let K be a smooth simple surface in $\mathbb{R}^3$. The area of K is defined to be*

9.1.5
$$\text{area } K = \iint_S \|v_\rho(u, v)\| \, du \, dv,$$

where $\rho : S \subseteq \mathbb{R}^2 \to \mathbb{R}^3$ is any simple parametrization of K.

9.1.6 Theorem. *Let $\rho : S \subseteq \mathbb{R}^2 \to \mathbb{R}^3$ and $\eta : T \subseteq \mathbb{R}^2 \to \mathbb{R}^3$ be simple parametrizations of a smooth simple surface K in $\mathbb{R}^3$. Then*

9.1.7
$$\iint_S \|v_\rho(u, v)\| \, du \, dv = \iint_T \|v_\eta(s, t)\| \, ds \, dt.$$

Proof. It follows from Theorem 8.1.10 that there is a 1–1 smooth function ϕ from S onto T such that $\rho = \eta \circ \phi$. By Theorem 8.2.7

9.1.8
$$\|v_\rho(u, v)\| = |\det J_{\phi, (u, v)}| \, \|v_\eta(\phi(u, v))\|.$$

Put $f = \|v_\eta\| : T \subseteq \mathbb{R}^2 \to \mathbb{R}$. Since $\phi : S \subseteq \mathbb{R}^2 \to \mathbb{R}^2$ is 1–1 and smooth on S, the Change of Variables Theorem 7.7.5 implies that

$$\iint_{\phi(S)} f(s, t) \, ds \, dt = \iint_S f(\phi(u, v)) \, |\det J_{\phi, (u, v)}| \, du \, dv.$$

Expression 9.1.7 now follows from 9.1.8.

Theorem 9.1.6 guarantees that Definition 9.1.4 is not ambiguous. The argument applies equally to smooth piecewise simple surfaces. We can therefore take 9.1.5 as a definition of the area of a smooth piecewise simple surface K, where $\rho : S \subseteq \mathbb{R}^2 \to \mathbb{R}^3$ is a simple parametrization of K whose domain S is a piecewise simple region.

9.1.9 Example. Let S be a simple region in $\mathbb{R}^2$ and let $\rho : S \subseteq \mathbb{R}^2 \to \mathbb{R}^3$ be given by $\rho(u, v) = (u, v, 0)$, $(u, v) \in S$. Then

$$\nu_\rho(u, v) = (0, 0, 1) = \mathbf{k} \qquad \text{and} \qquad \|\nu_\rho(u, v)\| = 1, \qquad (u, v) \in S.$$

The area of the planar surface $\rho(S)$ is therefore the same under 9.1.5 as it was given in Definition 7.4.4.

9.1.10 Example. Let S be a simple region in $\mathbb{R}^2$ and let $f : S \subseteq \mathbb{R}^2 \to \mathbb{R}$ be a C^1 function. Then $\gamma : S \subseteq \mathbb{R}^2 \to \mathbb{R}^3$, defined by

$$\gamma(x, y) = (x, y, f(x, y)), \qquad (x, y) \in S,$$

is a simple parametrization of the smooth simple surface K which is the graph of f in $\mathbb{R}^3$. In this case

$$\nu_\gamma(x, y) = \left(-\frac{\partial f}{\partial x}(x, y), \ -\frac{\partial f}{\partial y}(x, y), \ 1 \right),$$

and the area of K is

$$\text{area } K = \iint_S \sqrt{\left(\left(\frac{\partial f}{\partial x}\right)^2 + \left(\frac{\partial f}{\partial y}\right)^2 + 1 \right)} \, \mathrm{d}x \, \mathrm{d}y.$$

9.1.11 Example. Let K be the compact cylinder $x^2 + y^2 = 1$, $|z| \leqslant 1$ in $\mathbb{R}^3$. Let S be the piecewise simple region (annulus) in $\mathbb{R}^2$

$$S = \{(u, v) \in \mathbb{R}^2 \mid 1 \leqslant u^2 + v^2 \leqslant 9\}$$

and let $\rho : S \subseteq \mathbb{R}^2 \to \mathbb{R}^3$ be the simple parametrization of K (compare Exercise 8.4.1) given by

$$\rho(u, v) = \left(\frac{u}{\sqrt{(u^2 + v^2)}}, \ \frac{v}{\sqrt{(u^2 + v^2)}}, \ \sqrt{(u^2 + v^2)} - 2 \right), \qquad (u, v) \in S.$$

Then

$$\nu_\rho(u, v) = \left(\frac{-u}{u^2 + v^2}, \ \frac{-v}{u^2 + v^2}, \ 0 \right), \qquad (u, v) \in S$$

and so

$$\text{area } K = \iint_S \frac{1}{\sqrt{(u^2 + v^2)}} \, \mathrm{d}u \, \mathrm{d}v.$$

Applying the change of variables $(u, v) = (r \cos \theta, r \sin \theta)$, we obtain area $K = 4\pi$.

9.1.12 Remark. Let S be a simple region in $\mathbb{R}^2$ and let $\rho : S \subseteq \mathbb{R}^2 \to \mathbb{R}^3$ be a C^1 function which is 1–1 and smooth, except possibly on a null set. Then the argument of Theorem 9.1.6 also justifies using expression 9.1.5 to define the area of the set $K = \rho(S)$. This allows us to deal with surfaces like the sphere, the torus, the figure 8, and the Möbius strip which are all the image of a rectangle under functions which fail to be 1–1 on the boundary. Even more generally, if $K = \rho(S_1 \cup \cdots \cup S_k)$ is a surface in $\mathbb{R}^3$ as defined in Definition 8.5.1, then we may take the area of K to be

$$\text{area } K = \sum_{1}^{k} \iint\limits_{S_i} \|v_\rho(u, v)\| \, du \, dv.$$

9.1.13 Example. The cylinder K considered in Example 9.1.11 is naturally parametrized by the function $\eta : T \subseteq \mathbb{R}^2 \to \mathbb{R}^3$, where

$$T = \{(\theta, t) \in \mathbb{R}^2 \mid 0 \leq \theta \leq 2\pi, \qquad -1 \leq t \leq 1\}$$

and

$$\eta(\theta, t) = (\cos \theta, \sin \theta, t), \qquad (\theta, t) \in T.$$

Hence

$$v_\eta(\theta, t) = (\cos \theta, \sin \theta, 0), \qquad (\theta, t) \in T$$

and so

$$\text{area } K = \iint\limits_{T} 1 \, d\theta \, dt = 4\pi,$$

giving a simple confirmation of Example 9.1.11.

9.1.14 Example. Let $S = [0, \pi] \times [0, 2\pi]$, and let $\rho : S \subseteq \mathbb{R}^2 \to \mathbb{R}^3$ be the C^1 function defined, for $c > 0$, by

$$\rho(\phi, \theta) = (c \sin \phi \cos \theta, c \sin \phi \sin \theta, c \cos \phi), \qquad (\phi, \theta) \in S.$$

Then $\rho(S)$ is the sphere $x^2 + y^2 + z^2 = c^2$ in $\mathbb{R}^3$. In Example 8.5.11 we showed that for all $(\phi, \theta) \in S$,

$$\|v_\rho(\phi, \theta)\| = c^2 \sin \phi.$$

The function ρ is 1–1 and smooth on Int S. Since ∂S is a null set, the area of the sphere is given by

$$\iint\limits_{S} \|v_\rho(\phi, \theta)\| \, d\phi \, d\theta = c^2 \iint\limits_{S} \sin \phi \, d\phi \, d\theta$$

$$= c^2 \int_0^{2\pi} d\theta \int_0^{\pi} \sin \phi \, d\phi = 4\pi c^2.$$

9.1.15 Example.   Let $S = [-\pi, \pi] \times [-\pi, \pi]$, and let $\rho : S \subseteq \mathbb{R}^2 \to \mathbb{R}^3$ be the C^1 function defined, for $0 < r < R$, by

$$\rho(u, v) = ((R + r \cos v) \cos u, (R + r \cos v) \sin u, r \sin v)$$

for all $(u, v) \in S$. Then $\rho(S)$ is the torus K of Exercise 8.5.3. Since ρ is 1–1 and smooth on Int S the area of the torus may be found using the above technique (see Exercise 9.1.1).

9.1.16 Example.   Let K be the (bounded) circular cone of height h

$$K = \{(x, y, z) \in \mathbb{R}^3 \mid z = \sqrt{(x^2 + y^2)}, \, 0 \leqslant z \leqslant h\}.$$

The parametrization $\rho : S \subseteq \mathbb{R}^2 \to \mathbb{R}^3$ of K where $S = \{(u, v) \in \mathbb{R}^2 \mid u^2 + v^2 \leqslant h^2\}$ and

$$\rho(u, v) = (u, v, \sqrt{(u^2 + v^2)}), \qquad (u, v) \in S$$

is not differentiable at $(0, 0) \in S$. However, the function $\|v_\rho\|$ is continuous and bounded on $S \backslash \{(0, 0)\}$ (in fact it is constant). We may ignore the bad behaviour of ρ at the origin and conclude that the area of K is given by the formula 9.1.5.

In certain cases a limiting argument is useful in finding the area of a surface. See Exercise 9.1.3.

Exercises 9.1

1.   Find the area of the following surfaces in $\mathbb{R}^3$.
   (a)   The (bounded) cylinder of radius k and height h

$$\{(x, y, z) \in \mathbb{R}^3 \mid x^2 + y^2 = k^2, \, 0 \leqslant z \leqslant h\};$$

   (b)   the (bounded) circular cone of height h

$$\{(x, y, z) \in \mathbb{R}^3 \mid z = \sqrt{(x^2 + y^2)}, \, 0 \leqslant z \leqslant h\};$$

   (c)   the sphere $x^2 + y^2 + z^2 = c^2$;
   (d)   the ellipsoid $\dfrac{x^2}{a^2} + \dfrac{y^2}{b^2} + \dfrac{z^2}{c^2} = 1$;
   (e)   the spherical cap $x^2 + y^2 + z^2 = c^2$, $z \geqslant h$, where $0 < h < c$;
   (f)   the torus K given in Example 9.1.15;
   (g)   part of the paraboloid $z = 2 - (x^2 + y^2)$ above the x, y plane.

Answers: (a) $2\pi kh$; (b) $\sqrt{2}\pi h^2$; (c) $4\pi c^2$; (d) the integral formula cannot be simplified; (e) $2\pi c \sqrt{(c^2 - h^2)}$, use spherical coordinate or graph parametrization. In the latter case evaluate resulting integral over $S_h = \{(x, y) \in \mathbb{R}^2 \mid x^2 + y^2 \leqslant h^2\}$ by polar coordinates; (f) $4\pi^2 rR$; (g) $13\pi/3$.

2. Let $\rho:S\subseteq\mathbb{R}^2\to\mathbb{R}^3$ be a simple parametrization of a smooth simple surface K, and let $\rho^-:S^*\subseteq\mathbb{R}^2\to\mathbb{R}^3$ be the opposite parametrization to ρ (Example 8.3.12). Show that

$$\iint\limits_{S}\|v_\rho(u,v)\|\,du\,dv = \iint\limits_{S^*}\|v_{\rho^-}(s,t)\|\,ds\,dt.$$

Both integrals give the area of K.

3. Let K be the hemisphere $x^2+y^2+z^2=c^2$, $z\geq 0$. This is a smooth simple surface (Example 8.1.3). Show that the graph parametrization of K given by $\gamma:S\subseteq\mathbb{R}^2\to\mathbb{R}^3$, where $S=\{(x,y)\in\mathbb{R}^2\,|\,x^2+y^2\leq c^2\}$ and

$$\gamma(x,y)=(x,y,\sqrt{(c^2-x^2-y^2)}),\qquad (x,y)\in S$$

is not a simple parametrization. (Consider $\partial\gamma/\partial x$ and $\partial\gamma/\partial y$ on the boundary of S.)
   Prove that

$$\|v_\gamma(x,y)\|=\frac{c}{\sqrt{(c^2-x^2-y^2)}},\qquad (x,y)\in\text{Int }S.$$

Justify the formula for the area of K given by the *improper* integral

$$\text{Area }K=\iint\limits_{S}\frac{c}{\sqrt{(c^2-x^2-y^2)}}\,dx\,dy = 2\pi c^2.$$

(Note that the integrand is infinite on ∂S. The formula can be justified as the limiting case as $h\to 0$ of the spherical cap area, Exercise 1(e).)

4. Let K be that part of the square cone $z=|x|+|y|$ which lies between the planes $z=0$ and $z=1$. Set up simple parametrizations for each of the four triangular faces and hence find the area of K.

Answer: the face in the first octant, for example, can be parametrized by $\rho:S\subseteq\mathbb{R}^2\to\mathbb{R}^3$, where $S=\{(u,v)\in\mathbb{R}^2\,|\,u+v\leq 1,\,u\geq 0,\,v\geq 0\}$, and $\rho(u,v)=(u,v,u+v)$, $(u,v)\in S$. Each face has area $\frac{1}{2}\sqrt{3}$. This can be verified by a simple geometrical argument.

5. Show that the area of the Möbius strip K^* discussed in Example 8.5.5 is given by the integral

$$\text{Area }K^* = 2\int_0^{1/2\pi}d\theta\int_{-1}^1\sqrt{(t^2+4(k-t\cos\theta)^2)}\,dt.$$

Surprisingly this is not equal to the area $4\pi k$ of the paper strip S suggested for the construction of a model of K^*. The explanation is as follows: in the Möbius strip K^* the circle C running along the middle of the strip S is constrained to lie in the x, y plane. In order to fit the strip S on to K^* some stretching or compressing near the edges is necessary.

6. Let C be the graph in $\mathbb{R}^2$ of a C^1 function $g : [a, b] \subseteq \mathbb{R} \to \mathbb{R}$, and let K be the graph in $\mathbb{R}^3$ of a C^1 function $f : S \subseteq \mathbb{R}^2 \to \mathbb{R}$, where S is a simple region. Contrast the formulae for the length of C and the area of K (9.1.10):

$$l(C) = \int_a^b \sqrt{\left(\left(\frac{dg}{dx}\right)^2 + 1\right)} \, dx,$$

$$\text{Area } K = \iint_S \sqrt{\left(\left(\frac{\partial f}{\partial x}\right)^2 + \left(\frac{\partial f}{\partial y}\right)^2 + 1\right)} \, dx \, dy.$$

Consider in particular the case $S = [a, b] \times [c, d]$, and $f(x, y) = g(x)$, $(x, y) \in S$. The Area $K = (d - c)l(C)$. Illustrate this case with a sketch.

7. Find the area of part of the plane $2x + y + 2z = 16$ which is cut off by the planes $x = 0$, $y = 0$ and $x + y = 1$.

Answer: $\frac{3}{4}$.

9.2   Surface integral of a scalar field

The integral of a continuous scalar field (real-valued function) over certain subsets of $\mathbb{R}^2$ was defined in Chapter 6. We shall extend that definition to the integral of continuous scalar fields over surfaces. In doing so we follow the scheme adopted in Section 5.2 where we considered the integral of scalar fields along a path.

Imagine that a smooth simple surface K in $\mathbb{R}^3$ is made from heavy material of uniform thickness whose mass per unit area varies over K according to a continuous function $f : K \subseteq \mathbb{R}^3 \to \mathbb{R}$. What is the total mass of the material used? Let $\rho : S \subseteq \mathbb{R}^2 \to \mathbb{R}^3$ be a simple parametrization of K and let $\{R_{ij}\}$ be a grid of rectangles covering S. For each i and j the mass of $\rho(R_{ij})$ is approximately

$$f(\rho(\mathbf{p}_{ij})) \text{ area } \rho(R_{ij}),$$

where $\mathbf{p}_{ij}$ is a point in R_{ij}. But by 9.1.3, this expression is approximately

$$f(\rho(\mathbf{p}_{ij})) \, \|v_\rho(\mathbf{p}_{ij})\| \text{ area } R_{ij}$$

and therefore the mass of the material used is approximately

9.2.1 $$\sum_{i,j} f(\rho(\mathbf{p}_{ij})) \, \|v_\rho(\mathbf{p}_{ij})\| \text{ area } R_{ij}.$$

Expression 9.2.1 is a Riemann sum of the integral 9.2.3 which appears in the following definition.

9.2.2 Definition.   *Let K be a smooth simple surface in $\mathbb{R}^3$ and let $f : K \subseteq \mathbb{R}^3 \to \mathbb{R}$ be a continuous scalar field on K. The* surface integral of f over K *is defined to be*

9.2.3
$$\iint_K f \, \mathrm{d}A = \iint_S f(\rho(u, v)) \, \|v_\rho(u, v)\| \, \mathrm{d}u \, \mathrm{d}v,$$

where $\rho : S \subseteq \mathbb{R}^2 \to \mathbb{R}^3$ is any simple parametrization of K.

The reader should compare Definition 9.2.2 and Definition 6.1.1 for the line integral of a scalar field over a simple arc.

We justify Definition 9.2.2 by showing that the surface integral 9.2.3 is independent of the simple parametrization of K used.

9.2.4 Theorem.   *Let $\rho : S \subseteq \mathbb{R}^2 \to \mathbb{R}^3$ and $\eta : T \subseteq \mathbb{R}^2 \to \mathbb{R}^3$ be simple parametrizations of a smooth simple surface K in $\mathbb{R}^3$. For any continuous scalar field $f : K \subseteq \mathbb{R}^3 \to \mathbb{R}$,*

$$\iint_S f(\rho(u, v)) \, \|v_\rho(u, v)\| \, \mathrm{d}u \, \mathrm{d}v = \iint_T f(\eta(s, t)) \, \|v_\eta(s, t)\| \, \mathrm{d}s \, \mathrm{d}t.$$

Proof.   Follow the proof of Theorem 9.1.6.

9.2.5 Example.   The area of a smooth simple surface K is $\iint_K 1 \, \mathrm{d}A$. The mass of the material covering K in the discussion preceding Definition 9.2.2 is $\iint_K f \, \mathrm{d}A$.

9.2.6 Example.   Let S be a simple region in $\mathbb{R}^2$ and let $\rho : S \subseteq \mathbb{R}^2 \to \mathbb{R}^3$ be given by $\rho(u, v) = (u, v, 0)$, $(u, v) \in S$. Then $v_\rho(u, v) = (0, 0, 1)$ for all $(u, v) \in S$ and for any continuous scalar field $f : \rho(S) \subseteq \mathbb{R}^3 \to \mathbb{R}$,

$$\iint_{\rho(S)} f \, \mathrm{d}A = \iint_S f(u, v, 0) \, \mathrm{d}u \, \mathrm{d}v.$$

The integral of f over this subset $\rho(S)$ of the x, y plane is therefore essentially the same as that defined in Section 7.4.

9.2.7 Definition.   *Let $K = \rho(S_1 \cup \cdots \cup S_k)$ be the surface con-*

sidered in Definition 8.5.1. *For any continuous scalar field*
$f : K \subseteq \mathbb{R}^3 \to \mathbb{R}$ *we define*

9.2.8 $$\iint_K f \, dA = \sum_1^k \iint_{S_i} f(\rho(u, v)) \, \|v_\rho(u, v)\| \, du \, dv.$$

This definition is independent of the chosen parametrization of K.
It applies to most of the sets we have been considering—smooth
piecewise simple surfaces, sphere, cone, torus, Möbius strip, figure
8 sheet, and cube.

Once again one allows for bad behaviour on null sets.

9.2.9 Example.   The sphere K in $\mathbb{R}^3$ with centre $(0, 0, 1)$ and radius c is
given by the equation

$$x^2 + y^2 + (z - 1)^2 = c^2.$$

We can parametrize K by the function $\rho : S \subseteq \mathbb{R}^2 \to \mathbb{R}^3$ defined on the
simple region $S = [0, 2\pi] \times [-c, c]$ in $\mathbb{R}^2$, where

$$\rho(\theta, t) = (\sqrt{(c^2 - t^2)} \cos \theta, \ \sqrt{(c^2 - t^2)} \sin \theta, \ t + 1), \qquad (\theta, t) \in S.$$

Notice that ρ is 1–1 and smooth on Int S but that it is not 1–1, nor
differentiable, on ∂S.

We find that, for all $(\theta, t) \in$ Int S,

$$v_\rho(\theta, t) = (\sqrt{(c^2 - t^2)} \cos \theta, \ \sqrt{(c^2 - t^2)} \sin \theta, \ t)$$

and hence that

$$\|v_\rho(\theta, t)\| = c.$$

Define a continuous scalar field $f : K \subseteq \mathbb{R}^3 \to \mathbb{R}$ by

$$f(x, y, z) = x^2 + y^2 + z^2, \qquad (x, y, z) \in K.$$

Then since $\rho(\partial S)$ is a null set and the function $\|v_\rho\|$ is continuous and
bounded on Int S we may ignore the bad behaviour on ρ and conclude that
the surface integral of f over K is

$$\iint_K f \, dA = \iint_S f(\rho(\theta, t)) \, \|v_\rho(\theta, t)\| \, d\theta \, dt$$

$$= \int_0^{2\pi} d\theta \int_{-c}^c (c^2 + 2t + 1) c \, dt = 4\pi c^2 (1 + c^2).$$

It is interesting that whereas the C^1 function ρ on Int S cannot be
extended to a C^1 function on S, the function $\|v_\rho\| = \|\partial\rho/\partial u \times \partial\rho/\partial v\|$ can
be so extended.

In Section 7.4 we found that the integral of a function f over a subset of $\mathbb{R}^2$ was not affected by relaxing the continuity condition on a null set of $\mathbb{R}^2$. A similar result holds for surface integrals.

9.2.10 Definition. *Let K be a surface in $\mathbb{R}^3$ and let A be a subset of K. Then A is said to be a null set in K if $\rho^{-1}(A)$ is a null set in $\mathbb{R}^2$, where ρ is a parametrization of K.*

We leave it to the reader to show that this definition is independent of the parametrization ρ chosen.

9.2.11 Example. The edge ∂K of a simple surface K is a null set in K.

We extend the scope of Definition 9.2.7 by saying that the (not necessarily continuous) function $f : K \subseteq \mathbb{R}^3 \to \mathbb{R}$ is *integrable* over the surface K if the right-hand integral in 9.2.8 exists. The integral of f over K is then denoted by $\iint_K f \, dA$.

9.2.12 Theorem. *Let K be a surface in $\mathbb{R}^3$, let A be a null set in K and let $f : K \subseteq \mathbb{R}^3 \to \mathbb{R}$ be a scalar field which is continuous on $K \backslash A$. Then* **[i]** *f is integrable over K;* **[ii]** *for any scalar field $g : K \subseteq \mathbb{R}^3 \to \mathbb{R}$ which agrees with f on $K \backslash A$,*

$$\iint_K f \, dA = \iint_K g \, dA.$$

Proof. [*i*] Consider a parametrization ρ of K in the notation of Definition 8.5.1. Put $N = \rho^{-1}(A)$. Then N is a null set in $\mathbb{R}^2$. By Theorem 7.4.13, since the functions $f \circ \rho$ and $\|v_\rho\|$ are continuous on $S_i \backslash N$, the integrals on the right-hand side of 9.2.8 exist, so f is integrable over K.

(*ii*) This follows from Theorem 7.4.28.

Exercises 9.2

1. Let K be a surface in $\mathbb{R}^3$ with area $\mathscr{A} = \iint_K 1 \, dA$. Let $f : K \subseteq \mathbb{R}^3 \to \mathbb{R}$ be integrable over K. Provided $\mathscr{A} \neq 0$, the *mean value of f over K* is

defined to be the number

$$\bar{f} = \frac{1}{\mathscr{A}} \iint_K f \, dA.$$

In particular, the mean values of $f(x, y, z) = x$, $f(x, y, z) = y$ and $f(x, y, z) = z$ over K gives the coordinates $(\bar{x}, \bar{y}, \bar{z})$ of the *centroid* of K. (Compare with Exercise 5.2.4.)

Find the centroids of the following surfaces considered in Exercise 9.1.1:

(a)   the circular cone of height h

$$\{(x, y, z) \in \mathbb{R}^3 \mid z = \sqrt{(x^2 + y^2)}, \, 0 \leqslant z \leqslant h\};$$

(b)   the spherical cap $x^2 + y^2 + z^2 = c^2$, $z \geqslant h$, where $0 < h < c$;
(c)   the paraboloid $z = 2 - (x^2 + y^2)$, $z \geqslant 0$.

Answers: $\bar{x} = \bar{y} = 0$, by symmetry about the z-axis. (a) $\bar{z} = \frac{2}{3}h$;
(b) $\bar{z} = \frac{1}{2}(c + h)$; (c) $\bar{z} = 111\pi/30$.

2.   Let K be a surface in $\mathbb{R}^3$. The *second moment of area* about the z-axis of K is defined to be the integral

$$\iint_K \rho^2 \, dA,$$

where $\rho(x, y, z) = \sqrt{(x^2 + y^2)}$, the distance of the point (x, y, z) from the z-axis.

Find the second moment of area about the z-axis of the following surfaces. Express your answer in the form $\mathscr{A}\xi^2$, where $\mathscr{A}$ is the area of K. The number ξ is called the *radius of gyration* of K about the z-axis.
(a)   The cylinder $\{(x, y, z) \in \mathbb{R}^3 \mid x^2 + y^2 = k^2, \, 0 \leqslant z \leqslant h\}$;
(b)   the circular cone $\{(x, y, z) \in \mathbb{R}^3 \mid z = \sqrt{(x^2 + y^2)}, \, 0 \leqslant z \leqslant h\}$;
(c)   the sphere $x^2 + y^2 + z^2 = c^2$;
(d)   the paraboloid $z = 2 - (x^2 + y^2)$, $z \geqslant 0$.

Answers: the value of ξ^2 is (a) k^2; (b) $\frac{1}{2}h^2$; (c) $\frac{2}{3}c^2$; (d) 149/130.

3.   The electric potential of a unit point charge is $1/r$ at a distance r from the charge. Let K be a sphere of radius c.
(*a*)   Show that the mean value of the potential over K due to a point charge placed anywhere inside K is $1/c$. Remarkably the result is independent of the position of the charge inside K.

Hint: choose a coordinate system origin at the centre of K, with the point charge on one of the axes, say the z-axis, a distance $p < c$ from the origin. Parametrize K by

$$\rho(\theta, z) = (\sqrt{(c^2 - z^2)} \cos \theta, \sqrt{(c^2 - z^2)} \sin \theta, z), \quad (\theta, z) \in [0, 2\pi] \times [-1, 1].$$

Deduce that

$$\text{Mean potential over } K = \frac{1}{4\pi c^2} \int_0^{2\pi} \mathrm{d}\theta \int_{-1}^{1} \frac{c}{\sqrt{(c^2 \pm 2zp + p^2)}} \, \mathrm{d}z$$

and complete the integration.
(b) Show that the mean value of the potential over K due to a unit point charge placed outside K is $1/p$, where $p > c$ is the distance of the charge from the centre of the sphere.

4. (a) *Pappus' Theorem* for surfaces of revolution. A plane simple arc C is revolved about an axis in its plane which does not intersect the curve, except possibly at an end point. Prove that the area of the surface of revolution K so generated is equal to $2\pi d\, l(C)$, where $l(C)$ is the length of C and d is the distance of the centroid of C from the axis of rotation.

Hint: consider the curve C in the y, z plane and take the z-axis as axis of rotation. Let $\alpha:[a, b] \subseteq \mathbb{R} \to \mathbb{R}^2$ be a simple parametrization of C. Put

$$\alpha(t) = (y(t), z(t)), \qquad t \in [a, b],$$

where $y(t) \geq 0$ for all t. Hence obtain the parametrization $\rho:S = [a, b] \times [0, 2\pi] \subseteq \mathbb{R}^2 \to \mathbb{R}^3$ of K defined by

$$\rho(t, \theta) = (y(t) \cos \theta, y(t) \sin \theta, z(t)), \qquad (t, \theta) \in S.$$

Deduce that

$$\text{Area } K = \iint_S y(t) \, \|\alpha'(t)\| \, \mathrm{d}t \, \mathrm{d}\theta.$$

Hence show that Area $K = 2\pi \bar{y} l(C)$.
(b) Using Pappus' Theorem check the surface areas calculated in Exercises 9.1.1(a), (b), (c), (e) and (g). Note also the simple derivation of the surface area of the torus, Exercise 9.1.1(f).

5. Let K be the subset of the plane $x + y + z = 1$ in $\mathbb{R}^3$ enclosed by the triangle, with vertices $P(1, 0, 0)$, $Q(0, 1, 0)$, $R(0, 0, 1)$. Compute
(a) $\iint_K x \, \mathrm{d}A$ and (b) $\iint_K z \, \mathrm{d}A$.

Hint; (a) parametrize K by $\rho(x, y) = (x, y, 1 - x - y)$, $(x, y) \in T$, where T is the triangle in $\mathbb{R}^2$ with vertices $(0, 0)$, $(1, 0)$, $(0, 1)$.

Answers: (a) $\sqrt{3}/6$; (b) $\sqrt{3}/6$. Equal by symmetry.

6. Consider the following short solution of Exercise 5 by a symmetry argument:

$$\iint_K x \, \mathrm{d}A = \iint_K y \, \mathrm{d}A = \iint_K z \, \mathrm{d}A = \tfrac{1}{3} \iint_K 1 \, \mathrm{d}A,$$

and the area of triangle PQR is $\sqrt{3}/2$. Hence $\iint_K x \, \mathrm{d}A = \sqrt{3}/6$.

7. A bowl is in the shape of the hemisphere $z = \sqrt{(c^2 - x^2 - y^2)}$, $0 \leqslant z \leqslant c$. The thickness of the bowl tapers towards the rim, the mass density per unit area being given by $d(x, y, z) = kz$, where $k > 0$ is constant. Find the mass of the bowl.

Answer: $\pi k c^3$.

9.3 Surface integral of a vector field

In this section we consider the integration of vector fields over *oriented* surfaces in $\mathbb{R}^3$. We saw in Section 6.1 that the integral of a vector field along an oriented simple arc corresponds to the work done by the field in moving a particle along the arc. An important physical interpretation of a surface integral of a vector field concerns the rate of flow of a fluid across a surface. The orientation of the surface gives the direction in which the rate of flow is to be measured.

Let (K, N) be an oriented simple surface in $\mathbb{R}^3$ and let $F : \mathbb{R}^3 \to \mathbb{R}^3$ be the velocity field of a fluid moving in $\mathbb{R}^3$. At any point $\mathbf{q}$ in K the velocity of the fluid is $F(\mathbf{q})$ and the component of this velocity normal to K in direction $N(\mathbf{q})$ is $F(\mathbf{q}) \cdot N(\mathbf{q})$ (see Fig. 9.2).

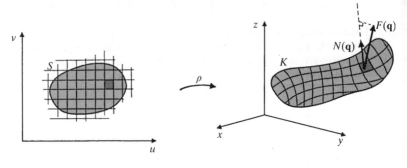

Fig. 9.2

Let $\rho : S \subseteq \mathbb{R}^2 \to \mathbb{R}^3$ be a simple parametrization of K and let $\{R_{ij}\}$ be a grid of rectangles covering S. For each i and j the rate of flow across $\rho(R_{ij})$, in the direction of the field, is approximately

$$|F(\rho(\mathbf{p}_{ij})) \cdot N(\rho(\mathbf{p}_{ij}))| \text{ area } \rho(R_{ij}),$$

where $\mathbf{p}_{ij}$ is a point of R_{ij}. By 9.1.3 this expression is approximately

9.3.1 $\qquad |F(\rho(\mathbf{p}_{ij})) \cdot N(\rho(\mathbf{p}_{ij}))| \, \|v_\rho(\mathbf{p}_{ij})\| \text{ area } R_{ij}.$

The above discussion applies equally to piecewise simple surfaces. Expression 9.3.1 is a Riemann sum of the integral contained in the following definition.

9.3.2 Definition. *Let (K, N) be an oriented piecewise simple surface in $\mathbb{R}^3$ and let $F : K \subseteq \mathbb{R}^3 \to \mathbb{R}^3$ be a continous vector field on K. The surface integral of F over (K, N) is denoted and defined by*

9.3.3 $\qquad \displaystyle\iint\limits_{(K,N)} F \cdot d\mathbf{A} = \iint\limits_{K} F \cdot N \, dA.$

In the left-hand integral 9.3.3 the symbol $d\mathbf{A}$ can be thought of as denoting an element of the *oriented* surface. In the right-hand integral the integrand $F \cdot N$ depends on the given orientation of K. Obviously the integral of F over (K, N) is the negative of the integral of F over $(K, -N)$ since

$$\iint\limits_{K} F \cdot N \, dA = -\iint\limits_{K} F \cdot (-N) \, dA.$$

The above discussion shows that if F is the velocity field of a fluid then the integral $\iint\limits_{(K,N)} F \cdot d\mathbf{A}$ measures the rate of flow of the fluid across the surface K in the direction determined at points on K by the unit normal field N. If length is measured in metres and time in seconds then the rate of fluid flow across K is given in cubic metres per second.

In Section 6.1 we considered the integral of a continuous vector field F along an oriented simple curve $C^+ = (C, T)$, where T is the unit tangent field to C^+. It is interesting to compare the relationships

9.3.4 $\qquad \displaystyle\int_{C^+} F \cdot d\mathbf{r} = \int_C F \cdot T \, ds \qquad \text{and} \qquad \iint\limits_{K^+} F \cdot d\mathbf{A} = \iint\limits_{K} F \cdot N \, dA,$

where we have put $K^+ = (K, N)$.

The following theorem is useful in computations.

9.3.5 Theorem. *Let (K, N) be an oriented piecewise simple surface in $\mathbb{R}^3$ and let $F : K \subseteq \mathbb{R}^3 \to \mathbb{R}^3$ be a continuous vector field over K.*

Then

$$\iint\limits_{(K,N)} F \cdot \mathrm{d}\mathbf{A} = \iint\limits_{S} F(\rho(u, v)) \cdot v_\rho(u, v) \, \mathrm{d}u \, \mathrm{d}v$$

for any simple parametrization $\rho : S \subseteq \mathbb{R}^2 \to \mathbb{R}^3$ *of* (K, N).

Proof.   Since

9.3.6 $$N(\rho(u, v)) = \frac{v_\rho(u, v)}{\| v_\rho(u, v) \|} \qquad \text{for all } (u, v) \in S$$

we find, using 9.2.3, that

9.3.7 $$\iint\limits_{K} F \cdot N \, \mathrm{d}A = \iint\limits_{S} F(\rho(u, v)) \cdot v_\rho(u, v) \, \mathrm{d}u \, \mathrm{d}v.$$

The result follows from 9.3.3.

9.3.8 *Example.*   Let K be the simple surface in $\mathbb{R}^3$

$$K = \{x, y, z) \in \mathbb{R}^3 \mid 2x + y + 2z = 4, \ x \geqslant 0, \ y \geqslant 0, \ z \geqslant 0\}.$$

Then K is a triangular piece of the plane $2x + y + 2z = 4$, and since grad $(2x + y + 2z) = (2, 1, 2)$ for all $(x, y, z) \in K$, we can define a continuous unit normal vector field $N : K \subseteq \mathbb{R}^3 \to \mathbb{R}^3$ by

$$N(x, y, z) = (\tfrac{2}{3}, \tfrac{1}{3}, \tfrac{2}{3}), \qquad (x, y, z) \in K.$$

To evaluate the integral of a continuous vector field F over (K, N) we look for a simple parametrization of (K, N) and then use Theorem 9.3.5.

Now K intersects the x, y plane in the line $z = 0$, $2x + y = 4$ and K is the graph $z = 2 - x - \tfrac{1}{2}y$ over the triangular subset of the x, y plane bounded by this line and the positive x and y axes. Let S be the corresponding triangular subset in $\mathbb{R}^2$. The parametrization $\rho : S \subseteq \mathbb{R}^2 \to \mathbb{R}^3$ of K defined by

$$\rho(u, v) = (u, v, 2 - u - \tfrac{1}{2}v), \qquad (u, v) \in S$$

leads to

$$v_\rho(u, v) = (1, \tfrac{1}{2}, 1), \qquad (u, v) \in S,$$

and so ρ is in fact a simple parametrization of (K, N).

For example, let $F : K \subseteq \mathbb{R}^3 \to \mathbb{R}^3$ be defined by

$$F(x, y, z) = (1 - 2z, 0, 2y), \qquad (x, y, z) \in K.$$

Then, by Theorem 9.3.5, the surface integral of F over (K, N) is

$$\iint\limits_{(K,N)} F \cdot d\mathbf{A} = \iint\limits_{S} F(\rho(u, v)) \cdot v_\rho(u, v) \, du \, dv$$

$$= \iint\limits_{S} (2u + v - 3, 0, 2v)(1, \tfrac{1}{2}, 1) \, du \, dv$$

$$= \iint\limits_{S} (2u + 3v - 3) \, du \, dv$$

$$= \int_0^2 du \int_0^{4-2u} (2u + 3v - 3) \, dv = 28/3.$$

For a sketch of (K, N) and further discussion, see Example 9.4.5.

The following example concerns an oriented piecewise simple surface that is not simple.

9.3.9 Example. Let K be the cylinder $\{(x, y, z) \in \mathbb{R}^3 \mid x^2 + y^2 = 1, 0 \leqslant z \leqslant 1\}$ and let $N: K \subseteq \mathbb{R}^3 \to \mathbb{R}^3$ be given by $N(x, y, z) = (-x, -y, 0)$. Then N is a continuous unit normal vector field on K pointing inwards.

The cylinder K is the image of the 1–1 smooth function $\rho: S \subseteq \mathbb{R}^2 \to \mathbb{R}^3$ where S is the (piecewise simple) annulus $\{(u, v) \in \mathbb{R}^2 \mid 1 \leqslant u^2 + v^2 \leqslant 4\}$ and

$$\rho(u, v) = \left(\frac{u}{\sqrt{(u^2 + v^2)}}, \frac{v}{\sqrt{(u^2 + v^2)}}, \sqrt{(u^2 + v^2)} - 1 \right), \qquad (u, v) \in S.$$

Since, for all $(u, v) \in S$,

$$v_\rho(u, v) = \left(\frac{-u}{u^2 + v^2}, \frac{-v}{u^2 + v^2}, 0 \right) = \frac{1}{\sqrt{(u^2 + v^2)}} N(\rho(u, v)),$$

the function ρ is a simple parametrization of the oriented piecewise simple surface (K, N).

If $F: K \subseteq \mathbb{R}^3 \to \mathbb{R}^3$ is given by $F(x, y, z) = (-y, x, z)$ then F flows tangentially to the cylinder and, as expected, $(F \cdot N)(x, y, z) = yx - xy = 0$. Therefore

$$\iint\limits_{(K,N)} F \cdot d\mathbf{A} = \iint\limits_{K} F \cdot N \, dA = 0.$$

If $G: K \subseteq \mathbb{R}^3 \to \mathbb{R}^3$ is given by $G(x, y, z) = (x, y, z)$ then G flows in the opposite direction to N and

$$\iint\limits_{(K,N)} G \cdot d\mathbf{A} = \iint\limits_{K} G \cdot N \, dA = \iint\limits_{K} (-x^2 - y^2) \, dA = \iint\limits_{K} (-1) \, dA = -2\pi.$$

If $H : K \subseteq \mathbb{R}^3 \to \mathbb{R}^3$ is given by $H(x, y, z) = (x(z + 1)^3, y(z + 1)^3, xy)$ then we may use the above parametrization ρ of K to obtain

$$
\iint\limits_{(K, N)} H \cdot \mathrm{d}\mathbf{A} = \iint\limits_S H(\rho(u, v) \cdot v_\rho(u, v) \, \mathrm{d}u \, \mathrm{d}v
$$

$$
= \iint\limits_S \left(u(u^2 + v^2), \, v(u^2 + v^2), \, \frac{uv}{u^2 + v^2} \right)
$$

$$
\cdot \left(\frac{-u}{u^2 + v^2}, \frac{-v}{u^2 + v^2}, 0 \right) \mathrm{d}u \, \mathrm{d}v
$$

$$
= \iint\limits_S (-u^2 - v^2) \, \mathrm{d}u \, \mathrm{d}v = -15\pi/2.
$$

The work of this section also applies to orientable surfaces. The following two examples illustrate the integration of a continuous vector field F over an oriented sphere, and over an oriented ellipsoid.

9.3.10 *Example.* Let (K, N) be the sphere $x^2 + y^2 + z^2 = c^2$ in $\mathbb{R}^3$ positively oriented by the unit normal vector field $N : K \subseteq \mathbb{R}^3 \to \mathbb{R}^3$ where $N(\mathbf{r}) = \mathbf{r}/r$, $\mathbf{r} \in K$. The vector field $F : \mathbb{R}^3 \backslash \{\mathbf{0}\} \to \mathbb{R}^3$ defined by $F(\mathbf{r}) = \mathbf{r}/r^3$, $\mathbf{r} \in \mathbb{R}^3 \backslash \{\mathbf{0}\}$ is continuous over K. The integral of F over (K, N) is

9.3.11 $$\iint\limits_{(K, N)} F \cdot \mathrm{d}\mathbf{A} = \iint\limits_K \frac{\mathbf{r} \cdot N}{r^3} \, \mathrm{d}A = \iint\limits_K \frac{1}{c^2} \, \mathrm{d}A = 4\pi,$$

since $r = c$ for all $\mathbf{r} \in K$ and the area of K is $4\pi c^2$. It is interesting and important to notice that the integral is independent of the radius of the sphere.

9.3.12 *Example.* Let (K, N) be the ellipsoid in $\mathbb{R}^3$

$$
\frac{x^2}{a^2} + \frac{y^2}{b^2} + \frac{z^2}{c^2} = 1
$$

positively oriented by the (outward facing) unit normal vector field $N : K \subseteq \mathbb{R}^3 \to \mathbb{R}^3$. Define $f : \mathbb{R}^3 \to \mathbb{R}$ by

$$
f(x, y, z) = \frac{x^2}{a^2} + \frac{y^2}{b^2} + \frac{z^2}{c^2}.
$$

Then the field N is given by

$$
N(x, y, z) = \frac{(\mathrm{grad}\, f)(x, y, z)}{\|(\mathrm{grad}\, f)(x, y, z)\|}, \qquad (x, y, z) \in K.
$$

In order to integrate a continuous vector field F over the oriented ellipsoid (K, N) we look for a suitable parametrization of K. Put $S = [0, \pi] \times [0, 2\pi]$, and let $\rho : S \subseteq \mathbb{R}^2 \to \mathbb{R}^3$ be the C^1 function given by

$$\rho(\phi, \theta) = (a \sin \phi \cos \theta, \, b \sin \phi \sin \theta, \, c \cos \phi), \qquad (\phi, \theta) \in S.$$

Then $\rho(S) = K$, and ρ is 1–1 on Int S. Since

$$\nu_\rho(\phi, \theta) = (bc \sin^2 \phi \cos \theta, \, ac \sin^2 \phi \sin \theta, \, ab \sin \phi \cos \phi), \qquad (\phi, \theta) \in S,$$

it follows that ρ is smooth on Int S. A straightforward calculation shows that at $\mathbf{q} = \rho(\tfrac{1}{2}\pi, 0)$ the normal vector $\nu_\rho(\tfrac{1}{2}\pi, 0)$ points in the direction of $N(\mathbf{q})$. Therefore ρ captures the orientation of (K, N).

Now let $F : K \subseteq \mathbb{R}^3 \to \mathbb{R}^3$ be an arbitrary *constant* vector field on K. Then

$$\iint\limits_{(K,N)} F \cdot \mathrm{d}\mathbf{A} = \iint\limits_S F(\rho(\phi, \theta)) \cdot \nu_\rho(\phi, \theta) \, \mathrm{d}\phi \, \mathrm{d}\theta$$

$$= \iint\limits_S (A \sin^2 \phi \cos \theta + B \sin^2 \phi \sin \theta + C \sin \phi \cos \phi) \, \mathrm{d}\phi \, \mathrm{d}\theta$$

for suitable constants A, B and C

$$= \int_0^\pi \mathrm{d}\phi \int_0^{2\pi} (A \sin^2 \phi \cos \theta + B \sin^2 \phi \sin \theta + C \sin \phi \cos \phi) \, \mathrm{d}\theta$$

$$= 0.$$

The reader is recommended to find both a physical and a geometrical explanation for this result.

9.3.13 Example. Let K be the cube whose faces lie in the planes $x = \pm 1$, $y = \pm 1$, and $z = \pm 1$, and let (K, N) be the positively oriented cube with outward facing unit normals. Let $F : \mathbb{R}^3 \to \mathbb{R}^3$ be the continuous vector field defined by $F(\mathbf{r}) = \mathbf{r}$, $\mathbf{r} \in \mathbb{R}^3$. The integral of F over (K, N) is the sum of the integrals of F over the six oriented faces of the cube K.

Let K_1 be the face $x = 1$, $-1 \leq y \leq 1$, $-1 \leq z \leq 1$. The unit normal vector field N on K_1 is given by

$$N(x, y, z) = (1, 0, 0), \qquad (x, y, z) \in K_1.$$

Hence

$$\iint\limits_{(K_1,N)} F \cdot \mathrm{d}\mathbf{A} = \iint\limits_{K_1} F \cdot N \, \mathrm{d}A = \iint\limits_{K_1} x \, \mathrm{d}A = 4.$$

By symmetry we see that each of the integrals of F over the other five oriented faces of the cube takes the value 4. Therefore

$$\iint\limits_{(K,N)} F \cdot \mathrm{d}\mathbf{A} = 24.$$

We have seen that the Möbius strip K (Example 8.5.13) is non-orientable. However, any parametrization $\eta : S \subseteq \mathbb{R}^2 \to \mathbb{R}^3$ of K leads to a unit normal vector field $N_\eta : K \subseteq \mathbb{R}^3 \to \mathbb{R}^3$ whose discontinuities are confined to the simple arc C_η in K which constitutes the 'join' of the twisted strip. Since C_η is a null set in K, the integral of a continuous vector field F over (K, N_η) can be defined, as in 9.3.3, by

9.3.14
$$\iint\limits_{(K, N_\eta)} F \cdot dA = \iint\limits_K F \cdot N_\eta \, dA.$$

In evaluating the integral, care has to be taken that the appropriate parametrization η of K is used, since the direction of the unit vector $N_\eta(\mathbf{q})$ at the point $\mathbf{q}$ in K depends on the choice of η.

9.3.15 *Example.* Let K be the Möbius strip considered in Examples 8.5.5 and 8.5.13. Define the smooth function $\eta : \mathbb{R}^2 \to \mathbb{R}^3$ by

$$\eta(\theta, t) = ((k - t \sin \theta) \cos 2\theta, (k - t \sin \theta) \sin 2\theta, t \cos \theta)$$

for *all* $(\theta, t) \in \mathbb{R}^2$. Let η^* be the function η restricted to the half-open rectangle

$$T = \{(\theta, t) \in \mathbb{R}^2 \mid 0 \leq \theta < \pi, -1 \leq t \leq 1\}.$$

Then η^* is 1–1, and $\eta^*(T) = K$. The associated unit normal field $N^* : K \subseteq \mathbb{R}^3 \to \mathbb{R}^3$ given by

$$N^*(\eta(\theta, t)) = \frac{v_\eta(\theta, t)}{\|v_\eta(\theta, t)\|} \qquad (\theta, t) \in T$$

is continuous, except at the points $\eta(0, t) = (k, 0, t)$ on K where a reversal in direction of the unit normals occurs.

Let $\eta^\dagger$ be the function η restricted to the half-open rectangle

$$W = \{(\theta, t) \in \mathbb{R}^2 \mid -\tfrac{1}{2}\pi \leq \theta < \tfrac{1}{2}\pi, -1 \leq t \leq 1\}.$$

Then $\eta^\dagger$ is also 1–1, and $\eta^\dagger(W) = K$. The associated unit normal field $N^\dagger$ is continuous, except at the points $\eta(-\tfrac{1}{2}\pi, t) = (-k - t, 0, 0)$.

Now let $F : K \subseteq \mathbb{R}^3 \to \mathbb{R}^3$ be the constant vector field $F(\mathbf{q}) = (0, 0, 1)$, $\mathbf{q} \in K$. Then by the method of Theorem 9.3.5,

$$\iint\limits_{(K, N^*)} F \cdot dA = \iint\limits_T F(\eta(\theta, t)) \cdot v_\eta(\theta, t) \, d\theta \, dt$$

$$= \int_0^\pi d\theta \int_{-1}^1 2(k - t \sin \theta) \sin \theta \, d\theta, \qquad \text{by Exercise 8.5.1(d)}$$

$$= 8k.$$

However,

$$\iint\limits_{(K,N^\dagger)} F \cdot d\mathbf{A} = \int_{-\pi/2}^{\pi/2} d\theta \int_{-1}^{1} 2(k - t \sin \theta) \sin \theta \, d\theta = 0.$$

It is not surprising that the integrals of F over (K, N^*) and over $(K, N^\dagger)$ are quite different, since the fields N^* and $N^\dagger$ agree on half the Möbius strip and are opposite on the other half.

There is another notation for the integral of a vector field over an oriented surface which matches that for the line integral of a vector field along an oriented simple curve. Let $\rho : S \subseteq \mathbb{R}^2 \to \mathbb{R}^3$ be a simple parametrization of an oriented simple surface (K, N) in $\mathbb{R}^3$. We can define three real valued C^1 functions, x, y and z, on S by the rule

$$\rho(u, v) = (x(u, v), y(u, v), z(u, v)), \qquad (u, v) \in S,$$

in which case (see 8.2.9), for each $(u, v) \in S$,

$$v_\rho(u, v) = \left(\frac{\partial(y, z)}{\partial(u, v)}, \frac{\partial(z, x)}{\partial(u, v)}, \frac{\partial(x, y)}{\partial(u, v)} \right),$$

where the Jacobians are evaluated at (u, v).

Hence for any continuous vector field $F = (F_1, F_2, F_3)$ on K, by Theorem 9.3.5,

$$\iint\limits_{(K,N)} F \cdot d\mathbf{A} = \iint\limits_{S} F(\rho(u, v)) \cdot v_\rho(u, v) \, du \, dv$$

$$= \iint\limits_{S} \left((F_1 \circ \rho) \frac{\partial(y, z)}{\partial(u, v)} + (F_2 \circ \rho) \frac{\partial(z, x)}{\partial(u, v)} \right.$$

$$\left. + (F_3 \circ \rho) \frac{\partial(x, y)}{\partial(u, v)} \right) du \, dv.$$

This expression leads to the following notation for the integral of F over (K, N)

9.3.16 $$\iint\limits_{(K,N)} (F_1 \, dy \, dz + F_2 \, dz \, dx + F_3 \, dx \, dy).$$

All the necessary ingredients are contained in 9.3.16. The simple surface K involved is parametrized by $(x, y, z) : S \subseteq \mathbb{R}^2 \to \mathbb{R}^3$, the orientation of K is the one determined by this parametrization, and the vector field involved is (F_1, F_2, F_3). In Example 9.3.8 for example we evaluated

$$\iint\limits_{(K,N)} (1 - 2z)\, dy\, dz + 2y\, dx\, dy,$$

where $x(u, v) = u$, $y(u, v) = v$ and $z(u, v) = 2 - u - \frac{1}{2}v$.

Exercises 9.3

1. Let $F: \mathbb{R}^3 \to \mathbb{R}^3$ be the vector field defined by $F(x, y, z) = (x, y, z)$. Find the integral of F over the following oriented surfaces:
    (a) the cylinder $x^2 + y^2 = k^2$, $0 \leqslant z \leqslant h$ with outward facing unit normals;
    (b) the circular cone of height h, $z = \sqrt{(x^2 + y^2)}$, $0 \leqslant z \leqslant h$ with outward facing normals (away from z-axis);
    (c) the paraboloid $z = 2 - (x^2 + y^2)$, $z \geqslant 0$, with inward facing normals (towards z-axis);
    (d) part of the plane $2x + y + 2z = 16$ which is cut off by the planes $x = 0$, $y = 0$ and $x + y = 1$, with normal having positive z-coordinate.

Answers: (a) $2\pi h k^2$; (b) 0 (draw a picture); (c) -6π; (d) 4.

2. Find the integral of the vector field $F(x, y, z) = (x^2, y^2, 2z^2)$ over the unit cube containing $[0, 1] \times [0, 1] \times [0, 1]$, oriented by outward facing normals.

Answer: 4.

3. The velocity field of a fluid is given by $F(x, y, z) = (\frac{1}{4}x, \frac{1}{4}, 0)$, with measurements in metres per second. Compute, in cubic metres per second,
    (a) the rate of fluid flow across the plane $y = 0$ through the unit square $-\frac{1}{2} \leqslant x \leqslant \frac{1}{2}$, $0 \leqslant z \leqslant 1$;
    (b) the rate of fluid flow across the plane $y = 1$ through the unit square $-\frac{1}{2} \leqslant x \leqslant \frac{1}{2}$, $0 \leqslant z \leqslant 1$;
    (c) the rate of fluid flow across the cylinder $x^2 + y^2 = k^2$, $0 \leqslant z \leqslant h$;
    (d) the rate of fluid flow across the sphere $x^2 + y^2 + z^2 = 1$.
    Illustrate with the help of Example 4.11.5 and Fig. 4.18.

Answers: (a) and (b) $\frac{1}{4}$. The x-component of velocity contributes zero to the flow; (c) $\frac{1}{4}\pi k^2 h$; (d) $\pi/3$, net rate of flow outwards across sphere, zero contribution from y-component.

4. *Heat flow.* Suppose that at a fixed time the temperature in a body is described by a C^1 function $T: W \subseteq \mathbb{R}^3 \to \mathbb{R}$. Heat tends to flow in the body in the direction of most rapid decrease in temperature. For a thermally isotropic body (which conducts heat uniformly), the rate of heat flow per unit area is given by the *temperature gradient field* $-\kappa \operatorname{grad} T: W \subseteq \mathbb{R}^3 \to \mathbb{R}$, where the constant κ is the *thermal conductivity* of the body.

If $T(x, y, z) = x^2 + y^2 + z^2$, $(x, y, z) \in \mathbb{R}^3$, and $\kappa = 1$, compute the rate of heat flow (heat flux) across the surface of the unit sphere $K : x^2 + y^2 + z^2 = 1$ from the formula

$$\iint\limits_{(K,N)} -\kappa \, \text{grad} \, T \cdot d\mathbf{A},$$

where K is positively oriented by the outward facing unit normal field N.

Answer: -8π. Heat flows into the sphere towards the centre.

5. The electric field of a point charge e placed at the origin is described by the vector function

$$E(\mathbf{r}) = -\frac{e}{r^3}\mathbf{r}, \qquad \mathbf{r} \neq \mathbf{0},$$

where $r = \|\mathbf{r}\|$. (It is the gradient of its scalar potential field $f(\mathbf{r}) = e/r$.) The *electric flux* due to the charge across an oriented surface (K, N) is defined to be the integral of its electric field E over (K, N).

Let (K, N) be a sphere of radius c oriented by outward facing normals.

(a) Show that the electric flux across (K, N) due to a point charge e inside is $-4\pi e$, irrespective of the radius of the sphere or the precise position of the charge.

(b) Show that the electric flux across (K, N) due to a point charge outside is 0.

Hint: parametrize K as in Exercise 9.2.3, and put point charge at $(0, 0, -p)$. See also Example 10.5.8.

6. Let $F : \mathbb{R}^3 \to \mathbb{R}^3$ be the constant vector field $F(x, y, z) = (1, 1, 1)$, $(x, y, z) \in \mathbb{R}^3$, and let K be the plane rectangular surface $z = 1$, $-1 \leqslant x \leqslant 1$, $0 \leqslant y \leqslant 1$.

(a) Show that the integral of F over (K, N) is 2 or -2, depending on the choice of continuous unit normal field N determining the orientation of K.

(b) Verify that $\mu : [0, 1] \times [-1, 1] \subseteq \mathbb{R}^2 \to \mathbb{R}^3$ defined by

$$\mu(u, v) = \begin{cases} (u^3, v^3, 1) & \text{when } 0 \leqslant u \leqslant 1, \quad 0 \leqslant v \leqslant 1 \\ (v^3, u^3, 1) & \text{when } 0 \leqslant u \leqslant 1, \; -1 \leqslant v < 0 \end{cases}$$

is a 1–1 C^1 parametrization of K, but that

$$\int_0^1 du \int_{-1}^1 F(\mu(u, v)) \cdot v_\mu(u, v) \, dv = 0.$$

Explain why the answer is different from that of part (a).

Hint: sketch the normals.

Answer: the parametrization μ of K is not smooth. A formal calculation
gives, in particular, that $v_\mu(u, 0) = (0, 0, 0)$. Moreover, for $u \neq 0$,
$v_\mu(u, v)$ points 'upwards' when $v > 0$ and 'downwards' when $v < 0$.
Hence μ does not lead to an orientation of K.

7. Show that the parametrizations η^* and $\eta^\dagger$ of the Möbius strip K of
    Example 9.3.15 give the same value for the area of K. This in in
    agreement with the conclusion (Section 9.1) that surface area is
    independent of the parametrizations used in its evaluation.

9.4   Stokes' Theorem

We now have enough information to state and prove another
classical theorem of vector analysis. Stokes' Theorem is essentially a
generalization to smooth simple surfaces of Green's Theorem for
simple regions. In the form given in 7.5.14, Green's Theorem
equates the line integral of a vector field around the oriented edge
∂S^+ of an oriented simple region S^+ and the integral of a related
vector field over S^+ itself.

A C^2 simple parametrization $\rho : S \subseteq \mathbb{R}^2 \to \mathbb{R}^3$ of a smooth simple
surface $K = \rho(S)$ transforms Green's Theorem on S^+ into Stokes'
Theorem on the oriented simple surface $K^+ = \rho(S^+)$.

9.4.1 *Theorem (Stokes' Theorem for oriented simple surfaces).*
*Let (K, N) be an oriented simple surface in $\mathbb{R}^3$ for which there is a
C^2 simple parametrization, and let $F : K \subseteq \mathbb{R}^3 \to \mathbb{R}^3$ be a C^1 vector
field on K. Then*

9.4.2 $$\int_{\partial(K,N)} F \cdot d\mathbf{r} = \iint_{(K,N)} \operatorname{curl} F \cdot d\mathbf{A}.$$

Proof.   The required result is obtained by adding the three identities
[*i*]

$$\int_{\partial(K,N)} (F_1, 0, 0) \cdot d\mathbf{r} = \iint_K \left(0, \frac{\partial F_1}{\partial z}, -\frac{\partial F_1}{\partial y} \right) \cdot N \, dA,$$

[*ii*]

$$\int_{\partial(K,N)} (0, F_2, 0) \cdot d\mathbf{r} = \iint_K \left(-\frac{\partial F_2}{\partial z}, 0, \frac{\partial F_2}{\partial x} \right) \cdot N \, dA,$$

[*iii*]

$$\int_{\partial(K,N)} (0, 0, F_3) \cdot d\mathbf{r} = \iint_{\tilde{K}} \left(\frac{\partial F_3}{\partial y}, -\frac{\partial F_3}{\partial x}, 0 \right) \cdot N \, dA.$$

We shall prove just the first identity; the other two are proved similarly.

Let $\rho : S \subseteq \mathbb{R}^2 \to \mathbb{R}^3$ be a C^2 simple parametrization of (K, N) and let $\alpha : [a, b] \subseteq \mathbb{R} \to \mathbb{R}^2$ be a simple parametrization of the counterclockwise oriented boundary ∂S^+. Then $\rho \circ \alpha$ is a simple parametrization of $\partial(K, N)$ and

$$(\rho \circ \alpha)'(t) = ((\rho_1 \circ \alpha)'(t), (\rho_2 \circ \alpha)'(t), (\rho_3 \circ \alpha)'(t)), \qquad t \in [a, b].$$

Hence

$$\int_{\partial(K,N)} (F_1, 0, 0) \cdot d\mathbf{r} = \int_a^b F_1(\rho(\alpha(t))(\rho_1 \circ \alpha)'(t) \, dt$$

and this in turn, by the Chain Rule, is equal to

$$\int_a^b F_1(\rho(\alpha(t))) \left(\frac{\partial \rho_1}{\partial u}(\alpha(t)) \frac{d\alpha_1}{dt}(t) + \frac{\partial \rho_1}{\partial v}(\alpha(t)) \frac{d\alpha_2}{dt}(t) \right) dt.$$

Therefore, since α parametrizes ∂S^+,

9.4.3 $$\int_{\partial(K,N)} (F_1, 0, 0) \cdot d\mathbf{r} = \int_{\partial S^+} (F_1 \circ \rho) \left(\frac{\partial \rho_1}{\partial u}, \frac{\partial \rho_1}{\partial v} \right) \cdot d\mathbf{r}.$$

Green's Theorem equates the right-hand side of 9.4.3 and

$$\iint_S \left(\frac{\partial}{\partial u} \left((F_1 \circ \rho) \frac{\partial \rho_1}{\partial v} \right) - \frac{\partial}{\partial v} \left((F_1 \circ \rho) \frac{\partial \rho_1}{\partial u} \right) \right) dA$$

which, since the mixed partial derivatives of the C^2 function ρ_1 are equal, is

9.4.4 $$\iint_S \left(\frac{\partial}{\partial u}(F_1 \circ \rho) \frac{\partial \rho_1}{\partial v} - \frac{\partial}{\partial v}(F_1 \circ \rho) \frac{\partial \rho_1}{\partial u} \right) dA.$$

Now, by the Chain Rule,

$$\frac{\partial}{\partial u}(F_1 \circ \rho) = \frac{\partial F_1}{\partial x} \frac{\partial \rho_1}{\partial u} + \frac{\partial F_1}{\partial y} \frac{\partial \rho_2}{\partial u} + \frac{\partial F_1}{\partial z} \frac{\partial \rho_3}{\partial u}$$

and

$$\frac{\partial}{\partial v}(F_1 \circ \rho) = \frac{\partial F_1}{\partial x}\frac{\partial \rho_1}{\partial v} + \frac{\partial F_1}{\partial y}\frac{\partial \rho_2}{\partial v} + \frac{\partial F_1}{\partial z}\frac{\partial \rho_3}{\partial v}.$$

It follows, after some cancellation, that the integrals in 9.4.4 and 9.4.3 are equal to

$$\iint_S \left(\frac{\partial F_1}{\partial z}\frac{\partial(\rho_3, \rho_1)}{\partial(u, v)} - \frac{\partial F_1}{\partial y}\frac{\partial(\rho_1, \rho_2)}{\partial(u, v)} \right) \mathrm{d}A$$

and therefore, by 9.3.7 and 8.2.9, to

$$\iint_K \left(0, \frac{\partial F_1}{\partial z}, -\frac{\partial F_1}{\partial y} \right) \cdot N \,\mathrm{d}A,$$

since ρ is a parametrization of (K, N). This establishes identity (i).

9.4.5 Example.   Let (K, N) be the triangular surface $2x + y + 2z = 4$, $x \geq 0$, $y \geq 0$, $z \geq 0$, oriented by the unit normal field $N(x, y, z) = (\frac{2}{3}, \frac{1}{3}, \frac{2}{3})$. (See Example 9.3.8.)

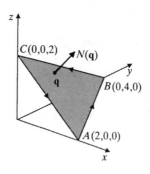

Fig. 9.3

Consider the continuous vector field $G : K \subseteq \mathbb{R}^3 \to \mathbb{R}^3$ defined by

$$G(x, y, z) = (x^2, 2xy + z^2, y + 2z), \qquad (x, y, z) \in K.$$

Then

$$(\mathrm{curl}\,G)(x, y, z) = (1 - 2z, 0, 2y) = F(x, y, z), \qquad (x, y, z) \in K$$

where $F : K \subseteq \mathbb{R}^3 \to \mathbb{R}^3$ is the vector field considered in Example 9.3.8.

Hence

9.4.6 $$\iint\limits_{(K,N)} \text{curl } G \cdot d\mathbf{A} = 28/3.$$

In order to verify Stokes' Theorem in this case, we need to find $\partial(K, N)$. Let A, B and C be the three vertices $(2, 0, 0)$, $(0, 4, 0)$, $(0, 0, 2)$ of K respectively. By Remark 8.3.25, reference to Fig. 9.3 shows that $\partial(K, N)$ is oriented from A to B to C to A.

We can evaluate the integral of G around the oriented edge of (K, N) by considering parametrizations of the oriented line segments AB, BC, CA which make up $\partial(K, N)$:

$\alpha:[0, 1] \subseteq \mathbb{R} \to \mathbb{R}^3$ defined by $\alpha(t) = (2, 0, 0) + t(-2, 4, 0)$, $t \in [0, 1]$;

$\beta:[0, 1] \subseteq \mathbb{R} \to \mathbb{R}^3$ defined by $\beta(t) = (0, 4, 0) + t(0, -4, 2)$, $t \in [0, 1]$;

$\gamma:[0, 1] \subseteq \mathbb{R} \to \mathbb{R}^3$ defined by $\gamma(t) = (0, 0, 2) + t(2, 0, -2)$, $t \in [0, 1]$.

We obtain

$$\int_\alpha G \cdot d\alpha = \int_\alpha x^2 \, dx + (2xy + z^2) \, dy + (y + 2z) \, dz$$

$$= \int_0^1 ((2 - 2t)^2(-2) + 2(2 - 2t)4t \,.\, 4) \, dt = 8.$$

Similarly

$$\int_\beta G \cdot d\beta = 8/3 \quad \text{and} \quad \int_\gamma G \cdot d\gamma = -4/3.$$

Hence

9.4.7 $$\int_{\partial(K,N)} G \cdot d\mathbf{r} = \int_\alpha G \cdot d\alpha + \int_\beta G \cdot d\beta + \int_\gamma G \cdot d\gamma = 28/3.$$

Comparison of 9.4.7 and 9.4.6 gives the expected result.

9.4.8 *Example.* Let S be a simple region in $\mathbb{R}^2$ with counterclockwise oriented boundary ∂S^+. Let K be the simple surface in the x, y plane of $\mathbb{R}^3$, parametrized by $\rho:S \subseteq \mathbb{R}^2 \to \mathbb{R}^3$, where

$$\rho(x, y) = (x, y, 0), \qquad (x, y) \in S.$$

Then K is a copy of S in the x, y plane in $\mathbb{R}^3$ and $N_\rho(\mathbf{p}) = (0, 0, 1) = \mathbf{k}$ for all $\mathbf{p} \in S$. The edge of $(K, \mathbf{k})$ is just a copy of ∂S^+ in the x, y plane in $\mathbb{R}^3$.

For any C^1 vector field $F:K \subseteq \mathbb{R}^3 \to \mathbb{R}^3$ of the form $F = (F_1, F_2, 0)$,

$$\iint\limits_{(K,\mathbf{k})} \text{curl } F \cdot d\mathbf{A} = \iint\limits_S \left(\frac{\partial F_2}{\partial x} - \frac{\partial F_1}{\partial y} \right) dA \quad \text{and} \quad \iint\limits_{\partial(K,\mathbf{k})} F \cdot d\mathbf{r} = \iint\limits_{\partial S^+} (F_1, F_2) \cdot d\mathbf{r}.$$

In this way we see that, when the simple surface lies in a plane, Stokes' Theorem reduces to Green's Theorem 7.5.14.

Since Green's Theorem is true for piecewise simple regions (Theorem 7.5.24) the argument of Theorem 9.4.1 is readily adapted to proving 9.4.2 when (K, N) is an oriented piecewise simple surface for which there is a C^2 simple parametrization, and $F: K \subseteq \mathbb{R}^3 \to \mathbb{R}^3$ is a C^1 vector field on K. This is a generalization of Theorem 9.4.1.

9.4.9 *Example*.   Consider the section of the parabolic bowl $z = x^2 + y^2$, $1 \leq z \leq 4$ with inward facing normals. The oriented piecewise simple surface (K, N) is sketched in Fig. 8.10 (Example 8.4.3).

Let S be the piecewise simple region

$$S = \{(u, v) \in \mathbb{R}^2 \mid 1 \leq u^2 + v^2 \leq 4\}$$

and let $\rho: S \subseteq \mathbb{R}^2 \to \mathbb{R}^3$ be the C^2 simple parametrization of (K, N) defined by

$$\rho(u, v) = (u, v, u^2 + v^2), \qquad (u, v) \in S.$$

The oriented edge of (K, N) is also sketched in Fig. 8.10. It consists of two circles

$$E_1^+ = \{(x, y, z) \in \mathbb{R}^3 \mid x^2 + y^2 = 1, z = 1\}$$

and

$$E_2^+ = \{x, y, z) \in \mathbb{R}^3 \mid x^2 + y^2 = 4, z = 4\}$$

with orientations determined by $\alpha_1: [0, 2\pi] \subseteq \mathbb{R} \to \mathbb{R}^3$ and $\alpha_2: [0, 2\pi] \subseteq \mathbb{R} \to \mathbb{R}^3$ respectively, where

$$\alpha_1(t) = (\cos t, -\sin t, 1) \quad \text{and} \quad \alpha_2(t) = (2 \cos t, 2 \sin t, 4), \qquad t \in [0, 2\pi].$$

Let $F: K \subseteq \mathbb{R}^3 \to \mathbb{R}$ be the C^1 vector field on K defined by

$$F(x, y, z) = (y, x^2, 3z^2), \qquad (x, y, z) \in K.$$

Then

$$\iint\limits_{(K,N)} \operatorname{curl} F \cdot d\mathbf{A} = \iint\limits_{K} (0, 0, 2x - 1) \cdot N \, dA$$

$$= \iint\limits_{S} (0, 0, 2u - 1) \cdot v_\rho(u, v) \, du \, dv$$

$$= \iint\limits_{S} (2u - 1) \, du \, dv = -3\pi.$$

But also

$$\int_{\partial(K,N)} F \cdot d\mathbf{r} = \int_{E_1^+} F \cdot d\mathbf{r} + \int_{E_2^+} F \cdot d\mathbf{r}$$

$$= \int_0^{2\pi} (-\sin t, \cos^2 t, 3) \cdot (-\sin t, -\cos t, 0)\, dt$$

$$+ \int_0^{2\pi} (2\sin t, 4\cos^2 t, 48) \cdot (-2\sin t, 2\cos t, 0)\, dt$$

$$= -3\pi.$$

9.4.10 *Example.* Let (K, N) be a cube in $\mathbb{R}^3$ positively oriented by outward facing normals.

For any C^1 vector field F over K and for each of the six faces $K_1, \ldots, K_6$

9.4.11 $$\iint_{(K_i,N)} \text{curl } F \cdot d\mathbf{A} = \int_{\partial(K_i,N)} F \cdot d\mathbf{r} \qquad i = 1, \ldots, 6.$$

The line integral in 9.4.11 involves integrating along four sides. Adding up for $i = 1, \ldots, 6$ these integrals cancel—see Fig. 8.14 p. 452. We find therefore that

$$\iint_{(K,N)} \text{curl } F \cdot d\mathbf{A} = 0.$$

In particular, since curl $(lz, mx, ky) = (k, l, m)$ for all $(x, y, z) \in \mathbb{R}^3$, it follows that the integral of a constant vector field over the positively oriented cube (K, N) is zero. This special case is easily proved without appealing to Stokes' Theorem.

We can now generalize Stokes' Theorem to oriented surfaces. Let $(K, [\rho])$ be an oriented surface, where the parametrization ρ satisfies the conditions of Definition 8.5.9. We define the *edge* of $(K, [\rho])$ to be the subset E of the curves in $\rho(\partial S_1) \cup \cdots \cup \rho(\partial S_k)$ which are traced just once when ρ acts on $\partial S_1^+ \cup \cdots \cup \partial S_k^+$, with the orientation inherited from this action. The oriented edge E^+ is denoted by $\partial(K, [\rho])$ or by $\partial(K, N)$, where N is the unit normal vector field determined by ρ. With this definition of oriented edge Stokes' Theorem 9.4.1 applies to oriented surfaces. The proof is based on the observation that the line integral of a vector field along an oriented arc changes sign with a reversal of orientation. Put

$K_i = \rho(S_i)$, $i = 1, \ldots, k$. Then

$$\iint\limits_{(K,N)} \operatorname{curl} F \cdot d\mathbf{A} = \sum_{i=1}^{k} \iint\limits_{(K_i,N)} \operatorname{curl} F \cdot d\mathbf{A}$$

$$= \sum_{i=1}^{k} \int_{\partial(K_i,N)} F \cdot d\mathbf{r} = \int_{\partial(K,N)} F \cdot d\mathbf{r}.$$

9.4.12 *Example.* The positively oriented cube and the positively oriented sphere are two examples of oriented surfaces with no (oriented) edge! For any such surface (K, N),

$$\iint\limits_{(K,N)} \operatorname{curl} F \cdot d\mathbf{A} = 0.$$

Our definition of the edge of an oriented surface turns out to be a natural one—the edge just indicated where the surface stops: if a curve lying in the surface crosses an edge, then it must leave the surface. Looked at in this way, a fold in the surface (for example in the cube) does not count towards its edge. Stokes' Theorem, simply stated, tells us that the integral of curl F over an oriented surface is equal to the line integrals of F along its oriented edge.

Difficulties arise in attempting to generalize Stokes' Theorem to non-orientable surfaces, as the following example illustrates.

9.4.13 *Example.* Let K be a Möbius strip. The natural definition of its edge is the simple closed curve E illustrated in Fig. 9.4.

We know that there is no orientation of K by a continuous unit normal vector field. However, we can consider K as the union of four oriented simple surfaces (K_i, N_i), $i = 1, \ldots, 4$ with orientation as shown in Fig. 9.4.

Let E^+ be the edge of the strip K with orientation inherited from the edges of the oriented simple surfaces, and let J^+ be the oriented curve shown running around the centre of the strip.

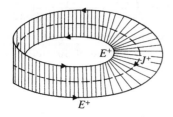
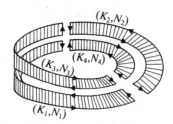

Fig. 9.4

Let $N^*: K \subseteq \mathbb{R}^3 \to \mathbb{R}^3$ be a vector field on K which agrees with N_i for each $i = 1, \ldots, 4$ except perhaps around the edge of K_i.

For any C^1 vector field $F: K \subseteq \mathbb{R}^3 \to \mathbb{R}^3$,

$$
\iint_{(K,N^*)} \text{curl } F \cdot d\mathbf{A} = \sum_{1}^{4} \iint_{(K_i,N_i)} \text{curl } F \cdot d\mathbf{A}
$$

$$
= \sum_{1}^{4} \int_{\partial(K_i,N_i)} F \cdot d\mathbf{r} \qquad \text{(by Stokes' Theorem)}
$$

$$
= \int_{E^+} F \cdot d\mathbf{r} + 2 \int_{J^+} F \cdot d\mathbf{r}.
$$

Here the line integral is not confined to the edge of the strip, and so our attempt to generalize Stokes' Theorem to the 'vectored surface' (K, N^*) has failed. Notice that the fields N_1 and N_3 point in opposite directions on either side of the curve J. So do the fields N_2 and N_4.

Exercises 9.4

1. Verify Stokes' Theorem

$$
\int_{\partial(K,N)} F \cdot d\mathbf{r} = \iint_{(K,N)} \text{curl } F \cdot d\mathbf{A},
$$

where $F(x, y, z) = (y, z, x)$ and (K, N) is the following oriented surface:

(a) the hemisphere $x^2 + y^2 + z^2 = 1$, $z \geqslant 0$ with outward facing normals (away from origin);

(b) the paraboloid $z = 2 - (x^2 + y^2)$, $z \geqslant 0$, with inward facing normals;

(c) part of the plane $2x + y + 2z = 16$ which is cut off by the planes $x = 0$, $y = 0$ and $x + y = 1$, with normal having positive z coordinate;

(d) the cylinder $x^2 + y^2 = k^2$, $0 \leqslant z \leqslant h$ with outward facing normals.

Answers: (a) $-\pi$; (b) 2π; (c) $-5/4$ ($\partial(K, N)$ is the oriented triangle $(0, 0, 8) \to (1, 0, 7) \to (0, 1, 7\frac{1}{2}) \to (0, 0, 8)$); (d) 0.

2. Let (K, N) be the oriented hemisphere $x^2 + y^2 + z^2 = 1$, $z \geqslant 0$, with outward facing normals, and let (K^*, N^*) be the oriented unit disc $x^2 + y^2 \leqslant 1$, $z = 0$ with unit normals $(0, 0, -1)$. Prove by Stokes' Theorem that, for any C^1 vector field F,

$$
\iint_{(K,N)} \text{curl } F \cdot d\mathbf{A} + \iint_{(K^*,N^*)} \text{curl } F \cdot d\mathbf{A} = 0.
$$

Hence obtain a quick method to check the answer of Exercise 1(a). Consider similarly Exercises 1(b) and 1(d).

3. Let $F(x, y, z) = (-y + z, x + yz, xyz)$. Using Stokes' Theorem compute the integral of curl F over the following oriented surfaces:
    (a) the hemisphere $x^2 + y^2 + z^2 = 1$, $z \geqslant 0$, outward normals;
    (b) the hemisphere $x^2 + y^2 + z^2 = 1$, $z \leqslant 0$, outward normals;
    (c) the hemi-ellipsoid $\frac{1}{2}x^2 + y^2 + 2z^2 = 1$, $z \geqslant 0$, inward normals;
    (d) the plane square region with corners $(0, 0, 0)$, $(0, 1, 0)$, $(1, 1, 0)$, $(1, 0, 0)$ and with normal $(0, 0, -1)$.

Answers: (a) 2π; (b) -2π; (c) -4π; (d) -2.

4. Let $F : D \subseteq \mathbb{R}^3 \to \mathbb{R}^3$ be a C^1 vector field defined on an open set D in $\mathbb{R}^3$. Let N be a *constant* unit vector field on $\mathbb{R}^3$. Given $\mathbf{p} \in D$, let K_ε be a circular disc radius ε, centre $\mathbf{p}$, having $N(\mathbf{p})$ as normal. Prove by Stokes' Theorem that

$$((\text{curl } F) \cdot N)(\mathbf{p}) = \lim_{\varepsilon \to 0} \frac{1}{\text{Area } K_\varepsilon} \int_{\partial(K_\varepsilon, N)} F \cdot d\mathbf{r}.$$

This generalizes 7.6.19. The expression provides a formula for the component of curl F at $\mathbf{p}$ in any desired direction, without reference to a particular coordinate system. It also shows how curl F at $\mathbf{p}$ is related to the circulation of F around small circuits close to $\mathbf{p}$.

5. The velocity field of a fluid is defined by $F(x, y, z) = (-z, 1, x)$. What is the circulation of F around the circle (a) $x^2 + y^2 = \varepsilon^2$, $z = 0$; (b) $x^2 + z^2 = \varepsilon^2$, $y = 0$; (c) $y^2 + z^2 = \varepsilon^2$, $x = 0$? Sketch the field and interpret your answers in terms of curl F.

Answers: (a) 0; (b) $\pm 2\pi\varepsilon^2$ depending on the orientation chosen; (c) 0.

6. Repeat Exercise 5 with the velocity field $F(x, y, z) = (-z, x, 1)$.

Answers: (a) $\pm \pi\varepsilon^2$; (b) $\pm \pi\varepsilon^2$; (c) 0.

7. *Electromagnetic Theory. Faraday's Law of Induction.* Faraday discovered that a change of magnetic flux across a (simple) surface K induces an electric current in a conducting wire lying along the edge ∂K. Specifically, Faraday's Law states that

$$-\iint_{(K,N)} \frac{\partial B}{\partial t} \cdot d\mathbf{A} = \int_{\partial(K,N)} E \cdot d\mathbf{r},$$

where $B(x, y, z, t)$ is the magnetic field and $E(x, y, z, t)$ the induced electric field. Deduce by Stokes' Theorem that

$$\text{curl } E = -\frac{\partial B}{\partial t}.$$

8. Let $F : D \subseteq \mathbb{R}^3 \to \mathbb{R}^3$ be a C^1 vector field defined on an open ball D in $\mathbb{R}^3$. Prove that a necessary and sufficient condition that F be a conservative field is that curl $F = \mathbf{0}$.

Hint: to prove sufficiency choose a base point $\mathbf{b} \in D$ and define
$f : D \subseteq \mathbb{R}^3 \to \mathbb{R}$ by

$$f(\mathbf{r}) = \int_\beta F \cdot \mathrm{d}\beta, \qquad \mathbf{r} \in D,$$

where β is the straight line path in D from $\mathbf{b}$ to $\mathbf{r}$. For a fixed $\mathbf{p} \in D$
show that $F(\mathbf{p}) = (\mathrm{grad}\, f)(\mathbf{p})$ by applying Stokes' Theorem to the
triangle formed by points $\mathbf{b}, \mathbf{p}, \mathbf{p} + t\mathbf{e}_i$ in D.

The above result is more generally true for simply connected open
sets $D \subseteq \mathbb{R}^3$.

Triple integrals in $\mathbb{R}^3$

10.1 Integral over a rectangular box

In this chapter we generalize the work of Chapter 7 to three dimensional integrals. We begin by considering integrals over a rectangular box (or rectangular parallelepiped) in $\mathbb{R}^3$.

Let R be the closed rectangular box

$$R = [a_1, b_1] \times [a_2, b_2] \times [a_3, b_3]$$
$$= \{(x, y, z) \in \mathbb{R}^3 \mid a_1 \leqslant x \leqslant b_1, a_2 \leqslant y \leqslant b_2, a_3 \leqslant z \leqslant b_3\},$$

and let $f : R \in \mathbb{R}^3 \to \mathbb{R}$ be a bounded function. Partitions of $[a_i, b_i]$, $i = 1, 2, 3$ lead to a partition $\mathscr{P}$ of R and lower and upper Riemann sums of f over $\mathscr{P}$ in an obvious generalization of 7.1.2 and 7.1.3. We define

10.1.1 $\qquad L(f, \mathscr{P}) = \sum_{i,j,k} m_{ijk} V_{ijk}$, the lower Riemann sum

and

10.1.2 $\qquad U(f, \mathscr{P}) = \sum_{i,j,k} M_{ijk} V_{ijk}$, the upper Riemann sum,

where V_{ijk} denotes the volume of the subrectangular region R_{ijk} in $\mathscr{P}$, and m_{ijk} and M_{ijk} are respectively the greatest lower bound and the least upper bound of f on R_{ijk}.

10.1.3 *Definition.* *The function $f : R \subseteq \mathbb{R}^3 \to \mathbb{R}$ is said to be* (Riemann) integrable *over the compact rectangular box R if the least upper bound of all lower sums is equal to the greatest lower bound of all upper sums. The common value is then called the* triple integral *of f over R, or simply the* integral *of f over R. It is denoted by*

10.1.4 $\qquad \iiint\limits_R f \, dV \qquad$ *or by* $\qquad \iiint\limits_R f(x, y, z) \, dx \, dy \, dz.$

Compare Definition 7.1.9. We leave it to the reader to check that the results corresponding to Theorems 7.1.10 and 7.1.12 are true of triple integrals.

10.1.5 Theorem. *If the function $f : R \subseteq \mathbb{R}^3 \to \mathbb{R}$ is continuous on the compact rectangular box R then f is integrable over R.*

Proof.   Adapt the proof of Theorem 7.1.14.

The work of Section 7.2 on partitions and null sets carries over to the three dimensional case with 'rectangle in $\mathbb{R}^2$' replaced by 'rectangular box in $\mathbb{R}^3$' and 'area' replaced by 'volume'.

The three sequences of numbers $x_0 < x_1 < \cdots < x_k$ and $y_0 < y_1 < \cdots < y_l$ and $z_0 < z_1 < \cdots < z_m$ lead to three families of grid lines in $\mathbb{R}^3$. These lines lead to a partition $\mathscr{P}$ of $\mathbb{R}^3$ into $(k+2)(l+2)(m+2)$ rectangular boxes of which klm are compact. For any subset A of $\mathbb{R}^3$, the sum of the volumes of these boxes of $\mathscr{P}$ whose intersection with A is non-empty is called the *contact* of $\mathscr{P}$ with A and is denoted by $\kappa(A, \mathscr{P})$.

10.1.6 Definition.   *A set $N \subseteq \mathbb{R}^3$ is said to be a null set in $\mathbb{R}^3$ if to each $\varepsilon > 0$ there corresponds a partition $\mathscr{P}$ of $\mathbb{R}^3$ such that $\kappa(N, \mathscr{P}) < \varepsilon$.*

We have two theorems which match Theorem 7.2.11 and Theorem 7.2.12.

10.1.7 Theorem.   *A surface is a null set in $\mathbb{R}^3$.*

10.1.8 Theorem.   *Let $f : R \subseteq \mathbb{R}^3 \to \mathbb{R}$ be a bounded function defined on a compact rectangular box R in $\mathbb{R}^3$. If the set of points at which f is discontinuous forms a null set in $\mathbb{R}^3$ then f is integrable over R.*

10.1.9 Example.   Let A be the spherical annulus

$$A = \{(x, y, z) \in \mathbb{R}^3 \mid 1 \leqslant x^2 + y^2 + z^2 \leqslant 8\}$$

and let $R = [-3, 3] \times [-3, 3] \times [-3, 3] \subseteq \mathbb{R}^3$. The function $f : R \subseteq \mathbb{R}^3 \to \mathbb{R}$ defined by

$$f(x, y, z) = \begin{cases} x^2 + \sin yz, & (x, y, z) \in A \\ 0 & (x, y, z) \in R \backslash A \end{cases}$$

is integrable over R. The set of discontinuities of f is the *boundary* ∂A of A. This is a null set made up of four simple (hemispherical) surfaces.

Exercises 10.1

1. Let $R = [0, 1] \times [0, 1] \times [0, 1] \subseteq \mathbb{R}^3$, and let $f : R \subseteq \mathbb{R}^3 \to \mathbb{R}$ be the continuous function defined by $f(x, y, z) = x + y + 2z$, $(x, y, z) \in R$. By considering $U(f, \mathcal{P})$ and $L(f, \mathcal{P})$ for suitable partitions of R prove that

$$\iiint\limits_R f \, dV = 2.$$

Hint: generalize the method of Exercise 7.1.1.

2. Let $R = [0, 1] \times [1, 2] \times [2, 3] \subseteq \mathbb{R}^3$. Prove that
    (a) $\iiint\limits_R x \, dx \, dy \, dz = \frac{1}{2}$,
    (b) $\iiint\limits_R y \, dx \, dy \, dz = \frac{3}{2}$,
    (c) $\iiint\limits_R z \, dx \, dy \, dz = \frac{5}{2}$.

3. Let $f : R \subseteq \mathbb{R}^3 \to \mathbb{R}$ and $g : R \subseteq \mathbb{R}^3 \to \mathbb{R}$ be integrable over a compact rectangular box R. Prove that the functions $f + g$, $f - g$ and cf, $c \in \mathbb{R}$ are also integrable over R.

    Prove also that if f and g take identical values except on a null set in R then $\iiint\limits_R f \, dV = \iiint\limits_R g \, dV$.

10.2   Repeated integrals

Let $f : R \subseteq \mathbb{R}^3 \to \mathbb{R}$ be an integrable function over a compact rectangular box R. The triple integral $\iiint\limits_R f \, dV$ is often conveniently evaluated as a repeated integral. We shall meet with two types.

Type 1:   a repeated integral that involves a double integral followed by a single integral, or vice versa: for example

10.2.1 $\displaystyle \int_c^d \left\{ \iint\limits_S f(x, y, z) \, dx \, dy \right\} dz \qquad and$

$$\iint\limits_S \left\{ \int_c^d f(x, y, z) \, dz \right\} dx \, dy,$$

where S is a suitably chosen rectangle in $\mathbb{R}^2$. The integrals in 10.2.1

are conveniently denoted by

10.2.2 $$\int_c^d dz \iint_S f(x, y, z) \, dx \, dy \qquad and$$

$$\iint_S dx \, dy \int_c^d f(x, y, z) \, dz.$$

In evaluating $\iint_S f(x, y, z) \, dx \, dy$, the 'variable' z is kept fixed and the integration is performed with respect to the variables x and y.

Type 2:   a threefold repeated (single) integral: for example

10.2.3 $$\int_{a_1}^{b_1} \left\{ \int_{a_3}^{b_3} \left(\int_{a_2}^{b_2} f(x, y, z) \, dy \right) dz \right\} dx$$

The repeated integral 10.2.3 is denoted by

10.2.4 $$\int_{a_1}^{b_1} dx \int_{a_3}^{b_3} dz \int_{a_2}^{b_2} f(x, y, z) \, dy.$$

At each stage one variable is 'integrated out' while the remaining variables are temporarily kept fixed.

10.2.5 *Example.*

$$\int_0^1 dx \int_1^3 dy \int_0^2 (x^2 + yz) \, dz = \int_0^1 dx \int_1^3 [x^2 z + \tfrac{1}{2} yz^2]_{z=0}^{z=2} \, dy$$

$$= \int_0^1 dx \int_1^3 (2x^2 + 2y) \, dy = \int_0^1 [2x^2 y + y^2]_{y=1}^{y=3} \, dx$$

$$= \int_0^1 (4x^2 + 8) \, dx = 28/3.$$

It will be found similarly that

$$\int_1^3 dy \int_0^2 dz \int_0^1 (x^2 + yz) \, dx = 28/3.$$

In fact, each of the six possible threefold repeated integrals of $x^2 + yz$ over the intervals $x \in [0, 1]$, $y \in [1, 3]$, $z \in [0, 2]$ takes the value 28/3. This is no accident, as we shall see in Theorem 10.2.9.

The following theorem shows that, subject to suitable conditions, a triple integral can be expressed as a type 1 repeated integral.

10.2.6 Theorem. *Let R be the compact rectangular box $[a_1, b_1] \times [a_2, b_2] \times [a_3, b_3]$ in $\mathbb{R}^3$, let $f : R \subseteq \mathbb{R}^3 \to \mathbb{R}$ be a bounded function which is integrable over R and let $S = [a_1, b_1] \times [a_2, b_2]$.*

   [i] *If for each fixed $(x, y) \in S$ the integral*

$$F(x, y) = \int_{a_3}^{b_3} f(x, y, z) \, \mathrm{d}z$$

exists, then the function $F : S \subseteq \mathbb{R}^2 \to \mathbb{R}$ is integrable, and

10.2.7 $$\iiint\limits_R f \, \mathrm{d}V = \iint\limits_S F(x, y) \, \mathrm{d}x \, \mathrm{d}y = \iint\limits_S \mathrm{d}x \, \mathrm{d}y \int_{a_3}^{b_3} f(x, y, z) \, \mathrm{d}z.$$

   [ii] *If for each fixed $z \in [a_3, b_3]$ the double integral*

$$g(z) = \iint\limits_S f(x, y, z) \, \mathrm{d}x \, \mathrm{d}y$$

exists, then the function $g : [a_3, b_3] \subseteq \mathbb{R} \to \mathbb{R}$ is integrable, and

10.2.8 $$\iiint\limits_R f \, \mathrm{d}V = \int_{a_3}^{b_3} g(z) \, \mathrm{d}z = \int_{a_3}^{b_3} \mathrm{d}z \iint\limits_S f(x, y, z) \, \mathrm{d}x \, \mathrm{d}y.$$

Proof.   Adapt the proof of Theorem 7.3.11.

Theorem 10.2.6 provides a half-way stage towards expressing the triple integral of f over R as a threefold repeated integral. We can achieve such expressions subject to further conditions on the function f. For example, in Theorem 10.2.6 part (i), if we assume in addition that the function F is integrable as a function of y for each fixed $x \in [a_1, b_1]$ then, by 10.2.7 and Theorem 7.3.11

$$\iiint\limits_R f \, \mathrm{d}V = \int_{a_1}^{b_1} \mathrm{d}x \int_{a_2}^{b_2} \mathrm{d}y \int_{a_3}^{b_3} f(x, y, z) \, \mathrm{d}z.$$

Alternatively, if we assume that F is integrable as a function of x for each fixed $y \in [a_2, b_2]$, then

$$\iiint\limits_R f \, \mathrm{d}V = \int_{a_2}^{b_2} \mathrm{d}y \int_{a_1}^{b_1} \mathrm{d}x \int_{a_3}^{b_3} f(x, y, z) \, \mathrm{d}z.$$

In the following important special case the six possible threefold repeated integrals of f are all equal.

10.2.9 Theorem. *Let $f: R \subseteq \mathbb{R}^3 \to \mathbb{R}$ be continuous on the compact rectangular box $R = [a_1, b_1] \times [a_2, b_2] \times [a_3, b_3]$. Then f is integrable over R and*

$$\iiint_R f \, dV = \int_{a_i}^{b_i} dx_i \int_{a_j}^{b_j} dx_j \int_{a_k}^{b_k} f(x_1, x_2, x_3) \, dx_k$$

where i, j, k is any arrangement of 1, 2, 3.

Proof. Exercise. Each of the six possibilities is established by application of Theorems 10.2.6 and 7.3.18.

10.2.10 Example. Let $R = [0, 1] \times [1, 3] \times [0, 2]$. Then by Theorem 10.2.9 and Example 10.2.5

$$\iiint_R (x^2 + yz) \, dx \, dy \, dz = 28/3.$$

Exercises 10.2

1. Let R be the rectangular box $[-1, 2] \times [-1, 1] \times [0, 1]$ in $\mathbb{R}^3$. For each of the following continuous functions $f: R \subseteq \mathbb{R}^3 \to \mathbb{R}$ evaluate the triple integral $\iiint_R f \, dV$

    (a) as the repeated integral $\int_{-1}^{2} dx \int_{-1}^{1} dy \int_{0}^{1} f(x, y, z) \, dz$;
    (b) as the repeated integral $\int_{0}^{1} dz \int_{-1}^{2} dx \int_{-1}^{1} f(x, y, z) \, dy$;
    (c) as one of the four remaining possible repeated integrals.
    (i) $f(x, y, z) = xy + z$; (ii) $f(x, y, z) = x^2 z \sin y$.

Answers: (i) 3; (ii) 0.

2. Define the function $f: R \subseteq \mathbb{R}^3 \to \mathbb{R}$ over the unit cube $R = [0, 1] \times [0, 1] \times [0, 1]$ by the rule

    $$f(x, y, z) = \begin{cases} 1 \text{ when } x = \tfrac{1}{2}, \ y \text{ rational or } y = \tfrac{1}{2}, \ x \text{ rational} \\ 0 \text{ otherwise.} \end{cases}$$

    Prove that f is integrable over R.

Hint: apply Theorem 10.1.8.

    Put $S = [0, 1] \times [0, 1] \subseteq \mathbb{R}^2$. Which of the following repeated integrals exist?
    (a) $\iint_S dx \, dy \int_0^1 f(x, y, z) \, dz$,   (b) $\int_0^1 dz \iint_S f(x, y, z) \, dx \, dy$,
    (c) $\int_0^1 dx \iint_S f(x, y, z) \, dy \, dz$,   (d) $\iint_S dy \, dz \int_0^1 f(x, y, z) \, dx$,
    (e) $\int_0^1 dx \int_0^1 dy \int_0^1 f(x, y, z) \, dz$,   (f) $\int_0^1 dx \int_0^1 dz \int_0^1 f(x, y, z) \, dy$.

Answer: all except (d)–(f) exist and take value 0.

10.3   Integrals over general subsets of $\mathbb{R}^3$

In this section we generalize the work of Section 7.4 on integrals over bounded subsets of $\mathbb{R}^2$. The basic definitions are easily adapted to integrals over bounded subsets of $\mathbb{R}^3$, essentially by replacing $\mathbb{R}^2$ by $\mathbb{R}^3$ and considering rectangular boxes in place of rectangles. We leave the reader to formulate definitions of a bounded subset of $\mathbb{R}^3$ and of the boundary of a set.

To define the integral of a bounded function $f : W \subseteq \mathbb{R}^3 \to \mathbb{R}$ on a bounded subset W of $\mathbb{R}^3$ we consider a compact rectangular box R in $\mathbb{R}^3$ containing W and the associated function $f^* : R \subseteq \mathbb{R}^3 \to \mathbb{R}$ defined by

$$f^*(\mathbf{p}) = \begin{cases} f(\mathbf{p}) & \text{if} \quad \mathbf{p} \in W \\ 0 & \text{if} \quad \mathbf{p} \in R \backslash W. \end{cases}$$

We say that f is *integrable over* W if f^* is integrable over R. The integral of f over W is then defined to be the integral of f^* over R and is denoted by

$$\iiint_W f \, \mathrm{d}V \quad \textit{or by} \quad \iiint_W f(x, y, z) \, \mathrm{d}x \, \mathrm{d}y \, \mathrm{d}z.$$

Hence

10.3.1
$$\iiint_W f \, \mathrm{d}V = \iiint_R f^* \, \mathrm{d}V.$$

10.3.2 Theorem.   *Let W be a bounded subset of $\mathbb{R}^3$ with null boundary, and let $f : W \subseteq \mathbb{R}^3 \to \mathbb{R}$ be a bounded function which is continuous everywhere except on a null set in $\mathbb{R}^3$. Then f is integrable over W.*

Proof.   Adapt the proof of Theorem 7.4.13.

Our first example concerns the centroid of a subset of $\mathbb{R}^3$.

10.3.3 Definition.   *Let W be a bounded subset of $\mathbb{R}^3$ whose boundary is a null set. The* volume *of W is defined to be*

$\mathscr{V} = \iiint_W 1 \, \mathrm{d}V$. *Let $f : W \subseteq \mathbb{R}^3 \to \mathbb{R}$ be integrable over W. Provided*

$\mathcal{V} \neq 0$, *the* mean value of f over W is defined to be the number

$$\bar{f} = \frac{1}{\mathcal{V}} \iint_W f \, dV.$$

In particular, the mean values of $f(x, y, z) = x$, $g(x, y, z) = y$ *and* $h(x, y, z) = z$ *over* W *give the coordinates* $(\bar{x}, \bar{y}, \bar{z})$ *of the* centroid *of* W.

10.3.4 *Example.* Find the centroid of the hemispherical subset of $\mathbb{R}^3$

$$W = \{(x, y, z) \in \mathbb{R}^3 \mid x^2 + y^2 + z^2 \leqslant c^2, z \geqslant 0\}.$$

See Fig. 10.1.

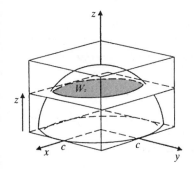 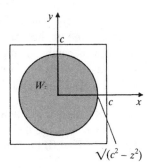

Fig. 10.1 Section W_z of hemisphere: $x^2 + y^2 \leqslant c^2 - z^2$

By symmetry, $\bar{x} = \bar{y} = 0$. We calculate

$$\bar{z} = \frac{1}{\mathcal{V}} \iiint_W z \, dV,$$

where $\mathcal{V} = \frac{2}{3}\pi c^3$, the volume of W (see Exercise 10.3.1). Consider the compact rectangular box R containing W given by $R = K \times [0, c]$, where $K = [-c, c] \times [-c, c]$.

Put $f(x, y, z) = z$, $(x, y, z) \in W$, and let f^* be the associated function defined over R by

10.3.5 $\qquad f^*(x, y, z) = \begin{cases} z & \text{when } x^2 + y^2 \leqslant c^2 - z^2 \\ 0 & \text{otherwise.} \end{cases}$

Note that for fixed $z \in [0, c]$ the inequality $x^2 + y^2 \leqslant c^2 - z^2$ specifies the circular disc W_z in $\mathbb{R}^2$ with centre the origin, radius $\sqrt{(c^2 - z^2)}$, and area

$\pi(c^2 - z^2)$. Therefore

$$
\begin{aligned}
\iiint_W z \, dV &= \iiint_R f^* \, dV && \text{by 10.3.1} \\
&= \int_0^c dz \iint_K f^*(x, y, z) \, dx \, dy && \text{by 10.2.8} \\
&= \int_0^c z \, dz \iint_{W_z} 1 \, dx \, dy \\
&= \int_0^c \pi z (c^2 - z^2) \, dz = \tfrac{1}{4}\pi c^4.
\end{aligned}
$$

It follows that $\bar{z} = \tfrac{3}{8}c$.

The above method of evaluating a triple integral is useful when, by fixing one of the variables, the double integral in the remaining variables is easily calculated. In practice it is usual to omit mention of the associated function and to deal directly for a fixed value of z with the double integral of f over the section W_z.

10.3.6 *Example.* Let W be the subset of $\mathbb{R}^3$ bounded by the parabolic bowl $z = x^2 + y^2$ and the planes $x = 0$, $y = 0$ and $z = a$, where $a > 0$ (See Fig. 10.2). Find **[i]** the volume of W; **[ii]** the z coordinate $\bar{z}$ of the centroid W.

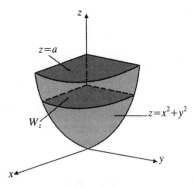

Fig. 10.2

[i]   Let W_z be the section obtained by slicing W by the plane $z = $ constant. Then W_z is a quadrant of a circle of area $\frac{1}{4}\pi(x^2 + y^2) = \frac{1}{4}\pi z$. By the method of Example 10.3.4 we obtain

$$\text{Volume of } W = \iiint\limits_W 1 \, dV$$

$$= \int_0^a dz \iint\limits_W 1 \, dx \, dy$$

$$= \int_0^a \tfrac{1}{4}\pi z \, dz = \tfrac{1}{8}\,\pi a^2.$$

[ii]   Similarly

$$\iiint\limits_W z \, dV = \int_0^a z \, dz \iint\limits_{W_z} 1 \, dx \, dy = \int_0^a \tfrac{1}{4}\pi z^2 \, dz = \tfrac{1}{12}\pi a^3.$$

Therefore $\bar{z} = \frac{2}{3}a$.

   In general there is no short cut to the value of a triple integral over a set W be slicing W parallel to just one of the coordinate planes. The following example illustrates a general procedure for expressing a triple integral as a threefold repeated integral.

10.3.7. *Example.*   Let W be the subset of $\mathbb{R}^3$ defined in Example 10.3.6. Calculate $\iiint\limits_W f \, dV$, where $f(x, y, z) = xyz$, $(x, y, z) \in W$.

   Let S be the projection of W onto the x, y plane (see Fig. 10.3). This is

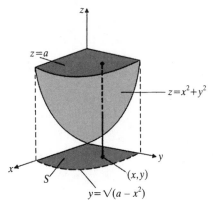

Fig. 10.3

the quadrant of a circle radius $\sqrt{a}$ given by

10.3.8 $S = \{(x, y) \in \mathbb{R}^2 \mid 0 \leqslant x \leqslant \sqrt{a}, \ 0 \leqslant y \leqslant \sqrt{(a - x^2)}\}.$

We first integrate with respect to z, keeping $(x, y) \in S$ fixed. By 10.2.7 applied to the associated function f^*,

$$\iiint_W f \, dV = \iint_S dx \, dy \int_{x^2+y^2}^a xyz \, dz.$$

By 10.3.8 and Theorem 7.4.21,

$$\iiint_W f \, dV = \int_0^{\sqrt{a}} dx \int_0^{\sqrt{(a-x^2)}} dy \int_{x^2+y^2}^a xyz \, dz.$$

A tedious calculation leads to $\iiint_W f \, dV = a^4/32.$

The procedure used in the above example illustrates the following general theory.

10.3.9 Definition.   *A bounded subset W of $\mathbb{R}^3$ is said to be an x, y simple region if there exists a simple region S in $\mathbb{R}^2$ and continuous functions $\lambda : S \subseteq \mathbb{R}^2 \to \mathbb{R}$ and $\tau : S \subseteq \mathbb{R}^2 \to \mathbb{R}$ such that*

10.3.10 $\lambda(x, y) < \tau(x, y)$ *for all* $(x, y) \in S \backslash \partial S,$

10.3.11 $W = \{(x, y, z) \in \mathbb{R}^3 \mid (x, y) \in S, \ \lambda(x, y) \leqslant z \leqslant \tau(x, y)\},$

and the graphs $z = \lambda(x, y)$ and $z = \tau(x, y)$ are both surfaces.

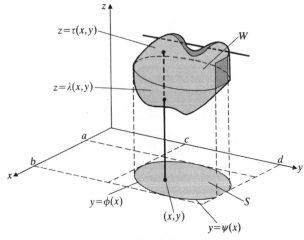

Fig. 10.4

Notice that in Definition 10.3.9 the simple region S is just the projection of W onto the x, y plane. See Fig. 10.4.

10.3.12 Example.   The set W in Example 10.3.6 is x, y simple. The projection S of W on to the x, y plane is given in 10.3.8. The functions λ, τ are defined by

$$\lambda(x, y) = x^2 + y^2, \qquad \tau(x, y) = a, \qquad (x, y) \in S.$$

In general, if W is the x, y simple region defined in 10.3.9 then, since the simple region S in $\mathbb{R}^2$ is x-simple, there exist continuous functions $\phi : [a, b] \subseteq \mathbb{R} \to \mathbb{R}$ and $\psi : [a, b] \subseteq \mathbb{R} \to \mathbb{R}$ with $\phi(x) < \psi(x)$ for all $x \in {]a, b[}$ such that

10.3.13 $W = \{(x, y, z) \in \mathbb{R}^3 \mid a \leqslant x \leqslant b,$

$$\phi(x) \leqslant y \leqslant \psi(x), \lambda(x, y) \leqslant z \leqslant \tau(x, y)\}.$$

10.3.14 Example.   The set W in Example 10.3.7 is given by

$$W = \{(x, y, z) \in \mathbb{R}^3 \mid 0 \leqslant x \leqslant \surd a, 0 \leqslant y \leqslant \surd(a - x^2), x^2 + y^2 \leqslant z \leqslant a\}.$$

10.3.15 Theorem.   *Let* $W \subseteq \mathbb{R}^3$ *be the* x, y *simple region given in* 10.3.13. *Then any continuous function* $f : W \subseteq \mathbb{R}^3 \to \mathbb{R}$ *is integrable over* W *and*

10.3.16 $$\iiint\limits_{W} f \, dV = \int_a^b dx \int_{\phi(x)}^{\psi(x)} dy \int_{\lambda(x, y)}^{\tau(x, y)} f(x, y, z) \, dz.$$

Proof.   Adapt the proof of Theorem 7.4.21. Briefly, since ∂W is a null set in $\mathbb{R}^3$, the function f is integrable over W by Theorem 10.3.2, and

10.3.17 $$\iiint\limits_{W} f \, dV = \iiint\limits_{R} f^* \, dV,$$

where R is a compact rectangular box containing W and f^* is the function on R associated with f. The theorem follows by an application of Theorems 10.2.6(i) and 7.4.21.

10.3.18 Corollary.   *With the notation of Theorem* 10.3.15, *let* U *be a subset of* W *whose closure is* W *and whose boundary is a null set. Then any continuous function* $f : W \subseteq \mathbb{R}^3 \to \mathbb{R}$ *is integrable over* U, *and*

$$\iiint_U f \, dV = \int_a^b dx \int_{\phi(x)}^{\psi(x)} dy \int_{\lambda(x,\,y)}^{\tau(x,\,y)} f(x, y, z) \, dz.$$

Proof.   Since $\iiint_U f \, dV = \iiint_W f \, dV$, the result follows.

A formula for $\iiint_W f \, dV$ alternative to that given in 10.3.16 is obtained if, by appealing to Theorem 7.4.22, we express the simple region S in $\mathbb{R}^2$ in the form

$$S = \{(x, y) \in \mathbb{R}^2 \mid c \leqslant y \leqslant d, \, \phi(y) \leqslant x \leqslant \psi(y)\}.$$

We then obtain

10.3.19 $$\iiint_W f \, dV = \int_c^d dy \int_{\phi(y)}^{\psi(y)} dx \int_{\lambda(x,\,y)}^{\tau(x,\,y)} f(x, y, z) \, dz.$$

See Exercise 10.3.3 for an illustration.

By interchanging the roles of y and z in Definition 10.3.9 we obtain the definition of an x, z simple region in $\mathbb{R}^3$. A y, z simple region is similarly defined. Theorem 10.3.15 is easily modified to apply to these cases.

10.3.20 *Definition.* *A simple region in* $\mathbb{R}^3$ *is a subset of* $\mathbb{R}^3$ *which is x, y simple and x, z simple and y, z simple.*

The set W sketched in Fig. 10.4 is not a simple region. Note the horizontal 'probe' parallel to the y-axis that pierces the boundary of W in more than two places. This shows that W is not x, z simple.

Note that if $f: W \subseteq \mathbb{R}^3 \to \mathbb{R}$ is continuous and if W is a simple region in $\mathbb{R}^3$ there are six ways of evaluating $\iiint_W f \, dV$ as a threefold repeated integral.

10.3.21 *Example.*   The set W defined in Example 10.3.6 is a simple region. Regarding it as an x, z simple region, for example, we obtain for any continuous function f,

10.3.22 $$\iiint\limits_{W} f \, dV = \int_0^{\sqrt{a}} dx \int_{x^2}^{a} dz \int_0^{\sqrt{(z-x^2)}} f(x, y, z) \, dy$$

$$= \int_0^{a} dz \int_0^{\sqrt{z}} dx \int_0^{\sqrt{(z-x^2)}} f(x, y, z) \, dy.$$

In particular, if $f(x, y, z) = xyz$, then $\iiint\limits_{W} f \, dV = a^4/32$.

Compare the method of Example 10.3.7 with the present shorter method for evaluating $\iiint\limits_{W} xyz \, dx \, dy \, dz$.

Exercises 10.3

1. (a) Let W be the hemispherical subset of $\mathbb{R}^3$ defined by

$$W = \{(x, y, z) \in \mathbb{R}^3 \mid x^2 + y^2 + z^2 \leqslant c^2, z \geqslant 0\}.$$

Draw supporting sketches for the expressions for the volume of W
   (i)   as the repeated integral

$$\int_0^{c} dz \iint\limits_{W_z} 1 \, dx \, dy,$$

where W_z is the section of W obtained by slicing W by the plane through $(0, 0, z)$ parallel to the x, y plane;
   (ii)   as the repeated integral

$$\int_{-c}^{c} dx \iint\limits_{W_x} 1 \, dy \, dz = \int_{-c}^{c} \tfrac{1}{2}\pi(c^2 - x^2) \, dx;$$

   (iii)   as the threefold repeated integral

$$\int_{-c}^{c} dx \int_{-\sqrt{(c^2-x^2)}}^{\sqrt{(c^2-x^2)}} dy \int_0^{\sqrt{(c^2-x^2-y^2)}} 1 \, dz.$$

In each case evaluate the integral.

Answer: $\tfrac{2}{3}\pi c^3$.

   (b)   Consider similarly the volume enclosed by the sphere $x^2 + y^2 + z^2 = c^2$.

2.   Sketch the tetrahedron W in $\mathbb{R}^3$ bounded by the plane $x + y + z = 1$, and the coordinate planes $x = 0$, $y = 0$ and $z = 0$. Find its volume $\mathcal{V}$ and the coordinates $(\bar{x}, \bar{y}, \bar{z})$ of its centroid.

Answer: consider the threefold repeated integral

$$\int_0^1 dx \int_0^{1-x} dy \int_0^{1-x-y} f(x, y, z) \, dz$$

for the cases $f(x, y, z) = 1$ and $f(x, y, z) = x$, $(x, y, z) \in \mathbb{R}^3$. Hence obtain $\mathcal{V} = 1/6$ and $\bar{x} = 1/4$. By symmetry, $\bar{y} = \bar{z} = 1/4$. Alternatively, consider

$$\int_0^1 dx \iint_{W_x} f(x, y, z) \, dy \, dz,$$

where W_x is a suitable section of the tetrahedron.

3.   Let $f: \mathbb{R}^3 \to \mathbb{R}$ be continuous. prove that

$$\int_0^1 dx \int_0^{1-x} dy \int_0^{1-x-y} f(x, y, z) \, dz = \int_0^1 dy \int_0^{1-y} dx \int_0^{1-x-y} f(x, y, z) \, dz.$$

Answer: both integrals are equal to $\iiint_W f \, dV$, where W is the tetrahedron of Exercise 2.

4.   (a)   Let $f: \mathbb{R} \to \mathbb{R}$ be continuous. Prove that

$$\int_0^a dx \int_0^x dy \int_0^y f(z) \, dz = \int_0^a \frac{(a-z)^2}{2} f(z) \, dz.$$

Hint: first interchange the order of integration with respect to y and z. Show that for a fixed x

$$\int_0^x dy \int_0^y f(z) \, dz = \int_0^x dz \int_z^x f(z) \, dy = \int_0^x (x-z)f(z) \, dz.$$

A sketch of the region of integration in the y, z plane for fixed x may be helpful.

(b)   Prove the generalization of part (a), that

$$\int_0^a dx_1 \int_0^{x_1} dx_2 \cdots \int_0^{x_{n-1}} f(x_n) \, dx_n = \int_0^a \frac{(a-x_n)^n}{n!} f(x_n) \, dx_n.$$

5.   Sketch the subset W of $\mathbb{R}^3$ relating to the repeated integral of Exercise 4(a). Calculate its volume.

Answer: the set bounded by the planes $z = 0$, $z = y$, $y = 0$, $y = x$, $x = 0$, and $x = a$. It is the tetrahedron with vertices at $(0, 0, 0)$, (a, a, a), $(a, a, 0)$, and $(a, 0, 0)$. Notice that for x constant, $0 \leqslant x \leqslant a$, the section W_x is the triangle with vertices (x, x, x), $(x, x, 0)$, and $(x, 0, 0)$ bounded in the plane $x = $ constant by the lines $z = 0$, $z = y$, $y = 0$, and $y = x$ (constant). Volume $W = 1/6$.

6. Let W be the subset of $\mathbb{R}^3$ bounded by the parabolic bowl $z = x^2 + y^2$ and the planes $x = 0$, $y = 0$ and $z = a$, where $a > 0$ (as in Example 10.3.6). Find suitable limits of integration for the expressions:

    (a) $\iiint_W f\,dV = \int dy \int dz \int f(x, y, z)\,dx$;

    (b) $\iiint_W f\,dV = \int dz \int dy \int f(x, y, z)\,dx$.

    Hence calculate $\iiint_W f\,dV$ for the cases (i) $f(x, y, z) = xyz$;
    (ii) $f(x, y, z) = xy^3$.

Answers: (i) $a^4/32$; (ii) $a^4/96$.

7. Sketch the cone $z = \sqrt{(x^2 + y^2)}$ and the sphere (of radius $\frac{1}{2}$ and centre $(0, 0, \frac{1}{2})$) $x^2 + y^2 + z^2 = z$. Find the volume of the subset of $\mathbb{R}^3$ lying inside the sphere and above the cone.

Answer: by the 'method of sections' illustrated in Example 10.3.6 the volume is given by the repeated integral

$$\mathcal{V} = \int_0^1 dz \iint_{W_z} 1\,dx\,dy$$

where

$$\text{Area } W_z = \begin{cases} \pi z^2 & \text{when } 0 \leqslant z \leqslant \frac{1}{2}, \\ \pi(z - z^2) & \text{when } \frac{1}{2} \leqslant z \leqslant 1. \end{cases}$$

Hence

$$\mathcal{V} = \int_0^{1/2} \pi z^2\,dz + \int_{1/2}^1 \pi(z - z^2)\,dz = \tfrac{1}{8}\pi.$$

8. A circular hole of radius b is drilled through a ball of radius $a > b$ along a diameter as axis. SHow that the volume of material removed from the ball is $\frac{4}{3}\pi(a^3 - (a^2 - b^2)^{3/2})$.

9. Find the volume of the subset W of $\mathbb{R}^3$ bounded by the parabolic bowl $z = x^2 + y^2$ and the plane $z = 2x$.

Answer: the volume can be expressed as

$$\int_0^2 dx \int_{-\sqrt{(2x-x^2)}}^{\sqrt{(2x-x^2)}} dy \int_{x^2+y^2}^{2x} dz,$$

which is equal to $\frac{1}{2}\pi$.

10. Find the volume bounded by the paraboloid $z = 4 - x^2 - y^2$ and the x, y plane.

Answer: 8π.

11.   Find the volume of a pyramid with horizontal base of area $\mathscr{A}$ and vertical height h. Find also the height above the base of its centroid.

Hint: the area of a section parallel to the base at a depth z from the apex is $\mathscr{A}z^2/h^2$.

Answers: volume $\frac{1}{3}\mathscr{A}h$; centroid $\frac{1}{4}h$ above base.

12.   Let W be a bounded subset of $\mathbb{R}^3$. Let $N \subseteq W$ be a null set, and let $f: W \subseteq \mathbb{R}^3 \to \mathbb{R}$ be a bounded function. Prove that f is integrable over W if and only if f is integrable over $W \backslash N$ and, when they exist, these two integrals are equal.

Hint: Adapt Theorem 7.4.28.

10.4   Change of variables in triple integrals

An important class of physical problems is concerned with spacial systems that display an element of spherical symmetry. Such problems are best considered in terms of spherical coordinates. Similarly, if axial symmetry exists then cylindrical coordinates are the appropriate ones to use. The following theorem describes how a change of variables

$$x = G_1(u, v, w), \qquad y = G_2(u, v, w), \qquad z = G_3(u, v, w)$$

from Cartesian x, y, z coordinates to curvilinear u, v, w coordinates affects the triple integral of a continuous function. Theorem 10.4.1 is a direct generalization of Theorem 7.7.5. Note that by Definition 4.3.24, a C^1 function $G: D \subseteq \mathbb{R}^3 \to \mathbb{R}^3$ defined on a open domain D is *smooth on D* is $\det J_{G,(u,v,w)}$ is non-zero for all (u, v, w) in D.

10.4.1 Theorem.   *Let $G: K \subseteq \mathbb{R}^3 \to \mathbb{R}^3$ be a C^1 function defined on a compact set K in $\mathbb{R}^3$, and let $D \subseteq K$ be an open subset of $\mathbb{R}^3$ such that*
  [i]   *$K \backslash D$ is a null set in $\mathbb{R}^3$, and*
  [ii]   *G is 1–1 and smooth on D.*
  Then for any bounded function $f: G(K) \subseteq \mathbb{R}^3 \to \mathbb{R}$ which is continuous on $G(D)$

10.4.2 $\displaystyle\iiint\limits_{G(K)} f(x, y, z) \, \mathrm{d}x \, \mathrm{d}y \, \mathrm{d}z$

$$= \iiint\limits_{K} f(G(u, v, w)) \, |\det J_{G,(u,v,w)}| \, \mathrm{d}u \, \mathrm{d}v \, \mathrm{d}w.$$

The relation 10.4.2 is commonly expressed in the form

10.4.3 $\displaystyle\iiint\limits_{G(K)} f(x, y, z)\,dx\,dy\,dz$

$$= \iiint\limits_{K} \tilde{f}(u, v, w) \left| \frac{\partial(x, y, z)}{\partial(u, v, w)} \right| du\,dv\,dw,$$

where $\tilde{f}(u, v, w) = f(G(u, v, w))$, $(u, v, w) \in K$.

The two most important applications of Theorem 10.4.1 concern spherical coordinates and cylindrical coordinates, and we consider these in turn.

The transition from spherical coordinates (r, ϕ, θ) to rectangular Cartesian coordinates (x, y, z) in three dimensional space is expressed by the rule $G(r, \phi, \theta) = (x, y, z)$, where

$$x = r \sin \phi \cos \theta, \qquad y = r \sin \phi \sin \theta, \qquad z = r \cos \phi.$$

(In some texts the roles of θ and ϕ are interchanged.) Clearly G is C^1, and an elementary calculation shows that

$$J_{G, (r, \phi, \theta)} = \begin{bmatrix} \sin \phi \cos \theta & r \cos \phi \cos \theta & -r \sin \phi \sin \theta \\ \sin \phi \sin \theta & r \cos \phi \sin \theta & r \sin \phi \cos \theta \\ \cos \phi & -r \cos \phi & 0 \end{bmatrix},$$

and hence that

$$\det J_{G, (r, \phi, \theta)} = r^2 \sin \phi.$$

In fact G is 1–1 and smooth on the open box

$$U = \{(r, \phi, \theta) \in \mathbb{R}^3 \mid r > 0, 0 < \phi < \pi, 0 < \theta < 2\pi\}.$$

See Fig. 10.5.

The image $G(U)$ consists of all points $(x, y, z) \in \mathbb{R}^3$ excluding the half-plane $\{(x, y, z) \in \mathbb{R}^3 \mid x \geq 0, y = 0\}$. The image $G(\bar{U})$ is the whole of $\mathbb{R}^3$.

The equation of a sphere, centre the origin, radius c takes the particularly simple form $r = c$ in spherical coordinates and the equation $\phi = \alpha$ where α is a constant and $0 < \alpha < \dfrac{\pi}{2}$ represents a cone with axis Oz.

10.4.4 Theorem. *Let* $G : \mathbb{R}^3 \to \mathbb{R}^3$ *be the spherical coordinate*

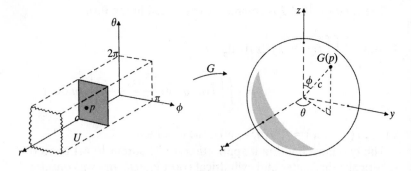

Fig. 10.5

transformation given by

10.4.5 $(x, y, z) = G(r, \phi, \theta)$
$$= (r \sin \phi \cos \theta, r \sin \phi \sin \theta, r \cos \phi).$$

Let D be a bounded open subset of $\mathbb{R}^3$ *with null boundary on which G is* 1–1 *and smooth, and let* $K = \bar{D}$. *Then for any bounded function* $f : G(K) \subseteq \mathbb{R}^3 \to \mathbb{R}$ *which is continuous on* $G(D)$,

10.4.6 $\displaystyle\iiint\limits_{G(K)} f(x, y, z) \, \mathrm{d}x \, \mathrm{d}y \, \mathrm{d}z$

$$= \iiint\limits_{K} \tilde{f}(r, \phi, \theta) r^2 \, |\sin \phi| \, \mathrm{d}r \, \mathrm{d}\phi \, \mathrm{d}\theta,$$

where $\tilde{f}(r, \phi, \theta) = f(G(r, \phi, \theta))$, $(r, \phi, \theta) \in K$.

Proof.   Apply Theorem 10.4.1.

10.4.7 *Example.*   Find the mass of a hemispherical bowl of inside radius a and outside radius b if the density at any point of the bowl is inversely proportional to the distance from the centre O.

Consider a Cartesian coordinate system relative to which the bowl is described (as sketched in Figure 10.6) by the set

$$W = \{(x, y, z) \in \mathbb{R}^3 \mid a^2 \leqslant x^2 + y^2 + z^2 \leqslant b^2, z \leqslant 0\}.$$

The density at (x, y, z) is given to be λ/r, where $r = \sqrt{(x^2 + y^2 + z^2)}$ and λ is

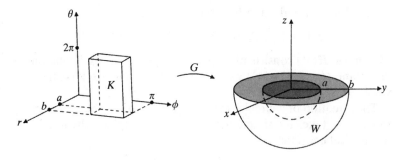

Fig. 10.6

a constant. Therefore the mass M of the bowl is

10.4.8
$$M = \iiint_W \frac{\lambda}{\sqrt{(x^2 + y^2 + z^2)}} \, dx \, dy \, dz.$$

The spherical symmetry of the problem suggests changing to spherical coordinates. Let $G : \mathbb{R}^3 \to \mathbb{R}^3$ be the spherical coordinate transformation 10.4.5, let $K = [a, b] \times [\tfrac{1}{2}\pi, \pi] \times [0, 2\pi]$ and let $D = \text{Int } K$. The conditions of Theorem 10.4.4 are satisfied and $W = G(K)$. Therefore applying 10.4.6 to determine 10.4.8 we find that

$$M = \iiint_K \frac{\lambda}{r} r^2 \, |\sin \phi| \, dr \, d\phi \, d\theta,$$

$$= \int_a^b dr \int_{\frac{1}{2}\pi}^{\pi} d\phi \int_0^{2\pi} \lambda r \sin \phi \, d\theta, \qquad \text{since } \sin \phi \geqslant 0 \text{ for } \phi \in [\tfrac{1}{2}\pi, \pi],$$

$$= \pi \lambda (b^2 - a^2).$$

The transition from *cylindrical coordinates* (ρ, θ, z) to rectangular Cartesian coordinates (x, y, z) in three-dimensional space is expressed by the rule $H(\rho, \theta, z) = (x, y, z)$, where

$$x = \rho \cos \theta, \qquad y = \rho \sin \theta, \qquad z = z.$$

Clearly H is C^1, and an elementary calculation shows that

$$J_{H,(\rho,\theta,z)} = \begin{bmatrix} \cos \theta & -\rho \sin \theta & 0 \\ \sin \theta & \rho \cos \theta & 0 \\ 0 & 0 & 1 \end{bmatrix}$$

and hence that

$$\det J_{H,(\rho,\theta,z)} = \rho.$$

In fact H is 1–1 and smooth on the open box

$$V = \{(\rho, \theta, z) \in \mathbb{R}^3 \mid \rho > 0, 0 < \theta < 2\pi\}.$$

The image $H(V)$ consists of all points $(x, y, z) \in \mathbb{R}^3$ excluding the half plane $\{(x, y, z) \in \mathbb{R}^3 \mid x \geqslant 0, y = 0\}$. The image $H(\bar{V})$ is the whole of $\mathbb{R}^3$.

The equation of a circular cylinder, whose axis is the z-axis and radius is a, takes the simple form $\rho = a$. This explains the term 'cylindrical coordinates'.

10.4.9 Theorem.   *Let $H : \mathbb{R}^3 \to \mathbb{R}^3$ by the cylindrical coordinate transformation given by*

10.4.10 $(x, y, z) = H(\rho, \theta, z) = (\rho \cos \theta, \rho \sin \theta, z),$ $\rho \geqslant 0.$

Let D be a bounded open subset of $\mathbb{R}^3$ with null boundary upon which H is 1–1 and smooth, and let $K = \bar{D}$. Then for any bounded function $f : H(K) \subseteq \mathbb{R}^3 \to \mathbb{R}$ which is continuous on $H(D)$,

10.4.11 $$\iiint\limits_{H(K)} f(x, y, z)\, \mathrm{d}x\, \mathrm{d}y\, \mathrm{d}z = \iiint\limits_{K} \tilde{f}(\rho, \theta, z)\rho\, \mathrm{d}\rho\, \mathrm{d}\theta\, \mathrm{d}z,$$

where $\tilde{f}(\rho, \theta, z) = f(H(\rho, \theta, z)), (\rho, \theta, z) \in K.$

Proof.   Apply Theorem 10.4.1.

10.4.12 Example.   Find the volume $\mathcal{V}$ bounded by the circular cone $z = \sqrt{(x^2 + y^2)}$ and the paraboloid $z = x^2 + y^2$ (see Fig. 10.7(i)).

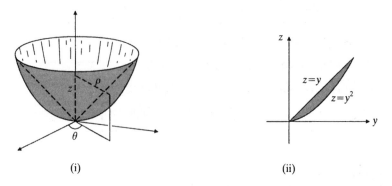

(i)   (ii)

Fig. 10.7

Symmetry about the z axis suggests the use of cylindrical coordinates. Let W be the set bounded by the given cone and paraboloid. If $H:\mathbb{R}^3 \rightarrow \mathbb{R}^3$ is the cylindrical coordinate transformation 10.4.10, then $W = H(K)$, where

$$K = \{(\rho, \theta, z) \in \mathbb{R}^3 \mid 0 \leqslant z \leqslant 1, \, 0 \leqslant \theta < 2\pi, \, z \leqslant \rho \leqslant \sqrt{z}\}.$$

Notice that $\rho = \sqrt{(x^2 + y^2)}$, and therefore for a fixed value of z the values of ρ such that $(\rho \cos \theta, \rho \sin \theta, z) \in W$ lie in the interval $[z, \sqrt{z}]$. Letting $D = \operatorname{Int} K$ the conditions of Theorem 10.4.4 are satisfied, and $W = H(K)$. Therefore, by 10.4.6,

$$\mathscr{V} = \iiint\limits_{H(K)} 1 \, dx \, dy \, dz = \iiint\limits_{K} \rho \, d\rho \, d\theta \, dz.$$

The last integral can be evaluated as a repeated integral. We obtain

$$\mathscr{V} = \int_0^1 dz \int_0^{2\pi} d\theta \int_z^{\sqrt{z}} \rho \, d\rho = \pi/6.$$

The set W sketched in Fig. 10.7(i) is an example of a *volume of revolution*. It is generated by revolving the plane region bounded by $z = y$ and $z = y^2$ in the y, z plane (Fig. 10.7(ii)) about the z-axis. See Exercise 10.4.5.

Exercises 10.4

1. Let $W = \{(x, y, z) \in \mathbb{R}^3 \mid x^2 + y^2 + z^2 \leqslant c^2, \, z \geqslant 0\}$ be the hemispherical subset of $\mathbb{R}^3$ considered in Example 10.3.4. Use (a) spherical coordinates, (b) cylindrical coordinates to show that

$$\iiint\limits_{W} z \, dV = \tfrac{1}{4}\pi c^4.$$

Compare with the method of Example 10.3.4.

2. Reconsider the examples and exercises of Section 10.3 and, where appropriate (in cases of spherical or cylindrical symmetry), solve them using spherical or cylindrical coordinates. For example, the volume bounded by the parabolic bowl $z = x^2 + y^2$ and the planes $x = 0$, $y = 0$ and $z = a > 0$ (Example 10.3.6) can be expressed simply in cylindrical coordinates as

$$\int_0^a dz \int_0^{1/2\pi} d\theta \int_0^{\sqrt{z}} \rho \, d\rho.$$

Again, the volume bounded by the hemisphere $x^2 + y^2 + z^2 = c^2$, $z \geqslant 0$ and the plane $z = 0$ (Exercise 10.3.1) can be expressed in

510 Triple integrals in $\mathbb{R}^3$

spherical coordinates as

$$\int_0^c dr \int_0^{1/2\pi} d\phi \int_0^{2\pi} r^2 \sin \phi \, d\theta.$$

3. Prove that the volume enclosed by the ellipsoid $x^2/a^2 + y^2/b^2 + z^2/c^2 \leqslant 1$ is $\frac{4}{3}\pi abc$.

Hint: apply 10.4.2, where $x = au$, $y = bv$, $z = cw$, and K is the unit ball $u^2 + v^2 + w^2 \leqslant 1$.

4. Let W be the unit ball $x^2 + y^2 + z^2 \leqslant 1$. Evaluate

$$\iiint_W e^{\sqrt{(x^2+y^2+z^2)}} \, dx \, dy \, dz.$$

Hint: use spherical coordinates.

Answer: $4\pi(e - 2)$.

5. (a) *Pappus' Theorem for volumes of revolution.* A plane region bounded by a simple closed curve is revolved about an axis in its plane which does not cross the region. Prove that the volume generated is equal to the product of the area of the region and the length of the path traversed in one revolution by the centroid of the region.

Hint: consider the curve in the y, z plane and take the z-axis as axis of rotation. Obtain the formula for the volume of revolution (in cylindrical coordinates)

$$\mathcal{V} = \int_0^{2\pi} d\theta \iint_{W_\theta} \rho \, d\rho \, dz.$$

(b) Using Pappus' Theorem, calculate:
(i) the volume of the torus generated by revolving the circular disc in the y, z plane $z^2 + (y - b)^2 \leqslant c^2$, $b > c > 0$ about the z-axis;
(ii) the volume generated by revolving the region bounded by the curves $z = y$ and $z = y^2$ in the y, z plane about the z-axis (Example 10.4.12);
(iii) the volume bounded by the cone $z = \sqrt{(x^2 + y^2)}$ and the plane $z = a$, $a > 0$.

Answers: (i) $2\pi^2 bc^2$; (ii) $\pi/6$; (iii) $\pi a^3/3$.

6. If the density of a body W in $\mathbb{R}^3$ is given by the density function $f : W \subseteq \mathbb{R}^3 \to \mathbb{R}$, then its mass is

$$M = \iiint_W f \, dV$$

and its centre of mass $(\bar{x}, \bar{y}, \bar{z})$ is defined by

$$\bar{x} = \frac{1}{M}\iiint_W xf\,dV, \qquad \bar{y} = \frac{1}{M}\iiint_W yf\,dV, \qquad \bar{z} = \frac{1}{M}\iiint_W zf\,dV.$$

Find the centre of mass of the hemispherical bowl of inside radius a and outside radius b with density function λ/r (see Example 10.4.7).

Answer: $\bar{z} = -\frac{1}{3}(a^2 + ab + b^2)/(a + b)$, and, by symmetry, $\bar{x} = \bar{y} = 0$.

7. The moment of inertia about the z-axis of a body W in $\mathbb{R}^3$ is defined to be the integral

$$I_z = \iiint_W \rho^2 f\,dV,$$

where f is the density function and $\rho(x, y, z) = \sqrt{(x^2 + y^2)}$, the distance of the point (x, y, z) from the z-axis.

Find the mass M and the moment of inertia about the z-axis of (a) a rectangular box

$$W = \{(x, y, z) \in \mathbb{R}^3 \mid -a \leqslant x \leqslant a, -b \leqslant y \leqslant b, -c \leqslant z \leqslant c\}$$

of uniform density λ; (b) a ball B, radius c, centre the origin, of density λr at distance r from the origin $0 \leqslant r \leqslant c$, where λ is a constant.

Answers: (a) $M = 8abc\lambda$, $I_z = \frac{1}{3}M(a^2 + b^2)$. Use rectangular coordinates.
(b) $M = \pi c^4 \lambda$, $I_z = \frac{4}{9}Mc^2$. Use spherical coordinates.

10.5   Gauss' Divergence Theorem

In introducing the concept of the divergence of a vector field F in Section 4.11 we explained that if F measures the velocity field of a gas then $(\operatorname{div} F)(\mathbf{r})$ measures the rate of expansion per unit volume of the gas at the point $\mathbf{r}$. We would expect that for a gas moving through a region in $\mathbb{R}^3$ there is a relationship between the net rate of flow across the boundary of the region and the total divergence of the field inside the region. This relationship is established in Gauss' Divergence Theorem.

We shall prove the Divergence Theorem for a simple region W in $\mathbb{R}^3$. Since W is simultaneously x, y simple, y, z simple, and x, z simple, its boundary ∂W consists of at most six graph-like surfaces suitably joined to form a closed surface. For example, a cube is made up of six simple surfaces; a sphere can be put together from two simple surfaces.

We define the *positively oriented boundary* $(\partial W, N)$ of W by
associating with each point **q** in W, a unit vector $N(\mathbf{q})$ normal to ∂W
at **q** and facing outwards from W, whenever such a vector exists
(Fig. 10.8). The Divergence Theorem equates the double integral of
a vector field F over $(\partial W, N)$ and the triple integral of the scalar
field div F over W. As such it corresponds to Green's Theorem in
the form given in Theorem 7.6.13.

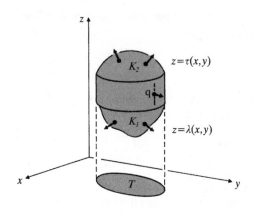

Fig. 10.8

10.5.1 Gauss' Divergence Theorem.   *Let W be a simple region in
$\mathbb{R}^3$ and let $(\partial W, N)$ be the positively oriented boundary of W. For
any C^1 vector field $F : W \subseteq \mathbb{R}^3 \to \mathbb{R}^3$*

10.5.2
$$\iint\limits_{(\partial W, N)} F \cdot d\mathbf{A} = \iiint\limits_W \operatorname{div} F \, dV.$$

Proof.   The result is established by adding together the three
identities

[i] $$\iint\limits_{\partial W} (F_1, 0, 0) \cdot N \, dA = \iiint\limits_W \frac{\partial F_1}{\partial x} \, dV,$$

[ii] $$\iint\limits_{\partial W} (0, F_2, 0) \cdot N \, dA = \iiint\limits_W \frac{\partial F_2}{\partial y} \, dV,$$

[iii]
$$\iint\limits_{\partial W} (0, 0, F_3) \cdot N \, \mathrm{d}A = \iiint\limits_{W} \frac{\partial F_3}{\partial z} \, \mathrm{d}V.$$

We shall prove just identity [iii]. The other two are proved similarly. Since W is a simple region it is in particular x, y simple in $\mathbb{R}^3$. Therefore there exists a simple region T in $\mathbb{R}^2$ and continuous functions $\lambda: T \subseteq \mathbb{R}^2 \to \mathbb{R}$ and $\tau: T \subseteq \mathbb{R}^2 \to \mathbb{R}$ such that

$$\lambda(x, y) < \tau(x, y) \qquad \text{for all } (x, y) \in T \backslash \partial T$$

and

$$W = \{(x, y, z) \in \mathbb{R}^3 \mid (x, y) \in T, \lambda(x, y) \leqslant z \leqslant \tau(x, y)\}.$$

See Fig. 10.8.

The boundary ∂W of W consists of a top (the graph of τ, a surface denoted by K_2) a bottom (the graph of λ, a surface denoted by K_1) and possibly a vertical collar V between them.

Since K_1 is a surface (Definition 8.5.1) there are simple regions $S_1, \ldots, S_k$ in $\mathbb{R}^2$ whose interiors are pairwise disjoint, and a C^1 function $\rho: S \subseteq \mathbb{R}^2 \to \mathbb{R}^3$ defined on $S = S_1 \cup \cdots \cup S_k$, such that ρ is 1–1 and smooth on

$$U = (\text{Int } S_1) \cup \cdots \cup (\text{Int } S_k),$$

and $\rho(S) = K_1$.

Since K_1 is the graph of λ it follows that

10.5.3 $\quad \rho_3(u, v) = \lambda(\rho_1(u, v), \rho_2(u, v)) \qquad$ for all $(u, v) \in S$.

We may choose ρ so that the normal function ν_ρ matches the positively orienting unit normal vector field N through

$$N(\rho(u, v)) = \frac{\nu_\rho(u, v)}{\|\nu_\rho(u, v)\|} \qquad \text{for all } (u, v) \in U.$$

Since K_1 is the graph of the bottom surface, the third coordinate of $N(\rho(u, v))$ is non-positive. Recall that

$$\nu_\rho = \left(\frac{\partial(\rho_2, \rho_3)}{\partial(u, v)}, \frac{\partial(\rho_3, \rho_1)}{\partial(u, v)}, \frac{\partial(\rho_1, \rho_2)}{\partial(u, v)} \right).$$

Hence for any C^1 vector field $F: W \subseteq \mathbb{R}^3 \to \mathbb{R}^3$

$$\iint_{K_1} (0, 0, F_3) \cdot N \, dA = - \iint_S F_3(\rho(u, v)) \left| \frac{\partial(\rho_1, \rho_2)}{\partial(u, v)}(u, v) \right| du \, dv$$

$$= - \iint_S F_3(\rho_1(u, v), \rho_2(u, v), \lambda(\rho_1(u, v), \rho_2(u, v)))$$

$$\times \left| \frac{\partial(\rho_1, \rho_2)}{\partial(u, v)}(u, v) \right| du \, dv$$

$$= - \iint_T F_3(x, y, \lambda(x, y)) \, dx \, dy,$$

since $x = \rho_1(u, v)$, $y = \rho_2(u, v)$ defines a C^1 function from S onto T which satisfies the conditions of the Change of Variables Theorem (Theorem 7.7.5).

Similarly

$$\iint_{K_2} (0, 0, F_3) \cdot N \, dA = \iint_T F_3(x, y, \tau(x, y)) \, dx \, dy.$$

On the other hand, since $(0, 0, F_3)$ is tangential to the vertical collar V,

$$\iint_V (0, 0, F_3) \cdot N \, dA = 0.$$

It follows that

$$\iint_{\partial W} (0, 0, F_3) \cdot N \, dA = \iint_{K_1} (0, 0, F_3) \cdot N \, dA + \iint_V (0, 0, F_3) \cdot N \, dA$$

$$+ \iint_{K_2} (0, 0, F_3) \cdot N \, dA$$

$$= \iint_T (F_3(x, y, \tau(x, y)) - F_3(x, y, \lambda(x, y))) \, dx \, dy$$

$$= \iint_T dx \, dy \int_{\lambda(x, y)}^{\tau(x, y)} \frac{\partial F_3}{\partial z} \, dz = \iiint_W \frac{\partial F_3}{\partial z} \, dV.$$

This establishes identity [iii].

10.5.4 *Example*. Let W be the spherical region $x^2 + y^2 + z^2 \leqslant c^2$ in $\mathbb{R}^3$ and let $(\partial W, N)$ be the positively oriented boundary of W. Consider the vector field $F : W \subseteq \mathbb{R}^3 \to \mathbb{R}^3$ defined by

$$F(x, y, z) = (0, 0, z), \qquad (x, y, z) \in \mathbb{R}^3.$$

Then div $F = 1$, and the Divergence Theorem implies that

10.5.5
$$\iiint\limits_{W} 1 \, dV = \iint\limits_{(\partial W, N)} F \cdot d\mathbf{A}.$$

The left-hand side of 10.5.5 is the volume of W. The right-hand side can be evaluated by taking the parametrization $\rho : [0, \pi] \times [0, 2\pi] \subseteq \mathbb{R}^2 \to \mathbb{R}^3$ defined by

$$\rho(\phi, \theta) = (c \sin \phi \cos \theta, c \sin \phi \sin \theta, c \cos \phi), \quad 0 \leqslant \phi \leqslant \pi, \quad 0 \leqslant \theta \leqslant 2\pi.$$

We find

$$\iint\limits_{(\partial W, N)} F \cdot d\mathbf{A} = c^3 \int_0^\pi d\phi \int_0^{2\pi} \sin \phi \cos^2 \phi \, d\theta = \tfrac{4}{3}\pi c^3.$$

This method of finding the volume of a simple region is developed in Exercise 10.5.4. It corresponds to the technique for finding the area enclosed by a simple closed curve established in Exercise 7.5.7.

The Divergence Theorem enables us to give an alternative definition of the divergence of a vector field F at a point in terms of the surface integrals of F over small spheres containing the point. See Exercise 10.5.6.

As in the case of Stoke's Theorem we can use the Divergence Theorem for simple regions to prove a corresponding theorem for composite regions in $\mathbb{R}^3$. A solid torus W for example can be cut up into four simple regions W_i, $i = 1, 2, 3, 4$, as shown in Fig. 10.9.

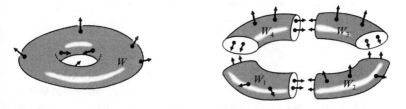

Fig. 10.9 Positively oriented boundary of solid torus and its decomposition

The positive orientation N of ∂W and of ∂W_i are indicated in the figure. Now

$$\iint\limits_{(\partial W, N)} F \cdot \mathrm{d}\mathbf{A} = \iint\limits_{\partial W_1} F \cdot N \, \mathrm{d}A + \iint\limits_{\partial W_2} F \cdot N \, \mathrm{d}A + \iint\limits_{\partial W_3} F \cdot N \, \mathrm{d}A$$

$$+ \iint\limits_{\partial W_4} F \cdot N \, \mathrm{d}A.$$

Where the simple regions join, the normals point in opposite directions. It follows that Expression 10.5.2 is also true of the torus W with positively oriented boundary ∂W. We could, for example, find the volume of the torus by using the technique of Example 10.5.4.

The general treatment of the Divergence Theorem for such composite regions is rather cumbersome and is best left to the intuition of the reader. The following important result provides an illustration of how it works.

10.5.6 Lemma (Gauss).   *Let W be a simple region in $\mathbb{R}^3$ with positively oriented boundary $(\partial W, N)$. Suppose that $\mathbf{0} \notin \partial W$. Let F be the vector field defined by*

$$F(\mathbf{r}) = \frac{\mathbf{r}}{r^3} \qquad \text{for all } \mathbf{r} \neq \mathbf{0},$$

where as usual $\mathbf{r} = (x, y, z)$ and $r = \|\mathbf{r}\|$. Then

$$\iint\limits_{(\partial W, N)} F \cdot \mathrm{d}\mathbf{A} = \begin{cases} 4\pi & \text{if} \quad \mathbf{0} \in W \\ 0 & \text{if} \quad \mathbf{0} \notin W. \end{cases}$$

Proof.   By Example 4.11.4, div $F(\mathbf{r}) = 0$, for all $\mathbf{r} \neq \mathbf{0}$. Hence, if $\mathbf{0} \notin W$ we obtain from the Divergence Theorem that

$$\iint\limits_{(\partial W, N)} F \cdot \mathrm{d}\mathbf{A} = \iiint\limits_{W} 0 \, \mathrm{d}V = 0.$$

If $\mathbf{0} \in W$ the Divergence Theorem cannot be applied directly to W since the vector field involved cannot be made continuous let alone C^1 at $\mathbf{0}$. Instead we apply it to the set $W^* = W \backslash U_\varepsilon$ where U_ε is a ball centre the origin and radius ε which is chosen to lie in W. The positively oriented boundary $(\partial W^*, N)$ is the disjoint union of

$(\partial W, N)$ and $(\partial U_\varepsilon, N)$ where the unit normals on ∂U_ε point towards **0**. Hence for $\mathbf{r} \in \partial U_\varepsilon$, $N(\mathbf{r}) = -\mathbf{r}/r$, and therefore

10.5.7 $\quad (F \cdot N)(\mathbf{r}) = -\dfrac{\mathbf{r} \cdot \mathbf{r}}{r^4} = -\dfrac{1}{r^2} = -\dfrac{1}{\varepsilon^2} \qquad$ for all $\mathbf{r} \in \partial U_\varepsilon$.

By the (extended form of) the Divergence Theorem,

$$\iint\limits_{(\partial W, N)} F \cdot d\mathbf{A} + \int_{(\partial U_\varepsilon, N)} F \cdot d\mathbf{A} = \iint\limits_{(\partial W^*, N)} F \cdot d\mathbf{A} = \iiint\limits_{W^*} 0 \, dV = 0.$$

It follows by 10.5.7 that

$$\iint\limits_{(\partial W, N)} F \cdot d\mathbf{A} = -\iint\limits_{\partial U_\varepsilon} F \cdot N \, dA = \iint\limits_{\partial U_\varepsilon} \frac{1}{\varepsilon^2} \, dA = 4\pi.$$

An important application of Lemma 10.5.6 is in Electric Field Theory, which we now briefly discuss.

10.5.8 *Example.* *Electric Fields and Potentials.* In Exercise 9.3.5 we observed **[i]** that the electric field due to a point charge e at the origin is given by

$$E(\mathbf{r}) = -\frac{e}{r^3}\mathbf{r} \qquad \mathbf{r} \neq \mathbf{0},$$

and **[ii]** that

$$E = \operatorname{grad} f, \qquad \text{where } f(\mathbf{r}) = \frac{e}{r}.$$

This means that E is a conservative field.

Let W be a simple region in $\mathbb{R}^3$ with positively oriented boundary $(\partial W, N)$. Then by Lemma 10.5.6

10.5.9 $\qquad \displaystyle\iint\limits_{(\partial W, N)} E \cdot d\mathbf{A} = \begin{cases} -4\pi e & \text{if the charge is inside } W, \\ 0 & \text{if the charge is outside } W. \end{cases}$

Formula 10.5.9 is also true when E is the electric field due to a point charge e placed anywhere in $\mathbb{R}^3 \backslash \partial W$. (Simply translate the coordinate axes.)

Hence if E is the electric field due to a finite number of point charges placed at various points in space (none of them on ∂W),

10.5.10 $\qquad \displaystyle\iint\limits_{(\partial W, N)} E \cdot d\mathbf{A} = -4\pi \text{ (total charge inside } W\text{)}.$

Consider now the electric field E due to a continuous charge distribution ρ in space. What can be said about E? It is reasonable to assume that E is conservative. Let its potential function be f, so

10.5.11 $E = \text{grad} f.$

Consider a fixed point $\mathbf{p} \in \mathbb{R}^3$ and let $U_\varepsilon = N(\mathbf{p}, \varepsilon)$. By a reasonable generalization of 10.5.10, we have

$$\iint\limits_{(\partial U_\varepsilon, N)} E \cdot d\mathbf{A} = -4\pi \text{ (charge inside } U_\varepsilon) = -4\pi \iiint\limits_{U_\varepsilon} \rho \, dV.$$

By the Divergence Theorem, it follows that

$$\iiint\limits_{U_\varepsilon} \text{div} \, E \, dV = -4\pi \iiint\limits_{U_\varepsilon} \rho \, dV.$$

Therefore, by a Mean-Value Theorem, there exist $\mathbf{p}_1$ and $\mathbf{p}_2$ in U_ε such that

$$(\text{Volume } U_\varepsilon) \times (\text{div} \, E)(\mathbf{p}_1) = -4\pi(\text{Volume } U_\varepsilon) \times \rho(\mathbf{p}_2),$$

that is

$$(\text{div} \, E)(\mathbf{p}_1) = -4\pi\rho(\mathbf{p}_2).$$

Letting $\varepsilon \to 0$, we obtain

$$(\text{div} \, E)(\mathbf{p}) = -4\pi\rho(\mathbf{p}).$$

But by 10.5.11

$$\text{div} \, E = \text{div grad} \, f = \nabla^2 f.$$

Hence we conclude that the electric potential function f due to a charge distribution ρ will satisfy the equation

10.5.12 $\nabla^2 f = -4\pi\rho.$

This is the important *Poisson's Equation*.

Exercises 10.5

1.   Verify the Divergence Theorem

$$\iint\limits_{(\partial W, N)} F \cdot d\mathbf{A} = \iiint\limits_{W} \text{div} \, F \, dV,$$

where $F(x, y, z) = (x^2, y^2, z^2)$, and W is the subset of $\mathbb{R}^3$ bounded by the surfaces:

(a)   the cylinder $x^2 + y^2 = k^2$ and the planes $z = h$ and $z = -h$;
(b)   the circular cone $z = \sqrt{(x^2 + y^2)}$ and the plane $z = h$, where $h > 0$;
(c)   the paraboloid $z = 2 - (x^2 + y^2)$ and the plane $z = 0$;

(d)   the planes $2x + y + 2z = 16$, $x = 0$, $y = 0$, $z = 0$, and $x + y = 1$;
(e)   the coordinate planes $x = 0$, $y = 0$, $z = 0$, and part of the
ellipsoid $x^2/a^2 + y^2/b^2 + z^2/c^2 = 1$ with $x \geq 0$, $y \geq 0$, $z \geq 0$.

Answers: (a) 0; note that $F(-x, -y, -z) = F(x, y, z)$, hence surface
integral is zero; (b) $\frac{1}{2}\pi h^4$; (c) $-4\pi/3$; (d) $31\frac{43}{48}$; (e) $\pi abc(a + b + c)/8$.

2.   Using the Divergence Theorem, obtain short solutions to Exercises 2,
3(d) and 5 of Section 9.3.

3.   Find by the Divergence Theorem the integral of the vector field
$F(\mathbf{r}) = \mathbf{r}/r$ over the surface of the sphere, centre the origin, radius c.
Interpret your answer.

Answer: $4\pi c^2$; surface area of sphere.

4.   Let W be a simple region in $\mathbb{R}^3$ and let $F(\mathbf{r}) = \mathbf{r}$ for all $\mathbf{r} \in \mathbb{R}^3$. Prove
that

$$\text{volume of } W = \tfrac{1}{3} \iint\limits_{(\partial W, N)} F \cdot d\mathbf{A},$$

where $(\partial W, N)$ is the positively oriented boundary of W.

5.   Let F be the vector field $F(x, y, z) = (x^2 y, y^2 z, z^2 x)$, and let B be a
ball in $\mathbb{R}^3$, with positively oriented boundary $(\partial B, N)$. Prove that

$$\iint\limits_{(\partial B, N)} \text{curl } F \cdot d\mathbf{A} = 0.$$

Hint: use the Divergence Theorem.

6.   Let $F : D \subseteq \mathbb{R}^3 \to \mathbb{R}^3$ be a C^1 vector field defined on an open set D in
$\mathbb{R}^3$. Given $\mathbf{p} \in D$, prove that

$$(\text{div } F)(\mathbf{p}) = \lim_{\varepsilon \to 0} \frac{1}{\text{Volume } U_\varepsilon} \iint\limits_{(\partial U_\varepsilon, N)} F \cdot d\mathbf{A},$$

where U_ε denotes an ε-neighbourhood of $\mathbf{p}$ in D, and ∂U_ε is oriented
by outward facing normals. If we accept that Volume U_ε is
independent of parametrization then this expression gives a
coordinate-free definition of div F. For the corresponding two-
dimensional expression see 7.6.18. Given that F is the velocity field of
a fluid, interpret the divergence of F at $\mathbf{p}$ as the rate of change of the
density of the fluid at $\mathbf{p}$ (assuming that there are no 'sources' or 'sinks'
where material is created or destroyed).

7.   Let W be a simple region in $\mathbb{R}^3$ with positively oriented boundary
$(\partial W, N)$. If $T : W \subseteq \mathbb{R}^3 \to \mathbb{R}$ is twice continuously differentiable,

prove, by the Divergence Theorem, that

$$\iint_{(\partial W, N)} \text{grad } T \cdot d\mathbf{A} = \iiint_W \nabla^2 T \, dV.$$

Hence obtain a short solution of Exercise 9.3.4.

8. *Green's Integral Theorems.* Let W be a simple region in $\mathbb{R}^3$ with positively oriented boundary $(\partial W, N)$. Let $f : W \subseteq \mathbb{R}^3 \to \mathbb{R}$ and $g : W \subseteq \mathbb{R}^3 \to \mathbb{R}$ be C^2 functions. Prove that

    (a) $\iint_{(\partial W, N)} f \nabla g \cdot d\mathbf{A} = \iiint_W (f \nabla^2 g + \nabla f \cdot \nabla g) \, dV;$

    (b) $\iint_{(\partial W, N)} (f \nabla g - g \nabla f) \cdot d\mathbf{A} = \iiint_W (f \nabla^2 g - g \nabla^2 f) \, dV.$

    Deduce from (a) that if f is *harmonic*, that is f satisfies Laplace's Equation $\nabla^2 f = 0$ in W, and if $f = 0$ on ∂W, then $f = 0$ throughout W.

Hint: put $f = g$ in (a).

9. *The Heat Equation.* Consider the heat flow in a thermally isotropic body W (Exercise 9.3.4). Let $U_\varepsilon = N(\mathbf{p}, \varepsilon) \subseteq W$ be a neighbourhood of a fixed point $\mathbf{p}$ in W. If with the passage of time the region U_ε suffers a net heat loss (gain) then the average temperature over U_ε will fall (rise). We must therefore regard the temperature T as a function of time as well as of position. The rate of heat loss over U_ε at time t_0 is given by the value of the integral

$$\iiint_{U_\varepsilon} -s\rho \frac{\partial T}{\partial t} \, dV$$

at $t = t_0$, where s is the specific heat of the material of the body and ρ is its density.

By Exercise 9.3.4

$$\iiint_{U_\varepsilon} -s\rho \frac{\partial T}{\partial t} \, dV = \iint_{(\partial U_\varepsilon, N)} -k \text{ grad } T \cdot d\mathbf{A}.$$

Deduce, by the Divergence Theorem, that under suitable differentiability conditions the temperature function T satisfies the *Heat Equation*

$$\frac{\partial T}{\partial t} = \frac{k}{s\rho} \nabla^2 T.$$

10. (a) Let K be a simple surface which intersects each line from the origin in at most one point. (Fig. 10.10). The solid cone of lines which meet K intersects the unit sphere centre the origin in a surface K^*.

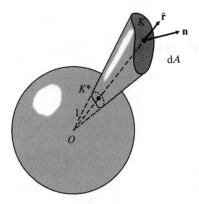

Fig. 10.10 Solid angle subtended by K

The area of K^* is called the *solid angle* subtended by K at the origin. Prove that the solid angle is given by

$$\iint\limits_{K} \frac{\mathbf{r} \cdot \mathbf{n}}{r^3}\, dA,$$

where $r = \|\mathbf{r}\|$ and K is oriented by the normals $\mathbf{n}$ making an acute angle with the position vectors $\mathbf{r}$.

Hint: apply the Divergence Theorem to the volume in the cone between K and K^*. The result is seen to be plausible by observing (i) that the area element dA on K subtends the same solid angle at $\mathbf{0}$ as its projection $\mathbf{r} \cdot \mathbf{n}\, dA$ onto the sphere of radius r, and (ii) that $(\mathbf{r} \cdot \mathbf{n}\, dA)/r^3 \approx dA^*$, since the area on a sphere cut off by a cone from the origin varies with the square of the radius.

(b) Prove that the solid angle subtended by a sphere at any point inside the sphere is 4π.

Hint: see Gauss' Lemma 10.5.6.

11. Verify the result of Example 9.3.13 by the Divergence Theorem.

Differential forms

11.1   Introduction

We have now completed the task begun in Chapter 5 of proving three major theorems of the integral vector calculus:

(1) the *Fundamental Theorem* for oriented simple arcs

$$\int_{C^+} \operatorname{grad} f \cdot d\mathbf{r} = f(\mathbf{q}) - f(\mathbf{p}),$$

where C^+ is a simple arc in $\mathbb{R}^n$ oriented from endpoint $\mathbf{p}$ to endpoint $\mathbf{q}$;

(2) *Stokes' Theorem* for oriented simple surfaces in $\mathbb{R}^3$

$$\iint_{(K,N)} \operatorname{curl} F \cdot d\mathbf{A} = \int_{\partial(K,N)} F \cdot d\mathbf{r};$$

(3) *Gauss' Divergence Theorem* for simple regions in $\mathbb{R}^3$

$$\iiint_W \operatorname{div} F \, dV = \iint_{(\partial W, N)} F \cdot d\mathbf{A}.$$

As a special case of Stokes' Theorem we have

(2′) *Green's Theorem* for simple regions in $\mathbb{R}^2$

$$\iint_S \left(\frac{\partial F_2}{\partial x} - \frac{\partial F_1}{\partial y} \right) dx \, dy = \int_{\partial S^+} F \cdot d\mathbf{r}.$$

It is remarkable that these classical theorems are all special cases of a far reaching generalization, the General Stokes' Theorem. In this chapter we shall give a brief introduction to this theorem.

The above four theorems have one common feature which suggests that they may be related—for suitable positive integer n each theorem expresses the equality of an n-fold integral of a vector field over a region and an $(n-1)$-fold integral of a related vector field over the boundary of the region. Of course the regions of integration and their boundaries have to be oriented, except when the vector field to be integrated is at the same time a scalar field.

In order to explain the General Stokes' Theorem we need the concept of differential forms. These generalize the differential of a real-valued function $f : D \subseteq \mathbb{R}^m \to \mathbb{R}$ which played such an important role in the earlier chapters of this book. We shall give formal functional meaning to the 'differentials' df, dx, dy, dz and also to their products such as $dx\, dy$, $dy\, dz$, ... paying due attention to problems of orientation, as the following example illustrates.

11.1.1 *Example.* Green's Theorem 7.5.14 was established as the combination of the two results

11.1.2
$$\int_{\partial S^+} F_1\, dx = -\iint_S \frac{\partial F_1}{\partial y}\, dx\, dy$$

and

11.1.3
$$\int_{\partial S^+} F_2\, dy = \iint_S \frac{\partial F_2}{\partial x}\, dx\, dy.$$

The asymmetry between these two expressions is related to the asymmetrical positioning of the x- and y-axes in the plane. In the equality

11.1.4
$$\int_{\partial S^+} f\, dx = -\iint_S \frac{\partial f}{\partial y}\, dx\, dy$$

a formal interchange of dx, dy, and $\partial/\partial x$, $\partial/\partial y$ leads to the formal equality

11.1.5
$$\int_{\partial S^+} f\, dy = \iint_S -\frac{\partial f}{\partial x}\, dy\, dx.$$

By 11.1.3 this is a true equality *provided that* $dy\, dx$ *is identified with* $-dx\, dy$. The theory of differential forms is designed to take this into account.

11.2   1-forms in $\mathbb{R}^m$

In Section 3.3 we saw how the differentiability of a real-valued function $f : D \subseteq \mathbb{R}^m \to \mathbb{R}$ at a point $\mathbf{p} \in D$ depended on approximating the difference function $\delta_{f,\mathbf{p}}$ by a linear function. We found that if f is differentiable at $\mathbf{p}$ then the approximating linear function $L_{f,\mathbf{p}} : \mathbb{R}^m \to \mathbb{R}$ is given by

11.2.1 $L_{f,\mathbf{p}}(\mathbf{h}) = \dfrac{\partial f}{\partial x_1}(\mathbf{p})h_1 + \cdots + \dfrac{\partial f}{\partial x_m}(\mathbf{p})h_m,$ $\mathbf{h} \in \mathbb{R}^m.$

So for a differentiable function $f : D \subseteq \mathbb{R}^m \to \mathbb{R}$ we have a *class* of linear functions $L_{f,\mathbf{p}}$ (approximating the difference functions), one for each $\mathbf{p} \in D$. We denote this class by

11.2.2 $df = \{ L_{f,\mathbf{p}} : \mathbb{R}^m \to \mathbb{R} \mid \mathbf{p} \in D \}$

and the individual member $L_{f,\mathbf{p}}$ by

11.2.3 $(df)_\mathbf{p} = L_{f,\mathbf{p}}.$

The class df is called the *differential* of f, and the function $(df)_\mathbf{p}$ is called the *differential of f at* $\mathbf{p}$. This agrees with Definition 3.3.15.

   The following definition of a 1-form on D in $\mathbb{R}^m$ concerns classes of *linear* functions, but omits reference to differentiability. For later use we first define a 0-form on D.

11.2.4 Definition.   *Let D be an open set in $\mathbb{R}^m$. A 0-form on D is a function $f : D \subseteq \mathbb{R}^m \to \mathbb{R}$.*

11.2.5 Definition.   *Let D be an open set in $\mathbb{R}^m$. A 1-form on D is a class of real-valued functions*

$$\omega = \{ \omega_\mathbf{p} : \mathbb{R}^m \to \mathbb{R} \mid \mathbf{p} \in D \}$$

such that the function $\omega_\mathbf{p}$ is linear for each $\mathbf{p} \in D$.

   Let ω be a 1-form on D in $\mathbb{R}^m$. For each $\mathbf{p} \in D$ the linear function $\omega_\mathbf{p} : \mathbb{R}^m \to \mathbb{R}$ will satisfy a linear relation of the form

11.2.6 $\omega_p(\mathbf{h}) = W_1(\mathbf{p}) h_1 + \cdots + W_m(\mathbf{p}) h_m,$ $\mathbf{h} \in \mathbb{R}^m,$ $\mathbf{p} \in D.$

In this way for each $\mathbf{p} \in D$ the linear function $\omega_\mathbf{p}$ is associated with a vector $W(\mathbf{p}) = (W_1(\mathbf{p}), \ldots, W_m(\mathbf{p}))$ in $\mathbb{R}^m$. Hence the 1-form is associated with the vector field $W : D \subseteq \mathbb{R}^m \to \mathbb{R}^m$ through the relation

11.2.7 $\omega_\mathbf{p}(\mathbf{h}) = W(\mathbf{p}) \cdot \mathbf{h}$ $\mathbf{h} \in \mathbb{R}^m, \quad \mathbf{p} \in D.$

Conversely, to any vector field $W : D \subseteq \mathbb{R}^m \to \mathbb{R}^m$ there corresponds through 11.2.6 a 1-form ω on D. So we have the following result.

11.2.8 Theorem.   *The relation 11.2.7 establishes a 1–1 correspondence between 1-forms ω on $D \subseteq \mathbb{R}^m$ and vector fields $W : D \subseteq \mathbb{R}^m \to \mathbb{R}^m.$*

11.2.9 *Notation.* We shall denote 1–1 correspondences by a double arrow $\leftrightarrow$. Thus in Theorem 11.2.8

$$\omega \leftrightarrow W.$$

A 1-form ω is said to be continuous (differentiable, C^1, C^2, ...) if the corresponding vector field W is continuous (differentiable, C^1, C^2, ...).

11.2.10 *Theorem.* *Let* $f : D \subseteq \mathbb{R}^m \to \mathbb{R}$ *be a differentiable function* (0-*form on* D). *Then the differential* df *defined in* 11.2.2 *is a* 1-*form on* D, *and*

$$\mathrm{d}f \leftrightarrow \mathrm{grad}\, f.$$

Proof.   Immediate from 11.2.3 and 11.2.1.

Let ω and γ be 1-forms and let f be a 0-form on D in $\mathbb{R}^m$. The *sum* $\omega + \gamma$ is defined by

11.2.11 $(\omega + \gamma)_{\mathbf{p}} = \omega_{\mathbf{p}} + \gamma_{\mathbf{p}},$ $\mathbf{p} \in D$

and the *product* $f\omega$ is defined by

11.2.12 $(f\omega)_{\mathbf{p}} = f(\mathbf{p})\omega_{\mathbf{p}}$ $\mathbf{p} \in D.$

Clearly $\omega + \gamma$ and $f\omega$ are again 1-forms on D, and properties of continuity and differentiability are inherited. If $\omega \leftrightarrow W$ and $\gamma \leftrightarrow G$ then $\omega + \gamma \leftrightarrow W + G$ and $f\omega \leftrightarrow fW$.

11.2.13 *Example.* Define the *projection* $x_i : \mathbb{R}^m \to \mathbb{R}$ by

$$x_i(p_1, \ldots, p_m) = p_i, \qquad (p_1, \ldots, p_m) \in \mathbb{R}^m,$$

so the function x_i singles out the *ith* coordinate of a vector $\mathbf{p}$ in $\mathbb{R}^m$. In this notation, for all $\mathbf{p} \in \mathbb{R}^m$

$$\frac{\partial x_i}{\partial x_j}(\mathbf{p}) = \begin{cases} 1 & \text{when } i = j \\ 0 & \text{when } i \neq j. \end{cases}$$

Hence by 11.2.3 and 11.2.1 the differential dx_i satisfies

11.2.14 $(\mathrm{d}x_i)_{\mathbf{p}}(\mathbf{h}) = h_i,$ $\mathbf{h} \in \mathbb{R}^m.$

For each $\mathbf{p} \in \mathbb{R}^m$ the differential of x_i at $\mathbf{p}$ acts as a projection on $\mathbb{R}^m$ given by 11.2.14. Notice that the result is independent of the choice of $\mathbf{p}$ in $\mathbb{R}^m$. In the notation

$$\mathbf{h} = (\Delta x_1, \ldots, \Delta x_m),$$

where Δx_i denotes an increment in the ith coordinate, 11.2.14 becomes

$$(dx_i)_{\mathbf{p}}(\mathbf{h}) = \Delta x_i.$$

Let ω be a 1-form on D in $\mathbb{R}^m$. If $\omega \leftrightarrow W$ then by 11.2.6 and 11.2.14 we have

$$\omega_{\mathbf{p}}(\mathbf{h}) = W_1(\mathbf{p})(dx_1)_{\mathbf{p}}(\mathbf{h}) + \cdots + W_m(\mathbf{p})(dx_m)_{\mathbf{p}}(\mathbf{h})$$

for all $\mathbf{h} \in \mathbb{R}^m$, $\mathbf{p} \in D$. Hence we obtain the equality of 1-forms

11.2.15 $\qquad \omega = W_1\, dx_1 + \cdots + W_m\, dx_m.$

11.2.16 Theorem. *A 1-form ω on D in $\mathbb{R}^m$ is uniquely expressible as the sum of 1-forms 11.2.15, where $\omega \leftrightarrow W$.*

Proof. Immediate.

We see from the expression 11.2.15 that every 1-form ω in $\mathbb{R}^m$ is expressible as a sum of forms involving the differentials $dx_1, \ldots, dx_m$. For this reason ω is called a *differential 1-form*. We shall continue to use the shorter description of ω.

11.2.17 Example. Let $f : D \subseteq \mathbb{R}^m \to \mathbb{R}$ be a differentiable function (0-form). Since $df \leftrightarrow \operatorname{grad} f$ we have the equality of 1-forms

$$df = \frac{\partial f}{\partial x_1}\, dx_1 + \cdots + \frac{\partial f}{\partial x_m}\, dx_m.$$

This commonly used formal expression has now been given a well defined meaning in terms of 1-forms.

11.2.18 Example. The expressions

$$(x + 4z)\, dx - 2xy^2z\, dy + (1 - e^x)\, dz$$

and

$$2xy^3z\, dx + 3x^2y^2z\, dy + x^2y^3\, dz$$

can be interpreted as 1-forms on $\mathbb{R}^3$. The second expression is the differential of the 0-form x^2y^3z, but the first is not the differential of a 0-form. A 1-form is said to be *exact* if it is the differential of a 0-form.

Exercises 11.2

1. For the following differential 1-forms ω on $\mathbb{R}^3$ evaluate $\omega_{\mathbf{p}}(\mathbf{h})$, where $\mathbf{p} = (a, b, c)$ and (a) $\mathbf{h} = (0.1, -0.1, 0.2)$; (b) $\mathbf{h} = (0.1, 0.2, -0.1)$;

  (c) $\mathbf{h} = (0.2, 0.1, -0.1)$:
    (i) $xy\,dx + (x - z)\,dy + 2xyz^2\,dz$,
    (ii) $2\,dx - 3\,dy + 4\,dz$.

Answers: (i) (a) $0.1ab - 0.1(a - c) + 0.4abc^2$, (b) $0.1ab + 0.2(a - c) - 0.2abc^2$; (ii) (c) -0.3, independent of a, b, c.

2.   Which of the following 1-forms are exact?
   (a) $yz\,dx + xz\,dy + xy\,dz$; (b) $xy\,dx + yz\,dy + zx\,dz$;
   (c) $(2xyz + z)\,dx + (x^2z + 1)\,dy + (x^2y + x)\,dz$.

Answers: (a) and (c) are exact, (b) is not. See Example 5.5.22 and Exercises 5.5. Notice that if a continuous 1-form $F_1\,dx + F_2\,dy + F_3\,dz$ is exact then the field F is conservative.

11.3 2-forms in $\mathbb{R}^m$

In defining a differential 2-form we shall take due regard of the comment made in Example 11.1.1, that in order to take account of the orientation of the x- and y-axes in the plane, the expression $dy\,dx$ should be identified with $-dx\,dy$. We shall presently define a product of two 1-forms which displays this 'skew-symmetry'.

11.3.1 Definition. *Let V be a vector space over $\mathbb{R}$. A function $f: V \times V \to \mathbb{R}$ is said to be* bilinear *(on $V \times V$) if it is linear in both its arguments, that is, if*

$$f(k\mathbf{u} + l\mathbf{v}, \mathbf{w}) = kf(\mathbf{u}, \mathbf{w}) + lf(\mathbf{v}, \mathbf{w})$$

and

$$f(\mathbf{u}, k\mathbf{v} + l\mathbf{w}) = kf(\mathbf{u}, \mathbf{v}) + lf(\mathbf{u}, \mathbf{w})$$

for all vectors $\mathbf{u}$, $\mathbf{v}$, $\mathbf{w} \in V$ and scalars k, $l \in \mathbb{R}$.

11.3.2 Definition. *Let $g: V \times V \to \mathbb{R}$ be a real-valued function. Then g is said to be* skew-symmetric *(on $V \times V$) if*

$$g(\mathbf{h}, \mathbf{k}) = -g(\mathbf{k}, \mathbf{h}) \qquad \text{for all } \mathbf{h}, \mathbf{k} \in V.$$

11.3.3 Example. For a fixed $a \in \mathbb{R}$ define the function $g: \mathbb{R}^2 \times \mathbb{R}^2 \to \mathbb{R}$ by

$$g(\mathbf{h}, \mathbf{k}) = a \det \begin{bmatrix} h_1 & h_2 \\ k_1 & k_2 \end{bmatrix} = a(h_1k_2 - h_2k_1).$$

Then g is bilinear and skew-symmetric on $\mathbb{R}^2 \times \mathbb{R}^2$.

11.3.4 *Example.*   For fixed a, b, $c \in \mathbb{R}$ define the function $g : \mathbb{R}^3 \times \mathbb{R}^3 \to \mathbb{R}$ by

$$g(\mathbf{h}, \mathbf{k}) = \det \begin{bmatrix} a & b & c \\ h_1 & h_2 & h_3 \\ k_1 & k_2 & k_3 \end{bmatrix}.$$

Then g is bilinear and skew-symmetric on $\mathbb{R}^3 \times \mathbb{R}^3$.

11.3.5 *Theorem.*   *The most general bilinear and skew-symmetric real-valued functions on $\mathbb{R}^2 \times \mathbb{R}^2$ and on $\mathbb{R}^3 \times \mathbb{R}^3$ are given in Examples 11.3.3 and 11.3.4.*

Proof.   See Exercise 11.3.1.

11.3.6 *Definition.*   *Let D be an open set in $\mathbb{R}^m$. A 2-form on D is a class of real-valued functions*

$$\phi = \{\phi_{\mathbf{p}} : \mathbb{R}^m \times \mathbb{R}^m \to \mathbb{R} \mid \mathbf{p} \in D\}$$

such that for each $\mathbf{p} \in D$ the function $\phi_{\mathbf{p}}$ is bilinear and skew-symmetric.

We now specialize to 2-forms in $\mathbb{R}^2$ and in $\mathbb{R}^3$. First let ϕ be a 2-form on D in $\mathbb{R}^2$. By Theorem 11.3.5, for each $\mathbf{p} \in D$ the bilinear function $\phi_{\mathbf{p}} : \mathbb{R}^2 \times \mathbb{R}^2 \to \mathbb{R}$ satisfies a bilinear relation of the form

11.3.7 $\phi_{\mathbf{p}}(\mathbf{h}, \mathbf{k}) = f(\mathbf{p})(h_1 k_2 - h_2 k_1),$ $\mathbf{h}, \mathbf{k} \in \mathbb{R}^2, \quad \mathbf{p} \in D.$

Next let ϕ be a 2-form on D in $\mathbb{R}^3$. By the same theorem, for each $\mathbf{p} \in D$ the function $\phi_{\mathbf{p}} : \mathbb{R}^3 \times \mathbb{R}^3 \to \mathbb{R}$ satisfies a relation

11.3.8 $\phi_{\mathbf{p}}(\mathbf{h}, \mathbf{k}) = \det \begin{bmatrix} F_1(\mathbf{p}) & F_2(\mathbf{p}) & F_3(\mathbf{p}) \\ h_1 & h_2 & h_3 \\ k_1 & k_2 & k_3 \end{bmatrix},$ $\mathbf{h}, \mathbf{k} \in \mathbb{R}^3, \quad \mathbf{p} \in D$

$= F(\mathbf{p}) \cdot \mathbf{h} \times \mathbf{k}.$

So we have the following result.

11.3.9 *Theorem.*   *The relations 11.3.7 and 11.3.8 establish 1–1 correspondences between 2-forms ϕ on $D \subseteq \mathbb{R}^m$ and [i] scalar fields $f : D \subseteq \mathbb{R}^m \to \mathbb{R}$ when $m = 2$, [ii] vector fields $F : D \subseteq \mathbb{R}^m \to \mathbb{R}^m$ when $m = 3$.*

A 2-form is said to be continuous (differentiable, C^1, C^2, ...) if the corresponding field is continuous (differentiable, C^1, C^2, ...).

Let ϕ and ψ be 2-forms and let f be a 0-form on D in $\mathbb{R}^m$. The *sum* $\phi + \psi$ is defined by

$$(\phi + \psi)_\mathbf{p} = \phi_\mathbf{p} + \psi_\mathbf{p}, \qquad \text{for all } \mathbf{p} \in D,$$

and the *product* $f\phi$ is defined by

$$(f\phi)_\mathbf{p} = f(\mathbf{p})\phi_\mathbf{p} \qquad \text{for all } \mathbf{p} \in D.$$

Clearly $\phi + \psi$ and $f\phi$ are again 2-forms on D, and properties of continuity and differentiability are inherited.

The following construction shows how two 1-forms can be combined to give a 2-form.

11.3.12 Definition. *Let ω and γ be 1-forms on D in $\mathbb{R}^m$. The* wedge product $\omega \wedge \gamma$ *of ω and γ is defined by*

$$(\omega \wedge \gamma)_\mathbf{p}(\mathbf{h}, \mathbf{k}) = \omega_\mathbf{p}(\mathbf{h})\gamma_\mathbf{p}(\mathbf{k}) - \omega_\mathbf{p}(\mathbf{k})\gamma_\mathbf{p}(\mathbf{h})$$

for all $\mathbf{h}, \mathbf{k} \in \mathbb{R}^m$, $\mathbf{p} \in D$.

We denote by 0 a form on D such that $0_\mathbf{p}$ is the zero function for each $\mathbf{p} \in D$. Given a form ϕ on D, the negative $-\phi$ is a form on D defined in the obvious way.

11.3.13 Theorem. *Let ω and γ be 1-forms on D in $\mathbb{R}^m$. Then*
  [i]   *the wedge product $\omega \wedge \gamma$ is a 2-form on D;*
  [ii] $\omega \wedge \gamma = -\gamma \wedge \omega$;
  [iii] $\omega \wedge \omega = 0$.

Proof.   Immediate from Definition 11.3.12.

The wedge product is said to be *anti-commutative* by property [ii] of Theorem 11.3.13. We leave the reader to check the *distributive* properties

11.3.14 $$\begin{cases} (\omega + \delta) \wedge \gamma = \omega \wedge \gamma + \delta \wedge \gamma \\ \gamma \wedge (\omega + \delta) = \gamma \wedge \omega + \gamma \wedge \delta, \end{cases}$$

where ω, δ and γ are 1-forms on D.

11.3.15 Example.   The wedge product of the differentials dx and dy on $\mathbb{R}^2$

gives the 2-form $dx \wedge dy$. By Theorem 11.3.13

$$dy \wedge dx = -dx \wedge dy$$

and

$$dx \wedge dx = dy \wedge dy = 0.$$

11.3.16 *Example.* Let ω and γ be the 1-forms on $\mathbb{R}^4$ defined by

$$\omega = 2x_2 \, dx_1 + x_1 x_2 x_4 \, dx_2 \qquad\qquad + x_3 \, dx_4$$
$$\gamma = \quad dx_1 - \qquad\quad dx_2 + x_1 \, dx_3 + \quad dx_4.$$

Bearing in mind that $dx_i \wedge dx_i = 0$ and that $dx_i \wedge dx_j = -dx_j \wedge dx_i$ for all i, j, we obtain

$$\omega \wedge \gamma = (2x_2 \, dx_1 + x_1 x_2 x_4 \, dx_2 + x_3 \, dx_4) \wedge (dx_1 - dx_2 + x_1 \, dx_3 + dx_4)$$
$$= (2x_1 - x_1 x_2 x_4) \, dx_1 \wedge dx_2 + \ 2x_1 x_2 \, dx_1 \wedge dx_3 + \quad (2x_2 - x_3) \, dx_1 \wedge dx_4$$
$$+ x_1^2 x_2 x_4 \, dx_2 \wedge dx_3 + (x_1 x_2 x_4 + x_3) \, dx_2 \wedge dx_4$$
$$- \qquad\qquad x_1 x_3 \, dx_3 \wedge dx_4.$$

This example illustrates the following general result.

11.3.17 *Theorem.* *Let ω and γ be 1-forms on D in $\mathbb{R}^m$, where $m \geq 2$. Then the wedge product $\omega \wedge \gamma$ is expressible in the form*

11.3.18 $$\omega \wedge \gamma = \sum_{1 \leq i < j \leq m} g_{ij} \, dx_i \wedge dx_j$$

for suitable functions $g_{ij} : D \subseteq \mathbb{R}^m \to \mathbb{R}$.

Proof. The result follows from Theorem 11.2.16 and the distributive and anti-commutative properties of the wedge product.

The wedge products $dx_i \wedge dx_j$ play a key role in the theory of differential forms. We describe their behaviour as 2-forms on $\mathbb{R}^m$. By Definition 11.3.12 and 11.2.14

11.3.19 $$\begin{cases} (dx_i \wedge dx_j)_{\mathbf{p}}(\mathbf{h}, \mathbf{k}) = (dx_i)_{\mathbf{p}}(\mathbf{h})(dx_j)_{\mathbf{p}}(\mathbf{k}) - (dx_i)_{\mathbf{p}}(\mathbf{k})(dx_j)_{\mathbf{p}}(\mathbf{h}) \\ \qquad\qquad\qquad = h_i k_j - k_i h_j. \end{cases}$$

Consider in particular the 2-form $dx \wedge dy$ on $\mathbb{R}^2$. Let $\mathbf{h} = (\Delta x, 0)$ and $\mathbf{k} = (0, \Delta y)$ be increments along the axes. Then for each $\mathbf{p} \in \mathbb{R}^2$

$$(dx \wedge dy)_{\mathbf{p}}(\mathbf{h}, \mathbf{k}) = \Delta x \, \Delta y$$

but

$$(dx \wedge dy)_{\mathbf{p}}(\mathbf{k}, \mathbf{h}) = -\Delta x \, \Delta y.$$

So the 2-form $dx \wedge dy$ on $\mathbb{R}^2$ picks out elements of area with orientation appropriate to the ordering of the coordinate axes.

11.3.20 Theorem. *Every 2-form ϕ on D in $\mathbb{R}^m$, $m \geq 2$, is uniquely expressible as a sum*

11.3.21
$$\phi = \sum_{1 \leq i < j \leq m} f_{ij} dx_i \wedge dx_j,$$

where $f_{ij} : D \subseteq \mathbb{R}^m \to \mathbb{R}$ are 0-forms on D. Conversely, every such expression defines a 2-form on D.

Proof. Suppose that ϕ is a 2-form on D in $\mathbb{R}^m$, $m \geq 2$. Consider any $\mathbf{p} \in D$ and $\mathbf{h}, \mathbf{k} \in \mathbb{R}^m$. Express $\mathbf{h}$ and $\mathbf{k}$ in terms of the standard basis $\mathbf{e}_1, \dots, \mathbf{e}_m$ of $\mathbb{R}^m$. Since $\phi_{\mathbf{p}}$ is bilinear,

$$\phi_{\mathbf{p}}(\mathbf{h}, \mathbf{k}) = \sum_{j=1}^{m} \sum_{i=1}^{m} h_i k_j \phi_{\mathbf{p}}(\mathbf{e}_i, \mathbf{e}_j).$$

Put $\phi_{\mathbf{p}}(\mathbf{e}_i, \mathbf{e}_j) = f_{ij}$. Then, since $\phi_{\mathbf{p}}$ is skew-symmetric,

$$\phi_{\mathbf{p}}(\mathbf{h}, \mathbf{k}) = \sum_{1 \leq i < j \leq m} f_{ij}(h_i k_j - k_i h_j).$$

It follows from 11.3.19 that ϕ is expressible as a sum 11.3.21. We leave the reader to prove the uniqueness of this expression. The converse is obvious.

By Theorem 11.3.20 the general 2-form on $\mathbb{R}^3$ is expressible as the sum

11.3.22 $F_1 \, dy \wedge dz + F_2 \, dz \wedge dx + F_3 \, dx \wedge dy,$

where $F = (F_1, F_2, F_3)$ is a vector field in $\mathbb{R}^3$. This expression bears a striking resemblance to the integrand in the integral of F over the oriented simple surface (K, N) given by 9.3.16:

11.3.23
$$\iint\limits_{(K,N)} F \cdot d\mathbf{A} = \iint\limits_{(K,N)} F_1 \, dy \, dz + F_2 \, dz \, dx + F_3 \, dx \, dy.$$

Here we have a strong hint of how we might define the integral of a continuous 2-form in $\mathbb{R}^3$ over an oriented simple surface (K, N) in $\mathbb{R}^3$. Before we give the definition let us take a closer look at the situation.

Let (K, N) be an oriented simple surface in $\mathbb{R}^3$, and let

$\rho : S \subseteq \mathbb{R}^2 \to \mathbb{R}^3$ be a simple parametrization of (K, N). Consider the point $\mathbf{q} = \rho(u, v)$ in K. Let Δu, Δv be positive increments in u, v. The vectors

$$\mathbf{h} = \frac{\partial \rho}{\partial u}(u, v)\, \Delta u \qquad \text{and} \qquad \mathbf{k} = \frac{\partial \rho}{\partial v}(u, v)\, \Delta v$$

are tangent vectors to K at $\mathbf{q}$. They determine a parallelogram which approximates to an element of area ΔA of K at $\mathbf{q}$. By Theorem 1.2.22

$$\Delta A = \|\mathbf{h} \times \mathbf{k}\| = \|v_\rho(u, v)\|\, \Delta u\, \Delta v.$$

The vectors $\mathbf{h}$, $\mathbf{k}$ *in that order* assign an orientation to the area element given by the direction of $\mathbf{h} \times \mathbf{k}$, that is, by the direction of $N(\mathbf{q})$. In this way we may regard the ordered pair $(\mathbf{h}, \mathbf{k})$ as determining an element of oriented area $\Delta \mathbf{A}$ of the oriented surface (K, N) at $\mathbf{q}$. Notice that the pair $(\mathbf{k}, \mathbf{h})$ determines an element of oriented area of $(K, -N)$.

We adopt the notation

$$\rho(u, v) = (x(u, v), y(u, v), z(u, v)).$$

Then by 11.3.19

$$(\mathrm{d}y \wedge \mathrm{d}z)_\mathbf{q}(\mathbf{h}, \mathbf{k}) = \left[\left(\frac{\partial y}{\partial u}\frac{\partial z}{\partial v} - \frac{\partial y}{\partial v}\frac{\partial z}{\partial u} \right)(u, v) \right] \Delta u\, \Delta v$$

$$= \frac{\partial(y, z)}{\partial(u, v)} \Delta u\, \Delta v,$$

with similar expressions for $(\mathrm{d}z \wedge \mathrm{d}x)_\mathbf{q}(\mathbf{h}, \mathbf{k})$ and $(\mathrm{d}x \wedge \mathrm{d}y)_\mathbf{q}(\mathbf{h}, \mathbf{k})$. Hence the 2-form ϕ given by 11.3.22 satisfies

$$\phi_\mathbf{q}(\mathbf{h}, \mathbf{k}) = \left(F_1(\mathbf{q}) \frac{\partial(y, z)}{\partial(u, v)} + F_2(\mathbf{q}) \frac{\partial(z, x)}{\partial(u, v)} + F_3(\mathbf{q}) \frac{\partial(x, y)}{\partial(u, v)} \right) \Delta u\, \Delta v,$$

where $\mathbf{q} = \rho(u, v)$. From Section 9.3 it is clear that this is a contribution to the Riemann sum of

$$\iint_S f(\rho(u, v)) \cdot v_\rho(u, v)\, \mathrm{d}u\, \mathrm{d}v = \iint_{(K, N)} F \cdot \mathrm{d}\mathbf{A}.$$

We have therefore justified the following definition.

11.3.24 Definition. *Let*

$$\phi = F_1\, \mathrm{d}y \wedge \mathrm{d}z + F_2\, \mathrm{d}z \wedge \mathrm{d}x + F_3\, \mathrm{d}x \wedge \mathrm{d}y$$

be a continuous 2-form on $\mathbb{R}^3$. For any oriented simple surface (K, N) in $\mathbb{R}^3$, the integral of ϕ over (K, N) is defined by

$$\iint\limits_{(K, N)} \phi = \iint\limits_{(K, N)} F \cdot d\mathbf{A},$$

where $F = (F_1, F_2, F_3)$.

A somewhat simpler argument leads to the definition of the integral of a 2-form on $\mathbb{R}^2$ over a simple region.

11.3.25 Definition.   *Let* $\phi = f\,dx \wedge dy$ *be a continuous 2-form on* $\mathbb{R}^2$. *The integral of* ϕ *over a simple region* S *in* $\mathbb{R}^2$ *is defined by*

11.3.26
$$\int_S \phi = \iint\limits_S f\,dA.$$

We require

$$\int_S f\,dy \wedge dx = -\int_S f\,dx \wedge dy,$$

so, strictly speaking, in 11.3.26 the integral of ϕ is taken over the positively oriented region S^+ (Definition 8.3.1), the meaning of 'counterclockwise orientation' being related to the ordering of the x, y axes.

The following result concerns a relationship between the wedge product of 1-forms in $\mathbb{R}^3$ and the cross product of vector fields in $\mathbb{R}^3$.

11.3.27 Theorem.   *Let* ω *and* γ *be 1-forms on* D *in* $\mathbb{R}^3$. *If* $\omega \leftrightarrow W$ *and* $\gamma \leftrightarrow G$ *then*

$$\omega \wedge \gamma \leftrightarrow W \times G$$

under the correspondence referred to in Theorems 11.2.8 *and* 11.3.9.

Proof.   We are given (by 11.2.6 and 11.2.14) that

$$\omega = W_1\,dx_1 + W_2\,dx_2 + W_3\,dx_3$$

and

$$\gamma = G_1\,dx_1 + G_2\,dx_2 + G_3\,dx_3.$$

Hence

$$\omega \wedge \gamma = F_1 \, dx_2 \wedge dx_3 + F_2 \, dx_3 \wedge dx_1 + F_3 \, dx_1 \wedge dx_2,$$

where

11.3.28 $F_1 = W_2 G_3 - W_3 G_2,$ $F_2 = W_3 G_1 - W_1 G_3,$

$$F_3 = W_1 G_2 - W_2 G_1.$$

It follows by 11.3.19 and 11.3.8 that

$$(\omega \wedge \gamma)_{\mathbf{p}}(\mathbf{h}, \mathbf{k}) = F(\mathbf{p}) \cdot \mathbf{h} \times \mathbf{k}.$$

Therefore

$$\omega \wedge \gamma \leftrightarrow F.$$

By comparing 11.3.8 with Definition 1.2.20 we see that $F = W \times G$.

Theorem 11.3.27 brings out a similarity between the wedge product and the cross product. However, they are quite different operations. Unlike the wedge product, the cross product is defined on $\mathbb{R}^3$ only. The cross product of two vectors is again a vector. In contrast, the wedge product of two 1-forms is not a 1-form but a 2-form. We shall see later that the wedge product is defined more generally as an associative operation on forms. The cross product is not associative.

Exercises 11.3

1.   Prove Theorem 11.3.5.

Hint: suppose that $g : \mathbb{R}^2 \times \mathbb{R}^2 \to \mathbb{R}$ is bilinear and skew-symmetric.

Put $\mathbf{h} = h_1 \mathbf{e}_1 + h_2 \mathbf{e}_2, \mathbf{k} = k_1 \mathbf{e}_1 + k_2 \mathbf{e}_2$. Then

$$g(\mathbf{h}, \mathbf{k}) = \sum_{j=1}^{2} \sum_{i=1}^{2} h_i k_j g(\mathbf{e}_i, \mathbf{e}_j).$$

Also $g(\mathbf{e}_i, \mathbf{e}_i) = 0$. Put $g(\mathbf{e}_1, \mathbf{e}_2) = a$.

2.   For the following differential 2-forms ϕ on $\mathbb{R}^3$ evaluate $\phi_{\mathbf{p}}(\mathbf{h}, \mathbf{k})$, where $\mathbf{p} = (a, b, c)$ and (a) $\mathbf{h} = (0.1, -0.1, 0.2)$, $\mathbf{k} = (0.3, 0.1, 0)$; (b) $\mathbf{h} = (0.3, 0.1, 0)$, $\mathbf{k} = (0.1, -0.1, 0.2)$:
   (i) $\phi = x \, dy \wedge dz + y \, dz \wedge dx + z \, dx \wedge dy$
   (ii) $\phi = 2 \, dy \wedge dz - 3 \, dz \wedge dx.$

Answers: (i) (a) $-0.02a + 0.06b + 0.04c$; (b) $0.02a - 0.06b - 0.04c$.
   (ii) (a) -0.22; (b) 0.22 independent of a, b, c.

3. Express the following wedge products of 1-forms in standard form 11.3.21:
    (a) $(dx + dy) \wedge (dx - dy)$;
    (b) $(x\, dx + y\, dy + z\, dz) \wedge (y\, dx + z\, dy + x\, dz)$;
    (c) $(dx_1 + 2\, dx_2 + dx_3 + dx_4) \wedge (dx_1 + dx_4)$.

Answers: (a) $-2\, dx \wedge dy$; (c) $-2\, dx_1 \wedge dx_2 - dx_1 \wedge dx_3 + 2\, dx_2 \wedge dx_4 + dx_3 \wedge dx_4$.

4. (a) Express the 2-form on $\mathbb{R}^2$

    $$x^2 y\, dx \wedge dy$$

    as the wedge product of two 1-forms on $\mathbb{R}^2$.
    (b) Express the 2-form on $\mathbb{R}^3$

    $$x\, dy \wedge dz + y\, dz \wedge dx + z\, dx \wedge dy$$

    as the wedge product of two 1-forms on $\mathbb{R}^3$.

Answers: there are many solutions, for example (a) $(x^2 dx) \wedge (y\, dy)$,

  (b) $\left(dx - \dfrac{x}{y} dy\right) \wedge (z\, dy - y\, dz)$ provided $y \neq 0$.

5. Prove that every 2-form on $\mathbb{R}^3$ is expressible as the wedge product of two 1-forms on $\mathbb{R}^3$.
    Is it true that every 2-form on $\mathbb{R}^4$ is expressible as the wedge product of two 1-forms on $\mathbb{R}^4$?

Hint: consider $dx_1 \wedge dx_2 + dx_3 \wedge dx_4$.

6. Prove that the only 2-form on $\mathbb{R}$ is $\phi = 0$.

7. Let $A = [a_{ij}]$ be a 2×2 matrix. Define the 1-forms

    $$\omega_j = a_{1j}\, dx_1 + a_{2j}\, dx_2, \qquad j = 1, 2.$$

    Prove that

    $$\omega_1 \wedge \omega_2 = \det A\, dx_1 \wedge dx_2.$$

8. Let ω be a continuous 1-form on $\mathbb{R}^m$. Let C^+ be an oriented simple arc in $\mathbb{R}^m$. Formulate and justify a definition of the integral $\int_{C^+} \omega$.

Hint: let $\omega = W_1\, dx_1 + \cdots + W_m\, dx_m$. Let $\mathbf{r}(t) = (x_1(t), \ldots, x_m(t))$, $t \in [a, b]$ be a simple parametrization of C^+. At $\mathbf{q} = \mathbf{r}(t)$ consider an incremental tangent vector $\mathbf{h} = (x_1'(t), \ldots, x_m'(t))\, \Delta t$. Interpret $\omega_{\mathbf{q}}(\mathbf{h})$ as a contribution to a Riemann sum of $\int_{C^+} \omega = \int_{C^+} W \cdot d\mathbf{r}$.

9. Evaluate $\int_S \phi$, where $\phi = (x^2 + y)\, dx \wedge dy$ and S is the rectangle $\{(x, y) \in \mathbb{R}^2 \mid 0 \leq x \leq 1, 1 \leq y \leq 3\}$.

Answer: 14/3.

10. Evaluate

$$\int_{(K,\,N)} x^2 \, dy \wedge dz + y^2 \, dz \wedge dx + 2z^2 \, dx \wedge dy,$$

where (K, N) is the positively oriented unit cube with opposite vertices $(0, 0, 0)$ and $(1, 1, 1)$ and faces in the coordinate planes.

Answer: 4.

11. Evaluate

$$\int_{(K,\,N)} x \, dy \, dz - 2z \, dx \, dy,$$

where (K, N) is the positively oriented unit sphere $x^2 + y^2 + z^2 = 1$.

Answer: $-4/3$.

12. Prove that the only 2-form on $\mathbb{R}$ is $\phi = 0$.

11.4   3-forms in $\mathbb{R}^m$

In this section we generalize the work of Section 11.3 on 2-forms. We could at this stage move to n-forms for any positive integer n, but it will suffice for our purposes (and lead to some simplification) if we consider just the case $n = 3$.

11.4.1 Definition.   *Let V be a vector space over $\mathbb{R}$. A function $f : V^3 = V \times V \times V \to \mathbb{R}$ is said to be* multilinear *if it is linear in each of its arguments. It is said to be* alternating *if an interchange of any two of its arguments leads to a change of sign.*

11.4.2 Example.   For a fixed $a \in \mathbb{R}$ define the function $g : \mathbb{R}^3 \times \mathbb{R}^3 \times \mathbb{R}^3 \to \mathbb{R}$ by

$$g(\mathbf{h}, \mathbf{k}, \mathbf{l}) = a \det \begin{bmatrix} h_1 & h_2 & h_3 \\ k_1 & k_2 & k_3 \\ l_1 & l_2 & l_3 \end{bmatrix}.$$

Then g is multilinear and alternating on $(\mathbb{R}^3)^3$. For example,

$$g(\mathbf{h}, \mathbf{k}, \mathbf{l}) = -g(\mathbf{l}, \mathbf{k}, \mathbf{h}) = g(\mathbf{k}, \mathbf{l}, \mathbf{h}).$$

11.4.3 Example.   For fixed $a, b, c, d \in \mathbb{R}$ the function $g : \mathbb{R}^4 \times \mathbb{R}^4 \times \mathbb{R}^4 \to$

$\mathbb{R}$ defined by

$$g(\mathbf{h}, \mathbf{k}, \mathbf{l}) = \det \begin{bmatrix} a & b & c & d \\ h_1 & h_2 & h_3 & h_4 \\ k_1 & k_2 & k_3 & k_4 \\ l_1 & l_2 & l_3 & l_4 \end{bmatrix}$$

is multilinear and alternating on $(\mathbb{R}^4)^3$.

Examples 11.4.2 and 11.4.3 give the most general multilinear alternating real-valued functions on $(\mathbb{R}^3)^3$ and $(\mathbb{R}^4)^3$ respectively.

11.4.4 Definition.   *Let D be an open set in $\mathbb{R}^m$. A 3-form on D is a class of real-valued functions*

$$\tau = \{\tau_\mathbf{p} : (\mathbb{R}^m)^3 \to \mathbb{R} \mid \mathbf{p} \in D\}$$

such that for each $\mathbf{p} \in D$ the function $\tau_\mathbf{p}$ is multilinear and alternating.

We leave the reader to formulate definitions of the sum of two 3-forms on D and the product $f\tau$ of a 0-form f and a 3-form τ on D.

The following construction shows how a 1-form and a 2-form can be combined to give a 3-form.

11.4.5 Definition.   *Let ω be a 1-form and ϕ a 2-form on D in $\mathbb{R}^m$. The* wedge products $\omega \wedge \phi$ *and* $\phi \wedge \omega$ *are defined as follows:*

$$(\omega \wedge \phi)_\mathbf{p}(\mathbf{h}, \mathbf{k}, \mathbf{l}) = \omega_\mathbf{p}(\mathbf{h})\phi_\mathbf{p}(\mathbf{k}, \mathbf{l}) - \omega_\mathbf{p}(\mathbf{k})\phi_\mathbf{p}(\mathbf{h}, \mathbf{l}) + \omega_\mathbf{p}(\mathbf{l})\phi_\mathbf{p}(\mathbf{h}, \mathbf{k}),$$

$$(\phi \wedge \omega)_\mathbf{p}(\mathbf{h}, \mathbf{k}, \mathbf{l}) = \phi_\mathbf{p}(\mathbf{h}, \mathbf{k})\omega_\mathbf{p}(\mathbf{l}) - \phi_\mathbf{p}(\mathbf{h}, \mathbf{l})\omega_\mathbf{p}(\mathbf{k}) + \phi_\mathbf{p}(\mathbf{k}, \mathbf{l})\omega_\mathbf{p}(\mathbf{h})$$

for all $\mathbf{h}, \mathbf{k}, \mathbf{l} \in \mathbb{R}^m$ and $\mathbf{p} \in D$.

Clearly $\omega \wedge \phi = \phi \wedge \omega$ for any 1-form ω and 2-form ϕ on D, and $\omega \wedge \phi$ is a 3-form, by Definition 11.4.4.

11.4.6 Example.   Consider the wedge product of the 2-form $dx_a \wedge dx_b$ and the 1-form dx_c on $\mathbb{R}^m$, where a, b, c are any three indices, not necessarily distinct, chosen from $1, 2, \ldots, m$. It follows by 11.3.19 and Definition 11.4.5 that for all $\mathbf{p} \in \mathbb{R}^m$

$$((dx_a \wedge dx_b) \wedge dx_c)_\mathbf{p}(\mathbf{h}, \mathbf{k}, \mathbf{l}) = \det \begin{bmatrix} h_a & h_b & h_c \\ k_a & k_b & k_c \\ l_a & l_b & l_c \end{bmatrix}$$

and that

11.4.7 $dx_a \wedge (dx_b \wedge dx_c) = (dx_a \wedge dx_b) \wedge dx_c.$

So the wedge product is *associative* on the differentials dx_i and, leaving out the brackets, the 3-form $dx_a \wedge dx_b \wedge dx_c$ has unambiguous meaning.

Notice that the interchange of two of dx_a, dx_b, dx_c in $dx_a \wedge dx_b \wedge dx_c$ leads to a change of sign. For example

11.4.8 $dx_a \wedge dx_b \wedge dx_c = -dx_b \wedge dx_a \wedge dx_c = dx_b \wedge dx_c \wedge dx_a.$

Also $dx_a \wedge dx_b \wedge dx_c = 0$ whenever two of a, b, c are equal.

11.4.9 Theorem. *Every 3-form τ on D in $\mathbb{R}^m$, $m \geq 3$, is uniquely expressible as a sum*

11.4.10 $\tau = \sum_{1 \leq i < j < k \leq m} f_{ijk}\, dx_i \wedge dx_j \wedge dx_k,$

where $f_{ijk} : D \subseteq \mathbb{R}^m \to \mathbb{R}$ are 0-forms on D. Conversely, every such expression defines a 3-form on D.

Proof.   Exercise.

11.4.11 Example.   Every 3-form τ on $\mathbb{R}^3$ is of the form

$$\tau = f\, dx \wedge dy \wedge dz.$$

The only 3-form on $\mathbb{R}^2$ or on $\mathbb{R}$ is the form $\tau = 0$.

Let τ be a continuous 3-form on $\mathbb{R}^3$, so $\tau = f\, dx \wedge dy \wedge dz$ for some continuous function $f : \mathbb{R}^3 \to \mathbb{R}$. We wish to define the integral of τ over a simple region W in $\mathbb{R}^3$. Take $\mathbf{q} \in W$ and consider a rectangular element of volume at $\mathbf{q}$ defined by the vectors

$$\mathbf{h} = (\Delta x, 0, 0), \qquad \mathbf{k} = (0, \Delta y, 0), \qquad \mathbf{l} = (0, 0, \Delta z)$$

based at $\mathbf{q}$. Then by Example 11.4.6

$$(dx \wedge dy \wedge dz)_{\mathbf{q}}(\mathbf{h}, \mathbf{k}, \mathbf{l}) = \det \begin{bmatrix} \Delta x & 0 & 0 \\ 0 & \Delta y & 0 \\ 0 & 0 & \Delta z \end{bmatrix} = \Delta x\, \Delta y\, \Delta z.$$

So for positive increments Δx, Δy, Δz the 3-form satisfies, for each $\mathbf{q} \in W$,

11.4.12 $\tau_{\mathbf{q}}(\mathbf{h}, \mathbf{k}, \mathbf{l}) = f(\mathbf{q})\, \Delta V,$

where $\Delta V = \Delta x \, \Delta y \, \Delta z$. Notice that 11.4.12 is a contribution to the Riemann sum of the integral of f over W. The following definition is therefore meaningful.

11.4.13 Definition.   *Let*

$$\tau = f \, dx \wedge dy \wedge dz$$

be a continuous 3-form on $\mathbb{R}^3$. *For any simple region W in $\mathbb{R}^3$ the integral of τ over W is defined by*

11.4.14 $$\int_W \tau = \int_W f \, dV.$$

We require

$$\int_W f \, dy \wedge dx \wedge dz = -\int_W f \, dx \wedge dy \wedge dz,$$

so strictly speaking in 11.4.14 the integral of τ is taken over the *positively oriented region* W^+, the orientation being determined by the ordering of the x, y, z axes.

Exercises 11.4

1.   Evaluate $(dx \wedge dy \wedge dz)_q(\mathbf{h}, \mathbf{k}, \mathbf{l})$ at $\mathbf{q} \in \mathbb{R}^3$, where

$$\mathbf{h} = (0, \Delta x, 0), \qquad \mathbf{k} = (\Delta y, 0, 0), \qquad \mathbf{l}(0, 0, \Delta z).$$

Answer: $-\Delta x \, \Delta y \, \Delta z$.

2.   Evaluate $\int_W xyz \, dx \wedge dy \wedge dz$, where W is the region in $\mathbb{R}^3$ bounded by the surface $z = x^2 + y^2$ and the planes $x = 0$, $y = 0$, and $z = 2$.

Answer: $\frac{1}{2}$.

3.   Let $A = [a_{ij}]$ be a 3×3 matrix. Define the 1-forms

$$\omega_j = a_{1j} \, dx_1 + a_{2j} \, dx_2 + a_{3j} \, dx_3, \qquad\qquad j = 1, 2, 3.$$

Prove that

$$\omega_1 \wedge \omega_2 \wedge \omega_3 = \det A \, dx_1 \wedge dx_2 \wedge dx_3.$$

Deduce that

$$\omega_1 \wedge \omega_2 \wedge \omega_3 = -\omega_2 \wedge \omega_1 \wedge \omega_3 = \omega_2 \wedge \omega_3 \wedge \omega_1.$$

11.5   The differential of a form

We have seen in Section 11.2 how a differentiable function (0-form) $f : D \subseteq \mathbb{R}^m \to \mathbb{R}$ gives rise to the 1-form df, the differential of f on D in $\mathbb{R}^m$. In Example 11.2.17 we obtained the expression

$$df = \frac{\partial f}{\partial x_1} \, dx_1 + \cdots + \frac{\partial f}{\partial x_m} \, dx_m$$

which relates df to the basic 1-forms $dx_1, \ldots, dx_m$ on $\mathbb{R}^m$. We now define the differential of a 1-form and a 2-form and show how these give rise to a 2-form and a 3-form respectively.

11.5.1 Definition.   *Let*

$$\omega = W_1 \, dx_1 + \cdots + W_m \, dx_m$$

be a differentiable 1-form on D in $\mathbb{R}^m$. The differential *of ω is defined to be the 2-form on D given by*

$$d\omega = dW_1 \wedge dx_1 + \cdots + dW_m \wedge dx_m.$$

11.5.2 Example.   Consider the differentiable 1-form on $\mathbb{R}^3$

$$\omega = xyz^2 \, dx + (x - y) \, dy + y \, dz.$$

Then

$$
\begin{aligned}
d\omega &= (yz^2 \, dx + xz^2 \, dy + 2xyz \, dz) \wedge dx + (dx - dy) \wedge dy + dy \wedge dz \\
&= dy \wedge dz + 2xyz \, dz \wedge dx + (1 - xz^2) \, dx \wedge dy.
\end{aligned}
$$

In evaluating this expression we freely used that $dx \wedge dy = -dy \wedge dx$, $dx \wedge dx = 0$, etc.

11.5.3 Definition.   *Let*

$$\phi = \sum_{1 \leqslant i < j \leqslant m} f_{ij} \, dx_i \wedge dx_j$$

be a differentiable 2-form on D in $\mathbb{R}^m$. The differential *of ϕ is defined to be the 3-form on D given by*

$$d\phi = \sum_{1 \leqslant i < j \leqslant m} (df_{ij}) \wedge dx_i \wedge dx_j.$$

11.5.4 Example.   The differential of the 2-form

$$\phi = x \, dy \wedge dz + (x + y) \, dz \wedge dx + z^2 \, dx \wedge dy$$

is the 3-form

$$d\phi = dx \wedge dy \wedge dz + (dx + dy) \wedge dz \wedge dx + 2z\,dz \wedge dx \wedge dy$$
$$= (2 + 2z)\,dx \wedge dy \wedge dz$$

since $dx \wedge dz \wedge dx = 0$ and $dy \wedge dz \wedge dx = dz \wedge dx \wedge dy = dx \wedge dy \wedge dz$.

The differential of a form is sometimes called its *exterior differential*.

In the following theorem we relate the general differentiable 0-, 1-, and 2-forms on D in $\mathbb{R}^m$ with their differentials.

11.5.5 Theorem. **[i]** *Let $f : D \subseteq \mathbb{R}^3 \to \mathbb{R}$ be a differentiable 0-form. Then*

$$df = F_1\,dx + F_2\,dy + F_3\,dz,$$

where

$$F = \operatorname{grad} f.$$

[ii] *Let ω be a differentiable 1-form on D in $\mathbb{R}^3$ given by*

$$\omega = W_1\,dx + W_2\,dy + W_3\,dz.$$

Then

$$d\omega = F_1\,dy \wedge dz + F_2\,dz \wedge dx + F_3\,dx \wedge dy,$$

where

$$F = \operatorname{curl} W.$$

[iii] *Let ϕ be a differentiable 2-form on D in $\mathbb{R}^m$ given by*

$$\phi = G_1\,dy \wedge dz + G_2\,dz \wedge dx + G_3\,dx \wedge dy.$$

Then

$$d\phi = g\,dx \wedge dy \wedge dz,$$

where

$$g = \operatorname{div} G.$$

Proof. [*i*] See Example 11.2.17.

[*ii*] If ω is as given, then by definition

$$d\omega = dW_1 \wedge dx + dW_2 \wedge dy + dW_3 \wedge dz.$$

A tedious calculation using $dx \wedge dx = 0$, $dy \wedge dx = -dx \wedge dy$, etc. shows that

$$d\omega = \left(\frac{\partial W_3}{\partial y} - \frac{\partial W_2}{\partial z}\right) dy \wedge dz + \left(\frac{\partial W_1}{\partial z} - \frac{\partial W_3}{\partial x}\right) dz \wedge dx$$

$$+ \left(\frac{\partial W_2}{\partial x} - \frac{\partial W_1}{\partial y}\right) dx \wedge dy.$$

The result follows.

[*iii*] This result is proved similarly.

Exercises 11.5

1. Evaluate $d\omega$ where ω is the 1-form:
   (a) $y\,dx - x\,dy$; (b) $y\,dx + x\,dy$; (c) $x\,dx + y\,dy + z\,dz$; (d) $yz\,dx + xz\,dy + xy\,dz$; (e) $(y^2 - z^2)\,dx + (z^2 - x^2)\,dy + (x^2 - y^2)\,dz$.

Answers: (a) $-2\,dx \wedge dy$; (b), (c), (d) 0; (e) $-2((y+z)\,dy \wedge dz + (z+x)\,dz \wedge dx + (x+y)\,dx \wedge dy)$.

2. Evaluate $d\phi$ where ϕ is the 2-form:
   (a) $x^2\,dy \wedge dz + y^2\,dz \wedge dx + z^2\,dx \wedge dy$;
   (b) $(x+y+z)\,dy \wedge dz + (x+y^2)\,dz \wedge dx + (1-z^2)\,dx \wedge dy$.

Answers: (a) $2(x+y+z)\,dx \wedge dy \wedge dz$; (b) $(1-2y-2z)\,dx \wedge dy \wedge dz$.

3. Let $\omega = W_1\,dx + W_2\,dy$ be a differentiable 1-form on $\mathbb{R}^2$. Show that
   $$d\omega = \text{rot}\,W\,dx \wedge dy.$$

4. Let f be a C^2 0-form on $\mathbb{R}^m$. Prove that
   $$d(df) = 0.$$
   Hence prove that for any C^2 1-form ω on $\mathbb{R}^m$
   $$d(d\omega) = 0.$$

5. Using the results of Exercise 4 and Theorem 11.5.5 show that for any C^2 functions $f:\mathbb{R}^3 \to \mathbb{R}$ and $F:\mathbb{R}^3 \to \mathbb{R}^3$
   $$\text{curl}(\text{grad}\,f) = 0$$
   and
   $$\text{div}(\text{curl}\,F) = 0.$$

6. Let f and g be differentiable 0-forms and let ω and γ be differentiable 1-forms on $\mathbb{R}^m$. Prove that
   (a) $d(fg) = (df)g + f\,dg$;
   (b) $d(f\omega) = df \wedge \omega + f\,d\omega$;
   (c) $d(\omega \wedge \gamma) = d\omega \wedge \gamma - \omega \wedge d\gamma$.

11.6   The General Stokes' Theorem

The stage is now almost set for the statement of the General Stokes'
Theorem. However, we first need to make a comment about
regions of integration. In the classical theorems listed in Section
11.1 the integrals concerned are evaluated over simple arcs, simple
closed curves, simple surfaces and simple regions in $\mathbb{R}^3$. Bearing in
mind the dimension of the vector space in which the domains of
their respective parametrizations lie, we shall call a simple arc in $\mathbb{R}^m$
a simple 1-region in $\mathbb{R}^m$, and a simple surface in $\mathbb{R}^3$ a simple
2-region in $\mathbb{R}^3$, and a simple region in $\mathbb{R}^3$ a simple 3-region in $\mathbb{R}^3$.

11.6.1 *The General Stokes' Theorem.*   *For $k = 0, 1, 2$ let λ be a
differentiable k-form on an open set D in $\mathbb{R}^3$ and let R^+ be an
oriented simple $(k + 1)$-region in D. Then*

11.6.2
$$\int_{R^+} d\lambda = \int_{\partial R^+} \lambda,$$

where ∂R^+ is the oriented boundary of R^+.

Let us see how the General Stokes' Theorem gives for $k = 0, 1, 2$
the three classical theorems of Section 11.1.

Case $k = 0$

λ is a differentiable 0-form on D, say $\lambda = f$,

$$d\lambda = \frac{\partial f}{\partial x}\, dx + \frac{\partial f}{\partial y}\, dy + \frac{\partial f}{\partial z}\, dz,$$

R^+ is a simple arc C in D oriented from end point $\mathbf{p}$ to end point $\mathbf{q}$,
$\partial R^+ = \{\text{ordered end points of } C^+\} = \{\mathbf{p}, \mathbf{q}\}$, $\int_{R^+} d\lambda = \int_{C^+} \operatorname{grad} f \cdot d\mathbf{r}$
(Exercise 11.3.8), $\int_{\partial R^+} \lambda = f(\mathbf{q}) - f(\mathbf{p})$, the difference in the values
of f at the end points. So 11.6.2 is the Fundamental Theorem in $\mathbb{R}^3$.

Case $k = 1$

λ is a 1-form on D, say

$$\lambda = \omega = W_1\, dx + W_2\, dy + W_3\, dz,$$
$$d\lambda = F_1\, dy \wedge dz + F_2\, dz \wedge dx + F_3\, dx \wedge dy$$

where $F = \operatorname{curl} W$, R^+ is an oriented simple surface (K, N),

$\partial R^+ = \partial(K, N)$,

$$\int_{R^+} d\lambda = \iint_{(K, N)} \operatorname{curl} W \cdot d\mathbf{A}$$

by Definition 11.3.24, and

$$\int_{\partial R^+} \lambda = \int_{\partial(K, N)} W \cdot d\mathbf{r}.$$

In this case 11.6.2 is Stokes' Theorem in $\mathbb{R}^3$.

Case $k = 2$

λ is a 2-form on D, say

$$\lambda = \phi = F_1 \, dy \wedge dz + F_2 \, dz \wedge dx + F_3 \, dx \wedge dy,$$
$$d\lambda = \operatorname{div} F \, dx \wedge dy \wedge dz,$$

R^+ is a simple region W in $\mathbb{R}^3$, $\partial R^+ = (\partial W, N)$, positively oriented,

$$\int_{R^+} d\lambda = \iiint_W \operatorname{div} F \, dV$$

by Definition 11.4.13, and

$$\int_{\partial R^+} \lambda = \iint_{(\partial W, N)} F \cdot d\mathbf{A}$$

by Definition 11.3.24. Here 11.6.2 is Gauss' Divergence Theorem.

We have not stated Stokes' Theorem in its most general form. The general theorem concerns the integration of differentiable k-forms in $\mathbb{R}^m$, where k and m are arbitrary. The interested reader is referred to more advanced texts, for example Spivak *Calculus on Manifolds*, or Corwin and Szczarba *Multivariable Calculus*.

Exercise 11.6

1. Interpret Stokes' Theorem 11.6.1 for a differentiable 1-form

$$F_1 \, dx + F_2 \, dy \qquad \text{in } \mathbb{R}^2.$$

Solution: the formula

$$\int_S d\omega = \int_{\partial S^+} \omega,$$

where S is a simple region in $\mathbb{R}^2$, is *Green's Theorem*.

Bibliography

The following list of books is intended for reference and as a guide to further reading.

Elementary analysis and linear algebra

Hadley, G., *Linear Algebra.* Addison-Wesley
Lang, S., *Linear Algebra.* Addison-Wesley
Reade, J. B., *An Introduction to Mathematical Analysis.* O.U.P.
Spivak, M., *Calculus.* W. A. Benjamin

Vector calculus

Bartle, R. G., *The Elements of Real Analysis.* Wiley
Flett, T. M., *Mathematical Analysis.* McGraw-Hill
Fulks, W., *Advanced Calculus.* Wiley
Lang. S., *Calculus of Several Variables.* Addison-Wesley
Marsden, J. E. and **Tromba, A. J.,** *Vector Calculus.* Freeman
Nickerson, H. K., Spencer D. C. and **Steenrod, N. E.,** *Advanced Calculus.* Van
  Nostrand
Rudin, W., *Principles of Mathematical Analysis.* McGraw-Hill
Williamson, R. E., Crowell, R. H. and **Trotter, H. F.,** *Calculus of Vector Functions.*
  Prentice-Hall

Applied mathematics

Bourne, D. E. and **Kendall, P. C.,** *Vector Analysis and Cartesian Tensors.* Nelson
Cole, R. J., *Vector Methods.* Van Nostrand Reinhold
Marder, L., *Vector Analysis.* Allen and Unwin
Smith, R. C. and **Smith, P.,** *Mechanics.* Wiley
Sowerby, L., *Vector Field Theory with Applications.* Longman

Advanced analysis

Apostol, T. M., *Mathematical Analysis.* Addison-Wesley
Corwin, L. J. and **Szczarba, R. H.,** *Multivariable Calculus.* Marcel Dekker.
Craven, B. D., *Functions of Several Variables.* Chapman and Hall
Flett, T. M., *Differential Analysis.* Cambridge University Press
Lang, S., *Analysis I and II.* Addison-Wesley.
Spivak, M., *Calculus on Manifolds.* W. A. Benjamin

Differential geometry

Gauld, David B., *Differential Topology,* Marcel Dekker
O'Neill, B., *Elementary Differential Geometry.* Academic Press
Schreiber, M., *Differential Forms, A Heuristic Introduction.* Springer
Thorpe, J. A., *Elementary Topics in Differential Geometry.* Springer
Willmore, T. J., *Introduction to Differential Geometry.* O.U.P.

Index